Analytical Techniques in the Oil and Gas Industry for Environmental Monitoring

Analytical Techniques in the Oil and Gas Industry for Environmental Monitoring

Edited by

Melissa N. Dunkle
The Netherlands

William L. Winniford
USA

WILEY

This edition first published 2020

Registered Office
John Wiley & Sons, Inc., 111 River Street, Hoboken, NJ 07030, USA

Editorial Office
111 River Street, Hoboken, NJ 07030, USA

For details of our global editorial offices, customer services, and more information about Wiley products visit us at www.wiley.com.

Wiley also publishes its books in a variety of electronic formats and by print-on-demand. Some content that appears in standard print versions of this book may not be available in other formats.

Library of Congress Cataloging-in-Publication Data applied for

Hardback ISBN: 9781119523307

Cover Design: Wiley
Cover Image: Courtesy of The Dow Chemical Company

Set in 9.5/12.5pt STIXTwoText by SPi Global, Chennai, India

10 9 8 7 6 5 4 3 2 1

MIX
Paper from responsible sources
FSC
www.fsc.org
FSC® C013604

Contents

Part I

Scope

1

Introduction

Melissa N. Dunkle[1] *and William L. Winniford*[2]

[1]*Analytical Science, Dow Benelux, BV, Herbert H. Dowweg 5, Building 446, Hoek, The Netherlands*
[2]*Analytical Science, The Dow Chemical Company, 230 Abner Jackson Parkway, Lake Jackson, TX, USA*

1.1 Introduction

The oil and gas industry, also known as the petroleum industry, is a global endeavor that encompasses exploration, extraction, refining, transportation, and marketing of petroleum products. The scope of this book is to provide readers with background knowledge on the petroleum industry, for example on how petroleum is sourced and refined, which will then be related to the multitude of analytical techniques required to characterize and quantify the various aspects of petroleum and petroleum-related products, combine with links to environmental impact.

This book has been designed in such a way as to provide a thorough background for students or novices in the industry, but also to provide enough detailed applications and analytical advancements to be useful for experts on the topic.

This book has been divided into six parts to simplify the complexity of the discussed topics for the readers:

- Part 1: Scope
- Part 2: Introduction to the Petroleum Industry
- Part 3: Analytical Techniques Utilized in the Petroleum Industry
- Part 4: Special Cases and Examples Related to the Petroleum Industry
- Part 5: Environmental Analysis
- Part 6: Future Trends in the Petroleum Industry

1.1.1 Petroleum Cycle

Whether we realize it or not, petroleum and petroleum-related products are part of our everyday lives. While petroleum is sourced as a raw material, it is further

Analytical Techniques in the Oil and Gas Industry for Environmental Monitoring, First Edition.
Edited by Melissa N. Dunkle and William L. Winniford.

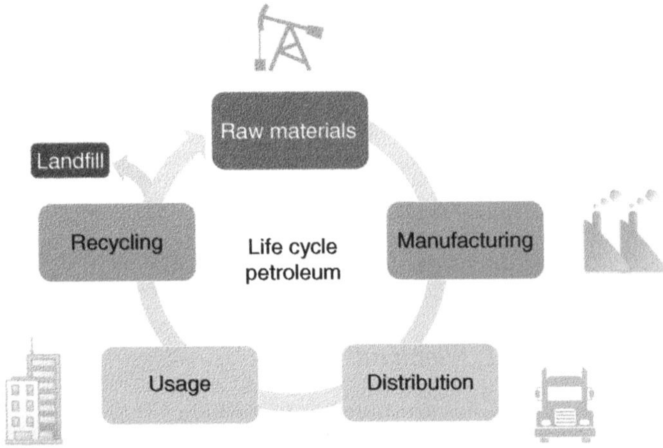

Figure 1.1 Life cycle of petroleum.

manufactured into useable materials, such as fuel for automobiles (e.g. gasoline and diesel) or into petrochemicals and chemicals that are further processed into everyday items (e.g. cosmetics, plastics, foams, construction and building materials, etc.). After use, these items are then (ideally) recycled back into the lifecycle to be manufactured into new products; Figure 1.1 shows a general schematic for the petroleum lifecycle.

At any point in the lifecycle, care must be taken to avoid environmental contamination, from the sourcing of petroleum, through to ensuring that we all do our part to recycle and avoid contributing to landfill waste. The authors realize that not all materials can be recycled; however, we do want to stress environmental awareness to the reader. Two examples of environmental contamination originating from different phases of the petroleum lifecycle will be discussed before providing an overview of the content of this book.

1.1.2 Well-Known Cases of Environmental Contamination

1.1.2.1 Oil-Drilling Rig Deepwater Horizon

Deepwater Horizon is one of the most well-known oil spills originating from the United States. On 20 April 2010, the Deepwater Horizon oil drilling rig exploded and sank in the Macondo Prospect in the Gulf of Mexico while sourcing crude oil (EPA 2017). In total, 11 workers lost their lives, and more than 4 million barrels of oil spilled into the Gulf of Mexico over a period of 87 days before the Macondo well was able to be capped. The Gulf Coast ecosystem was severely affected and altered by this event, and as a result, the Resources and Ecosystems Sustainability, Tourist Opportunities, and Revived Economies of the Gulf Coast States Act

(RESTORE Act) was signed into law (RESTORE 2015). The RESTORE Act was aimed at restoring the long-term health of the natural ecosystem and economy of the Gulf Coast Region. Additionally, the oil and gas industry identified several learnings from this disaster; firstly, the blowout preventer, which failed in the Deepwater Horizon operation, was in need of improvement, and secondly, there is a need for better deepwater sub-sea intervention capabilities (Griffin 2010).

Even as the spill was still active, cleanup activities were underway. Oil dispersants were used in an attempt to keep the surface oil from reaching the coastline by breaking down the hydrocarbons, and this was the first example where oil dispersants were used underwater. While oil dispersants should not be any more toxic than the oil they are intended to disperse, it was noted in a study by Rico-Martinez that for selected aquatic species (*Brachionus manjavacas*), the combination of the oil dispersants used in the Deepwater Horizon spill and the Macondo oil showed a synergetic effect for increased toxicity (Rico-Martínez et al. 2013). Even two years after the event, the full impact on the environment, including marine and human life was not fully known. In 2012, a study of the sediment and pore-water from coastal marshes was performed using both gas chromatography coupled to mass spectrometry (GC-MS) and isotope ratio mass spectrometry (IRMS), which showed that oil from the Deepwater Horizon spill was still present (Natter et al. 2012). In 2013, GC-MS was used to analyze tar balls washing up on the Alabama coastline; through chemical fingerprinting, it was determined that these tar balls originated from the Deepwater Horizon spill (Mulabagal et al. 2013). It should be noted that National Oceanic and Atmospheric Administration (NOAA) and the Food and Drug Administration (FDA) took great measures to ensure that no contaminated seafood reached consumers (Astaiza 2012). As a result of the Deepwater Horizon spill, the Gulf of Mexico Research Initiative coordinated scientific research to understand the environmental impact and consequences of oil spills better. Recently, an offering of books have been published on the environmental impact of the Deepwater Horizon oil spill and the scientific advancements and learnings attained (Murawski et al. 2020a,b).

As a result of the Deepwater Horizon spill, several improvements have been implemented for offshore oil drilling: sturdier wells, improved blowout preventers, remotely operated vehicles (ROVs) on every oil rig, improved preparedness for future blowouts, and improved technology for tracking and controlling spilled oil (Kiger 2012). In terms of sturdier wells, the US Federal Government now requires an engineer to certify that cementing used in offshore drilling can meet expected pressures (DOI 2012), and new stricter specifications for blowout preventers have been published (DOI 2019). The Bureau of Safety and Environmental Enforcement (BSEE), part of the US Department of the Interior (DOI), is the leading agency in the United States responsible for implementing these requirements and enforcing that the requirements are met with offshore drilling.

1.1.2.2 Sanchi Oil Tanker Collision

More recently, on 6 January 2018, the Sanchi oil tanker collided with the CF Crystal cargo ship in the East China Sea around 160 nautical miles from Shanghai, China. Immediately following the collision, the Sanchi broke out in fire; none of her 32 crew survived. The Sanchi was carrying 960,000 barrels (130,000 MT) of a natural-gas condensate, plus nearly 2000 MT of fuel to make the commute from Iran to South Korea when the collision occurred, and on 14 January 2018, the Sanchi sank. Satellite remote sensing (SRS) and the Visible Infrared Imaging Radiometer Suite (VIIRS) Nightfire were used to track the tanker both day and night from the collision to sinking (Sun et al. 2018).

Learnings from this example include the need for better safety training for shipping crews, but also that cleanup technologies need to improve in order to minimize the damages to the environment (Wan and Chen 2018). With the navigation technology available at the time, this collision should have been avoidable; however, human error from both the Sanchi crew and the CF Crystal crew led to the collision. In fact, nearly 80% of all offshore incidents are a result of human error (Bea et al. 1997).

The environmental impact of this spill is different than that of an oil spill, as natural-gas condensate is more volatile than crude oil. Due to this fact, the greatest environmental concern was focused around immediate toxic effects, rather than deposition of the condensate on sediments or polluting beaches (Carswell 2018). As such, focus was placed on analyzing the more toxic components of natural-gas condensate, such as heavy metals and polyaromatic hydrocarbons (PAHs), in open fishing areas. Heavy metals were analyzed by inductively coupled plasma mass spectrometry (ICP-MS), and GC-MS was used to quantify PAHs, and in seawater, sediments and seafood in the fishing grounds connecting China and Japan, Pb and low molecular weight PAHs were found at high levels (Wang et al. 2019). Additionally, as the Sanchi was ablaze, it was unclear as to how much of the condensate would burn off, how much would evaporate, and how much would go down with the ship. As such, a model was created to track the spread of condensate from the Sanchi spill; refer to Figure 1.2 (Qiao et al. 2019). With the issues observed utilizing dispersants to breakdown the hydrocarbons from crude oil spills, for example, at the Deepwater Horizon spill, research into the use of foams to separate water and the oil spilled was conducted (Chen et al. 2019).

1.1.3 Summary

There are many examples of environmental contamination originating from the oil and gas industry. Our goal is not to detail each case; rather, our intention is to provide the reader with information on analytical developments that are being

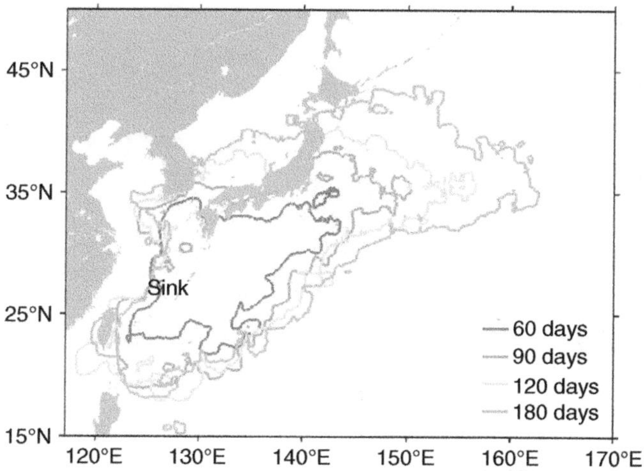

Figure 1.2 Map of the visitation frequency for oil particles based on the assumption that the bunker oil leaks continuously for two months. Source: Qiao et al. (2019). Reproduced with permission, Copyright (2019), Elsevier.

used in the oil and gas industry and how these developments, when applicable, are being exploited for environmental monitoring.

The remainder of this chapter is devoted to providing the reader with a thorough overview of the content contained within this book.

1.2 Petroleum

Petroleum is a naturally occurring material that can be found in geological formations in the Earth. A full overview of petroleum and the petroleum industry is given in Chapter 2; however, a brief overview is given here. Figure 1.3 shows a heat map of the 2018 global petroleum reserves as billions of barrels (EIA 2019b). As can be seen, Venezuela and Saudi Arabia have the largest petroleum reserves.

The petroleum from these reserves is then sourced, typically through drilling. Figure 1.4 shows the 2018 total global petroleum and other liquids production in thousands of barrels per day (EIA 2019b).

The sourced petroleum is then refined and separated via distillation into light (roughly C_4 to C_{12}), middle (up to C_{20}), and heavy distillates ($>C_{20}$). Light distillates include liquefied petroleum gas (LPG), liquefied natural gas (LNG), and gasoline. Middle distillates consist of naphtha, kerosene, jet fuel, and diesel. Heavy distillates are made up of heavy fuel oil, asphalt, bitumen, lubricating oils, and waxes. Many of the products from the distillation of refined petroleum will

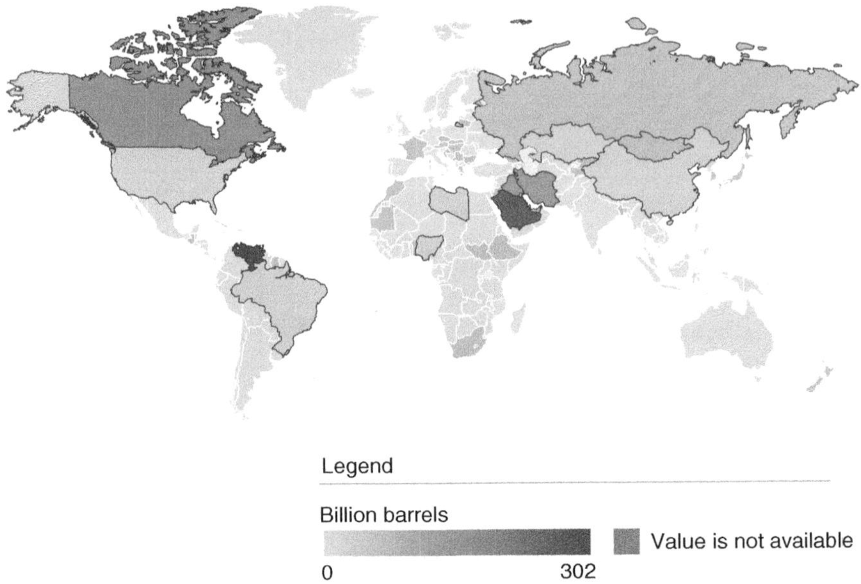

Figure 1.3 Heat map of the 2018 global petroleum reserves. Source: From EIA (2019b).

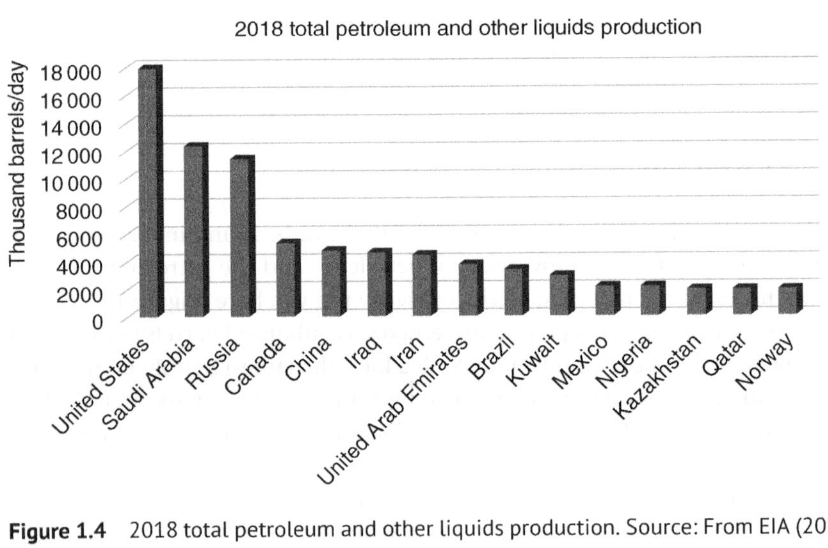

Figure 1.4 2018 total petroleum and other liquids production. Source: From EIA (2019b).

be recognizable to all readers, such as the various types of fuel for the transport industry, which also includes personal vehicles. What may be less well known to certain readers is that from the refined petroleum distillates, further processing can be performed to produce petrochemicals. The most common petrochemicals include olefins (ethylene and propylene) and aromatics (benzene, toluene, xylenes). Global ethylene and propylene production are around 150 million MTs and 80 million MTs per annum, respectively (Zimmermann and Walzl 2009), and combined aromatics global production is around 95 million MTs per annum (Bender 2013). As can be deduced, petrochemicals are not end products; however, they are the building blocks to a wide array of end products used in people's daily lives. Products produced from petrochemicals come from various sectors, including health, hygiene, housing and construction, and food packaging, to name a few.

1.3 Analytics

Throughout the years, as the petroleum industry has evolved, so has the analytical industry. Most significantly, the developments made in gas chromatography (GC) have gone hand-in-hand with the analysis of petroleum, including instrumentation, column developments, and detector technologies. A comprehensive overview of these developments is given in Chapter 3 of this book. As analytical technology advanced, so did the development of standardized methods for the evaluation of petroleum and petroleum distillates; numerous standardized methods have been published and are widely used today. Typically, these methods have been vetted by the American Society for Testing and Materials (ASTM) and are known as ASTM methods, or Universal Oil Products (UOP) standard guidelines and methods.

While conventional standardized methods are available, it should not be assumed that research and development are not present. The remaining chapters in Part 3 of this book will describe in detail the advancements that have taken place as well as non-conventional analyses that are available for the various aspects of the petroleum industry.

Advancements in GC include multidimensional GC as well as coupling the GC separation to non-typical GC detectors, such as IRMS and vacuum ultra-violet (VUV) detection. In multidimensional GC, both heart-cutting and comprehensive analyses have been developed. Both techniques have their merits, while heart-cutting would be beneficial for impurity or contamination identification and even quantification, comprehensive two-dimensional GC gives group type spatial separation, which has been applied to various middle and heavy petroleum distillates. Examples of the application of multidimensional GC in the petroleum industry will be discussed in detail in Chapter 4.

Using gas chromatography–isotope ratio mass spectrometry (GC-IRMS) allows source/origin identification for collected samples. While IRMS has not typically been applied to the petroleum industry, the capability to trace a spill or contamination to the original well, factory, or other location could be possible. The VUV detector, combined with either a one-dimensional or a two-dimensional GC separation, provides identification and quantification of the various hydrocarbon species in petroleum and petroleum distillates using the obtained spectra.

In additional to GC, other separation techniques can also be applied to different aspects of the petroleum industry. Liquid chromatography (LC) is one such technique. LC can be used as a pre-separation method for sample cleanup or sample simplification prior to GC or GC × GC analysis. LC and LC × LC can be combined with different detector technologies, including high-resolution mass spectrometry (HRMS) for the analysis of environmental samples (e.g. soil extracts, groundwater, etc.) for contamination tracking and identification. Chapter 5 gives a detailed overview of the application of LC in the petroleum and petrochemical industries.

Supercritical fluid chromatography (SFC) can be applied using either capillary columns or packed columns. Capillary SFC (cSFC) equipment resembles GC equipment; however, the technique can be applied to heavier petroleum samples, such as vacuum oil distillates. On the other hand, packed column SFC (pSFC) equipment is based on LC equipment. pSFC can be applied to the analysis of surfactants and additives that are utilized in the petroleum industry. The reader will find more information on SFC applied to the petroleum industry in Chapter 6.

A vast array of chromatographic techniques can be applied to the samples originating from the petroleum industry. However, since the 1950s, GC has been the primary tool for analyzing petroleum products and distillates. Two of the most common examples are discussed briefly.

One example where a chromatographic separation was able to replace a labor-intensive distillation often used in the petroleum industry (ASTM-D86 2018) is simulated distillation (SIMDIST). SIMDIST uses a GC separation to quickly and accurately determine the true boiling point distribution of petroleum and petroleum distillates. When GC is used, SIMDIST analysis can be applied to petroleum products and distillates with a final boiling point of 538 °C (ASTM-D2887 2019); however, high temperature GC can also be applied to analyze samples with final boiling points up to 700 °C (Vickers 2002). On the other hand, cSFC can also be utilized for SIMDIST analysis, extending the range of petroleum products and distillates to a final boiling point of around 900 °C (Dulaurent et al. 2007). For additional information on SIMDIST by GC and SFC, the readers are referred to Chapters 3 and 6, respectively.

The quantification of the various hydrocarbon groupings, for example, the paraffins (P), isoparaffins (I), olefins (O), naphthenes (N), and aromatics (A), or PIONA for short, is applied to petroleum distillates in order to better characterize the material. Typically, GC is applied for PIONA quantification, where either a single column or a multidimensional approach can be applied. Due to the inherent complexity of petroleum distillates, especially when olefins and oxygenates are present, the multidimensional approach is the qualified technique for PIONA quantification (ASTM-D6839 2018). When single column approaches are used, the separation is often combined with advanced detector technology. For example, the Pro-Q PIONA Analyzer from Global Analyser Solutions (GAS) combines a single high-resolution GC separation with mass spectrometric detection (GAS 2018). Quantification is performed using dedicated software that utilizes a specific MS spectral library, retention indices, and automated peak deconvolution to provide automated identification, quantification, and reporting. While this single-column PIONA analysis can be utilized, it is not a standardized method. For that, one must look to the gas chromatography–vacuum ultra-violet (GC-VUV), which has recently come out with an ASTM method for single-column PIONA quantification (ASTM-D8071 2019). For additional information on PIONA analysis, the reader is referred to Part 3.

The analytical techniques discussed thus far have focused on lab-scale equipment. However, it is important to mention that online or process instrumentation is also available. An online analyzer is connected to the source of the sample, where the sample stream is allowed to flow from the process equipment to the analyzer. With such a setup, no sample is required to be taken and transported to an analytical lab for analysis, which can have several benefits, including real-time analysis during the process, as well as no degradation or contamination of the sample due to the sample handling steps can occur.

Processes in the petroleum industry are typically run at high temperatures and pressures, and zone classifications are given in the plant for the different hazard levels. It should be noted that differences in zone classification exist between the United States and Europe, which will be described in more detail in Chapter 7. As a result of these zone classifications, it is required that the analytical instrumentation installed online is appropriate for the zone where the installation will occur. GC instrumentation for online SIMDIST or PIONA quantification is constructed very differently from that of a lab instrument. Typical online GC analyzers consist of an explosion-proof outer casing, where the analytical instrumentation is built into that casing. Additionally, due to the high temperature and pressures used in petroleum processes, the introduction of the sample into the online analyzer is a challenging but crucial step. Instead of direct injection into the GC column, a valve injection is performed, for example.

Online GC analyzers are just one example of possible online analyses; some other examples include but are not limited to the following:

- Online chlorine analysis in crude, water, and process streams (ASTM D7536)
- Online viscosity measurements
- Online Freeze, Clouds, and Pour Point Analyzer

- Online Oil in Water Analyzer (Laser-Induced Fluorescence)
- Online Sulfur Analyzer (EDXRF)
- Process Analyzer for Oil/total Fat Composition (FTNIR)
- SIMDIST Process Analyzer (GC-FID)
- PIONA Process Analyzer (GC-TCD/FID)

A wide variety of analytical techniques (off-line and online) and advancements discussed in Part 3 of this book can be utilized in environmental monitoring. While this topic is briefly mentioned with the description of various analytical techniques, several chapters have been dedicated to various aspects of the environment and the impact of the petroleum industry.

1.4 Reservoir Tracers

Oil geologists have used tracers to gain better insight into the geological formations of where drilling occurs for over 60 years. Tracers can be utilized with both conventional and non-conventional reservoirs; conventional reservoirs are those that are straightforward to tap (e.g. pumpjack or wellhead compressor is sufficient), and non-conventional reservoirs are those requiring hydraulic fracking. By using reservoir tracers, valuable knowledge on the fluid flow information is obtained, accurately measuring the oil, gas, and water flow. Such information can help maximize the oil yield to be recovered from a reservoir.

Tracers can be water-soluble, oil-soluble, or even gaseous, depending on the geology and type of reservoir of interest. Early tracers were radioactive; however, as one can imagine, many different hazards were associated with these tracers, especially concerning safe handling. Nowadays, chemical tracers are used, where thiocyanate (SCN–) is the most common water tracer, perfluorocarbons, specifically SF_6, is the most commonly used gas tracer, and naturally occurring isotopes are being evaluated for tracer applications. The reader is guided to Chapter 8 for a detailed explanation of reservoir tracers.

1.5 Emissions from the Petroleum Industry

Global warming is a topic of great environmental concern. The surface temperature of the planet has risen by 0.8 °C since 1980, which is a result of greenhouse gas (GHG) emissions and air pollution (Shaftel et al. 2019). Government policies have been implemented to reduce GHG emissions and air pollution in order to slow down and eventually prevent further increase in the surface temperature.

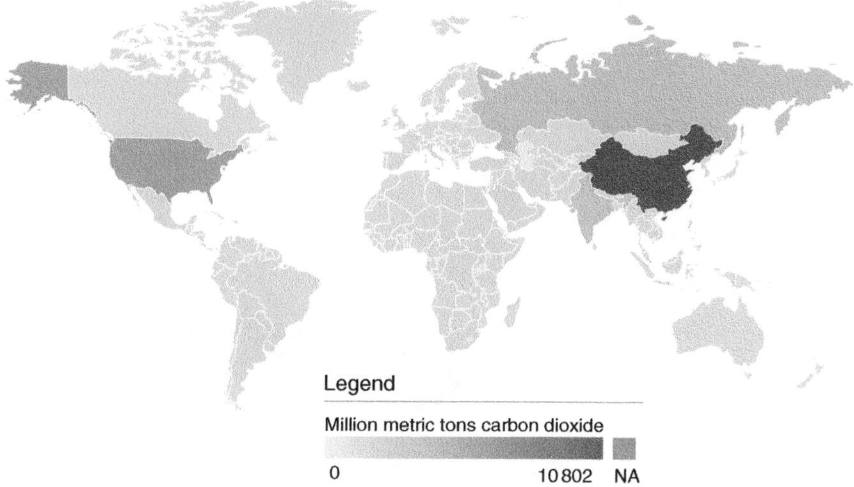

Figure 1.5 2017 Global CO_2 emissions per million MTs. Source: From EIA (2019a).

In the context of the petroleum industry, emissions refer to the release of air pollutants. Air pollutants can arise from exhaust gas, flue gas, and GHG emissions. Exhaust gas is any gas generated by fuel combustion and is typically related to transportation, flue gas is any gas exiting to the atmosphere by a flue, and GHGs are a specific class of gases that absorb and emit radiant energy. The most prevalent GHG is carbon dioxide (CO_2), making up over three-quarters of all GHG emissions; as such, CO_2 emission data is typically used to track GHG emissions. The other GHGs included methane, nitrous oxide, and fluorinated gases. Figure 1.5 shows the 2017 global CO_2 emissions per million MTs (EIA 2019a).

Air pollutants include the GHGs and many other harmful substances, including sulfur oxides (SO_x), nitrogen oxides (NO_x), volatile organic compounds (VOCs), particulate matter (PM), persistent free radicals, ammonia, odors, and radioactive pollutants. Some of these air pollutants will contribute to acid rain, such as SO_x and NO_x, while others will contribute to smog, such as NO_2 and PM. PM is classified by its size; PM10 corresponds to particles less than 10 μm, and PM2.5 corresponds to particles less than 2.5 μm. Both PM10 and PM2.5 are inhalable particles. In addition to sources from the oil and gas industry (e.g. fossil-based fuels and fossil-based electricity), PM can also originate from other industrial and agricultural emissions, the transportation industry, residential heating and cooking, the manufacture and distribution of chemicals, and forest fires. Figure 1.6 shows the global modeled annual mean for PM less than 2.5 μm for 2016 (Chan 2017). When exposed to moderate to high levels of PM2.5, which is above 51 on the air

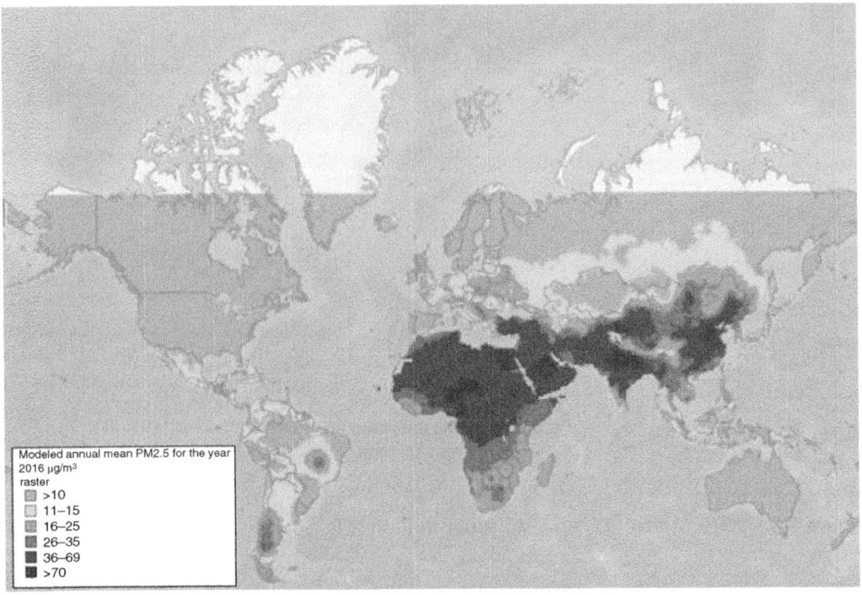

Figure 1.6 Map of global air pollution, modeled annual mean for PM 2.5 for 2016 (μg/m^3). Source: Reproduced from Chan (2017), World Health Organization. License: CC BY-NC-SA 3.0 IGO.4

quality index (AQI), then negative health effects can be observed. AQI is a unitless measure; an AQI score for PM2.5 of 51 is around 12 μg/m^3, which corresponds to the light gray (11–15 μg/m^3) regions (or moderate PM2.5 levels) in Figure 1.6.

Chapter 9 is dedicated to two examples from the oil and gas industry, where both European and US regulatory committees have implemented stringent policies to reduce GHG emissions and air pollution. Not only can regulations differ between the United States and Europe, but regulations can also vary by state in the United States and by country in Europe. The first example focuses on diesel emissions and the improvements made to both the diesel engine and diesel fuel over the years to reduce emissions. The second example focuses on the energy sector, specifically fossil fuel-based energy (e.g. coal-fired power stations and natural gas-fired power stations), and the improvements made to reduce emissions.

1.6 Environmental Analysis and Monitoring

The need for environmental monitoring seems logical, as the risks of spills originating from the well, spills occurring during transport, issues during and arising from refining and processing, and the risk of air emissions at any point in the process are of real concern. Environmental monitoring covers a broad scope,

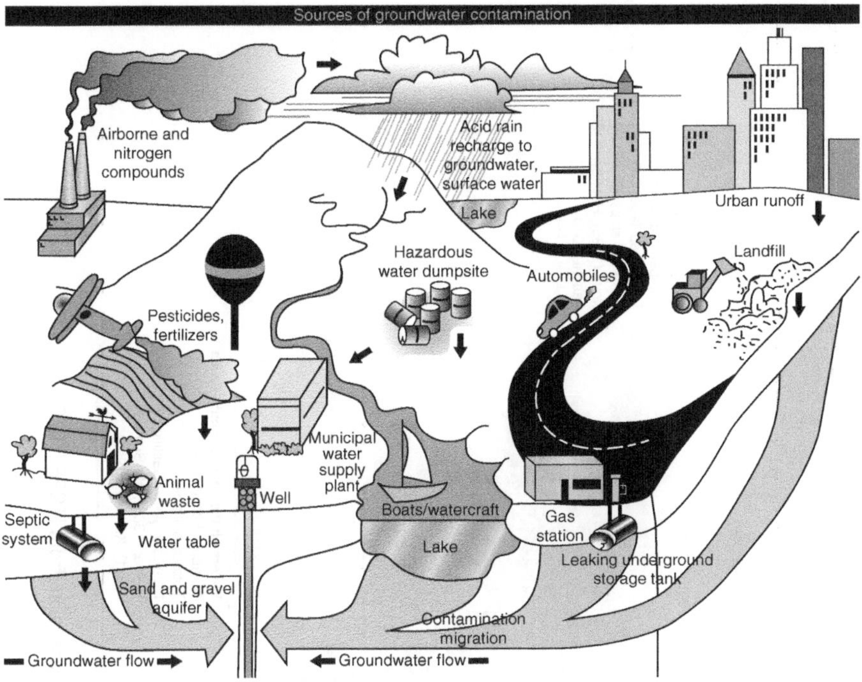

Figure 1.7 Diagram of potential sources for groundwater contamination. Source: Reproduced with permissions, Copyright (2019), Groundwater (2019).

including air analysis, water analysis, and soil/ground analysis and is discussed in detail in Chapter 10. The reader might ask why, with all the analytical developments discussed herein, is there a specific focus on environmental analysis? This comes down to the importance and care that must be taken during the sampling of environmental samples.

Care must be taken when collecting environmental samples, as the sampling technique should not have an influence on the material or affect the results. A sample can be affected by dilution, weathering, and traversing great distances from the original source location, which adds to the complexity of environmental monitoring. Figure 1.7 contains a diagram of potential sources for air pollution and groundwater contamination (Groundwater 2019). Airborne pollutants from industry and transportation not only contribute to air pollution but also create acid rain, which recharges groundwater and surface water. Urban runoff, runoff from agriculture, and ground contamination from industry, landfills, and other sources are all possible points of origin for groundwater contamination.

However, groundwater is not the only water that can be contaminated; of specific concern for the petroleum industry is the risk for tanker oil spills. Figure 1.8 shows a world map of selected oil tanker spills, including the top 20 largest volume

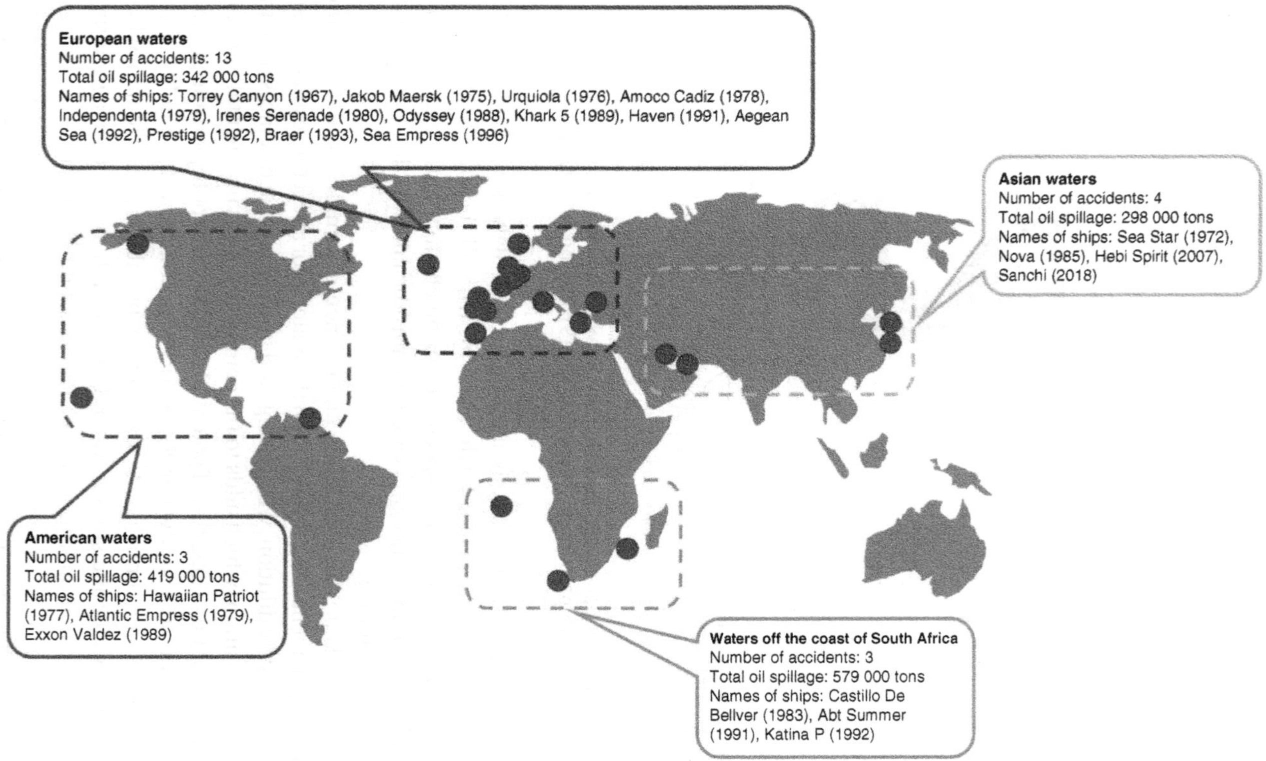

Figure 1.8 Map of major global tanker oil spills since 1967. Source: Data taken from ITOPF (2019).

spills from the last 50 years (ITOPF 2019). While it can be seen that the highest number of tanker spills have occurred in European Waters, these waters have by no means the largest volume of oil spilled. Once spilled, the oil disperses along the surface of the oceans, seas, and other waterways, where dilution and weathering occur.

The focus of Chapter 10 is to illustrate the importance of environmental sample collection and handling and to detail examples compiled from the literature. Examples of total petroleum hydrocarbons (TPHs), VOCs, and non-volatile components, to name a few, are discussed, including the necessary sample preparation steps. In addition to the use of laboratory-scale analytical equipment for environmental analyses, portable measuring devices, such as portable GCs, along with the benefits and drawbacks of such equipment, will also be mentioned.

1.7 Conclusions

The Scope detailed herein provides the reader with a brief overview of the material presented in this book. The topics range from background information on petroleum and the petroleum industry, analytical techniques utilized for various aspects of petroleum and petroleum-related products, and finally, making the connection to environmental impact.

With this book, the authors have attempted to compile a comprehensive overview of analytics utilized in the oil and gas industry with specific emphasis on environmental monitoring. From conventional analytical tools to advanced multidimensional techniques, a wide range of analytics have been covered; however, this book is not all-encompassing.

In addition to the analytical techniques available, the authors have provided the readers with some insight into the future trends we envision for the petroleum industry (Chapter 11), as well as gaps in the current analytical toolbox.

References

Astaiza, R. (2012). 19 months later, here's what we've learned from the Deepwater Horizon oil spill. *Business Insider* (4 December). Retrieved from https://www.businessinsider.com.

ASTM-D2887. (2019). *Standard test method for boiling range distribution of petroleum fractions by gas chromatography*. West Conshohocken, PA: ASTM International. Retrieved from https://www.astm.org.

ASTM-D6839. (2018). *Standard test method for hydrocarbon types, oxygenated compounds, and benzene in spark ignition engine fuels by gas chromatography*. West Conshohocken, PA: ASTM International. Retrieved from https://www.astm.org.

ASTM-D8071. (2019). *Standard test method for determination of hydrocarbon group types and select hydrocarbon and oxygenate compounds in automotive spark-ignition engine fuel using gas chromatography with vacuum ultraviolet absorption spectroscopy detection (GC-VUV)*. West Conshohocken, PA: ASTM International. Retrieved from https://www.astm.org.

ASTM-D86. (2018). *Standard test method for distillation of petroleum products and liquid fuels at atmospheric pressure*. West Conshohocken, PA: ASTM International. Retrieved from https://www.astm.org.

Bea, R., Holdsworth, R., and Smith, C. (1997). Human and organizational factors in the safety of offshore platform. In: *1996 International Workshop on Human Factors in Offshore Operations* (eds. R.G. Bea, R.D. Holdsworth and C. Smith), 109–126. New York, NY: American Bureau of Shipping.

Bender, M. (2013). Global aromatics supply – today and tomorrow. *Ocean Gas European Magazine* 39: 209–212.

Carswell, C. (2018). Unique oil spill in East China Sea frustrates scientists. *Nature* 554: 7690.

Chan, M. (2017). A global health guardian: climate change, air pollution, and antimicrobial resistance. In: *Ten Years in Public Health: 2007–2017* (ed. M. Chan), 136–142. France: World Health Organization.

Chen, X.Y., Wang, L., Nagamine, S., and Ohshima, M. (2019). Study oil/water separation property of PE foam and its improvement by in situ synthesis of zeolitic–imidazolate framework (ZIF-8). *Polymer and Engineering Science* 59: 1354–1361. https://doi.org/10.1002/pen.25118.

DOI (2012). *Oil and Gas and Sulphur Operations on the Outer Continental Shelf–Increased Safety Measures for Energy Development on the Outer Continental Shelf (4310-VH-P)*. Online. US Department of the Interior (DOI). Retrieved from https://www.bsee.gov.

DOI. (2019). BSEE finalizes improved blowout preventer and well control regulations (Press release). Retrieved from https://www.doi.gov (accessed 19 March 2020).

Dulaurent, A., Dahan, L., Thiebaut, D. et al. (2007). Extended simulated distillation by capillary supercritical fluid chromatograph. *Oil & Gas Science and Technology – Rev. IFP* 62: 33–42. https://doi.org/10.2516/ogst:2007003.

EIA (2019a). *CO_2 Emissions*. Online. US Energy Information Administration (EIA). Retrieved from https://www.eia.gov.

EIA (2019b). *Crude Oil Including Lease Condensate Reserves 2017*. Online. US Energy Information Administration (EIA). Retrieved from https://www.eia.gov.

EPA. (2017). Deepwater Horizon – BP Gulf of Mexico oil spill. Retrieved from https://www.epa.gov (accessed 19 March 2020).

GAS (2018). *ProQ-PIONA Analyser: Single column GCMS method.* Application Note: 213WA1409D. Breda, The Netherlands: Global Analyser Solutions (GAS).

Griffin, J. (2010). Getting to grips with the Gulf of Mexico spill. *OPEC Bulletin* 7–8: 20–23.

Groundwater. (2019). Sources of groundwater contamination. Retrieved from https://www.groundwater.org (accessed 19 March 2020).

ITOPF (2019). *Oil Tanker Spill Statistics 2018.* Online. International Tanker Owners Pollution Federation, Ltd.

Kiger, P.J. (2012). *5 Improvements in Offshore Oil Drilling.* How Stuff Works?. Retrieved from https://science.howstuffworks.com/environmental/energy/5-improvements-offshore-oil-drilling.htm.

Mulabagal, V., Yin, F., John, G.F. et al. (2013). Chemical fingerprinting of petroleum biomarkers in Deepwater Horizon oil spill samples collected from Alabama shoreline. *Marine Pollution Bulletin* 70: 147–154. https://doi.org/10.1016/j.marpolbul.2013.02.026.

Murawski, S.A., Ainsworth, C.H., Gilbert, S. et al. (2020a). *Scenarios and Responses to Future Deep Oil Spills: Fighting the Next War.* Basel, Switzerland: Springer.

Murawski, S.A., Ainsworth, C.H., Gilbert, S. et al. (2020b). *Deep Oil Spills Facts, Fate, and Effects.* Basel, Switzerland: Springer.

Natter, M., Keevan, J., Wang, Y. et al. (2012). Level and degradation of Deepwater Horizon spilled oil in coastal marsh sediments and pore-water. *Environmental Science and Technology* 46: 5744–5755. https://doi.org/10.1021/es300058w.

Qiao, F., Wang, G., Yin, L. et al. (2019). Modelling oil trajectories and potentially contaminated areas from the Sanchi oil spill. *Science of the Total Environment* 685: 856–866. https://doi.org/10.1016/j.scitotenv.2019.06.255.

RESTORE (2015). *Department of the Treasury Regulations for the Gulf Coast Restoration Trust Fund.* Washington, DC: Department of the Treasury.

Rico-Martínez, R., Snell, T.W., Shearer, L., and T. L. (2013). Synergistic toxicity of Macondo crude oil and dispersant Corexit 9500A® to the *Brachionus plicatilis* species complex (Rotifera). *Environmental Pollution* 173: 5–10. https://doi.org/10.1016/j.envpol.2012.09.024.

Shaftel, H., Jackson, R., and Callery, S. (2019). Global temperature. Retrieved from https://climate.nasa.gov/vital-signs/global-temperature (accessed 19 March 2020).

Sun, S., Lu, Y., Liu, Y. et al. (2018). Tracking an oil tanker collision and spilled oils in the East China sea using multisensor day and night satellite imagery. *Geophysical Research Letters* 45: 3212–3220. https://doi.org/10.1002/2018GL077433.

Vickers, A.K. (2002). *Higher-Temperature Simulated Distillation with DB-HT Sim Dis Columns*. USA: Agilent Technologies. https://www.agilent.com.

Wan, Z. and Chen, J. (2018). Human errors are behind most oil-tanker spills. *Nature* 560: 161–163. https://doi.org/10.1038/d41586-018-05852-0.

Wang, Q., Peng, F., Chen, Y. et al. (2019). Heavy metals and PAHs in an open fishing area of the East China Sea: multimedia distribution, source diagnosis, and dietary risk assessment. *Environmental Science and Pollution Research* 26 https://doi.org/10.1007/s11356-019-05355-z.

Zimmermann, H. and Walzl, R. (2009). Ethylene. In: *Ullmann's Encyclopedia of Industrial Chemistry* (ed. B. Elvers). Weinheim: Wiley.

Part II

Introduction to the Petroleum Industry

2

Petroleum: From Wells to Wheels

Clifford C. Walters, Steven W. Levine, and Frank C. Wang

ExxonMobil Research and Engineering Company, 1545 Route 22 East Clinton, Annandale, NJ 08801, USA

2.1 Introduction

The petroleum/petrochemical industry is complex. It is composed of interconnected operations involved in the exploration and production of oil and gas (upstream), transport (midstream), refining to liquid fuels and lubes (downstream), and in the conversion to chemical products and consumer products (Figure 2.1). The companies at each stage range in size from very small, private businesses to large companies that focused on only one aspect, to nationally owned operations, to very large integrated and globally positioned corporations. Additionally, each location has its own history and evolution, so one site may not look like another for a variety of reasons.

In this chapter, we briefly describe the basic principles and practices that comprise each stage of the industry. Our goal is to provide a broad overview of key elements that will serve as a foundation for this volume, rather than a complete description or comprehensive literature survey. The reader is encouraged to use the general references listed at the end of the chapter for more information on specific topics.

2.2 Petroleum in the Ancient World

It is tempting to mark the beginnings of the petroleum industry with modern benchmarks such as the drilling of the 1859 Drake well in Pennsylvania. But to do so ignores millennia of human innovation. Practically every aspect of the petroleum industry – exploration, production, transportation, and refining – has its roots in the ancient world. Taking a grand view, petroleum has influenced

Analytical Techniques in the Oil and Gas Industry for Environmental Monitoring, First Edition.
Edited by Melissa N. Dunkle and William L. Winniford.

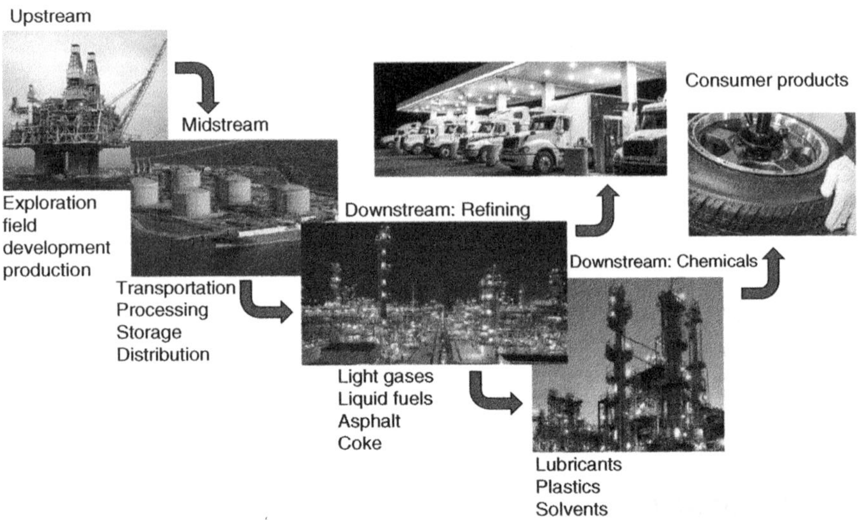

Figure 2.1 Petroleum industry value chain from wells to wheels. Source: Photo courtesy of ExxonMobil.

all cultures to some degree, and the evolution of many of today's processes and applications can be traced through history.

The earliest known use of petroleum is not as fuel but as a glue and sealant (Connan 1999; Speight 2014). Prehistoric discoveries at a ~70,000-year-old site in Syria (Boëda et al. 2008) and in a ~40,000-year-old cave in Romania (Cârciumaru et al. 2012) show that bitumen was used to adhere stone tools to hafts. The Chumash in the Channel Islands of California waterproofed their seagoing canoes with bitumen ~10,000 years ago (Brown 2016). Other prehistoric finds reveal that bitumen was used as a glue for repairing broken statues and pottery and to attach decorative façades of minerals to ornamental items.

Abundant deposits of surface bitumens and seeping oil contributed to the rise of the earliest city-states of Mesopotamia. Along the banks of the Euphrates in southern Mesopotamia, the "Ubaids" (early Sumerians) lacked stone and constructed their houses of reeds and mud that eroded away with the annual floods. Their solution was to heat locally available surface bitumen with the clay and straw for bricks that were then cemented with more bitumen resulting in water-resistant, permanent structures (Connan 1999). This technology dates back to ~4500 BCE and was used for millennia by the succession of Mesopotamian cultures, the Sumerians, Babylonians, and Assyrians, in the construction of paved roads, palaces, and ziggurats. Bitumen mixed with gypsum, limestone, clays, and other minerals form asphalt mastic, an ancient petrochemical with many properties of modern

Figure 2.2 Examples of petroleum bitumen used in the ancient world. (a) Bitumen mastic found at Susa. Fragment of bas-relief, the Spinner, from the Neo-Elamite period (around 750 BCE). Source: Raux (2010). Reproduced with permission, Copyright (2010), RMN-Grand Palais (Musée du Louvre) / Franck Raux. (b) Bitumen mortar used in the construction of the ~third century BCE Ekur ziggurat at Nippur. Source: Al-Sibahy and Edwards (2017). Reproduced with permission, Copyright (2017), Elsevier. (c) Extensively blackened with bitumen adult male XXI dynasty mummy (1064–948 BCE). Source: Clark et al. (2016). Licenced under CC BY 4.0.

plastics. Excavations near the Tigris and Euphrates rivers have found entire walls of asphalt mastic constructed for flood protection. The early Sumerians also constructed boats made of bundled reeds that were waterproofed with bitumen, allowing them to venture in the Arabian Sea (Connan et al. 2005). Other examples of bitumen as a waterproofing sealant are known throughout the ancient world. Bitumen mortar was used in the construction of the Great Bath built ~2500 BCE in Mohenjo-daro (Pakistan), and the Olmecs (~1200–400 BCE) used bitumen to waterproof structures and boats. Figure 2.2 shows three examples of petroleum bitumen used in the ancient world: a bas-relief, mortar, and coating a sarcophagus.

Petroleum has also shaped man's religious beliefs in other cultures. The bitumen deposits of Hit were so critical to the Babylonian and Assyrian kingdoms that the city was considered sacred. Prehistoric natives of California (Brown 2016) and Mexico (Argáez et al. 2011) applied tar to bodies in ritual mourning and burial ceremonies. The natural gas seeps in the Absheron Peninsula inspired the Zoroastrian concept of the eternal flame, and the intoxicating hydrocarbon vapors gave visions to the Oracle of Delphi (Etiope 2015).

The earliest use of crude oil as a combustible fuel is lost in antiquity. Solid bitumen and even oil-soaked soil will burn, but not cleanly, so its use as a fuel for lamps and cooking was restricted. Only the lightest crudes are suitable, and these are rare at the surface, occurring only in a few areas with active seepage. The Medes may be the first oil explorationists sourced, using the occurrence of natural seeps around the southern shores of the Caspian Sea as indicators for where to hand-dig shallow wells for oil as early as the seventh century BCE. In 329 BCE, while camped on the Caspian shores near the Oxus River (modern-day Amu Darya), Alexander the Great marveled at natives burning a yellowish oil for light. Ancient writings of Herodotus, Plutarch, Pliny the Elder, Marcus Vitruvius, and others document the occurrence of other seeps in various locations throughout the Mediterranean and the Middle East.

Abundant surface seeps of oil in the Absheron Peninsula, near the modern-day city of Baku, gave rise to one of the earliest centers of industrial-scale oil production and export. There is archeological evidence that high-quality Baku oil was collected and traded throughout Asia as early as the third and certainly by the ninth century. By the twelfth century, fields consisting of hand-dug shafts were producing segregated white and brown oils, establishing the concept of variable oil pricing based on composition. Marco Polo visited the Baku oil field in the late thirteenth century, noting its use for lighting and in medicines. By the sixteenth century, as much as 60,000 kg (~380 barrels) of oil were being produced and transported by camel caravans per day.

The Chinese can take credit for inventing drilling for oil and gas (Feng et al. 2013). Around 800 BCE, a thriving salt industry was established in the Zhejiang Province based on the solar evaporation of seawater. By 350 BCE, wells were being dug by hand to produce subsurface brines that contained a much higher salt content. A percussive drilling technology was developed that, by ~250 BCE, were reaching depths up to 140 m. The technical solutions the Chinese developed to drill and maintain these wells foreshadowed many modern methods. The wells were drilled using iron bits of several different designs to deal with different rock types that were attached to bamboo piping that was wound around a large wheel, essentially equivalent to modern coiled tubing (Khun 2004). Various downhole tools were developed to lift out the pulverized rock, repair cave-ins with cement, and remove scale build-up. Around 350 BCE, oil recovered from hand-dug wells began to be burned to speed up evaporation. In 347 BCE, the Chinese adapted the percussive method long used for the production of deep brines to drill a well specifically to recover oil.

Most crude oils will not burn cleanly in oil lamps or cookstoves; for that, the kerosene or naphtha fraction needs to be separated from the heavier components. The Babylonians had a primitive method where a hide was stretched over boiling bitumen, and the condensed liquid was squeezed out. By the twelfth century, more advanced methods of distillation using an alembic, consisting of a boiling pot, condenser, and collection vessel, were described by Muhammad al-Razi, a Persian physician and chemist. Crude oil was being distilled at Baku prior to shipment to remove the foul odor of light organic acids produced by biodegradation and is probably the first example of oil quality upgrading at the site of production. By the twelfth century, petroleum refining reached an artisanal scale in Europe and industrial scale in Egypt and Damascus. A fire at the residence of the Fatimid ruler is said to have destroyed 100 tons of refined naphtha.

Petroleum has always been a fungible resource, with its value determined by principles of economic supply and demand. From around 1100 BCE, when the supply of aromatic resins used in ritual mummification became depleted, the Egyptians began to substitute bitumen (Clark et al. 2016), primarily from the Dead Sea, mixed with pine resins, beeswax, and fats. The Egyptian demand for bitumen for mummification provided great wealth to the Nabataeans, a nomadic tribe that roamed the Arabian Desert and controlled the lucrative Dead Sea bitumen trade for centuries. They fought many battles with the Greeks, Macedonians, and Romans, occasionally losing control but later regaining authority. The Nabataean trade empire collapsed after the Octavian banned mummification, instantly cutting demand. The term mummy is derived from Medieval Latin *mumia*, which itself is from the Arabic word *mumiyah*, meaning an embalmed body, which was taken from the Persian word *mum* meaning wax (AHDictionary 2020). Medieval Persian physicians formulated bitumen into *mūmiyā*, a medicine that was used as a salve to heal cuts and fractures and taken internally for stomach ulcers, tuberculosis, pleurisy, and edema. Crusaders learned of this miraculous preparation, and by the twelfth century its demand in Europe outstripped supplies. This shortage was first met by recycling the bitumen harvested from Egyptian mummies, then later by grinding up the entire remains. Renaissance scholars finally realized that the drug was ineffective and the market again collapsed. The notion that the application and consumption of crude oil have beneficial medicinal properties returned many times and was a widely held belief in nineteenth century America. Prior to founding America's first refinery in 1853, Samuel Kier sold oils distillate as a patent medicine. Belief in the curative power of crude oil is still held today in rural West Africa and by some alternative medicine practitioners.

Oil is not only a resource that was fought over for millennia, but it was also used as a weapon in battle, likely one of the earliest uses of petroleum was as a combustible fuel. The Assyrians are known to have used oil-soaked flaming arrows and incendiary bombs made of oil-filled clay pots as early as ~900 BCE. Historic texts tell of Persians using oil-soaked flaming arrows in the siege of Athens in 480 BCE, and the Romans later launched incendiary bombs from catapults. Perhaps the most effective and feared oil-based weapon is the flamethrower. In 424 BCE, the Boeotians, allies of Sparta, used a flamethrowing device to route the Athenians from Delium during the Peloponnesian War, though the nature of the fuel is not known. The Byzantines used Greek fire shot from ship-based flamethrowers to break the Arab naval siege of Constantinople in 672 BCE. This technology was declared a state secret, and the exact formulation of Greek fire remains unknown today but was most likely a mixture of crude or refined oil mixed with plant resins, resulting in a flammable liquid with properties similar to napalm. The Chinese invented a flamethrower in the tenth century that used the naphtha distilled from crude oil. The Northern Song Dynasty (960–1126) established the "fierce oil workshop," the world's first military-industrial complex to produce the flammable liquid.

2.3 The Petroleum System

An integrated view of the Earth and its geologic and geochemical processes is required to understand how oil and gas accumulations form. Hydrocarbons are found throughout Earth's atmosphere, waters, and all recent and ancient sediments. Even igneous and metamorphic rocks can contain trace amounts of petroleum preserved in fluid inclusions. However, petroleum must be concentrated to be exploited economically, and such focusing occurs only under specific conditions. For conventional subsurface plays, the key elements are the presence of organic-rich source rocks, faults, and permeable strata that allow for petroleum migration, highly porous reservoir rocks to store the migrated petroleum, and impermeable sealing rocks or structures that limit further migration, trapping the petroleum. The mere presence of these geologic elements within a basin is insufficient to guarantee petroleum reserves. Timing of the geologic processes is critical as the elements must be present as the source rocks are generating and expelling petroleum, and traps, once charged, must maintain their integrity. In unconventional plays, such as tight shales and coals, source rocks themselves serve as the reservoir and seal. Collectively, these elements and processes constitute the *Petroleum System*.

2.3.1 Source Rocks

Petroleum originates from sedimentary organic matter derived from biological precursors that generate oil and gas when buried and heated. This concept is nearly universally accepted. As early as 1757, Mikhailo Lomonosov advocated that coals, known from fossil evidence to be derived from plant remains, generated oil when buried. The concept that marine organisms also could contribute to oil generation was advanced by Belsazar de la Motte Hacquet in 1793 and Charles Hatchett in 1798. Mounting evidence for the biogenic hypothesis accumulated in the nineteenth and early twentieth centuries. In 1936, Alfred Treibs established a link between chlorophyll in living organisms and porphyrins in petroleum (Treibs 1936) (in German). Additional geochemical evidence followed with the discoveries that low to moderate maturity oils still retained molecules with optical activity, evidence for biogenic carbon isotopic fractionation, and hydrocarbon structures that can be traced back to specific biological precursors. These are chemical fossils (biomarkers) that can be used to correlate produced oils to specific organic-rich sediments, the source rocks.

A few scientists, primarily several noted astronomers and enclaves in Eastern Europe, have advocated for a different origin, one where oil and gas are generated deep within the Earth and migrate to the surface. The abiotic theory can be traced to Dmitri Mendeleev, who, in 1877 reasoned that petroleum results from the reaction of water with metallic carbides forming acetylene that could then condense further into larger hydrocarbons (Mendeleev 1877) (in French). Indeed, abiotic hydrocarbons are generated via rock–water interactions involving serpentinization of ultramafic rocks, thermal decomposition of siderite in the presence of water, and during magma cooling as a result of Fischer–Tropsch type reactions (Etiope 2015). Vast quantities of hydrocarbons also are known to occur in many planetary bodies and in the interstellar media. However, the contribution of abiotic hydrocarbons to the global crustal carbon budget on Earth's surface is considered by most geoscientists to be inconsequential. There are no known commercial accumulations of oil or gas that can be shown to be definitively abiogenic.

Source rocks are water-deposited sedimentary rocks that contain sufficient amounts of biogenic organic matter to generate oil and/or gas when heated. Three factors determine the hydrocarbon generative potential of a source rock (Figure 2.3); for example, how much organic matter it contains and the chemical nature of that organic matter: the influx of organic matter (primary production and secondary transport), the rate of its destruction before lithification, and the influx of inorganic minerals that dilute the amount of preserved carbon (Bohacs et al. 2005).

With a few minor exceptions, the biochemicals produced and comprising living organisms do not directly form petroleum. Rather they are chemically altered, joining together into a geopolymer termed *kerogen* as unconsolidated sediments are transformed into rocks. The chemical nature of kerogen, whether it generates oil or gas primarily, is determined by the type of biotic input and the conditions upon deposition. Biologic input can be from land plants, marine organisms, or single-celled microbes. Land plants are mostly composed of cellulose and lignin.

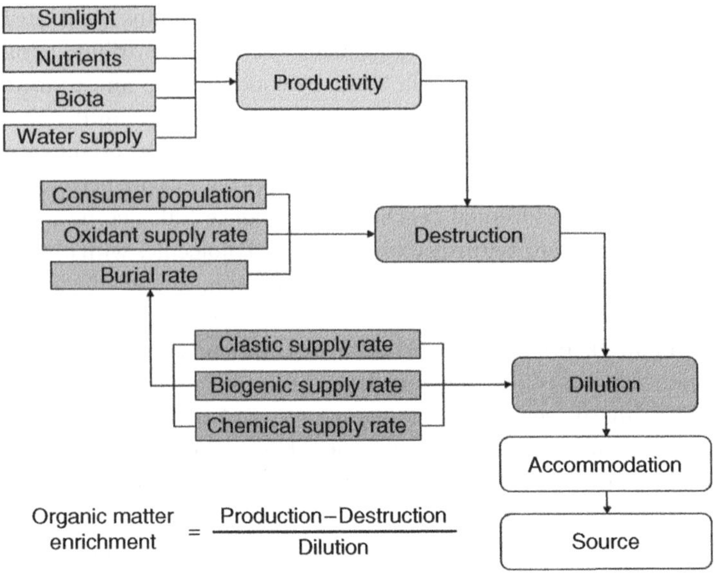

Figure 2.3 Major factors that determine the amount of organic matter preserved in sedimentary rocks.

Cellulose and related polymers of sugars are relatively easy to break down, while lignin, an aromatic-rich cross-linked polymer with mostly methyl groups, is highly resistive to decomposition. Consequently, when land plant biomass is contributed to sediments, it tends to form aromatic-rich kerogens that are prone to generate mostly methane upon heating. Marine organisms, such as algae and plankton, tend to be much more enriched in long aliphatic lipids, giving rise to aliphatic-rich kerogens.

Organic matter can be produced in the water column above the sediment (autochthonous) or transported via wind or rivers to the site of sedimentation (allochthonous). The amount of autochthonous organic matter is determined by the degree of primary biologic productivity, which in turn is controlled by the nutrient supply, solar radiance, and other oceanographic factors. The open ocean is nutrient-poor. Environments such as coastal upwelling zones where deepwater nutrients are carried to the surface or coastal runoff from fluvial systems promote primary productivity. Allochthonous organic matter is mainly derived from near-shore plants or kerogens and coals eroded from older sedimentary rocks. Eroded organic matter is usually highly oxidized and inert, and while contributing to the amount of total organic carbon (TOC), it has no generative potential.

Preservation is key for source rock formation. Most primarily, biotic organic matter is formed in shallow water where there is sufficient sunlight to promote photosynthesis. This organic matter must fall through the water column to reach the sediment–water interface. If the water column contains oxygen, the organic matter will be chemically oxidized and consumed by aerobic organisms. In high energy environments, oxygen can persist below the sediment–water interface and be further consumed by macro- and micro-fauna. Oxic conditions result in poor preservation, and most of the organic matter is converted to CO_2, with the residual carbon being depleted in hydrogen. Organic matter is preserved in depositional environments where the water above the sediments transitions from oxic to anoxic (no oxygen) conditions. Anaerobic microbes will degrade the organic matter, but at a much slower rate than aerobic organisms. When sulfate is present, common to all marine environments and in some onshore settings, bacterial sulfate reduction will occur, producing H_2S that can react with the organic matter forming sulfur-rich kerogens. Euxinia occurs when free H_2S is present in the water column (Pettijohn 1957). The interface between the oxic surface water and the underlying euxinic water promotes the growth of anaerobic green and sulfur photosynthetic bacteria that utilize H_2S instead of H_2O as the terminal reductant.

The rate of deposition of inorganic minerals is the third major factor in source rock formation. If the flux of inorganic minerals greatly exceeds that of organic matter, the carbonaceous material will be diluted, and the resulting rock will have low carbon concentrations. On the other hand, high sedimentation rates can enhance the preservation of organic matter as more material buries the carbonaceous matter, removing it from oxic conditions. River deltas are such environments. The ideal conditions for source rock deposition are those involving high primary productivity, anoxic bottom waters for enhanced preservation, and relatively slow rates of sedimentation of fine-grained minerals (e.g. clays or carbonate muds) or slow chemical precipitation of minerals (e.g. carbonates, cherts).

Certain environments are prone to give rise to source rocks. These can be generally grouped into lake systems, marine carbonates, marine shale, river deltas, and coastal swamps. These environments promote high primary productivity and preservation, but by different means.

Lacustrine source rocks form in lakes – ephemeral bodies of water that occur on the continents in topographic lows – mostly due to local tectonics or the glacial retreat. Large tectonic lake systems can form along continental rifts. Lake waters can be completely oxic or contain anoxic bottom water. Stratification can occur in very deep lakes where the bottom waters cannot mix with the surface or when water evaporation exceeds recharge resulting in a salinity density gradient. Biotic input into lake systems includes mainly land plants from the surrounding terrain and freshwater algae, which can produce seasonal mats that sink into the sediments, eventually producing fine lamina of kerogen. Lakes typically have low sedimentation rates and are devoid of sulfate, which, when combined with high productivity and anoxic bottom water conditions, can result in oil shales, source rocks that contain very high concentrations of organic carbon that is hydrogen-rich, and prone to generate oil.

Marine carbonate source rocks form in restricted, low latitude embayments, or sags, and on shelves during transgressions. Both are considered to be low energy environments conducive to high productivity that promote preservation with low rates of sedimentation. During the Mesozoic era, the Middle East was situated at the Equator and was largely a carbonate platform. Vast areas contained salinity and/or temperature stratified anoxic to euxinic water. Clastic input was very low, and most of the carbonate formed from evaporation. Similar conditions produced the Smackover and Sunniland carbonate source rocks in the United States. Marine carbonates also form during periods of high sea level as marine waters flood the shallow continental shelf. These conditions can result in nutrient-rich surface waters with high primary productivity overlaying stratified anoxic bottom water. These source rocks can be thick and widespread (e.g. Duvernay and La Luna).

Upwelling can give rise to marine shales. Regions, where persistent winds blow the surface water off the coast, cold, nutrient-rich deep ocean water, can be transported upward, promoting high primary productivity and anoxic conditions. Sedimentation is slow as the clastic influx from the onshore is inhibited and may be dominated by chemical precipitation. Many marine source rocks throughout Earth's history are associated with upwelling zones, including the Lower Silurian Qusaiba "hot shale" that was deposited on the northern margin of Gondwana as it was positioned in a high southern latitude. Here, the upwelling zone was extensive, and Silurian source rock was deposited across much of northern Africa and the Middle East.

Marine shales can be deposited offshore, past the shelf, on the slope, and in deep-water basins (Figure 2.4). When high productivity in the shallow waters results in a high influx of organic matter into the water column, aerobic organisms metabolizing the settling organic matter can consume the available oxygen, creating oxygen-minimum layers that intersect the continental slope. The development of oxygen-minimum zones can be seasonal and is most common in areas such as the Gulf of Mexico, where ocean circulation is somewhat restricted. Anoxic conditions can also develop in sub-basins, localized topographic lows that allow for salinity or temperature stratification. Throughout Earth's history, the amount of oxygen in the oceans has been determined by global atmospheric temperature. During periods of high global atmospheric temperature, the amount of dissolved oxygen in the oceans was much lower than today. During periods in the Jurassic and Cretaceous, the oceans become largely anoxic. These oceanic anoxic events (OAEs) may have been triggered by tectonic positioning of the continents, impeding global circulation and increasing ocean nutrients from enhanced weathering that promoted high productivity and acceleration and expansion of oxygen minimum zones. Many of the world's marine shale source rocks correlate with OAEs.

Figure 2.4 Lower Jurassic Blue Lias Formation, Kilve, Somerset, UK. This major lithostratigraphical unit with a regional extent comprises alternations of limestone (white) and beds of gray marl and organic-rich laminated black shale, which variably contain terrigenous clay minerals and marine- and terrestrially derived organic matter. Source: Photo courtesy of ExxonMobil.

Coastal swamps and lagoons are ideal for forming coals. Terrestrial land plants growing near low-lying wetlands contribute large quantities of organic matter to the soil. Here, the influx of plant debris can be so high that it exceeds the rate of its biodegradation, which can be slowed by acidic water due to tannins and by the development of anoxic conditions from stagnation. Vast peatland can develop, which upon burial transforms into lignite and higher grade coals. This type of kerogen is hydrogen-poor and typically considered gas-prone or non-source, while rare depositional settings can give rise to oil-prone coals. Cannel coals are shales that occur at the top or bottom of large coal seams and contain large amounts of spores, pollen, and resins. Some shales, particularly from the Tertiary in Southeast Asia, can contain facies enriched in these oil-prone macerals.

Large petroleum systems are common to many of the world's major river deltas (e.g. Assam-Barail, Beaufort-Mackenzie, Mississippi, Mahakam, and Niger). However, the sources of the petroleum in these systems are difficult to determine and are often mixtures from older marine source rocks lying below the delta sediments, intra- and pro-delta shales with a high content of terrigenous organic matter, and coastal hydrogen-rich coals. The intra-deltaic shales typically have low TOC that has been poorly preserved and are considered as having only marginal potential to generate gas. Pro-deltaic shales generally have higher generative potential; they are deposited farther from the river outflow, receive less allochthonous matter and large-grain clastics, and more marine input and fine-grain clays. Preservation is enhanced as the deeper water is more likely anoxic and clastic dilution is lower.

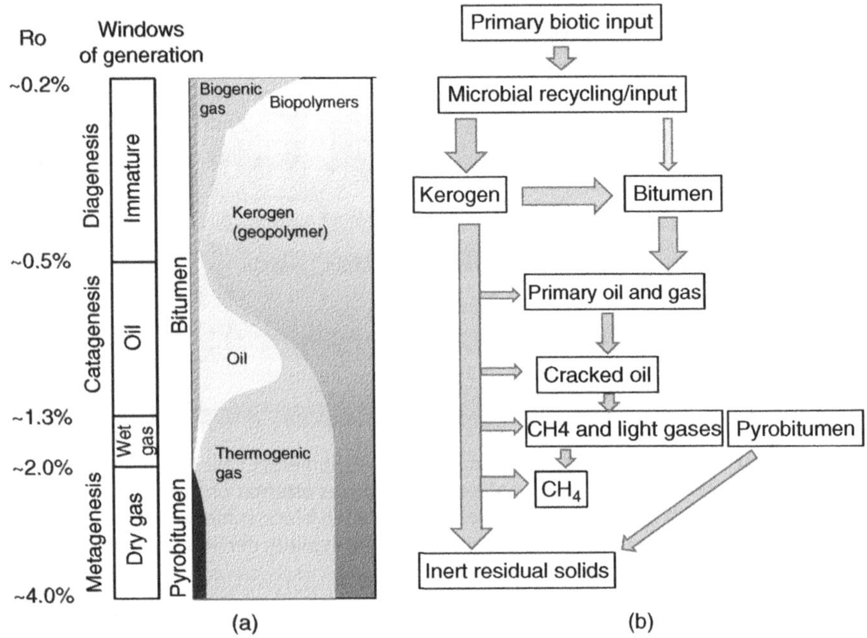

Figure 2.5 (a) Source rocks past through stages of diagenesis, catagenesis, and metagenesis as they are exposed to increasing temperature. The lithological stages correspond to the "windows" of oil, wet gas, and dry gas generation (b). Values of vitrinite reflectance (%Ro), an organic petrographic measurement, are approximate as they are dependent on the kerogen type and geothermal history.

2.3.2 Generation of Petroleum

Upon burial, sediments become source rocks as they are exposed to increasing temperature and pressure (Figure 2.5). Lithification occurs as the unconsolidated loose grains compact, expel water, and undergo reactions, whereby unstable minerals convert to more stable ones, and cements are formed from dissolution and precipitation. Similarly, the organic matter is altered as the preserved biochemicals are broken apart and rearranged into kerogen. These processes occur during *diagenesis*, the stage when the source rock has not generated or expelled any significant amount of petroleum. Such source rocks are termed *immature*.

As the source strata are buried deeper and exposed to higher temperatures and pressure, undergoing *catagenesis*. Here, the kerogen begins to decompose from thermal cracking, producing first *bitumen*, a proto-oil enriched in heteroatomic species. As the source rock *matures*, kerogen and bitumen further crack to yield hydrocarbons. The source rock is said to have entered the window of oil generation. Additional heating will accelerate the thermal processes until the rate of oil generation is maximized, and the source is said to be at peak oil generation. Eventually, the generative potential of the remaining kerogen and bitumen drops until no additional liquids can be formed – the end of the oil window. Secondary reactions occur as larger hydrocarbons crack to smaller species and heteroatoms are eliminated. These reactions are slow during oil generation but accelerate and become dominant afterward.

The end of catagenesis and the onset of metagenesis are marked by the absence of liquid hydrocarbons. Source rocks at this level of thermal alteration are said to be overmature. During catagenesis, oil-prone kerogen is being converted from an aliphatic-rich, disorganized mass comprised of small naphtheno-aromatic moieties typically with four or fewer rings, to a highly condensed aromatic structure that is becoming more ordered in structure. Although hydrogen-depleted, the residual kerogen and insoluble pyrobitumen can still generate small amounts of methane as they undergo additional rearrangement to form graphite.

As oil and gas are being generated, they are first absorbed into the kerogen matrix. The absorptive capacity of the kerogen changes as it matures, and prior to the oil window, can no longer retain all products (Pathak et al. 2017). Molecular solubility and confirmation come into play, and the least polar compounds, the saturated and small aromatic hydrocarbons, are the first to be expelled out the kerogen matrix, leaving behind the more polar, heteroatomic species. The expelled fluids then fill the available pore spaces within the source rock. Petroleum is then completely expelled from the source rock into the surrounding strata as the volume of petroleum products exceeds the source's total pore space. The net effect of this expulsion is a chemical fractionation between the heteroatom-rich bitumen that is retained in the source and the hydrocarbon-rich petroleum that is now free to migrate.

2.3.3 Migration and Accumulation

Expelled petroleum enters into associated strata, typically sandstones and carbonates with much higher porosity and permeability. These strata serve as carrier beds, allowing for the petroleum to migrate away from their source as it is driven by buoyancy and capillary pressure. The buoyancy force is produced by the density difference between hydrocarbons and water in the pore space. Capillary pressure is the difference in pressure that exists across any curved interface between two immiscible fluids. The interface must be small to be significant. The water-saturated carrier bed resists being filled as capillary pressure prevents petroleum from entering through the pores. Within an increasing petroleum fill, buoyancy force increases and eventually exceeds the capillary pressure of the carrier bed, allowing petroleum to migrate upward. Unlike hydraulic flow, migrating petroleum does not have to fill all pore spaces. Rather, water coats rock surfaces and is preferentially held in smaller pores while the non-wetting hydrocarbon phases occupy the central space of larger pores. Petroleum can migrate through carrier beds along narrow conduits and travel long distances with a minimum volume. Migration through faults can be even more efficient.

To be exploited, petroleum must accumulate in the reservoir rocks. These are also strata typically with high porosity and permeability and, in fact, may have also served as the carrier bed. A trap is required to allow petroleum to pool within a geologic closure. There are many different trap geometries that can be grouped into structural and stratigraphic traps (Figure 2.6). Structural traps are subsurface deformations resulting from tectonic movement. The most common are anticlines, where the reservoir is under a concave flexure, sealing faults, where the reservoir rock is split along a fault that is closed by clay fill, or types of cement, and salt domes, where the upward migration of deep salts (also driven by buoyancy) deforms the surrounding rock, forming structural features, where petroleum migration is trapped against the mobile salt. Many

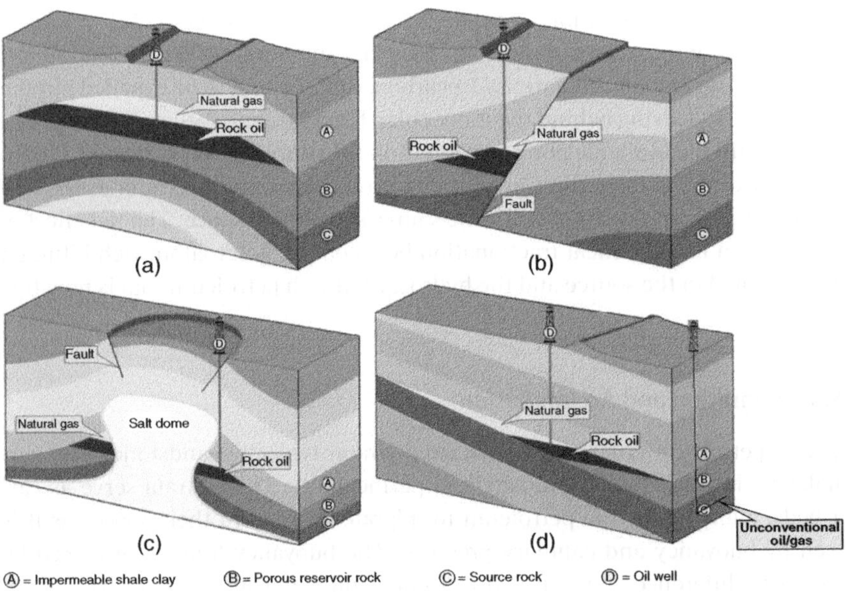

Figure 2.6 Geologic traps for oil and gas. Structural traps involve tectonic forces that fold strata into domed (anticlinal) features or juxtapose strata via faults. Stratigraphic traps are reservoirs sealed by sedimentary unconformities or develop through the movement of salt. These are conventional petroleum accumulations where petroleum migrates from source rocks through porous strata that are sealed with impermeable rock to present leakage. In unconventional plays, organic-rich shales/marls act as a source, reservoir, and seal. (a) Anticlinal trap. (b) Fault trap. (c) Salt dome trap. (d) Stratigraphic trap. Source: Green (2014a,b,c, 2017). Licenced CC BY-SA 3.0.

of the early exploration discoveries were made from projections of surface expressions of anticlines and salt domes. Because structural traps are relatively easily distinguished on seismic surveys, the presence of a trap is assigned a low risk if detected. Stratigraphic traps form reservoirs that are confined because of changes in porosity and permeability with a stratum rather than the result of structural deformations. These changes may result from the deposition of the reservoir rock as a discontinuous layer or from heterogeneous diagenetic alteration within the strata. Because they are difficult to seismically image, most stratigraphic traps were discovered by accident. Even with today's technology, stratigraphic traps are difficult to predict with certainty, and prospects based on their presence are assigned as high risk. Most targeted stratigraphic traps are actually combination traps, where structural thinning and/or faults assist in accumulation. For example, the recently discovered oil accumulations in the Guyana-Suriname Basin are expressed in both structural, stratigraphic, and combination traps.

While oil may pool in traps, buoyancy forces are still in play, and unless there are seals that impede further migration, the oil charge will leak out. Shale and salt, with their very low permeability and small pores, make excellent sealing rocks. Cemented sandstones and carbonates also can serve as seals. Sealing rocks fail when buoyancy forces exceed their capillary entrance pressure. Consequently, the volume of petroleum able to be retained in a trap is determined by the strength of the weakest seal.

2.4 The Upstream

The upstream oil industry is divided into three operations: (i) exploring for petroleum (oil and/or natural gas), (ii) development of newly discovered reserves, and (iii) long-term production. A company may focus on one or all of these operations. Each requires specialized skills and knowledge in science and engineering; however, there is an underlying geologic framework that is common across all aspects of the upstream.

2.4.1 Exploration

Exploring for petroleum is a risky business. Each year, billions of dollars are spent on acquiring acreage, developing plays and prospects, and drilling test wells. The exact methods that companies use to minimize risk are closely guarded secrets, but the overall approach is universal.

An exploration company first needs to purchase mineral rights from individual landowners or bid on onshore lands or offshore tracts being offered by governments. Alternatively, a company may purchase such rights through mergers or acquisitions. Nationally owned companies, such as Sinopec, China National Petroleum, and Saudi Aramco, and semi-public companies, such as Petrobras and Pemex, benefit from having first rights to lands within their country. Acquiring acreage requires an evaluation of potentially producible reserves and an economic evaluation of whether such reserves will be profitable. All of the tools and techniques described earlier are used in the technical evaluation to predict what resources are present. Risk is assigned to the individual key elements and processes of the petroleum system, and an overall score is developed. This is not an exact science, and the technical evaluation of prospective acreage can vary greatly depending on the available data, experience, and biases of the evaluators and a company's tolerance for accepting risk. Sophisticated economic models are then applied that take into account a multitude of factors, including the terms and agreements and projections of future resource value, development and infrastructure costs, political stability, market availability and deliverability, short- and long-term financial liability, and environmental impact. Ultimately, a value is assigned to the targeted acreage and the company tries to obtain the acreage below the target to maximize future profit.

2.4.1.1 Play and Prospect Evaluation

Once new acreage is obtained, decisions are made as to which prospects will be drilled first. In raw frontier areas, the earliest exploration wells are usually selected to give the best chance for proving that the projected reserves are actually present. The first successful discovery well typically leads to subsequent additional exploratory wells that evaluate the limits of the initial discovery or new prospects within the acreage. If significant new resources are found that can be economically produced, reserves can be "booked," and the value of acreage is no longer speculative. Prospects become plays only through the application of some of the most advanced science and technology, coupled with experience and personal knowledge.

2.4.1.1.1 Geology

Exploration using geoscience is largely a twentieth-century concept. Prior to this, the presence of petroleum surface seeps was the only tool available. Oils seeps are noted in many ancient texts. By the tenth century, the Arabian polymath Al-Mas'udi recorded oil seeps across much of the known world from the Mediterranean, throughout the Middle East and Asia Minor, to India and Sumatra. Exploring for seeps remained the primary geologic practice for determining where to drill into the nineteenth century, and once found, the fields were expanded using trial and error, following the paths of surface creeks, topography, and trends of the successfully completed wells.

Even with the emergence of practical geology and geophysics, exploration using seeps continued as a primary tool. Despite the ancient records, the Arabian Peninsula had long been considered unviable for exploration. While fighting at Gallipoli in the First World War, Frank Holmes heard about oil seeps in the Arabian Gulf. With a leased concession from the sheik of Bahrain and drilling support from the Standard Oil Company of California, he discovered oil in 1932. The detection of oil seeps remains a valuable exploration tool to this day. It is mostly conducted in deepwater, frontier basins where the petroleum system is not yet fully understood and relies on sophisticated sampling and analytical methods to detect and characterize trace quantities of oil and gas.

Modern geology starts with James Hutton's landmark publication *Theory of the Earth*, where he described sedimentary rocks as forming from repeated cycles of erosion and deposition, occurring over hundreds of million-year timeframes (Hutton 1788). Charles Lyell advanced these concepts in the 1830s with the publication of his three-volume masterpiece *Principles of Geology*, which cemented the concepts of "deep time" and a gradualistic view that the present is the key to the past (Lyell 1830–1833). Shortly thereafter, geologists began to recognize that there was an association of anticlinal folded structures and oil and gas accumulations. The earliest such reports were made in 1842 by William Edmond Logan, founder of the Geological Survey of Canada, on an oil seep in Quebec (Logan 1842, 1843), in 1855 by Thomas Oldham on hand-dug oil wells in Burma (Yule and Oldman 1855), and Ebenezer Andrew in 1861 on productive structures in the Appalachian Basin. Logan encouraged his assistant, Thomas S. Hunt, to study the area around western Ontario, and in 1861, Hunt reported that oil at Enniskillen was associated within a folded anticline (Hunt 1861). With further study, he concluded in 1863 that the organic matter in some North American Paleozoic rocks must be derived from marine vegetation or marine animals and that the transformation of this organic matter to bitumen must be similar to the processes involved in coal formation (Hunt 1863). This publication essentially establishes the modern concept of a petroleum system requiring source, reservoir, and sealing formations. It took several decades for the anticlinal theory to be widely accepted. The first discovery in Persia in 1908 was drilled over an anticline, and other successes soon followed, such that drilling the "hills" became standard practice.

The association between petroleum and salt has long been exploited in China and documented later by Hacquet in 1793, who developed his theories on biogenic origins of oil through observations made in a salt mine. This relationship was not adopted by modern exploration until the discovery at Spindletop, Texas, in 1901. Pattillo Higgins noticed gas seeps around Spindletop, a small hill in the flat plains near Beaumont. Convinced that this feature was an anticline, he began drilling in 1892. After six dry holes over a six-year span, Higgins was heavily in debt (Sternbach et al. 2017). He teamed up with Captain Anthony Lucas, who previously supervised operation at the salt mine in Louisiana, where he observed the association with salt, sulfur, and petroleum. Lucas was convinced that salt domes were ideal structures for oil accumulations and that Spindletop was not an anticline but, in fact, a surface expression of a salt dome. The 1901 gusher sparked a new era in exploration, and within 10 years, 36 other salt domes on the margins of the Gulf of Mexico were drilled.

The modern petroleum geologist involved in exploration is required to implement the techniques and tools needed to acquire and interpret data used to evaluate basin and reservoir architecture to identify new resources. A petroleum geologist may be a specialist in a specific discipline, such as sedimentology, stratigraphy, and tectonics, or a generalist who integrates data provided by the specialized geologists and other geoscientists.

2.4.1.1.2 Geophysics

Surface expressions of anticlines and salt domes are limited. Explorationists needed a tool that would allow them to see these structures deep within the Earth. Reflection seismology provided the key technology to do this. It is based on a simple and well-understood principle, that a portion of a compressional sound wave will reflect when it encounters a change in impedance. In air, these are called echoes. In the Earth, these are seismic reflections. Starting in the mid-nineteenth century, experiments and observations in the detection of seismic waves focused on earthquakes. The use of artificially generated acoustic waves for subsurface imaging was developed by Ludger Mintrop, who successfully detected a salt dome in Germany in 1914. Similar inventions were independently made by researchers in Canada and the United States shortly thereafter. Commercial application for petroleum exploration soon followed with the first field tests in 1921 in Oklahoma, followed by a discovery well in 1925. The first offshore field trial was conducted by Shell in 1938, four miles of the coast of Louisiana. While refined and expanded, the process of collecting seismic data remains the same today; an energy source directs a sound wave into the Earth, and a string of microphones listens to the reflections.

The physics needed to construct an image from seismic reflections has been long known, but the ability to perform the necessary computations exceeded the available technology for most of the twentieth century. In the 1950s, analog machines were built for simple analysis and calculation of dipping events. Digital recorders were developed in the mid-1950s, and computers began crude processing of seismic data in the early 1960s. Only compressional reflections (P-waves) were used for image processing, as the interpretation of shear reflections (S-wave) proved exceedingly difficult. By the 1980s, mathematicians developed new algorithms for seismic processing of multicomponent (both P- and S-wave) reflections, specifically P-to-S conversion and applications to 3D data cubes. Considerable effort was made to enhance P-wave data, but these techniques taxed the limits of the then-available computers. This changed in the 1990s as computers became more powerful, and seismic processing could be distributed across multiple processors. Cluster computers finally provided the horsepower to process large 3D seismic surveys. Most of the algorithms in use today for routine data processing were developed during this period, including a battery of programs designed for noise attenuation, correction of source and receiver acquisition irregularities, and conditions specific to marine surveys. However, these routine methods still use only a portion of the seismic data.

Modern supercomputers have finally advanced to the point that the algorithms developed in the 1980s can be fully used. Full wavefield inversion (FWI) uses all data acquired from a seismic survey not only to produce high-definition images but also to define rock properties, including the presence of hydrocarbons. In routine seismic image processing, subsurface properties that determine its seismic response are inferred, and deviations from the real and input parameters result in error. FWI begins with this geologic model of rock properties, a seismic response is computer-generated, and the simulated results are compared with the field acquired data. Differences between the modeled and the real images flag parameters that need to be adjusted, and the process for comparing simulated seismic with reality is repeated until they converge. The result is not only enhanced imaging, but also the rock properties are reliably predicted and the occurrence of petroleum accumulations can be directly observed. Figure 2.7 shows the evolution of seismic processing over the years, from around 1990 to the present (ExxonMobil 2018).

4D seismic imaging is now practical for monitoring changes that occur in subsurface reservoirs as they are produced. Here, individual 3D surveys are taken over time, and changes in the seismic response of the reservoir rock are related to changes in pressure, temperature, and fluid saturation and composition. 4D

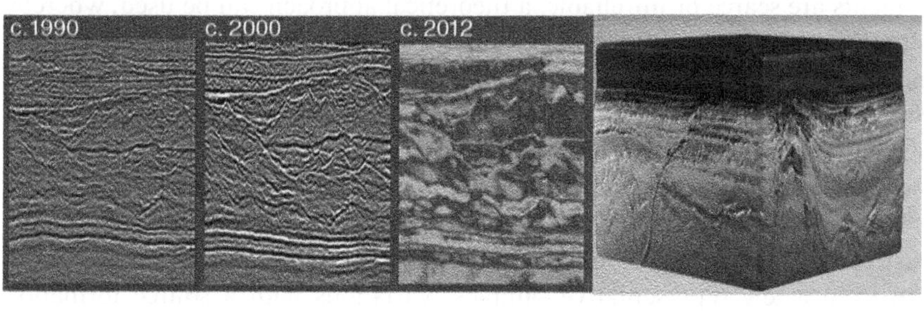

Figure 2.7 Advances in seismic processing over the last three decades. Full wave inversion of a 3D seismic survey. Different rock properties range from light to dark gray. Source: ExxonMobil (2018). Photo courtesy of ExxonMobil.

studies are increasingly being used in fields undergoing thermal treatment and enhanced oil recovery (EOR), where they can better define distinct compartments and map out fluid flow pathways.

2.4.1.1.3 Basin Modeling

Geologists and geochemists have developed models that predict the occurrence of the individual elements, parameters, and processes of the petroleum system. When integrated into one computer simulation, the complete history of a basin can be simulated, allowing explorationists to evaluate the risk of a play or prospect better.

The first step is the burial history reconstruction. This is done by integrating lithologic, stratigraphic and fossils, and/or isotopic age-dating descriptions of well cores and cutting, or inferred descriptions from well logs and seismic images. Each formation present today is projected back to its time of deposition, with its depth interval and porosity corrected for compaction, incorporating paleobathymetry. Simple "layer-cake" geology is rare and provisions for periods of non-deposition (hiatus) or where strata were deposited, but uplifted and eroded strata must be accounted for. Sometimes these missing layers occur elsewhere in the basin, but more often, they are extrapolated so that measured and modeled maturation parameters match. To better understand the tectonic processes influencing basin evolution and changes in sea-level over time, a "back-stripping" technique is used, and the effects of subsidence caused by sediment loading is removed.

Next, the thermal history of the basin is reconstructed. One approach relies on direct measurements of the local heat flux and geothermal gradient in wells, maturation parameters that reflect total thermal exposure (e.g. vitrinite reflectance), and indicators of maximum paleo-temperatures (e.g. fluid inclusions and apatite fission tracks, mineralogy). Using these data, along with complementary information (e.g. lithographic thermal conductivity, tectonic and structural models), the thermal history of the basin can be determined. In frontier basins where measurements are scarce or unreliable, a theoretical approach can be used, whereby a physical model of Earth's lithosphere and upper mantle is constructed, and the tectonics resulting in a basinal sag, such as rifting, is inferred.

Strata with petroleum source potential are identified, and key properties are assigned. These include the TOC, initial generative potential, and its kinetics of generation. All can be measured directly on immature source rocks or inferred from measurements taken on more mature samples. However, it is often impossible to capture the complex heterogeneity of a source formation across an entire basin from a few representative samples. Variations with a source formation can be predicted to some degree from knowledge gleaned from well logs and sequence stratigraphy. The kinetics of oil and gas generation depends on the chemical composition of the kerogen. Sulfur-rich kerogen (Type IIS), commonly found in evaporates and carbonates, generate oil at fairly low temperatures (onset at ~90 °C) as carbon–sulfur bonds are easily broken. Kerogens from lacustrine (Type I) and marine shales (Type II) generate oil at higher temperatures (onset at ~110 °C), while kerogens from terrigenous organic matter (Type III) are more refractory and mostly generate gas at still hotter temperatures. Kinetic models of kerogen decomposition can be determined using laboratory simulations, where the source rock is heated to temperatures that rapidly generate hydrocarbons (~350–500 °C) that are then used to infer generation at lower geologic conditions. The timing, quality, and quantity of generated petroleum can then be modeled using the geothermal history. This is coupled to an expulsion model that determines the petroleum product that leaves the source strata. Figure 2.8 shows an example of a total petroleum system events chart.

Movement of the expelled petroleum is then modeled as the basin evolves. Permeable carry beds form migration pathways that can be modeled using tomography, percolation theory, or explicit Darcy flow calculations. Accumulations occur when the migrating fluids charge reservoir rocks confined with a trapping mechanism and a seal. As the simulation models all processes as the basin evolves, the elements of the petroleum system can be tracked through time. Viable prospects require that the trap and seal be present at the time a petroleum charge occurs and that they remain intact and in place to the present day.

2.4.1.2 Predicting Petroleum Quantity and Quality

There are two basic questions essential for play and prospect evaluation: how much petroleum has accumulated in a trap, and what is its value. Models of petroleum generation exist that predict not only the timing and amount of petroleum migrating from the source, but also its composition. The complexity of such product models ranges from simple oil versus gas to detailed molecular descriptions. Knowing the composition of the expelled fluids allows one to predict their phase behavior as they migrate through carrier beds and into reservoirs (i.e. whether the fluid is single phase or has separated into distinct oil and gas phases). Basin models can couple reservoir conditions (e.g. volume, geometry, sealing capacity) with the arrival of fluids over time and predict the composition currently in place. For example, assuming a reservoir is receiving fluid from a single source that first expels liquid oil and then gas as it matures. The reservoir may be filled initially with a single-phase oil, but as it receives the gas charge, the fluid in the reservoir becomes two-phase and the gas displaces the oil.

Risk assessment requires knowing the composition of the accumulated petroleum to assign an economic value. Gas may or may not be economical, depending on current pricing and the cost of bringing it to market. An offshore prospect predicted to be filled with oil but proven to be filled with gas may be left stranded regardless of the size of the accumulation. Oil, too, has a variable

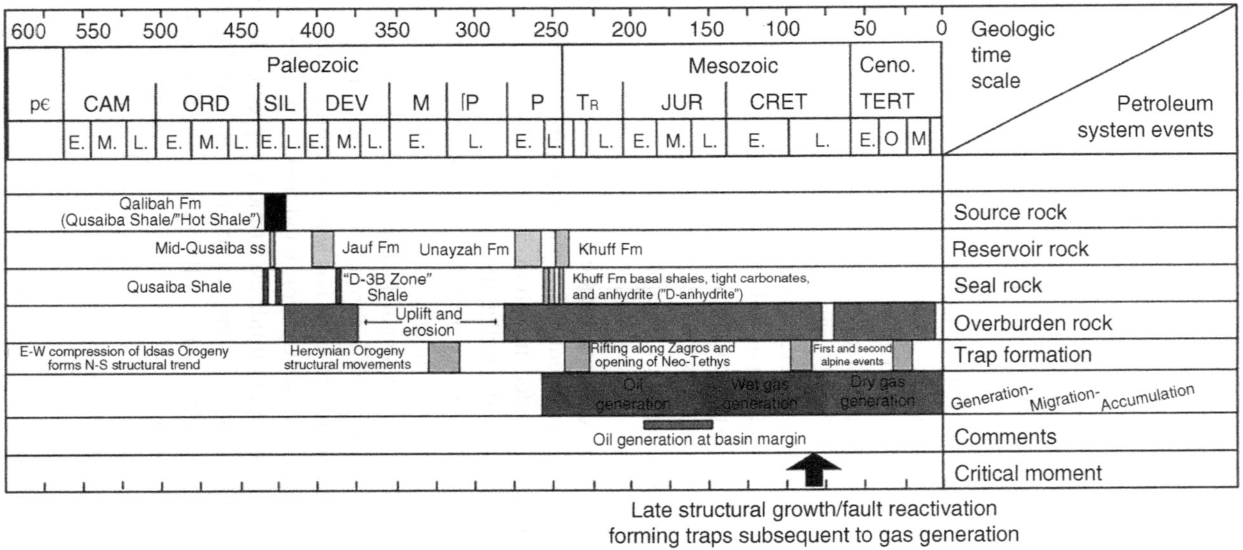

Figure 2.8 Total petroleum system events chart for Central Arabia. Source: Reprinted from Pollastro (2003).

value. The composition of an oil charge is dependent on the type of source rock and its thermal maturity at the time of expulsion. High-sulfur crude generated from sulfur-rich kerogen at low maturity has a lower value than low-sulfur oil, as it requires more energy and refinery infrastructure to refine. After filling the reservoir rock, oil can undergo several processes that further alter its composition. Oil cracking can occur if the reservoir is very hot, resulting in the formation of pyrobitumen and lighter hydrocarbons. Biodegradation can occur if the reservoir is relatively cool (c. <80 °C), resulting in low value, heavier crude with higher acid content. Asphaltenes can precipitate either from gravity segregation or by changes in temperature, pressure, gas addition, or mixing of incompatible fluids. While asphaltene precipitation improves the quality of the liquid oil, it can also form barriers within the reservoir that prevents efficient drainage. Thermochemical sulfate reduction (TSR) is a chemical reaction that occurs in hotter reservoirs (c. >100 °C), where oil is oxidized by sulfate converting hydrocarbons to CO_2, and the sulfate is reduced to H_2S. TSR is more likely to occur in carbonate reservoirs. The manner in which explorationists evaluate the risks of reservoir alterations varies greatly by company, from simple assignments of the probability of their occurrence, to complex chemical reaction models that attempt to predict changes in molecular composition given the initial oil charge and the reservoir conditions over time.

2.4.2 Drilling

Reserves remain speculative until proven when petroleum flows to the surface. This requires drilling into the Earth, penetrating the reservoir, determining what is present, and ultimately bringing petroleum to the surface above a minimum rate and total quantity that justifies the expense. Today, frontier exploration continues into challenging plays, such as drilling in the Arctic and in increasingly deeper water that demands the deployment of new technologies. The development of unconventional resources involving horizontal drilling and hydraulic fracturing on a massive scale demands careful planning and logistical control.

The task of drilling a well begins with a plan that will specify the surface position of the well and the trajectory of the hole to target. Using all available geologic and geophysical information, the technical details of the well mechanics are optimized. Wells are usually not just vertical holes but steered to follow a specific path to reach a target. The well trajectory is planned so that it avoids hitting nearby wells or problematic faults or formations. Factors such as pressure, temperature, porosity, permeability, lithology, and rock mechanics can affect borehole stability, hydraulic fracturing, and sand/clay production. The drill plan includes specifying the engineering and materials needed to drill, such as the casing design, mud system, bottom hole assembly (BHA), and drill bit selection. The BHA consists of various tools positioned above the bit to assure optimal penetration, steer the drill string (geosteering), or conduct logging and measurement while drilling for real-time formation evaluation. For petroleum to flow to the surface in a controlled and safe manner, a completion design is prepared where the wellbore is fully stabilized with steel and cement, the productive formations are stimulated via perforation, and in tight formations, hydraulic fracturing and undesired inorganic fines are screened out. All of this needs to be conducted safely and in compliance with environmental standards. Logistics and contingency plans are made to finish the well in the shortest time and with the lowest cost. This involves juggling the schedule of as many as 10–30 or more service companies, synchronizing the delivery of pipe, drilling muds, chemicals, proppant sand and other material, and continuous monitoring of well conditions. Onshore wells drilled in the United States range in cost from about US$ 3 million to over US$ 25 million. The cost of drilling on land in a foreign country varies greatly but typically ranges from US$ 25 to US$ 40 million. Offshore drilling is immensely more expensive, ranging from a low of about US$ 200 million to over US$ 1 billion, and averaging out about US$ 650 million. With offshore rig costs of US$ 200,000 per day, any delay due to faulty planning is not tolerated.

2.4.2.1 Development of Drilling Technology

Drilling involves piercing through the non-productive overburden so that the targeted reservoir rock is penetrated. The task becomes more difficult with increasing depth.

The oldest exploited subsurface reserves were very shallow and tapped by digging wells by hand, a very hazardous practice. Nevertheless, such methods persisted into the early twentieth century in the Baku oil fields, with some reaching over 35 m. Drilling is needed to tap deeper reservoirs.

Salt, percussive drilling, and oil have been entangled for centuries in China, and the relationship continued into the modern age. A claim for the first American oil well is one drilled in 1829 that was seeking brine but instead found high pressured gas and oil that shot up into the air. The earliest wells intentionally drilled for oil in the western world, the Bibi-Heybat well drilled in 1844 near Baku and the Edwin Drake well drilled in 1859 at Oil Creek, Pennsylvania, both used percussive drilling to tap shallow accumulations that were only ~20 m deep. Two years later, Henry Rouse drilled the first free-flowing well, a "gusher." Although claimed as major achievements that ushered in a new age, the Chinese were drilling as deep as 1000 m by 1835 to bring brine and gas up from Triassic strata (Khun 2004).

Rotary drilling technology, utilizing the spinning of a drill bit to cut through rock, was known for centuries and is attractive, in theory, as it permits deeper drilling with wider boreholes. In practice, the length and size of the well is limited by the material strength of the drill shaft. With increasing depth, the weight of the drill string and torque on the drill bit increases, and the shaft will eventually fail.

Leonardo da Vinci invented a complete rotary drilling system, but like many of his inventions, he lacked the materials for successful implementation. The success of rotary drilling essentially paces that of innovations in the iron and steel industries. The artisan water well at Grenelle, drilled using a dry rotary auger turned by horse, was started in 1833, and after four years, reached a depth over 380 m when the drilling rod broke. The well was finally completed in 1841 after spending over a year to clear out the fallen drill string (Azais 1845). The French are also credited with inventing drilling with water circulation in 1845 and the diamond core drill in 1863 (Mullin 1885). Continued improvement in drill bits and the circulation of drilling muds continued through the late nineteenth century, culminating in the 1901 gusher at Spindletop, Texas, that was drilled with a steam-driven rotary rig. Howard Hughes invented the conical rolling cutter bit, leading to the creation of the Hughes Tool Company in 1908. By 1930, rotary rigs had replaced cable-tools except for drilling very shallow wells.

Today, most well are drilled using a turbodrill, a 1922 invention of Matvei Kapelyushnikov that was first used to drill a well in Russia two years later. Here, the drill bit is rotated by a fluid-powered turbine using the hydraulic energy of the mud being pumped from above. This places much less torque on the drill string, allowing for faster rotation and use of diamond embedded bits that facilitate drilling through hard rock at faster penetration rates. In 1956, Neyrpic, a French company later named Neyrfor and now part of Schlumberger, purchased the rights to the Russian design and brought turbo drilling to the western world. Neyrfor introduced the first steerable motor in 1982, and the continued technological improvements that have been made in pump technology, bit design, adjustable-gauge stabilizers, and sensors have allowed for efficient and controlled directional drilling. This is critical when drilling unconventional wells with long laterals that need to be navigated within a thin productive pay zone.

Technological advances in drilling technology allow for wells to be drilled into ever deeper reservoirs. After a series of gas discoveries in very deep strata in the Anadarko Basin, the Lone Star Producing Co. drilled the 1–27 Bertha Rogers well in 1974 to a depth of 9,583 m, where it discovered a pool of molten sulfur (Gray and Tschirley 1975). This record depth was broken by the Soviet Union's Kola scientific borehole that began coring operations in 1970 and eventually reached a depth of 12,262 m (Popov et al. 1999), where it encountered temperatures exceeding 180 °C. Plans have been proposed for new exploration wells in the Caspian Sea to drill to depths of nearly 15 km. The record for the longest wells is held by ExxonMobil, where a specialized drilling rig sitting on Sakhalin Inland directionally drilled the extended reach Odoptu OP-11 well to a total drill string length of 12,345 m (Walker 2012).

Technological advances in drilling technology also pushed the frontiers of E&P offshore to ever-deepening waters. The first "offshore" rig, a wooden pier in the

Santa Barbara Channel, was drilled in 1897. The well successfully tapped into an oil play discovered onshore, and continued development placed over 150 "off-shore" wells by 1902, all connected to the shore by piers. Similar land-connected piers were used to drill in Lake Erie and Caddo Lake. The first piers constructed of steel appeared in 1921 at Rincon and Elwood. The Soviet Union developed much of the nearshore region of the Caspian Sea around Baku using piers. Starting in 1947 with the development of Neft Dashlari, more than 1300 fixed platforms connected by 450 km of piers were constructed over the next two decades. The Russians also were first to build artificial land to support drilling. The Bibi-Heybat field, site of the first drilled well, was extended with landfill into the bay. The project started in 1909 and was completed in 1932. Artificial islands remain attractive for shallow water development with recent construction in offshore Abu Dhabi and proposed structures for Arctic drilling off Prudhoe Bay.

The first drilling platforms not connected to land were constructed in the early 1930s in the swamps of Louisiana using barges that were partially sunk and supported by pilings driven into the muck. The first offshore platform not within sight of land was a barge set up in 1947 by Kerr-McGee that was partially sunk in 15 ft of water in the Gulf of Mexico. The first mobile offshore drilling rig was the *Mr. Charlie* that drilled its first well in 1954 at a water depth <40 ft.

Technological advances have allowed drilling at ever-increasing water depths (Figure 2.9). Fixed platforms sit on the seafloor and are typically used for water depths <200 m. Shell's Bullwinkle platform, one of the largest in the world, pushed this technology to a water depth of 412 m. Compliant tower platforms are tall narrow structures that sit on the seafloor with a smaller footprint than fixed platforms (Sterling et al. 1989). Amerada Hess' Baldpate compliant tower structure

Figure 2.9 Type of offshore oil and gas structures. (1, 2) Conventional fixed platforms; (3) compliant tower; (4, 5) vertically moored tension leg and mini-tension leg platform; (6) Spar; (7, 8) semi-submersibles; (9) floating production, storage, and offloading facility; (10) sub-sea completion and tie-back to host facility. Source: NOAA (2010). Image courtesy of NOAA ESRL Global Monitoring Division, Boulder, Colorado, USA.

is positioned at a water depth of 502 m. Tension leg platforms are floating and tethered by cables, the tallest being Chevron Big Foot, which was constructed in 2018 at a water depth of 1581 m. Spar platforms use a large cylinder to support the floating deck from the seafloor. The added stability of spar platforms makes them attractive for use in water depths >1000 m; the Shell Perdido spar platform, built in 2010, sits in 2383 m of water. Floating production storage and offloading (FPSO) systems have become the latest technology for deepwater drilling and production; Shell's Turritella FPSO, built in 2016, produces from a 2896 m water depth. The current record for deepwater drilling is held by the *Maersk Venturer* drillship that completed an exploratory well in 2016 at a water depth of 3400 m.

2.4.2.2 Modern Drilling Practices

The first step in drilling is site preparation. For an onshore well, this typically involves building roads and laying power and water lines to reach the site, constructing a pad for the rig, and digging out pits or setting up large metal bins to store the used mud and cuttings. Drilling offshore exploration wells involves contracting a floating rig that is temporarily anchored to the seafloor. Space is tight, and equipment and staff are transported on and off the rig by boat or helicopter as needed. More permanent, fixed platforms are set up on the seafloor or above sea level when a field enters into development. These topside facilities can be very large as they serve as platforms not only for drilling, but also to accommodate equipment for fluid separation, treatment, and storage.

A drilling rig consists of infrastructure for pipe management, a power train that drives the drill string and bit, and the mud circulation system. The familiar derrick is only the outer frame for the mechanics needed to store, move, position, and rotate the drill pipe. Electric motors, usually powered by local diesel-fueled generators, drive the drill string that is composed either of screwed-together straight segments 27–45 ft (8–14 m) long or continuous coiled tubing coupled with the BHA and drill bit. Pumps circulate drilling mud down the well pipe to the drill bit, providing the power for turbodrills, then back out through the annulus, the space between the drill pipe and the well casing, where it both lubricates a rotating drill string and supports an open hole from collapse. Once returned to the surface, the mud is cleaned and ready to be recirculated.

Drilling mud consists of minerals, such as barite, siderite, hematic, and calcium carbonate that are dispersed in water, oil, diesel fuel, or synthetic organic blends of esters, olefins, and paraffins. The pressure is continually monitored while drilling so that the mud density can be adjusted to balance against formation pressure. This is particularly critical while drilling into petroleum reservoirs that may be overpressured relative to the lithostatic load. Too low a mud density and the reservoir fluid may unexpectedly enter the wellbore; too high a density and the mud will be pushed into the reservoir, potentially damaging the rock, hindering fluid production. When a well penetrates rocks with high permeability or into fractures, a considerable amount of drilling mud may be lost. If not the target formation, drillers will add just about anything to the mud that is solid and can plug off the zone to prevent further mud loss. Cement, coarse minerals, nutshells, plant fibers, shredded tires, and flaked plastics have all been added as lost-circulation materials. Chemicals are also added to prevent steel corrosion and microbial growth, scavenge H_2S, control emulsion and foam formation, regulate pH conditions, reduce viscosity, counteract cement contamination, and disperse and flocculate particulates.

The borehole of a well is usually not stable and needs to be reinforced to prevent failure and to control fluid flow. Upon spudding a well, a large-diameter steel sleeve, known as the drive pipe or conductor casing, is driven into the onshore surface soil or offshore unconsolidated sediment and cemented in place. The conductor case acts like a cofferdam providing a stable, very shallow (typically 40–300 ft or 12–91 m) initial hole that is typically 30–42 in. (76–107 cm) in diameter for offshore wells and 16 in. in diameter for onshore wells. The structural casing is then set as the well is deepened. Running about 600–1000 ft (183–305 m) from the surface, the structural casing is a smaller diameter, stronger steel tube that is placed within the conductor casing. The structural casing isolates well fluids from contaminating shallow aquifers, maintains hole integrity through unconsolidated sediment, mitigates pressure kicks while drilling through shallow gas zones, supports the weight of all deeper casing strings, and serves as the attachment point for blowout preventers and wellhead systems. For deep wells, an intermediate casing may be required to maintain wellbore integrity or to isolate lost-circulation or abnormal pressure zones until the target reservoir is reached. The intermediate casing is cemented within the structural casing and multiple intermediate strings may be nested within each other, decreasing in diameter. Since the early 1990s, a technique termed "casing while drilling" involving tubulars that are not cemented has been developed for higher efficiency and lower drill cost. A final production casing may be set into the target reservoir at the final depth. At this point, the hole may be only 6 in. (15 cm) or less in diameter. Fluid production is achieved by perforating the steel casing with explosives. More often, a liner that is not cemented in place but hung from the intermediate casing, or the hole penetrating the reservoir is left open with no casing or liner. Figure 2.10 provides a schematic of casing placement within a well.

With the drilling then completed, production tubing is run from the reservoir to surface, and the annular space between the production tubing and the casings is filled with cement. For very deep wells at high temperatures and pressures or in corrosive environments, specially formulated types of cement are needed. The rig is removed and a wellhead assembly is bolted on and connected to a pipeline. The wellhead assembly, also known as the Christmas tree, consists of multiple safety

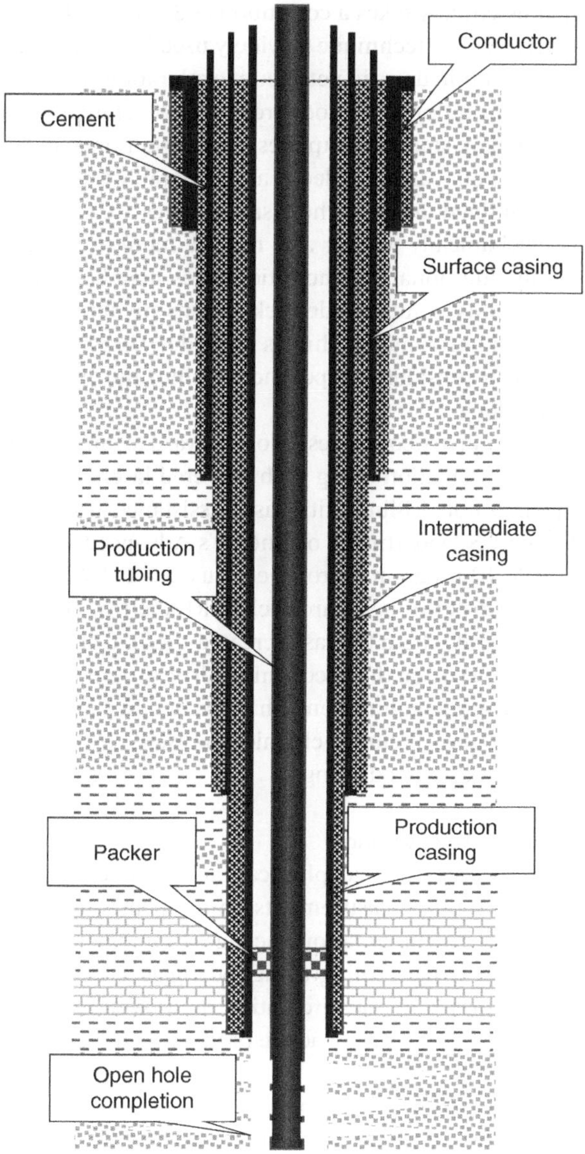

Figure 2.10 Schematic of well design showing casing placement.

valves, pressure and temperature gauges, chokes, and openings where samples can be collected or chemicals injected. Subsea wellheads can be remotely controlled.

Although technologic advances have allowed for success under increasingly difficult conditions, the basic concepts in rotary drilling have not changed drastically for over a hundred years. Slim-hole drilling is an alternative drilling method that essentially takes a continuous ~3–7 in. (8–18 cm) diameter core to the target reservoir. The technique is widely used by the mining industry on hard rocks but has had limited use in petroleum exploration. The advantages of slim-hole drilling over conventional methods are that it has a much smaller footprint and the needed equipment, staff, and supplies can be positioned in very remote locations via helicopter, wells can be drilled at a considerably lower cost, and the recovered core can be examined directly. The disadvantages are that the smaller, thinner drill string is mechanically weaker and may be more prone to failure, limiting the drilling depth; additionally, penetration rates may be slower, borehole stability may be compromised, future sidetrack options are limited, directional drilling is challenging, and horizontal drilling is limited to short lateral distances. Slim-hole drilling has been considered experimental since the 1980s and still is considered a niche technology.

The industry continues to look toward technology to provide better ways to drill. Full rock ablation using high powered lasers remains impractical for petroleum drilling, but drilling with laser assistance was demonstrated in 2012. Funded by the US Department of Energy's Advanced Research Projects Agency-Energy (ARPA-E) program, Foro Energy used a 20 kW laser coupled to a special optical fiber to drill through hard rock (ARPA 2016). The system was a laser-mechanical drill hybrid where the laser created thermal shocks in the rock, creating spalling that allowed for the mechanical drill to remove the disrupted rock with much lower energy requirements than needed in conventional drilling. Other emerging technologies include chemical-assisted hydrothermal, plasma-assisted, and vibration-assisted drilling.

2.4.2.3 Well Logging

Valuable geologic, geophysical, and geochemical information can be obtained by conducting measurements on the penetrated strata. These tests are made by lowering instrumentation packages (wireline tools) after the drill string has been removed, while drilling using tools placed in the BHA, or on material returned to the surface by the circulating mud. Well logging is also performed after well completion and during active production to determine well integrity.

2.4.2.3.1 Wireline Well Logs

The oldest and most common practice is the wireline resistivity well log. In the early 1920s, Conrad and Marcel Schlumberger developed the concept that the surface resistivity techniques they developed for the mining industry would be effective for characterizing the subsurface. Since hydrocarbons, freshwater, and most rocks have high resistivity, whereas saline connate water is conductive, porous reservoir rock filled with hydrocarbons will exhibit higher electrical resistance than if water-filled. This concept was tested for the first time in 1927 when the resistivity was measured in a shallow well over a 140 m interval. This was done by slowly lowering a tool on a cable that was connected to a power supply and potentiometer. The Schlumberger brothers also invented the resistivity dipmeter that determined the dip direction of well stratified beds by comparing the response of several horizontally spaced sensors.

While the measurements and interpretations have evolved since the first Schlumberger well log, the basic mechanics of slowly lowering a tool connected to cables that supply power to the tool and return data to the surface remain the same. Modern wireline tools typically contain a package of sensors that provide multiple measurements from a single wireline test. The most common are electrical, acoustic, and gamma ray.

Resistivity and induction sensors can measure electrical conductivity at various lengths from the well wall. This is necessary to overcome the invasion of drilling fluids into permeable rock and the use of oil-based drilling mud. Small differences in electrical potential occur spontaneously by the interaction of the formation water, drilling fluid, and ion-containing minerals. Measuring these produce a spontaneous potential (SP) log that is used to differentiate porous reservoir rock from impermeable clays and shales. Today, well logging companies have developed various sensor arrays that provide high-resolution electrical measurements that work in cased and open wellbores.

The acoustic well logging tool piezoelectric transducers and receivers generate sonic vibrations and detect their reflections. Porosity, lithology, pore-fluid composition and saturation, pore pressure, and temperature all influence the returned acoustic signal. The porosity and the presence of gas-bearing strata can be inferred by accounting for these effects. The ΔLogR method estimates TOC using resistivity and sonic signals. Acoustic well logs also provide high-quality velocity data at a much higher resolution than seismic surface reflection and are run to evaluate the integrity of casing-cement bonding. As with electric logging tools, service companies offer various configurations involving multiple types and arrays of acoustic sensors.

Gamma-ray logging reflects the amounts of potassium, thorium, and uranium that occur naturally in the rock. High gamma-ray counts are associated with organic-rich shales, as clays generally contain more potassium than other minerals, and uranium and thorium are enriched by mobilization and stabilization processes associated with reducing conditions. Developed in the 1930s, gamma-ray well logging using a modified Geiger–Müller counter was offered as a service in 1940. Potassium, thorium, and uranium emit gamma rays at different energies that can be individually measured with a scintillation counter. As different clays tend to contain varying amounts of these elements, and potassium-containing feldspars may be associated with quartz, spectral gamma-ray logs are used to differentiate clay types and sandy-shales.

Nuclear logging tools also rely on active sources of gamma-rays or neutrons. Density logging relies on a gamma-ray source, typically ^{137}Cs, and a gamma-ray detector to measure the amount of Compton scattering and photoelectric absorption. The response is a function of rock porosity and lithology. Neutron-emitting sources, commonly plutonium-beryllium, were first used in 1940 and counted gamma rays emitted when the neutrons were captured by hydrogen and chlorine. Neutron porosity logging tools rely on an artificial neutron source, typically americium-beryllium, and two detectors that measure the falloff of scattered neutrons. Neutrons penetrate further into rock than gamma rays, and neutron density logs are considered to be more reliable. Spectroscopic detectors can measure the gamma-ray energy and determine the elemental composition that can be interpreted to yield the mineral assembly, TOC, and petroleum saturation.

A variation is the pulsed neutron lifetime (PNL) log that relies on an accelerator to produce high energy neutrons that are largely unaffected by steel; hence, these tools are preferred for use in cased wells. PNL tools are used to measure well integrity and zonal isolation, by flow velocity as oxygen in water is activated by the source and the time it takes to be detected at a detector. Residual oil saturation and permeability can be determined using PNL by taking a baseline measurement, injecting brine at a different salinity as the formation, and measuring the difference.

The potential for using nuclear magnetic resonance (NMR) for well logging was recognized long before it became technically practical. Salinity, lithology, texture, pore fluid characterization, and fluid displacement are all possible measurements, and considerable research has been devoted toward inventing an NMR for downhole use. Such a device was demonstrated by Chevron in 1958, and logging companies have offered a form of NMR logging since the early 1960s, but technical limitations stifled acceptance. Practical applications emerged in the 1990s with more advanced NMR instruments and a better understanding of the science involved with porous media. The technique remains in limited use due to its high cost.

Numerous other wireline tools have been developed for specialized measurements. Cameras and ultrasonic imagers can return visual and UV fluorescence images of the borehole, providing information on reservoir geometry, faults and fractures, petrophysical parameters, and oil saturation. Cross-well tomography can provide high-resolution seismic images by lowering an array of acoustic sources down one hole and an array of microphones in a nearby well. Time-lapsed images can provide a direct view of the effectiveness of EOR flooding.

2.4.2.3.2 Mud Logging, Measurements While Drilling (MWD), and Logging While Drilling (LWD)

Just the act of drilling provides the rig operator with some information, such as the rate of penetration; however, far more data is needed for optimal and safe drilling efficiency, steering the bit to target, and making decisions that deviate from planned activities, such as which zones to test. To do this, information on direction (hole inclination, azimuth, tool facing), trajectory, drill assembly conditions (temperature, pressure, torque, weight-on-bit, revolutions per minute [RPM], severity of downhole vibration, and mudflow), the penetrated strata (hydrocarbon pay zones, resistivity, and other wireline-type log data), and geomechanics (placement of casing points, mud-weight management, cement and wellbore integrity) need to be made available as the well is being drilled. Sensors must be placed near the bit, and the collected data needs to be made available as the well is being drilled.

Gases, oil, and rock cuttings are carried to the surface with the circulating drilling mud, and analyzing these provides a window into the subsurface as it is being penetrated. In 1939, the Barnsdall Research offered the first mud logging services, consisting of a suction line over the shale shaker that carried gas to a hotwire gas detector to measure the total amount of flammable gases and a microscope to examine the lithology of the cuttings that were pulled off the shaker when a new drill pipe was added to the string. The volume of cuttings correlated with the rate of penetration. By being on site and able to provide real-time data, the mud logger could communicate the findings directly to the driller. Over the decades, mud logging has evolved in sophistication. One of the earliest applications of gas chromatography in the 1950s was C_1–C_4 hydrocarbon analysis of gases desorbed from the mud. Geopressure measurements were added in the 1960s. Recent advances in automated sampling, instrumentation, and data processing have greatly expanded mud logging capabilities (Ablard et al. 2012); Figure 2.11 shows an excerpt from a typical mud log. Service companies now offer continuous extended C_1–C_8 hydrocarbon and non-hydrocarbon gas analysis, carbon isotope ($\delta^{13}C$) measurements of methane, description of lithofacies using infrared spectroscopy, X-ray diffraction, digital microscopy and X-ray fluorescence, and hydrocarbon analysis of fluid inclusions.

Measurements while drilling (MWD) and logging while drilling (LWD) are not new concepts. Early wells drilled with cable tools could collect such data by placing a logging tool in series with the bit as the cable provided a way to supply power and telemetry. Rotary drilling, however, posed exceptional technical challenges. In 1932, Geophysical Services, Inc. (GSI) patented a system that incorporated a conducting rod into each section of drill pipe and published the first LWD electric log in 1935. Around the same time, an alternative system was developed by Amerada Research Corporation. Though demonstrated, neither system was adopted, as they required a specially designed pipe; this prompted research into alternative methods to receive bottomhole data. Patents were issued in the 1940s and 1950 that used electromagnetic waves, both direct and from relay stations placed along the drill string, acoustics through the drill pipe, and radioactive or chemical tracers released in the mud system – none of which were practical. In 1950, the Arps brothers patented a method where a tape recorder stored logged data that could

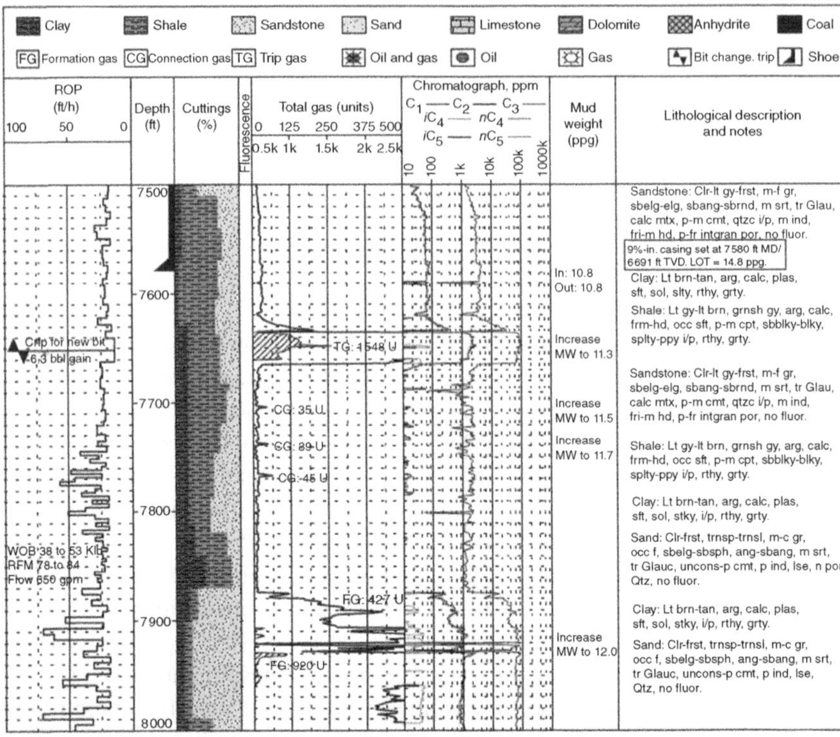

Figure 2.11 Excerpt from a typical mud log. ROP, rate of penetration; TVD, true vertical depth; LOT, leakoff test; SSP, static spontaneous potential. Source: Ablard et al. (2012). Reproduced with permission, Copyright (2012), Schlumberger.

then be retrieved when the drill string was pulled out. Such systems were used in the 1960s–1970s, with the major problem being a reliance on batteries that could not operate above 100 °C.

In the late 1950s–1960, the Arps brothers invented and patented a subsurface telemetry system around modulating the mud stream pressure using a valving device placed above the bit. Data could be transmitted in pressure pulses, similar to Morse code, and received and decoded at the surface. The system worked with normal pipe, and the short modulations, with a pressure differential of only 50–100 psi, did not impact the overall hydraulics of the mud system. Further development of pressure-pulse telemetry (positive/negative pulse and continuous wave modulation) led to its widespread use in the 1970s. Its main limitation was its very slow data rate, which limited the number of MWD sensors that could be transmitted. Mud-pulse telemetry is currently in widespread use with bandwidth up to 40 bit/s for ideal shallow wells and as low as 0.5–3 bit/s in very deep wells.

Technology has also caught up to make electromagnetic telemetry practical. Using the drill string and the Earth as a dipole antenna, low frequency (2–12 Hz) radio waves are modulated and detected as a voltage difference at the surface. The transmission rate may be higher than mud-pulsed, particularly in shallow wells, falling off with depth. The old concept of specially designed drill pipe has also made a comeback. In 2016, National Oilwell Varco offered the IntelliPipe with electronic components embedded along the drill string and the IntelliServ system that can provide two-way communication up to 1 million bit/s.

2.4.2.4 Development

Plans for field development take place in the exploration phase as the cost of materials and engineering needed for long-term exploitation enter into the risk assessment. These plans, however, enter into a new phase once the discovery and delineation of the new play are made. Field development focuses on how a reservoir will be produced for maximum yield and profit.

A field reservoir model is constructed using the available geologic, geochemical, and geophysical data. The model is similar to a basin simulation used by exploration, but at a much finer scale with a greater emphasis on lithographic variance within the reservoir formation, fluid migration, phase behavior, and processes measured in days and years instead of millennia. Modern reservoir modeling requires a fine-mesh of interconnected cells, each defined by parameters that characterize the temperature, pressure, and physical and chemical properties of the reservoir rock and the contained fluid. Porosity and permeability are the minimal rock properties needed for volumes and fluid flow modeling. Lithology/mineralogy is included for reactive fluid models involving mineral dissolution and cementation. Standard petroleum properties, such as density and viscosity under reservoir conditions and composition, are needed for calculating its phase behavior, such as its bubble point (the pressure at which a single-phase petroleum separates into two-phases and forms a gas cap) and asphaltene precipitation. The positions of the current oil/gas water contacts and water saturation are entered, as is the wettability of the reservoir rock mineral surfaces, which determine relative permeability and capillary entrance pressures, critical factors in modeling secondary oil recovery processes. Many input parameters depend on pressure, temperature, and saturation, and their effects on production can be simulated over future production. The reservoir model is updated and improved as new wells are drilled and additional information becomes available. Input parameters are continually tweaked so that the prediction results better match actual production. An example of a simulation model of a mature carbonate oil field is presented in Figure 2.12 (Beicip 2019).

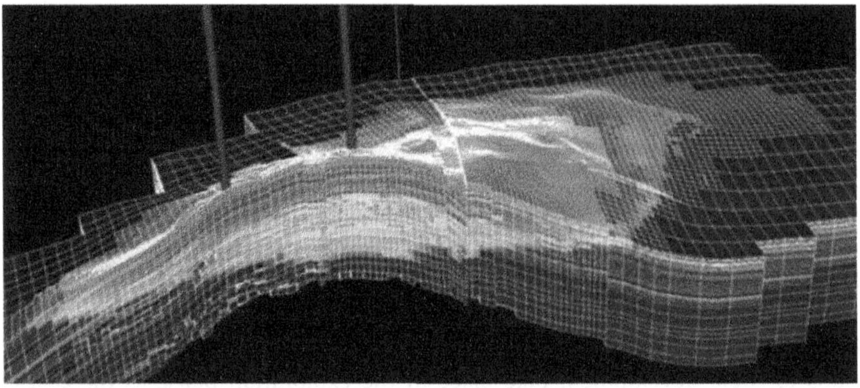

Figure 2.12 Simulation model of a mature carbonate oil field obtained through the integration of all static and dynamic data available. Source: (Beicip 2019). Image courtesy of Beicip-Franlab.

2.4.3 Production

Production is the final stage of upstream operations, where petroleum reserves are converted into a marketable product. The goal of a production field manager is not necessarily to maximize the amount of petroleum recovered but to extract oil and gas within the parameters of an economic model. This may necessitate setting high production rates at the sacrifice of total recoverable reserves, or settling for an inferior method of recovery if a superior method fails to meet the set cost/benefit criteria. Production geologists and engineers are responsible for drilling new wells that expand established fields, test reservoirs and structures that were not included in the initial field development plans, and maintain flow rates of wells under production. In addition, they also treat wellbores that are closing from scale, wax, or asphaltene precipitation, maintain the integrity of subsurface wellbore and pumps and surface facilities (e.g. separators, gathering systems, storage tanks), manage water disposal, and assure that the produced oil reaches the mid-stream.

2.4.3.1 Primary, Secondary, and Tertiary Production

Field production of oil and gas driven by natural forces is termed *primary* production. These include natural water drive, solution gas drive, gas cap drive, gravity drainage drive, and compaction drive. All are pressure-driven. Natural water drive is the most common and occurs when a water aquifer is in contact with the oil or gas. The contact may be partial (edge water drive) or complete (bottom water drive). Reservoirs under natural water drive are relatively easy to produce as the water pressure maintains reservoir pressure as the petroleum is withdrawn. Solution gas drive occurs when a pressure drop due to oil production lowers the reservoir pressure so that the gas dissolved in single-phase oils expands, forcing oil into the wellbore. Solution gas drive can be maintained only if the oil remains above its bubble point. Should this occur, solution drive is lost and the gas-to-oil ratio (GOR) of the produced fluid greatly increases. In conventional reservoirs, pressure above the bubble point can be maintained by water flooding or gas injection. Unconventional shale and tight sand oil wells are all produced initially by solution gas drive and are generally not considered suitable for pressure maintenance (though this is changing); consequently, their productive lifetimes are short and only ~5–10% of the oil in place is recovered. Gas cap drive occurs where the fluid in a reservoir is already two-phase, and a gas cap exists. As oil is withdrawn from the liquid leg and reservoir pressure drops, the gas cap expands, maintaining a pressure drive on the oil. Primary production by gravity drainage requires a specific geometry where the reservoir is tilted. The pressure is maintained as the production of oil from the deeper portion is offset by oil draining from the updip portion. Compaction drive occurs only under geologic conditions where weak reservoir rock collapses upon the removal of oil. Primary pressure may be insufficient to push oil to the surface, and pumps are needed to provide artificial lift. Pumping has little effect on the reservoir pressure and can only assist in the production of oil to and out of the wellbore. Electrically driven submersible and gas lift pumps are used on high flow rate wells common in the offshore. Mechanical pumps, such as sucker rod pumps, are used for low flow rate stripper wells that produce less than 10 barrels/d.

Secondary production involves methods that restore or maintain reservoir pressure by artificial means. The most common practice performed on almost all fields with primary natural water drive is water flooding. A field may be placed under water flooding early during primary production or at a later time. One or more wells are drilled into the aquifer, and water is injected to maintain pressure as the oil is produced from the reservoir above. The chemistry of the injected water is adjusted to prevent mineral precipitation and microbial growth and to modify the wettability of the reservoir rock. Gas injection is an alternative method for maintaining pressure, whereby hydrocarbon gases collected at the separator are re-injected into the reservoir.

Fields, where the reservoir rock is being manipulated directly, are under *tertiary* production. Primary and secondary production never completely drains a reservoir. The amount of oil left behind depends on the reservoir conditions and petroleum composition and can be 20–95% of the oil in place. A host of EOR techniques have been used to recover the unproduced oil. The most common EOR methods used on conventional reservoirs are similar to water flooding, where only water is injected directly into reservoir rock. Water chemistry is again adjusted to prevent mineral precipitation and microbial growth and to modify the oil–water interface either by changing the wettability of the mineral surfaces through chemical reactions or by the addition of surfactants. The water injection wells are distributed to create a sweep of the reservoir using immiscible water to push the trapped oil into production wells. Reservoir rocks are heterogeneous in flow properties, and the injected water will preferentially flow through zones of high permeability instead of a continuous front. Consequently, high permeability zones will be swept clean of oil, while low permeability zones will be untouched. More efficient water-EOR can be engineered by adding chemicals, such as polymers, that plug the high permeability zones, forcing injected water to sweep the oil-bearing low permeability zones. Chemically assisted EOR can be effective, but the cost of the chemicals limits its use. Microbial-enhanced oil recovery (MEOR) methods have been tested, but their effectiveness is disputed. MEOR can be viewed as a form of chemical assistance where nutrients, such as molasses, are added to promote the growth of co-injected microbial strains or indigenous microbes already present in the subsurface to produce biosurfactants or modify reservoir flow in a beneficial manner. An extreme form of MEOR is the production of methane via the conversion of residual oil or coals.

Instead of water, gases can be used for miscible flooding where the injected gas not only adds to the reservoir pressure but dissolves into the residual oil, lowering the oil's viscosity and improving its interfacial tension with water. CO_2, N_2, light hydrocarbon gases, and even air have been used for miscible flooding, but supercritical CO_2 behaves very similarly to a hydrocarbon solvent. CO_2–EOR has emerged recently as a particularly attractive form of EOR as it has the potential for CO_2 sequestration, but like much of petroleum science, its roots date back to earlier decades. Studied in the lab since the late 1950s, Chevron conducted the first, though limited, commercial-scale CO_2 flood in 1972 on the Sacroc Unit in the Permian Basin. The Sacroc Unit passed through several owners until purchased by Kinder Morgan, who implemented a major CO_2 injection program in 2000 that proved extremely successful, with benefits still seen today. While not all fields are good candidates for CO_2–EOR, many are but remain untested due to the cost and limited availability of CO_2. The designs of CO_2–EOR projects that are underway or in development vary considerably. Flooding methods include continuous CO_2 injection, mixed CO_2/N_2 or hydrocarbon gases, pulsed CO_2 and N_2 or hydrocarbon gases, continuous CO_2 followed by water, conventional water-alternating-gas (WAG) injection followed by water, WAG followed by CO_2, and tapered WAG where the volume of CO_2 is gradually decreased with each WAG cycle.

The production of very heavy oil and oil sand in the subsurface requires different EOR methods. The most common is steam "huff-and-puff," where steam laced with surfactants is pumped into a heavy oil reservoir, allowed to "soak" into the oil, thereby lowering its viscosity by thermal and chemical means so that it flows out the wellbore. The volume of rock that can be accessed by "huff-and-puff" is limited so that oil production declines after several cycles. Steam-assisted gravity drainage (SAGD) is a more advanced system. Here, two parallel wells are drilled through the heavy oil sands, one over the other separated by about 4–6 m. Steam is injected into the upper well, mobilizing the heavy oil that collects in the bottom well. Both steam processes can be enhanced with the addition of solvents. In situ combustion is a more drastic form of thermal enhanced recovery. After creating a chamber stripped of oil using steam, air is pumped down the well, and the reservoir is ignited, creating a front of hot gases. The gases ideally then move laterally as a firefront through the heavy oil sands, transferring heat to heavy oil forward of the combustion zone. The mobilized and partially cracked oil then drains downward into a lateral borehole connected to a vertical well. The method is termed toe-to-heel air injection (THAI), with the toe being the most distant portion of the lateral and the heel being the point where the lateral transitions to a vertical borehole. Technically challenged, THAI was demonstrated by Petrobank on its properties in Canada but was stopped in 2011 after limited success.

2.4.3.2 Surface Oil Sands

About 10% of the world's oil reserves are locked up in oil sands in Alberta, Canada. These deposits were formed when oil migrated to the surface and saturated unconsolidated sands. The surface conditions were ideal for microbial activity, and biodegradation oxidized most of the hydrocarbons into CO_2. The mobile oil converted into highly viscous tars composed of high molecular weight polynuclear aromatic (PNA) hydrocarbons and polar/asphaltene species, containing a relatively high percentage of sulfur, nitrogen, oxygen, and trace metals. About 20% of Alberta's oil sands are close enough to the surface that they can be mined. The oil sand is removed from open pits on an industrial scale, then trucked to an extraction plant where it is crushed and treated with hot caustic water laced with chemicals. The slurry is piped to a separation vessel where the bitumen froth floats and is removed, while the extracted coarse sand is returned to the mine and the water containing fine clays is held in large tailing ponds. The recovered bitumen is too viscous for pipeline transport and is either thinned with light hydrocarbons or upgraded into synthetic crude.

2.4.3.3 Unconventional Resources

Unconventional resources are present in low permeability rocks that cannot be economically produced by conventional means without fracturing the rock. Such rocks include coals, shales, microporous carbonates, and low permeability tight sandstones. In the United States, many conventional wells were drilled long before a basin was considered to have unconventional prospects. All elements of a viable petroleum system are still present; however, the source, reservoir, and seal may be within a single formation, and oil and gas migration may be limited to short distances. Since much of the information to define plays already exist, exploitation of unconventional resources is thought of as an engineering production challenge and not one of discovery. This is certainly not true in basins outside of the United States, where information from conventional wells is not as dense, and even in mature basins, the identification of "sweet spots" is critical in acquiring profitable acreage.

2.4.3.3.1 Coal Bed Methane (CBM)

Methane desorbing from coal mines was recognized as a safety issue in the nineteenth century and was captured as an exploitable resource from a few mines in the early twentieth century. The first deliberately drilled wells into coal were drilled in the early 1950s in the San Juan Basin in New Mexico. There, the Fruitland coals were considered a hazard for drilling into deeper conventional reservoirs, and the methane was produced to relieve the gas pressure of the coal beds. The energy crisis of the 1970s encouraged a few producers to capture the methane from the San Juan and Black Warrior Basins. A change in the 1980 US tax code granted credits for unconventional fuel sources, precipitating a boom in drilling wells into coal beds and producing the desorbed gas. The tax credit was so generous that companies could make money even if the wells were not economical. The credit was eliminated in 1992, but coal bed methane (CBM) continued to be developed in the United States, contributing ~1.8 TCF/yr, or ~10%, of the total gas production by the mid-2000s. Only a few CBM projects have been developed outside of the United States. With the recent dominance of shale gas and the corresponding drop in value, few new CBM wells are being drilled in the United States, but the older wells continue to produce slightly less than 1 TCF/yr.

Coals suitable for CBM development contain high volumes of methane and a well-developed cleat system. Methane, produced by the thermal alteration or by microbes, is physically absorbed within the coal matrix that possesses a very large surface area due to a profusion of sub-nanometer sized pores. Cleats are fractures within the coal that are naturally formed or induced by the injection of gas or water. A well drilled for CBM will have high initial water production from the dewatering of the cleats. After the water is drawn down, methane will desorb from the coal matrix into the cleat system and eventually into the wellbore. Gas desorption is a slow process, and the flow rate of produced CBM gas is very low compared with conventional resources.

2.4.3.3.2 Shale Gas

The exploitation of shales as unconventional shale gas is not a new concept. In the early 1820s, wells penetrating eastern US Paleozoic shales produced methane for commercial use. In the 1970s, the US government launched the Eastern Gas Shale Project to study this resource in response to energy shortfalls. The full resource potential of gas shales was finally recognized with the development of the Barnett Shale play starting in the late 1990s (Steward 2007). In 2000, shale gas, mostly from the Fort Worth Basin, contributed only 2% to the total US production of natural gas. Today, shale gas contributes ~24 TCF/yr or ~65% of the total. Shale gas production is currently mostly limited to the United States, Canada, China, and Argentina, but its global potential is now apparent. Tests are being conducted on all continents, although full development in many locations may be hindered by the lack of infrastructure or environmental/climatic impact concerns.

Just how gas is generated, stored, and transported in productive shales has been a topic of debate since the 1930s. At first, natural fractures were deemed required for transport, but it was unclear if the productive gas resided only in these fractures or if gas stored in the less permeable matrix could contribute. Despite strong arguments for a matrix contribution, the issue only settled when gas was produced out of Barnett Shale with no natural fractures. In the 1980s, the matrix contribution was generally considered to reside as free gas within small inorganic pores, and although a high abundance of organic matter was considered a necessity for gas generation, it was not viewed as necessary for storage. The recognition that shale productivity depends on a variety of factors, including the amount of organic carbon and level of thermal maturity, became accepted criteria as petrophysical measurements were developed for fine grained rocks. Again, the hypothesis for shale gas storage changed slowly, and gas absorption into the organic matter in a manner similar to coals was recognized as more reliable when production history data became available along with the non-detection of the required inorganic porosity. This view of shale gas storage and transport was held until the recent accelerated exploitation of this resource initiated new research. When conventional reservoir characterization methods failed to locate a significant porosity in the productive Barnett Shale, new techniques in sample preparation and imaging were applied by researchers at the University of Texas in Austin (Reed and Loucks 2007). The shale, when ion-milled and examined by scanning electron microscope (SEM), revealed that nearly all of the observed porosity was located within the organic matter. The images revealed physical meso- and macro-scaled pores in carbonaceous material, which could be either kerogen or thermally altered solid bitumen. The correlation seen between TOC and total porosity is now well established in many organic-rich shales. Figure 2.13 highlights the organic porosity of the carbon-rich Marcellus shale by helium ion microscopy.

Many basic questions concerning the origin of the organic porosity remain unanswered. It is unclear whether the porous material is spent kerogen, retained and thermally cracked bitumen, or both. Knowledge of the thermal conditions for the onset of porosity formation and development into the wet gas, as well as porosity preservation at high levels of catagenesis, are key elements in expanding unconventional resources. The size distribution of the pores and pore throats and the manner in which they are connected are factors in determining the amount of gas in place and its production.

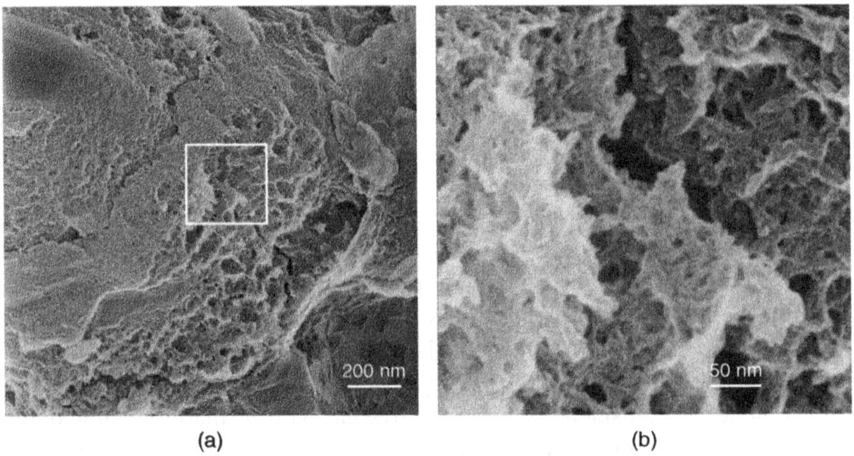

Figure 2.13 Helium ion microscopy of high maturity, carbon-rich Marcellus shale showing organic porosity (a), with zoom (b) of the highlighted region. Source: Eberle et al. (2016). Reproduced with permission, Copyright (2016), American Chemical Society.

The modern boom in developing unconventional resources relies on controlled horizontal drilling and hydraulic fracturing. Claims for the first directional wells include one drilled in Texon, Texas, in 1929, and one drilled in Franklin, Pennsylvania, in 1944. Drilling true horizontal wells, those having a lateral bore coming off the vertical hole at deviations around ~70°–90°, only became possible in the 1970s with the development of downhole "mud" motors (turbo drilling) that turned the drill bit using circulating mud while the vertical shaft remained stationary. Coupled with technology advances in downhole telemetry while drilling, horizontal drilling of long laterals within a specific production zone became practical. Today, wells drilled in the Permian Basin routinely drill to vertical depths >3000 m and continue with laterals that can exceed 4500 m.

The stimulation of shales for enhanced gas production is not a new concept. It was first tested in the early 1820s when exploding gunpowder in the wellbore of vertical wells penetrating eastern US Paleozoic shales produced higher rates of gas flow. The Roberts Petroleum Torpedo Company, one of the earliest oil field service companies, patented a design in 1866 that was essentially a canister of gunpowder or nitroglycerine rigged with first a percussion cap that was set off by dropping an iron weight, then later with a blasting cap and fuse. The device and its successors remained in service until 1990.

The concept of stimulating production by the injection of fluids was first proposed in the 1930s. However, it was not proven effective until tests conducted by Stanolind Oil and Gas in 1947 on the Hugoton gas field in Kansas. The research was patented in 1949, and an exclusive license was granted to Halliburton, who followed up that year with commercial tests in Texas and Oklahoma. Small-scale hydraulic fracturing of eastern Paleozoic shales was conducted in the mid-1960s. Hydraulic fracturing was not widely used until the 1970s, when it was applied to stimulate conventional oil and gas reservoirs penetrated by vertical wells. The energy crisis prompted numerous tests of fracturing shale using government funds. The shale gas boom started with Mitchell Energy, when in 1997, a horizontal well in the Barnett Shale conceived the "slickwater frac job," a technique where chemical friction reducers are added to increase fluid flow.

All unconventional horizontal wells are "fracked" using pumps to send a mixture of water, sand, and chemicals under extremely high pressure into shale or other tight reservoir rocks. First, placement of the fractures is done using a perforating gun, a device that is lowered by wire, and fires a charge that perforates the steel casing and cement at a given position in the lateral. The hydraulic process occurs in stages: the initial spearhead stage pumps a dilute acid solution to clear debris out of the wellbore, the pad stage pumps water and chemicals without proppant to initiate fracturing, then non-compressible materials are added in the proppant stage, and finally, the flush stage simply pumps down to remove excess proppant. The chemicals are added to reduce friction, to increase viscosity and suspend the proppant, and to prevent microbial growth, scale formation, and corrosion. Most wells use sorted and unsorted sand as proppant; other natural and synthetic materials have been used and may, in fact, be more effective proppants but are economically challenged compared to sand.

The goal of developing an unconventional gas or oil field is to maximize the volume of the stimulated reservoir using the fewest number of wells. This requires optimizing the placement, orientation, and lateral length of each well, and designing a perforation pattern for the lateral. Ideally, wells and their fracture stages are placed as close together as possible so that no rock volume is missed while remaining isolated from each other. The application of advanced geomechanical models that incorporate rock mechanical and fluid flow properties, the orientation of maximum geologic stress, placement within the target zone, and other parameters such as the presence of natural faults, can be used to optimize drilling and fracturing, but companies mostly use learnings gained from field experience. Various multistage techniques have been developed using packers, plugs, and sleeves to control the fracturing process; however, few if any wells are stimulated in an ideal manner. Worse case scenarios include fractures prorogating outside of target formation into permeable rock or into open fractures from a nearby well.

2.4.3.3.3 Unconventional Liquids

The production of liquids from unconventional reservoirs basically uses the same drilling and well completion techniques developed for gas shales. Gases produced from lower maturity shales can contain appreciable amounts of dissolved ethane, propane, and butane that can be condensed at the surface.

Liquids can be produced from oil-bearing formations with very low permeability using long laterals and hydraulic fracturing. Tight oils are produced from formations that are both reservoir and its own source or from non-source rocks that have been charged with migrated oil. Although also called shale liquids, if the producing formation is predominately shale, most if not all of the liquid production is from hybrid systems, with natural fractures and microporous lithologies.

Hybrid systems consist of source shales incorporating or in close proximity to thin conventional reservoir rock. The Bakken Shale is an example of a layered hybrid system, where organic-rich shales in the Upper and Lower Bakken serve as source and seals for oil that migrated into the Middle Bakken, which is a non-source member with highly variable lithology (cemented siltstones and sandstones) and having low porosity (1–15%) and permeability (0–20 mD) with a maximum thickness of 75 ft. The Middle Bakken can be viewed as a poor quality conventional reservoir with limited production using vertical wells as far back as the early 1960s. Innovations in precise geosteering permitted drilling laterals that threaded through the Middle Bakken; when coupled with fracking, this created an oil boom in the 2000s that continues to today. The Niobrara Formation is another hybrid system producing liquids. Here, organic-rich shales, which serve as source and seal, are interbedded with limestones, which serve as reservoirs.

Formations with high microporosity (5–10%) but low permeability are the latest tight oil plays to be developed. These are mainly marine sandstones and carbonates that contain migrated oil generated from a completely separate source or sources. They are essentially part of a conventional oil system and are only defined as unconventional in that horizontal drilling and hydraulic fracturing are needed for economic production. Examples of tight oil plays occur in the Anadarko Basin, with production out of the Upper Cleveland sandstones and Mississippi limestones that were charged with migrated oil generated by older shales. The Austin Chalk reservoir in southeastern Texas is a microporous carbonate with karsts and natural fractures, which were charged with oil that migrated from the underlying Eagle Ford Shale source. The fractures can be efficiently produced through horizontal drilling, while recovering oil trapped in the microporous matrix requires fracturing.

2.4.3.4 Plug and Abandonment

Once a well is exhausted, it is permanently plugged with cement. The entire borehole may not be filled, but cement is usually placed across any open reservoirs containing hydrocarbons, casing points, freshwater aquifers, and the uppermost section near the surface. Once the integrity of the cement is proven, the wellhead and all associated subsurface flowlines and surface equipment (e.g. storage tanks, separators) no longer in operation are removed. The well is cut several feet below grade, and an identifying marker is welded on. The site then undergoes remediation according to government regulation. Similar processes are used on offshore platforms when they are completely decommissioned.

2.5 Mid-Stream

The role of Upstream Exploration and Production ends once the oil is brought to the surface and ready for shipment. Mid-stream operations then transport oil and gas from the field to a refinery via pipeline, ocean tanker, or rail/truck, and construct and maintain loading and storage facilities.

2.5.1 Transportation

In ancient times, global trade of oil and bitumen stored in amphora, animal skin bottles, and wooden barrels were transported by pack animals and boats, a process that remained nearly unchanged for centuries. The early oil wells drilled in Pennsylvania gave rise to another boom, as coopers could barely keep up making barrels that were carried by wagons or rafts to railroad depots, ports, and refineries. We still use a "barrel" as a standard unit of oil volume. With production out of the Pennsylvania wells topping 1000 barrels/d, it soon became evident that a different technology was needed to overcome the midstream bottleneck. The first oil pipeline was constructed in 1861 in Erie between the rail depot and the refinery. Attempts to construct longer pipelines from the Oil Creek wells beginning in 1862 were largely unsuccessful and were destroyed by teamsters, whose lucrative and monopolistic transportation business was threatened. In 1865, Samuel Van Syckel succeeded in building a 5 mi long, 2 in. wrought iron pipeline from Oil Creek to the nearby railroad stop. A complete gathering system of interconnected pipelines soon followed, as did a war between the teamsters and the pipeline owners. Oil pipeline technology continued to advance in the twentieth century, becoming longer, wider, and safer. The first offshore pipelines, nearly 70 km long, were laid in 1944 across the English Channel to support the allied war effort. The first commercial offshore gas pipeline was laid in 1954 in the Gulf of Mexico, and the Trans-Alaskan Pipeline System (TAPS), started in 1974 and completed in 1977, is one of the longest and most technologically advanced.

Tankers, ships dedicated to the transport of petroleum, are a fairly modern invention. Some of the more well-known tankers are portrayed in Figure 2.14. Two sailing vessels were built in 1863 for service along the Tyne River. The *Vaderland* was the first steam-powered tanker, built in 1873 for the Belgium Red Star Line, but its use was curtailed over safety concerns. The Ludvig Nobel

Figure 2.14 Historic oil tankers. (a) Zoroaster photo. Source: From Zoroaster (1908) under Creative Commons Attribution License (CC BY-SA 3.0). (b) Batillus photo. Source: Reprinted from Batillus tanker in Saint-Nazaire (1976). (c) Exxon Valdez photo. Source: Courtesy of ExxonMobil.

had the first oil tanker, the *Zoroaster*, built in 1878 with safety features included in the design, which included separating the engines from the cargo holds and providing for improved ventilation. The system worked until a sister ship, the *Nordenskjöld*, became the first oil tanker disaster. It exploded in 1881 while being filled with kerosene that spilled into the engine room, prompting Nobel to invent an improved flexible, leak-proof loading pipe. Soon a fleet of single-hulled tankers driven by internal combustion engines was traveling between Baku and Europe, which incorporated additional innovations following the designs of Colonel Henry F. Swan. The *Glückauf* was Swan's crowning achievement, incorporating many features of today's tankers, controlling the fluid flow within the tanker and the ability to load seawater ballast when empty of oil. After World War II, the size of the tankers continued to increase with incremental improvements. Single-hull ships remained the norm until the *Exxon Valdez* ran aground in 1989, spilling its cargo into Prince William Sound. This prompted the passage of the Oil Pollution Act of 1990, which required oil tankers docking at US ports to be double-hulled.

The design of a modern oil tanker depends on its intended use. Small vessels and barges are used to offload oil from larger vessels that cannot dock at a port. Larger ocean-going tankers are built for specific routes. Panamax tankers that need to pass through the Panama Canal were limited by a beam of 32.3 m and a draft of 12 m. The 2016 expansion of the canal now allows Neo-Panamax tankers a beam of 52.2 m and a draft of 15.2 m. The four largest supertankers were built in the Daewoo shipyards in 2002–2003, with a cargo capacity of 3,166,353 barrels. Two of these ships are still in service, and the others have been converted into floating storage and offloading (FSO) vessels. Ships are designed as floating platforms for production, storage, and offloading (FPSO). The largest ship ever constructed is Shell's floating gas liquefaction platform, the *Prelude*, designed to process gas in the waters off the northwest coast of Australia.

Transporting natural gas has also progressed from onshore pipelines to specialized ships. In China, natural gas had long been co-produced with the brine from deep wells drilled to support the salt industry. Early on, the Chinese invented a system to separate the gas, which could contain toxic H_2S. But, it was only in the sixteenth century that the Chinese began to exploit natural gas, and a transportation system involving above-ground pipelines constructed of bamboo was invented. The bamboo system continued to evolve, delivering gas to both industry and residences into the twentieth century.

The modern gas pipeline system involves a distribution network that involves gathering the gas from producing fields, transporting the gas for purification to remove gas liquids and non-hydrocarbon gases, and distributing the clean methane to individual factories and domiciles. Methane is colorless and odorless, so trace amounts (1–10 ppm) of a mercaptan (usually 2-propanethiol) is added as an odorizer for safety. The gas pipeline systems in the United States and Europe are highly integrated networks that are extensively regulated. They are made from high-quality steel ranging from 6″ to 48″ in diameter that are welded together as they are laid into shallow (typically 2–5 ft deep) trenches. Valving is installed at defined points that allow operators to isolate flow in a section of pipe for pressure testing. Gas pipelines are operated at transmission pressures from ~500 to 1400 psi, and the pressure is closely monitored for indications of leakage. Russia exports huge volumes of gas to Europe via its pipeline systems, tapping gas fields in Western Siberia. Regulation and maintenance of their pipelines are less rigorous, and an infamous disaster occurred in 1989, when sparks from two passing passenger trains ignited gas leaking from a nearby pipeline. The pipeline operators had noticed that pressure was dropping in the line but increased the pressure to maintain flow instead of trying to find and repair the leak. Conditions of the Gazprom network have improved but are estimated to still leak 5–7% of the transported gas.

Transporting natural gas via ship involves completely different technology than oil tankers. Economical transport of gas requires it first to be liquefied and then transported on specially designed ships capable of maintaining cryogenic temperatures. The maiden voyage of the first liquefied natural gas (LNG) carrier, the

Methane Pioneer, took place in 1959, with a shipment from the Louisiana Gulf Coast to England. As with oil tankers, LNG ships continued to increase in size and incorporate innovations for improved safety, and the largest now transport up to ~95 million ft^3 (~2.7 million m^3) of cryogenically cooled methane.

Although LNG tankers are fairly small vessels (200–350 m in length) that could easily pull up to a dock, the potential of a catastrophic explosion requires that LNG tankers be loaded and unloaded at offshore mooring berths. The berths have articulated arms that connect the LNG carrier to onshore facilities. If empty of LNG, loading first requires that the tanks first be purged of oxygen. LNG is then introduced, which vaporizes as it cools the tanks, and eventually reaches less than −140 °C while the gas is captured and returned to shore. Loading of the LNG cargo can then commence. Unloading is essentially the reverse operation; LNG is pumped out of the tanks to the terminal, and the empty space is filled with gas sent from the terminal, or some of the LNG cargo is allowed to volatilize. Usually, about 5–10% of the LNG cargo is retained in the tanks so that they are not exposed to air and are already cool to receive the next load.

Rail and truck transportation offers the greatest flexibility for inland transport but at a higher cost. Rail transport is drafted into use when pipelines are unavailable. The recent surge in production out of the North Dakota Bakken Shale outpaced pipeline availability, and almost 1 million barrels/d were transported by rail during its peak production (2010–2014). Rail transport is not without risk. A 2013 derailment of a freight train carrying Bakken oil in Lac-Mégantic, Quebec, resulted in the explosion of multiple tank cars resulting in 47 deaths and the destruction of 30 buildings in the center of town. A series of similar, though less disastrous events occurred in the following months and years in Canada and the United States. The accidents involved the DOT-111A, a thin-walled unpressured tank car that already had been redesigned in 2011 to improve crashworthiness. In 2015, the Federal Railroad Administration and Transport Canada agreed that the design of DOT-111A was inadequate and that all newly constructed tank cars would have to meet DOT-117 specifications. Oil can still be transported in DOT-111A cars with restrictions until 2025, when all of the older oil cars need to be retrofitted to meet DOT-117 standards or be retired. Nevertheless, rail remains viable, if not a preferred option, as new pipeline construction is facing citizen opposition in many areas.

2.5.2 Storage

Safe storage of oil, gas, and LNG has different engineering requirements. These can be underground, on the surface, or floating. For stripper wells with only a few barrels/day production, a single above-ground storage tank will be sufficient.

Refineries, utilities, and pipeline operators may need to construct massive tank farms or use underground voids.

Above-ground tanks operate at low, atmospheric, and high pressure depending on the oil or refinery product being stored. Circular atmospheric tanks are used to store non-volatile heavy oils, bitumen, and residual fuel oil. These circular tanks have a fixed, conical roof that is intentionally built with a weld weaker than the sides that will fail first should pressure build up in the tank. Oils and volatile liquid products are stored in circular tanks with flat, floating roofs, which may be open or capped by a fixed conical roof. The height of the floating roof is adjusted to be in contact with the stored oil to minimize evaporation. Gases are stored above ground in large atmospheric pressure circular tanks called gasometers. These consist of an inverted tank containing the gas that sits inside an outer tank containing water. The water provides a seal for the gas and is lowered to accommodate gas storage or raised to pump the gas out. Refined gases such as propane and butane are stored in high-pressure containers that may be spheroids or cylinders.

Underground storage can be either buried tanks, such as used at gas stations, home propane heating, or geologic voids. Gas, but not oil, can be stored underground in depleted gas reservoirs with formations with high porosity and permeability. Static aquifers can also be used to store gas where depleted wells are not available. Aquifer gas storage is mostly performed seasonally by utility companies that purchase gas during the summer when prices are lower than during the winter. Salt caverns are ideal structures for underground storage of gas or oil. They are formed by pumping hot water into a layer of subsurface salt and pumping out the brine, leaving behind a void encapsulated by nearly impermeable rock. The underground salt caverns are heated by the geothermal energy that will induce convection currents in stored oils or petroleum products, assuring that the material remains mixed and homogeneous. The US Strategic Petroleum Reserve (SPR) is a series of caverns up to 1000 m deep that were made within Luann salt domes at four sites on the Texas/Louisiana Gulf Coast. Originally specified to hold 1 billion barrels, the SPR's maximum capacity was 727 million barrels and was filled completely in 2009. The facilities store both sweet and sour crudes isolated in individual caverns. Oil has been released at times of storages due to hurricanes and currently holds 641 million barrels.

Ships are intentionally designed to serve as floating platforms for oil FPSO. Decommissioned oil tankers can also be converted for offshore storage. Occasionally, active oil tankers are pressed into service for storage. This situation can arise if tanker shipments become uncoordinated due to storms or if a terminal shuts down unexpectedly due to an accident or political unrest.

LNG terminals have unique features to handle LNG storage. Cryogenic tanks must be capable of maintaining −160 °C to keep the LNG in its liquid form. The inner steel tank is surrounded by a thick wall of pre-stressed reinforced concrete that acts as insulation. Heat still enters the inner tank so a small amount of LNG will continuously volatilize. The boil-off gas is captured and converted back to LNG using a compressor and recondensing systems.

2.6 Downstream

Petroleum refining processes convert crude oil feedstocks into more valuable products with specific physical and chemical properties through separation and conversion processes. An understanding of the composition of the crude feed and products, as well as reaction mechanisms and kinetics, is key to optimizing the refining conditions.

2.6.1 Evolution of Modern Refining

Distillation, the process of heating crude and condensing the vapors, has been used for millennia for separating oil into highly volatile naphtha, middle cuts of flammable kerosene, and nonvolatile residuum. By the fourteenth century, kerosene distillate was commonly used for lighting and heating throughout China and the Middle East. Kerosene was distilled in Europe, but whale oil was preferred as it burned more cleanly. By the mid-1800s, with the invention of the flat-wick kerosene lamp, and with whale oil becoming increasingly scarce as the whale population was decimated, the demand for kerosene surged globally, and large-scale petroleum refineries were constructed in Baku, Poland, Romania, Canada, United States, and soon throughout the world. Most of the early refineries distilled crude oil in batches, and the produced kerosene was contaminated with naphtha or residuum. Continuous distillation using a series of connected stills emerged in the 1880s, and in 1903, John Van Dyke and William Irish developed the tower still at the Standard Oil Point Breeze refinery in Philadelphia, paving the way for the multi-staged tray towers used today.

Until the twentieth century, naphtha was considered a waste product. This all changed with the growth of the auto industry and internal combustion engines that run best on gasoline. Gasoline is the lightest fraction of naphtha, but it comprises only a small fraction of most crude oils, and very little can be recovered by distillation. The demand for gasoline spurred refiners to find ways to convert heavier petroleum components into lighter hydrocarbons. The problem was first tackled commercially at the Standard Oil refinery at Whiting, Indiana. With construction starting in 1889, the Whiting Refinery was designed to distill oil at a massive scale and was soon producing 20% of the total US kerosene production. The Whiting Refinery is first noteworthy as it processed sulfur-rich "sour crude" from the Lima, Ohio oilfields, considered useless by others. The process using copper oxide to remove the sulfur was developed by Herman Frasch and implemented by John Van Dyke. Twenty years later, as Standard Oil was broken apart, the demand for gasoline prompted Amoco scientists William Burton and Robert Humphreys to conduct research on converting kerosene into gasoline. After many failures at using catalysts, in 1913, they invented a non-catalytic thermal cracking process using both high heat and pressure. The Burton Distillation Unit ushered in the modern era of petroleum refining, and the original unit is preserved at the Smithsonian. It was quickly refined over the next decade into a continuous flow process involving "soaker" units where distillate fractions could be moved on- and off-stream to undergo thermal cracking. Innovations in the internal combustion engine demanded refineries to produce fuels with specific properties. Higher performance for gasoline-driven engines requires fuel that resists detonation upon compression (knocking) while diesel engines prefer fuels that undergo autoignition. Gasoline in the early twentieth century had very low octane ratings that limited engine performance. After years of testing, General Motors discovered in 1921 that adding tetraethyl lead could increase octane rating, and soon refineries were adding this inexpensive compound even though lead was known to be a health hazard. Looking for a way to produce lead-free, high-performance gasoline, Socony Vacuum and Sun Oil companies collaborated with Eugene Houdry, who had conducted research on the conversion of lignite into coal tars using silica–alumina catalyst. Humphreys' idea of using chemical catalysts for crude upgrading was finally realized in 1937 when the Sun Oil Marcus Hook refinery put into operation the "Houdry Unit." Since then, a variety of thermal reforming, polymerization, alkylation, and isomerization refinery processes were developed to produce higher octane gasoline and to convert more of the barrel into quality fuels. Many of the processes were spurred on by the military needs of World War II. Today's refineries convert crude feedstock into roughly 40% gasoline, 22% diesel, and 8% kerosene jet fuel. The remainder can be converted to a variety of products ranging from high-value lubricants to low-value asphalt and coke.

2.6.2 Modern Refinery Processes

The purpose of refining crude oil is to convert the crude feed into a set of products with maximum economic value. Refining starts with the delivery of crude feedstock and finishes with the sale of refined products. In between are serial and parallel processes involving separation, conversion, and purification to convert the crude oil into valued commodities (Figure 2.15). The highest value molecules are those in naphtha, gasoline, kerosene, and diesel; consequently, a major portion of the refinery processing involves converting both smaller and larger

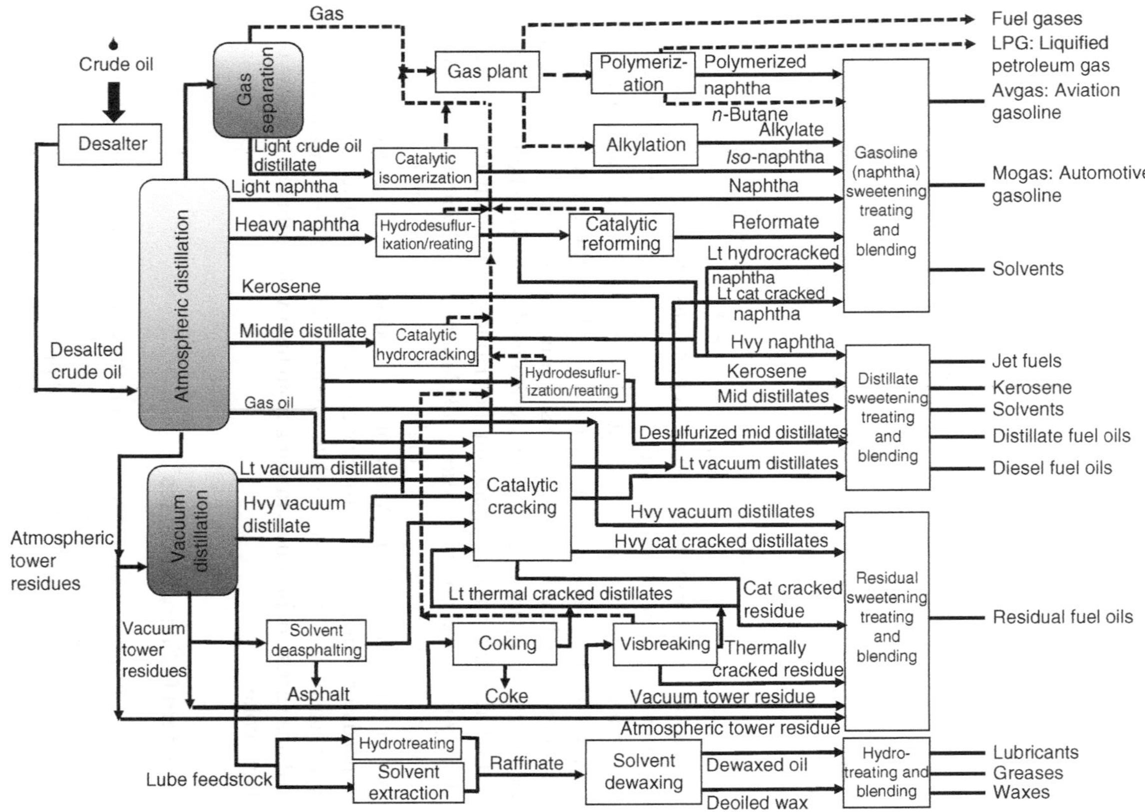

Figure 2.15 Overview of petroleum refining processes. Source: OSHA (1999).

molecules into those with a higher value. Refining employs a wide variety of processes, and the flow scheme is largely determined by the molecules in the crude oil feed and the chosen slate of refined products. More than 2500 distinct products may be produced, including various grades of liquefied petroleum gas (LPG), gasoline, kerosene, aviation fuel, diesel fuel, fuel oils, lubricating oils, and feedstock for the petrochemical industry. Each product is defined by its performance specifications, which are determined by the molecular composition.

2.6.2.1 Crude Oil Pretreatment

The crude oil feedstock, as delivered to the refinery, needs to be cleaned. Non-hydrocarbon gases (H_2S, CO_2, N_2), water, inorganic salts, suspended solids, and water-soluble trace metals may be dissolved or suspended within the crude. In addition to natural materials co-produced with the crude oil, synthetic substances may have been intentionally added during drilling or transport. The removal of dissolved salts is critical, as HCl can form from chloride during the high-temperature processing of crude causing metal corrosion. Other impurities can be detrimental to the refinery, as they may impede fluid flow and poison catalysts in the processing units. The amount of material removed from crude feedstocks varies considerably but typically ranges from 0.1% to 2% volume.

Pretreatment begins with one or more separators driving off gases, which are treated in the refinery as a separate feed from liquid crude. The liquid petroleum feed is then mixed with pH-adjusted hot water that is laced with chemical surfactants (demulsifiers) and oxygen scavengers.

The purpose of desalting (also known as dehydration) pretreatment is to reduce or completely remove these inorganic molecules from the petroleum gas, liquid, and/or solid phases. The process begins when pH adjusted hot water, either fresh or recycled, that is laced with chemical surfactants (demulsifiers) is mixed into the crude feedstock. Salts and other solid impurities are extracted by the water, which is then allowed to settle out from the petroleum phase. Mildly heating (typically between 150 and 350 °F or between 66 and 177 °C) can reduce the viscosity and improve mixing and separation. The application of a high-voltage electrostatic charge is used to promote the concentration of salts in the bottom of the settling tank. Wastewater is discharged from the bottom of the settling tank, and the cleaned crude is sent for refining as a continuous process. The desalter wastewater is polluted with whole crude, added chemicals, and dissolved and suspended solids and is enriched in the more water-soluble species in petroleum (e.g. light aromatics, phenols, and small fatty acids).

2.6.2.2 Separation

The crude oil is a complex mixture of organic molecules, mostly hydrocarbons. Each molecule has an economic value that varies depending on its association

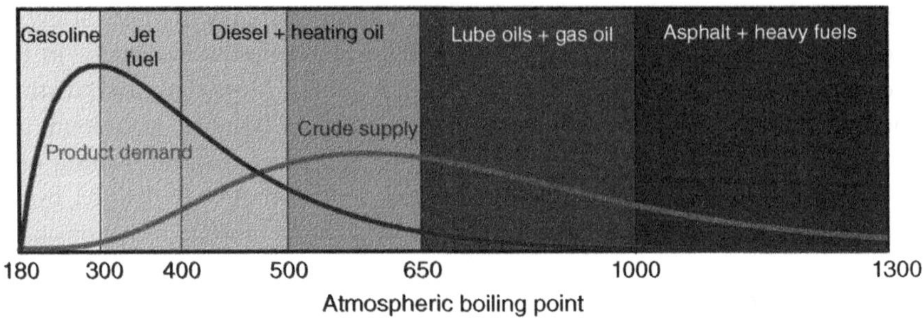

Figure 2.16 Boiling point distribution of a typical crude feedstock and converted products.

with other molecules. The boiling point is one of the major physical properties that determine the value of individual molecules in crude oil. In general, higher-value products are composed of lighter, more volatile hydrocarbon mixtures with the desired flow and combustion properties.

Crude feeds may contain species ranging from methane (bp −258.7 °F/−161.5 °C) to C_{100+} waxes (bp > 1320 °F/715 °C); a typical boiling point curve for a typical crude feed is shown in Figure 2.16. These are mostly paraffinic, naphthenic, and aromatic hydrocarbons. Molecules containing heteroatoms, such as sulfur, nitrogen, oxygen, and chelated metals (e.g. Ni and V), are generally of lower value and, depending on the quality of the crude, may be minor to major components. Heteroatomic species range from the acidic gases, CO_2 (bp −109.3 °F/−78.5 °C) and H_2S (−76 °F/−60 °C), to non-volatile asphaltenes that thermally degrade ($\gtrsim$450 °C) before boiling.

Distillation is one of the most robust ways to selectively separate high-value molecules present with the crude feedstock from other molecules of lower worth. There are two major distillation processes: atmospheric distillation and vacuum distillation.

2.6.2.2.1 Atmospheric Distillation

Atmospheric distillation is the first process of separation in crude oil refining (Figure 2.17). The desalted crude oil is preheated using recovered process heat. The crude oil then flows to a direct-fired crude oil charge heater, then fed into the vertical atmospheric distillation tower just above the bottom, at pressures slightly above atmospheric and at temperatures ranging from ~650 to 700 °F (~345 to 370 °C). Heating crude oil above these temperatures may cause undesirable thermal cracking and coke formation. The tower is hottest at the bottom, where liquids are heated and vaporized, and coolest at the top. All but the heaviest molecules are vaporized, and as molecules travel up the tower, they condense at different temperatures. Unlike the simple single-stage distillations performed in

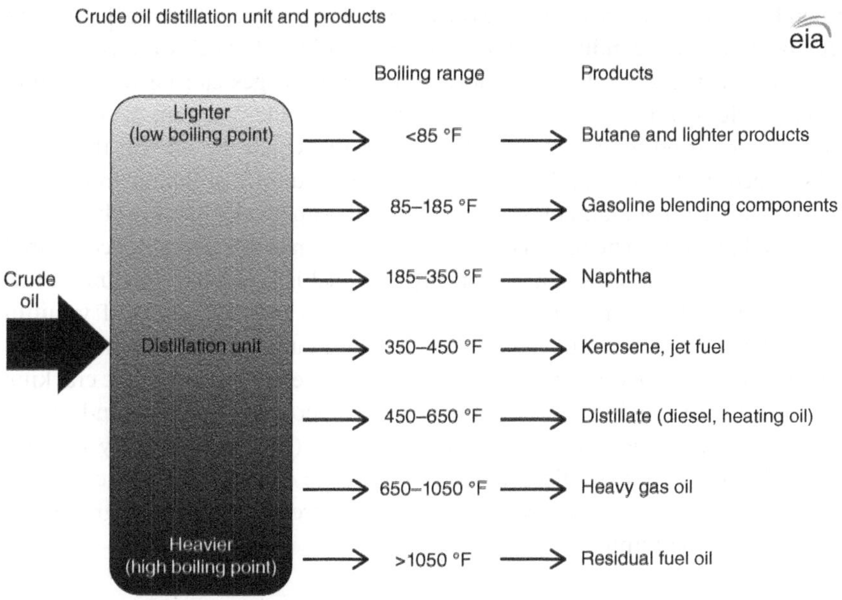

Figure 2.17 Atmospheric distillation tower. Source: EIA (2012).

a freshman organic chemistry class, the distillation tower contains a set of trays that capture molecules within a range of boiling points. At each tray, vapors from lower boiling temperature areas enter perforations and bubble caps. They allow the vapors to pass through the liquid on the tray, promoting condensation at the temperature of that tray. An overflow line drains the excess condensed liquids from each tray back to the lower boiling area, where the higher temperature causes re-evaporation. The evaporation, condensing, and scrubbing operation are repeated many times until a dynamic equilibrium is reached, and the molecules with the desired boiling temperature range are isolated. Side streams are then taken off from certain trays to obtain the desired fractions. The distillation process separates the major constituents of crude oil into so-called straight-run products defined as naphtha, gasoline, kerosene, diesel, and heavy gas oil that gather at the different tray levels, leaving heavy fuel oil or asphalt residue at the bottom of the distillation tower. Light gases are flared or captured from the top of the distillation tower. Sometimes crude oil is "topped" by distilling off only the lighter fractions, leaving a heavy residue that is distilled further under a high vacuum.

2.6.2.2.2 *Vacuum Distillation*

Vacuum distillation operation is conducted at very low pressures (50–100 mmHg). Under near-vacuum conditions, most molecules will volatilize at a lower temperature than at atmospheric pressure (760 mmHg). By conducting distillation under near-vacuum conditions, the relative volatility of species increases, and the separation process refines as more equilibrium stages can be reached and heavier species can be recovered from the residual crude. Vacuum distillation is operated at reduced temperature (~400 °C), above which the oil will thermally crack, impeding the distillation process. These vacuum distillation towers are designed with a larger diameter so the comparable vapor velocities are obtained at the reduced pressure. The internal design of a vacuum tower is simpler compared with atmospheric columns, since typically, the feed is separated only into a few fractions unless lubricant cuts are pulled. The two main products of vacuum distillation are termed vacuum gas oil (VGO) and vacuum resid. VGO includes all products that were volatilized and condensed and is fed to fluid catalytic cracking (FCC) and hydrocracking units for further conversion into gasoline and diesel. VGO is often sub-divided into light vacuum gas oil (LVGO) and heavy vacuum gas oil (HVGO), and specialized units may produce gas oils and lubricating oil base stocks. The resid is fed to visbreaker or hydrotreater units for upgrading or to coker and asphalt units.

2.6.2.2.3 Light Gases

Crude oil entering the refinery is largely a liquid under standard conditions. However, it also contains small amounts of propane, butanes, and pentanes, which are dissolved but highly volatile. Additionally, other light gases can be present in small amounts, including hydrogen sulfide and low molecular weight thiols and mercaptans, which lend some crudes a strong odor. Besides these small amounts of light hydrocarbons coming in with the crude, most conversion steps in the refinery also generate light cracked products and collectively create a large amount of light gas that the refinery must manage.

When cracked gases enter the gas stream, they introduce compounds that are not typically present in virgin feeds. The most common of these are olefins (alkenes) corresponding to the dehydrogenated versions of the molecules present in natural light gases. Dienes and alkynes, typically just acetylene, methyl acetylene, and propadiene, may be formed in the more extreme processing steps. As the cracked product's molecular weight grows, it tends to convert into aromatics and styrenics, particularly in the steam cracking process.

Where economical, these mixed gas streams may be sent to an olefins recovery unit where the olefins are captured for use in high-value chemical feedstocks for polyolefin manufacture. Additionally, contaminants such as H_2S and ammonia are scrubbed. The remaining gas is burned to supply heat to the refineries various processes, either directly in fired furnaces or indirectly in boilers that provide steam throughout the site.

2.6.2.2.4 Lube Extraction and Solvent Dewaxing

The VGO stream from the vacuum pipestill contains historically valuable molecules for use as lubricants. Without additional processing, VGO is typically not suitable for conversion into a marketable product. Lubricant basestocks are categorized industry-wide via API-1509, Appendix E, into five grades (API 2019). Group I basestocks, the lowest quality, are derived from separations and purifications of the VGO stream. Higher quality Group II and III basestocks are derived from hydroprocessing of VGO-range streams. Group IV and V basestocks are synthetic and derived chemically from common petrochemical feedstocks such as ethylene. The basestocks are blended and mixed with additives, often heavily, to meet performance specifications. Additization usually occurs at a blending plant, which may be at a refinery site by a lubricant manufacturer.

Of critical importance for nearly all basestocks is its viscosity, generally defined by ASTM-D97. Refineries equipped for making basestocks often employ additional draws on their vacuum pipestills so that specific boiling point ranges can be obtained, which in turn correlate to specific viscosity ranges based on the crude(s) being distilled. These draw streams may be stored and/or reblended to achieve a specific desired viscosity. Basestock manufacture is done in blocked operation on a single viscosity grade at a time.

In many lubricant applications, especially engine lubrication, the viscosity needs to remain as constant as possible over a wide range of possible operating temperatures. As a result, the temperature dependence of viscosity is as important as the viscosity itself. This dependence is quantified via a viscosity index (VI), as defined in ASTM-D2270, and must meet a minimum threshold of 80 in order to qualify as a Group I basestock (ASTM-D2270 2016).

To meet this specification, the VGO must be depleted of aromatic and naphthenic compounds, which have an inherently stronger viscosity–temperature response than paraffins, resulting in a low VI. This is accomplished through a continuous liquid–liquid extraction process with one of a number of solvents. Most commonly, *N*-methyl pyrrolidone (NMP) is used, but some extraction is also performed with furfural or phenol. The solvent is recovered and reused via a closed loop. The raffinate, a term used for a liquid from which impurities have been removed by solvent extraction, is enriched in saturates, while the solvent extract stream is depleted. The extract stream is of low value and typically reprocessed in the refinery or blended into various fuel oil grades.

Although paraffins have superior resistance to viscosity change as a function of temperature, they are prone to precipitation with decreasing temperature. When this crystallization begins at the wax appearance temperature (WAT), viscosity can dramatically increase. From a basestock standpoint, the pour point (e.g. ASTM-D97 or ASTM-D5950) is a second critical specification of the basestock (ASTM-D97 2017; ASTM-D5950 2014). A non-flowing lubricant at cold temperatures can lead to catastrophic failure of equipment.

To mitigate this aspect of the raffinate stream, it must be dewaxed. Historically, this was done using solvent and refrigeration. In solvent dewaxing, the raffinate is first diluted under high mixing with a chilled solvent such as methyl ethyl ketone (MEK) and/or methyl isobutyl ketone (MIBK). The mixture is further chilled in specially designed scraped-surface heat exchangers, which create a slurry of wax particles in the diluted raffinate. The degree of chilling determines the yield of wax crystallized and can be used to achieve the desired pour point. The wax particles are then removed from the slurry in a continuous rotating filter. The solvent is recovered and reused from both the dewaxed raffinate and the wax stream. Solvent dewaxing is still used today, but a more modern practice involves hydroprocessing using proprietary catalysts that add methyl branches to linear alkyl chains without cracking to lower molecular weight species.

The dewaxed product and the wax may each be optionally further mildly hydrotreated to remove some remaining sulfur depending on subsequent use of the stream. The dewaxed fraction is sold as basestock and additized, and the wax is sold as a byproduct, known as slack wax, used in many applications such as candles, cosmetics, and packaging.

A product known as brightstock is a specific basestock from the solvent lubes process. Brightstock is the heaviest basestock grade and is made from vacuum resid that has subsequently been deasphalted with a light paraffin such as propane or butane (see deasphalting). It is then fed to the solvent lubes process, resulting in a thick basestock that fluoresces (hence the term brightstock) and is a component in some heavy lubes and greases.

2.6.2.2.5 Solvent Deasphalting

The solvent deasphalting (SDA) process separates heavy oil vacuum residues into relatively high hydrogen deasphalted oil (DAO) product and a pitch product that contains the majority of undesirable components, including metals and coke-forming material (i.e. Conradson carbon residue [CCR]). The first step in the process mixes heavy oil residue (feedstock) with four to eight times the volume of light (C_3 to C_5) paraffinic solvents. Heavier vacuum residue requires a higher solvent to residue (S/R) ratio for effective separation of asphalt. DAO and asphalt precipitate are separated, and each stream is purified and flashed to recover and recycle the light paraffins. The four unit operations of deasphalting are the precipitation of asphaltenes, asphalt decanting and separation, asphalt washing to remove entrained oil, solvent recovery, and recycle.

2.6.2.2.6 Other Distillation

In other refining processes described in the following text, feed molecules are converted into products with molecules spanning a range of boiling points, where separation is then required to obtain products within specification. Most times, this separation will be accomplished by fractional distillation, similar to atmospheric distillation and vacuum distillation conducted on the crude feedstock. A refinery will contain many smaller distillation towers, called fractionators, designed to separate specific and unique products. Fractionators all work on the same distillation principles. For example, a fractionator after steam cracking is a distillation unit designed to separate ethylene and propylene as well as other lighter gases from other liquid by-products and heavier components.

2.6.2.3 Conversion

Only a fraction of crude oil feeds are composed of the high-value molecules (naphtha, kerosene, jet fuel, gasoline, diesel, and lubricant base stock). The purpose of the conversion process is to react and alter low-value molecules to produce those of higher value. Different conversion processes are required to convert molecules into gasoline, jet fuel, and diesel fuel from components such as residual oils, fuel oils, and light ends. Cracking, coking, and visbreaking processes are used to convert large petroleum molecules into smaller ones. Polymerization and alkylation processes combine smaller petroleum molecules into larger ones, while isomerization and reforming processes are applied to rearrange the structure of petroleum molecules to produce higher-value molecules of similar molecular size.

2.6.2.3.1 Thermal Cracking

The two basic types of cracking are thermal cracking, using heat to break chemical bonds, and catalytic cracking, using catalysts to control the type or position in the molecules to selectively break and/or rearrange chemical bonds.

The basic principle of thermal cracking is to apply energy via heat and pressure to break down large, low-value molecules into smaller, higher-value molecules. Any chemical bond can be broken if enough heat is applied to overcome the bond dissociation energy. However, not every chemical bond within a molecule has equal strength. The bond that will be preferred to break first depends on molecular structure, reaction kinetics, and decomposition reaction mechanisms. Only the breakage of carbon–carbon or carbon–heteroatom bonds will transform large molecules into smaller ones. Breaking carbon–hydrogen bonds will not achieve this purpose. In general, thermal cracking offers little control over which bond will break within a molecule, and many different bond-breaking reactions compete with each other. Several reactions can occur that are undesirable; the most important ones are dehydrogenation (breakage of the carbon–hydrogen bond) and aromatization (conjugated double bonds further lose hydrogen to form aromatic rings). These undesired reactions produce a large amount of a solid, low-hydrogen carbon material called coke. Coke formation was one of the first refining problems to be recognized, as it plugs pipes and coats catalysts. The first thermal cracking process, developed around 1913, involved heating distillate fuels and heavy oils in large pressurized drums until they cracked into smaller molecules within the range of naphtha, kerosene, jet fuel, gasoline, and diesel. However, this method also produced large amounts of unwanted coke. This early process has evolved into the following applications of thermal cracking: visbreaking, steam cracking, and coking that can potentially convert more than 50% of the incoming feed into naphtha and light diesel oils.

Visbreaking In order to control the thermal cracking process to preferentially follow desired reaction pathways and to maximize energy efficiency, it is important to heat the crude feed evenly. Uneven heating may result in some molecules exposed to insufficient energy to promote cracking or over-heated so that undesired reaction pathways produce undesired products.

Visbreaking, as its name indicates, refers to the fact that it reduces the viscosity of heavy oil to achieve uniform thermal cracking. The visbreaking process reduces the viscosity of the feed stream while increasing the quantity of middle distillate products from the thermal cracking of heavy oil. Visbreaking units typically convert about 15% of the feedstock to naphtha and diesel oils and produce a lower-viscosity residual fuel.

Visbreaking is achieved using one of two technologies: coil visbreaking and soaker visbreaking. In coil visbreaking, the cracking process occurs in furnace tubes (or "coils"), and the extent of the cracking reaction is controlled by regulation of the temperature and the flow speed of the heavy oil feed. Typical conditions for coil visbreakers are ~885–930 °F (473–499 °C) for one to three minutes and 3–10 bar pressure. The cracked heavy oil exiting the furnace is quenched to halt the cracking reactions, frequently by heat exchangers through which the virgin material being fed to the furnace is preheated, improving overall energy efficiency. Coil visbreaking products are passed to a fractionator where they are separated and recovered. In soaker visbreaking, the cracking reaction occurs not in the furnace but in a drum, "the soaker." Compared with coil visbreakers, the heavy oil feeds are cracked at a lower temperature (~ 800–830 °F/427–443 °C) for a pre-determined longer period of time (five to eight minutes). The cracked heavy oils then pass to a fractionator for further refining.

Steam Cracking The production of olefins (alkenes) from petroleum streams has gained widespread popularity as they can be used to make a wide variety of high-performance, low-cost polyolefin materials. Of all the olefins, ethylene is the most ubiquitous and versatile and is the most important industrial chemical building block today in terms of volume and variety of end uses. Propylene, butenes, pentenes, and aromatics are also of very high value and end up in plastics, synthetic rubbers, adhesives, and numerous chemical intermediates.

Often, ethane alone is the primary feedstock as it produces the highest selectivity to ethylene and is also very abundant in North America. However, naphtha cracking is also very common around the world. While still producing a high yield of ethylene, it also produces an array the larger olefins and aromatics that are also in demand. In some cases, even heavier portions of the crude barrel, well into the VGO range, will be steam cracked (Corma et al. 2018).

Steam cracking is a carefully controlled pyrolysis process (Figure 2.18). The feed stream is diluted with steam to dilute intermolecular reactions between the radicals that are formed. The feed stream enters the convective section of a furnace where the bulk of the heating occurs, but not quite to the point where substantial reaction begins. The heated feed then enters the radiant section of the furnace, where it is divided into numerous hot tubes that enable very short (<50 ms) residence times and very high (over 1500 °F/816 °C) temperatures that push the boundaries of metallurgy. At these temperatures, the hydrocarbons decompose, generally cracking to smaller, hydrogen-deficient molecules and

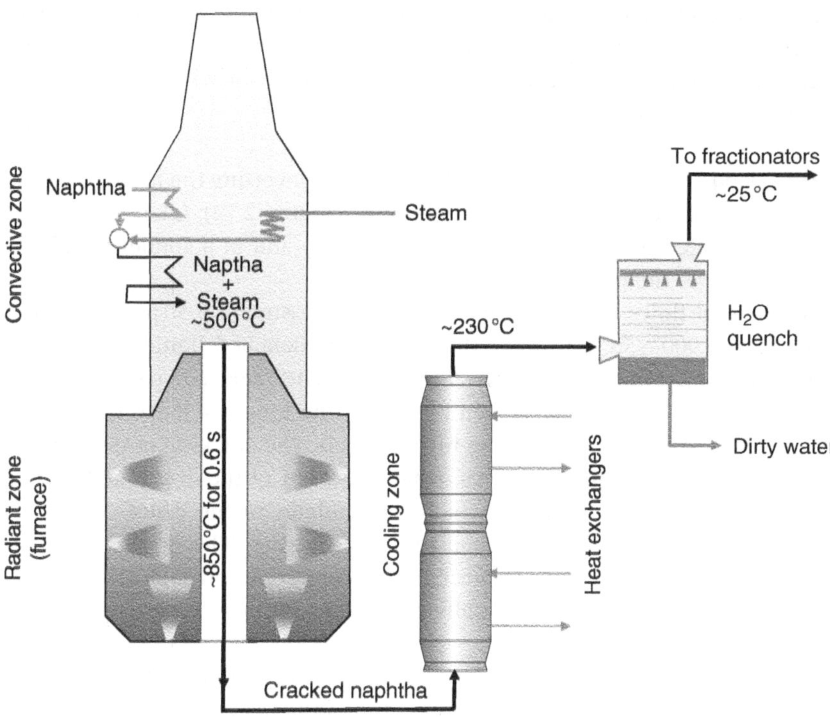

Figure 2.18 Schematic of a steam cracker. The feed is typically ethane or the naphtha fraction, but some refineries can process heavier fractions.

free hydrogen. The reaction is quenched in specialized heat exchangers, and the resultant hydrocarbons are sent to a primary fractionator, where olefin-rich gases are sent for purification. As with a pipestill, the product liquids are also separated, where steam-cracked naphtha (SCN), C_6–C_8 aromatics, steam cracked gas oil (SCGO), and a heavy fraction call steam cracked tar (SCT) are separated from one another. All of the liquid products are highly aromatic and have very little paraffinic character.

The C_4 gases leave the "hot ends" of the steam cracker and flow to the "cold ends." Here, the mixture of light alkanes, alkenes, alkynes, water, hydrogen, and other byproducts flow to a highly heat-integrated series of cryogenic distillation steps, where the olefins are purified to a high degree, typically >99.999% purity, for subsequent polymerization. Hydrogen and alkanes are recycled back to either re-crack, be used in other processes, or burned for fuel. Alkynes and dienes are lightly hydrogenated to convert into more valuable olefins.

The mixture of C_4 hydrocarbons (1-butene, *cis*-2-butene, *trans*-2-butene, isobutylene, and butadiene) are sent for additional purification. In particular, butadiene is recovered through a reactive distillation where they form a heavy complex with dimethyl sulfoxide, which enables them to be easily distilled. Subsequent distillation extracts the other isomers, each of which is a valuable chemical feedstock for derivative products.

Delayed Coking One of the primary techniques for converting the heaviest part of the barrel into lighter products is delayed coking (Figure 2.19). Like steam cracking, it is controlled pyrolysis, but at a lower severity and in a condensed phase, which encourages the production of cracked liquids.

Vacuum resid flows through a furnace at low pressure (30–50 psig), where it is brought to ~900–950 °F (~482–510 °C). The unquenched liquid is then sent into a large drum where it continues to pyrolyze. The "delayed" part of delayed coking relates to the reaction happening during the long residence time in the drum rather than immediately in the furnace itself. The cracking of the vacuum resid creates lighter liquids and gases, which flow out of the drum, and coke, a solid hydrogen-deficient material, which accumulates in the drum. When the drum is full of coke, flow is switched to another drum. The original drum is cooled, and then the solid coke is mechanically removed and sold as a product. The cyclic nature of the process usually entails four drums that are operated in fill-soak-empty series.

The liquid product is sent to a fractionation tower, much like a pipestill. Products usually require additional hydroprocessing due to the olefin content. Any unreacted residual material is recycled to the furnace until extinction.

The solid coke can take on different morphologies (needle, shot, and sponge) depending on the specific blend of crudes fed to the unit. As coke originates from

Figure 2.19 Fifty thousand barrel per day delayed coker units commissioned in 2018 at the ExxonMobil Antwerp refinery designed to convert heavy, higher sulfur residual oil into marine gasoline and diesel. Source: Photo courtesy of ExxonMobil.

the petroleum resid that will not boil or pyrolyze, it is enriched in sulfur and inorganic particulate contaminants that were not removed by the desalting process. High sulfur crudes tend to yield high sulfur cokes that can be burned with suitable emission controls. Low sulfur and low metal/ash cokes can be calcined and further upgraded into a high purity (>98%) carbon that is sold to the aluminum and steel industry for use as an anode in the electrochemical reduction of metal ore to its metallic form.

Fluidized Coking Fluidized coking serves the same purpose as delayed coking, but instead of producing coke as a saleable product, it is combusted as the heat source in a highly integrated manner. At the heart of a fluidized coker are two specialized vessels. The first, a regenerator, contains fine, hot coke particles with a stream of air passing upward, causing the coke particles to fluidize rather than form a solid mass. These particles incompletely combust in the presence of the air and release heat and CO-rich flue gas. These hot coke particles are transferred into a second reactor vessel. In this vessel, vacuum resid is sprayed onto the hot coke particles, where it coats them and pyrolyzes due to the heat. Cracked products are volatile and escape as gases, and coke is formed that deposits on the particle and increases its size. These particles are then recycled back to the regenerator to complete the process. With careful control of the reaction and regeneration rates and conditions, a heat balance is established between the two vessels, as well as a coke formation and combustion equilibrium.

The hydrocarbon products that come from the reactor are fractionated as with the delayed coker and have a similar composition. The fluidized coker also introduces a low-heating value flue gas, which may be burned elsewhere in the refinery. Unlike the delayed coker, there is no solid coke product. However, occasionally carbonaceous and ash-containing solids may build up in the system and need to be purged as a waste stream.

2.6.2.3.2 Catalytic Cracking

Catalytic cracking is a conversion process used throughout a refinery. A catalyst is defined as a substance that lowers a reaction's activation energy without being consumed in the process. Catalysts are deployed in refineries not only to speed up thermal cracking reactions but also to promote specific reactions toward desired products selectively. Thermal cracking reactions are free radical reactions that have low product selectivity and yield gasoline with a low octane number. Catalytic cracking reactions are ionic reactions that can be tuned for high product selectivity and yield gasoline enriched in isoparaffins and aromatics with a higher octane number. Both natural substances and synthetic materials are used as refinery catalysts. By definition, a catalyst participates but emerges unaltered after promoting a reaction and, in theory, should maintain its efficiency with continual use. In practice, catalysts have a finite number of cycles before their efficiency is no longer acceptable and need to be replaced.

Fouling of catalysts occurs when they are coated with coke. They can be regenerated by burning off the carbon residue, but regenerated catalysts are never completely restored to their original effectiveness. This is due to several other deactivation processes. Sulfur, nitrogen, and metals can poison a catalyst by irreversibly absorbing onto active sites, reacting to modify the catalytic surface, blocking access to the reactive sites, or hindering diffusion. Vanadium is particularly problematic as it can convert into H_3VO_4, a strong acid that can destroy a zeolite's framework. Catalysts can be degraded thermally if the reaction temperature exceeds its operating limits, which can occur in hot spots if the reactor temperature is not controlled to be uniform via adequate mixing and flow.

Refineries have used several catalytic cracker designs to maintain a flow of feedstock over a catalyst bed and a flow of product out of the reactor while controlling pressure and temperature and a means to regenerate the catalyst. The earliest catalytic cracker utilized a fixed-bed of catalyst that required shutting off the feed for regeneration within the reactor. These were replaced with moving-bed crackers, where the spent catalyst in the form of beads or pellets is moved from the reactor to a separate kiln for regeneration. The fluid catalytic cracker is deemed to be the most efficient design and is in common use in modern refineries.

The fluid catalytic cracker, more commonly known as a "cat cracker" or just FCC, is a highly heat-integrated process for converting heavier hydrocarbons into lighter ones, especially gasoline (also known as mogas = motor gas) (Figure 2.20). It is typically fed with VGO streams for conversion to distillate and gasoline, but its feed window can sometimes be expanded to consume residua as well. A primary feed attribute for FCC feed characterization is its propensity to form coke. This is determined by measuring the microcarbon residue (MCR), the amount of carbon residue left after laboratory evaporation and pyrolysis in an inert atmosphere (ASTM-D4530 2015). It is also known as the Conradson carbon (CCR) using an earlier testing method (ASTM-D189 2019).

The FCC is comprised of a reactor vessel and a regenerator vessel. The FCC gets its name from the powdered catalyst that flows between these two vessels in a fluidized state. The catalysts are typically silica–alumina based; and, in fact, the original catalysts used were acid-treated clays. The yield slate of the FCC can be adjusted by the introduction of other catalysts, such as zeolites. The flow rate of catalyst in the FCC can be surprisingly large, on the order of 40 tons/min.

Feed enters near the bottom of the reactor vessel into a tall vertical tube known as a riser, where it is sprayed and mixed with a hot powdered catalyst. The endothermic reaction cracks the feed into smaller molecules, typically cleaving alkyl chains from aromatic rings and leaving methyl groups behind. Highly aromatic molecules will grow in molecular weight and form coke on the catalyst surface during this process. The trip up the riser can last a second, and the uppermost riser temperature after completion of the reactions is around 900 °F (~480 °C). At the top of the riser, the mixture opens into the larger reactor vessel, where superficial velocities are drastically reduced, and the solids disengage from the gases. The gases leave through the top of the reactor through a series of cyclones that recover any entrained catalyst particles. From there, the gases are distilled and proceed to additional treating units and product blend pools.

The catalyst solids, now coated with coke, collect in the bottom of the reactor, still fluidized by the addition of steam. Additional steam strips any remaining hydrocarbons. The spent catalyst is then transferred to the regenerator vessel. Here, air, and possibly additional oxygen, is fed into the fluidized bed of the spent catalyst, and the coke is combusted. This results in the creation of a flue gas stream, and in some cases, incomplete combustion is preferred, resulting in a CO-containing low heating value stream. The combustion heats up the catalyst, recovering the heat lost due to reaction endothermicity in the reactor riser. The hot, regenerated catalyst is then sent back to the reactor for another cycle.

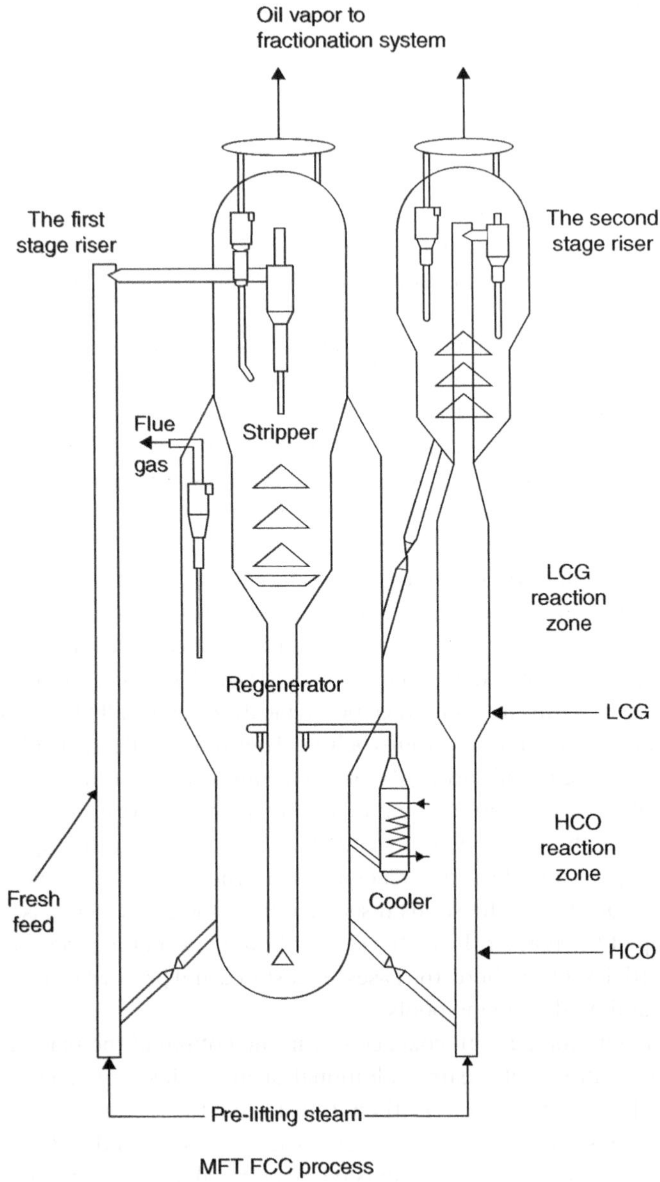

Figure 2.20 A schematic of a catalytic cracker that converts vacuum gasoils and residuals into gasoline. MFT, multifunctional two-stage riser; LCG, light cracking gasoline; HCO, heavy cycle oil. Source: Zhang et al. (2014). License (CC BY).

In the FCC riser, a number of chemical transformations are occurring (Sadeghbeigi 2012), beyond thermal cracking that is common to the pyrolysis reactions mentioned in the steam cracker and delayed coker. The rates of these are enhanced by the presence of the powdered catalyst.

As the name implies, one of the predominant chemistries is cracking, where paraffins are cracked to smaller paraffins, olefins are cracked to smaller olefins, and side chains are cracked from rings (the dealkylation reactions typically leaving a methyl on the ring). Naphthenic rings may also be cracked into olefins. These cracking reactions significantly reduce the molecular weight (and therefore boiling point) of the heavy feed molecules, bringing them into distillate and gasoline range.

Isomerization chemistry also takes place, where the olefin double bonds can shift, or methyl branches can be introduced to otherwise normal paraffins or olefins. Additionally, hydrogen transfer reactions, olefin cyclization, dehydrogenation of paraffins to olefins, and cycloaromatization reactions are occurring, resulting in the formation of aromaticity. These reactions also typically increase the octane number of the produced fuel, making the FCC an ideal process for gasoline production.

The cyclone system in the FCC is effective at removing and recovering the catalyst from the dilute phases at the top of the reactor and regenerator. However, there is some small leakage in the system. This is most important on the reactor side, where small amounts of catalyst fines will end up in the product fractionator, and leave with the heaviest stream, known at the "cat bottoms" or main column bottoms (MCBs).

Other commercial higher-temperature variants of the FCC are starting to appear at some refineries. These are known as deep catalytic cracking (DCC) units and catalytic pyrolysis process (CPP) units. The primary difference in these units is that the riser operates at a higher temperature. This results in more thermal cracking reactions and significantly higher production of olefins, approaching 40–50%. These units start to bridge the space between traditional FCC and steam cracking when it comes to the spectrum of fuels and olefin production for petrochemicals.

2.6.2.3.3 Reforming

Reforming is also called catalytic reforming, as it uses one or more catalysts to transform light hydrocarbons into hydrocarbon molecules of higher value. This process accepts feeds from various refinery streams, such as the heavy straight-run naphtha from atmospheric/vacuum distillation, and converts low octane naphtha molecules into higher octane hydrocarbons such as light aromatics (e.g. benzene, toluene, ethylbenzene, and xylene, known as BTEX). These reformates are used in gasoline blending and petrochemicals. Hydrogen and LPG are byproducts. The growing demand for high-octane number aromatic molecules assures that catalytic reforming will continue to be one of the most important unit processes in the petroleum and petrochemical refinery.

Reforming represents the total effect of numerous reactions such as cracking, polymerization, dehydrogenation, and isomerization taking place simultaneously. Depending on the molecules in the naphtha feedstock (as measured by the paraffin, olefin, naphthene, and aromatic content) and the catalysts used, reformates can be produced with very high concentrations of benzene, toluene, ethylbenzene, and xylene (BTEX) and other aromatic molecules. Hydrogen, a significant by-product, is separated from the catalytic reforming products for recycling and use in other processes.

There are many different commercial types of catalytic reforming processes, including Rheniforming (Chevron), Powerforming (ExxonMobil), Ultraforming (British Petroleum), Magnaforming (Atlantic Richfield), and Platforming (Universal Oil Products [UOP]). Each involves different treatments of the feed, reactor configuration, and use of specific catalysts under varying operating conditions. For example, in the Platforming process shown in Figure 2.21 (Honeywell 2013), the first step is to remove impurities from the feed to reduce the potential of catalyst degradation. The purified naphtha feed is then mixed with hydrogen, vaporized, and passed through a series of alternating furnaces and fixed-bed reactors that contain an appropriately defined catalyst. The effluent from the last reactor is cooled and sent to a separator for the removal of the hydrogen-rich gas stream from the top of the separator for recycling. The reformate, the liquid product from the bottom of the separator, is sent to a fractionator for further refining. Butanes and lighter gases are sent to the saturated gas plant.

The operating conditions of catalytic reformers vary considerably. Some operate at low pressure (50–200 psi), while others operate at high pressures (up to 1000 psi). Some catalytic reforming systems continuously regenerate the catalyst, whereas in other systems, one reactor is taken off-stream for catalyst regeneration while the others continue to operate, or the unit is shut down while the catalysts in all reactors are regenerated.

2.6.2.3.4 Alkylation

Alkylation combines low-molecular-weight olefins to form larger molecules. The reactants are isobutene and a mixture of propylene and butylenes, which are the by-products from isobutane and C_3–C_4 olefins produced from FCC process and other catalytic and thermal conversion processes in the refinery. In the presence of a catalyst, these small molecules are converted into larger molecules, called alkylates, which are composed of a mixture of high octane, branched-chain paraffinic hydrocarbons. Alkylate is a premium blending stock because it has exceptional antiknock properties and is clean burning. The octane number of alkylates will depend mainly upon the kind of olefins used and operating conditions.

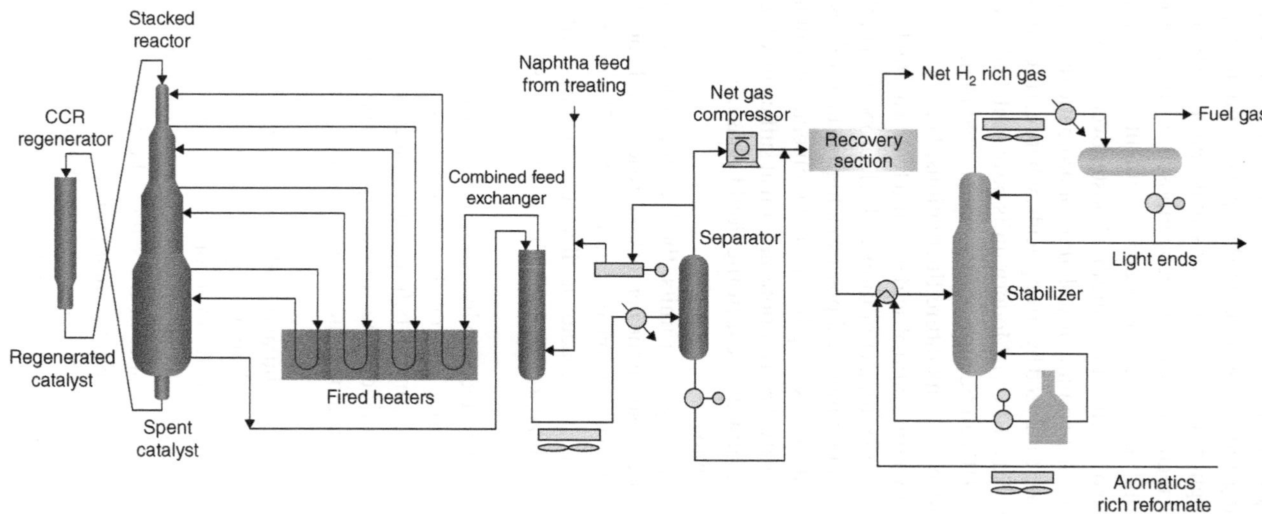

Figure 2.21 UOP CCR Platforming™ Process. *Note*: ™ Trademark of Honeywell UOP. Source: Courtesy of Honeywell UOP (Honeywell 2013). Copyright 2020 UOP LLC.

The alkylation process was developed in the 1930s and 1940s to produce high-octane aviation gasoline (Avgas) initially (PSU 2018), but later it became important for producing motor gasoline as the spark ignition engines became more powerful with higher compression ratios that require higher octane fuel. Recent environmental regulations have been imposed, restricting the amount of benzene and the total aromatic hydrocarbon content of gasoline. As such, alkylation has gained favor as an octane number booster over catalytic reformates, because the alkylate does not contain any olefinic or aromatic hydrocarbons.

Alkylation reactions are catalyzed by strong acids, where early commercial units typically used sulfuric acid (H_2SO_4). Modern units have switched to hydrofluoric acid (HF) alkylation, as HF can be more easily regenerated than H_2SO_4, and HF alkylation is less sensitive to temperature fluctuations. For both H_2SO_4 and HF processes, roughly a 1 : 1 volume ratio of acid to liquid hydrocarbon feed is utilized. By careful selection of the operating conditions, a high proportion of the products is within the gasoline boiling range and possesses high-octane values. Important operating variables include acid strength, reaction temperature, isobutane/olefin ratio, and olefin space velocity. The reactions are run at sufficiently high pressures to keep the hydrocarbons and the acid in the liquid phase, and good mixing is essential for high conversions.

In sulfuric acid (H_2SO_4) alkylation units, the feedstock (propylene, butylene, 2-methyl-2-butene, and fresh isobutane) enters the reactor and contacts the concentrated sulfuric acid catalyst (in concentrations of 85–95% for good operation and to minimize corrosion). The reactor is divided into zones, with olefins fed through distributors to each zone, and the sulfuric acid and isobutanes flowing over baffles from zone to zone.

The reactor effluent is separated into hydrocarbon and acid phases in a settler, and the acid is returned to the reactor. The hydrocarbon phase is hot-water washed with caustic for pH control before being successively depropanized, deisobutanized, and debutanized. The alkylate products obtained from the deisobutanizer can be used in motor fuel blending. The isobutane is recycled to the feed.

In HF alkylation, olefin and isobutane feed streams are dried to remove water before they are mixed with the isobutane recycle stream; this mixture is then fed to the reactor, where it is mixed with an incoming stream of acid catalyst. Conversion of feeds to high-quality alkylate happens in a very short time, and the mixture flows up to the settling zone. In the settler, the catalyst is separated out as a bottom phase and flows through the catalyst cooler and returns to the reactor. The hydrocarbon phase from the settler containing propane, recycled isobutane, *n*-butane, and alkylate is collected from the top of the settler and transferred to the main fractionator. High-purity propane is sent overhead to pass through the HF-propane stripper, defluorinator, and potassium hydroxide (KOH) treater before it is recovered. Recycled isobutane is drawn from the side of the fractionator and returned to the entrance of the reactor after it is mixed with the dried olefin and isobutane feed. The *n*-butane product is taken from the side of the fractionators as vapor, condensed, and KOH-treated before recovery. The alkylate product is obtained from the bottom of the fractionator. The HF catalyst is regenerated onsite in the regeneration section, where heavy oils (tars) are removed from the catalyst.

Transporting and working with concentrated acids pose safety risks. In particular, HF tends to form a vapor cloud that is difficult to disperse and is highly toxic. A newly designed method by UOP (UOP 2020), a major licensor of the HF alkylation process, uses a solid catalyst called Alkylene®.[1] The advantages of this new process over traditional HF alkylation processes (liquid acid technology) include no acid transportation, no acid spills, no corrosion, and reduced maintenance cost. Efforts to develop alternative processes that use solid acid catalysts instead of concentrated HF for alkylation are underway.

2.6.2.3.5 Polymerization

The term "polymerization" as used in the refinery process is really an "oligomerization" reaction from a broader polymer chemistry point of view. Polymerization is the process of oligomerizing light olefin gas molecules, including ethylene, propylene, and butylene, into hydrocarbons of higher molecular weight and higher octane number that can be used as gasoline. Polymerization combines two or more identical olefin molecules to form a single larger molecule with the same elements in the same proportions as the original molecules. Polymerization may be accomplished thermally or in the presence of a catalyst at lower temperatures. The polymerization process was used extensively in the 1930s and 1940s, but it was replaced to a large extent by the alkylation process after the 1950s.

The olefinic feed is purified to remove sulfur and other undesirable compounds. In the catalytic process, the feed molecules are either passed over a solid phosphoric acid catalyst or come in contact with liquid phosphoric acid, where an exothermic reaction occurs. This reaction requires cooling water and the injection of cold feedstock into the reactor to control temperatures between 300 and 450 °F at pressures from 200 to 1200 psi. The reaction products are sent to stabilization and/or fractionation systems to separate saturated and unreacted gases from the oligomerized molecules. Commonly licensed polymerization processes include the UOP polymerization process, which uses phosphoric acid as catalyst and the Institut Français du Pétrole (IFP)\Dimersol process, which produces dimers from propene or butene using a homogeneous aluminum alkyl catalyst.

1 ® Registered Trademark of Honeywell UOP.

2.6.2.3.6 Isomerization

Isomerization converts straight-chain paraffin molecules, *n*-butane, *n*-pentane, and *n*-hexane into their respective isoparaffin molecules that have substantially higher octane numbers. The component carbon atoms are the same but are arranged in a different geometric structure. Isomerization conversion of *n*-butane into isobutane provides feedstock for alkylation units, while the branched isomers from the conversion of *n*-pentanes and *n*-hexanes can be used directly for gasoline blending. Butane isomerization and pentane/hexane isomerization require separate processes.

In butane isomerization, aluminum chloride catalyst plus hydrogen chloride are used in a low-temperature reaction, while platinum or another metal catalyst is used for the higher-temperature processes. In a typical low-temperature process, hydrogen is added to *n*-butane or a mixed butane isomer feed to inhibit olefin formation and is passed to the reactor at appropriate temperature and pressure. Hydrogen is flashed off in a high-pressure separator, and the hydrogen chloride is removed in a stripper column. The butane mixture is sent to a fractionator (de-isobutanizer) to separate *n*-butane from the isobutane product.

In a typical pentane/hexane isomerization process, feeds from hydrotreated light straight-run naphtha or natural condensates are dried and desulfurized, and then mixed with a small amount of organic chloride and recycled hydrogen. They are passed over a supported-metal catalyst in a temperature-controlled reactor. Isomerization product yields depend on the flow configuration and feed properties but are typically high. The feed then goes to the isomerization reactor, where the paraffins are catalytically isomerized to isoparaffins. The reactor effluent is then cooled and subsequently separated in the product separator into two streams: a liquid product and a recycle hydrogen-gas stream. The isomerization products are washed, acid-stripped, and stabilized before use in gasoline blends.

2.6.2.3.7 Natural Gas Conversion via Steam Methane Reforming

The various hydrotreating, hydrocracking, and hydroisomerization processes within a refinery are consumers of hydrogen gas. The aromatics reforming process is a large net producer of hydrogen gas, which feeds into the hydroprocessing operations. However, it is often the case where hydrogen from reforming does not meet the hydrogen needs of the refinery, and additional hydrogen must be supplied. This is done via the steam methane reforming (SMR) of natural gas.

In the SMR process, the methane is mixed with superheated steam (900 °C) over a nickel-based catalyst. This produces a synthesis gas at about 370 °C due to the endothermicity of the reaction (Eq. (2.1)). This heat is supplied by burning methane. Then, in the shift converter, the CO in the syngas is converted via the water-gas shift reaction (Eq. (2.2)). The produced CO_2 is removed via scrubbing with monoethanolamine. The net chemical reaction of the SMR process is that half of the hydrogen produced is from the natural gas, while the other half comes from the steam (Eq. (2.3)).

$$CH_4 + H_2O \rightleftharpoons CO + 3H_2 \quad (2.1)$$

$$CO + H_2O \rightleftharpoons CO_2 + H_2 \quad (2.2)$$

$$CH_4 + 2H_2O \rightarrow CO_2 + 4H_2 \quad (2.3)$$

2.6.2.4 Purification

Petroleum feedstocks and intermediate products contain undesirable molecules that must be removed during refining. These are non-hydrocarbon molecules containing one or more heteroatoms, such as sulfur, nitrogen, and oxygen, or organic complexes containing metals, such as vanadium, nickel, and iron. Their removal is essential for assuring efficient refinery operations (e.g. flow assurance, prevention of catalysis poisoning) or meeting final product specifications (e.g. sulfur content in fuels). Removal of these undesired components can be accomplished by purification via separation and segregation, or by treating via chemical conversion using processes such as hydrodesulfurization (HDS) and hydrotreating.

2.6.2.4.1 Hydrotreating

Hydrotreating is a ubiquitous process in refineries. A subset of the more general hydroprocessing processes, its focus is on the removal of impurities from its feed streams. These impurities and the levels to which they must be removed depend on the particular hydrotreating service. Commonly, the impurities are sulfur compounds, nitrogen compounds, organometallics, and aromatics.

For finished fuels and specialties, there are specifications on a number of impurities addressable by hydrotreating, which may be imposed by federal, state, or pipeline agencies. Many of these specifications have become more stringent over the past 30 years, increasing the reliance on hydrotreating operations for refining. For gasoline, testing includes color (ASTM-D4176), aromatics (ASTM D5769-10), benzene (ASTM D3606-10), olefins (ASTM D1319-13), and sulfur (ASTM D2622-10). For diesel fuel, these include color (ASTM D1500) and sulfur (various). For jet fuel, these include acidity (ASTM D3242), color (ASTM D156), smoke point (ASTM D1322), aromatics (ASTM D1319 or ASTM D6379), density (ASTM D4052), and sulfur (various methods). Lubricant basestocks also have sulfur, nitrogen, and aromatics requirements, which are commonly addressed by hydrotreating. Recent International Maritime Organization (IMO) specifications on marine fuels are driving new sulfur requirements in traditionally modestly processed refinery products. These marine fuel limits are driving a current wave of hydrotreating investment to produce fuels that can meet the United States and worldwide specifications (Figure 2.22).

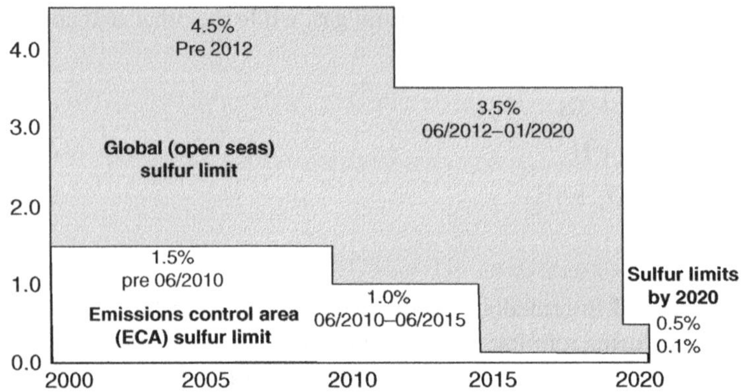

Figure 2.22 Marine fuel sulfur limits (wt%).

Besides finished fuels, hydrotreating also is used as an intermediate process to prepare feed for other units. For example, feed to the FCC is often hydrotreated in order to remove a large portion of the nitrogen, which can act as a catalyst poison as it titrates the acidic sites on the catalyst. Resid streams may be hydrotreated to not only reduce aromaticity and sulfur but also remove organometallics such as nickel and vanadyl porphyrins, which can deactivate catalysts in downstream units.

In hydrotreating, the hydrocarbons flow over a catalyst at moderate to high temperatures (550–700 °F, or 290–370 °C) along with hydrogen gas at high pressure, ranging from 200 to 2000 psi, depending on the feedstock and necessary treatment. Under most conditions, the hydrotreating reactions are highly exothermic, and gas or liquid quenches are added along the reactor to help control reactor temperature. The catalyst is a supported sulfide catalyst typically based on nickel, cobalt, tungsten, molybdenum, and combinations thereof.

One of the primary chemistries occurring in hydrotreating is sulfur removal, which is known as hydrodesulfurization or HDS. Sulfur is found across all boiling point ranges of petroleum in the form of organic compounds, including mercaptans, sulfides (mono-, di-, tri-sulfides), thiophenes, benzothiophenes, dibenzothiophenes (DBTs), and many higher aromatic compounds as molecular weight increases. The sulfur is ultimately removed from the stream in the form of H_2S, which is removed from the gas phase downstream of the reactor and sent to the Claus process, where it is converted into elemental sulfur. The amount of sulfur can be considerable, as crudes can have anywhere from wppm levels of sulfur for some sweet crudes, up to 7 or 8 wt% for certain heavy crudes. Different sulfur compounds can require greatly varying hydrotreating severity to remove. Sulfides and mercaptans are typically easy to remove, whereas aromatic sulfur compounds become increasingly difficult with increasing molecular complexity.

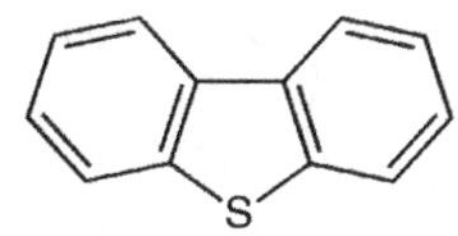

Figure 2.23 Structure of dibenzothiophene (DBT).

The sulfur specifications on fuels, particularly for diesel fuel, experienced a dramatic tightening in the mid-1990s to early 2000s due to air quality concerns arising from motor vehicles. For on-road motor diesel fuel in the United States, the specification is currently 15 wppm (Code of Federal Regulations, CFR §80.520). This drove a surge in high-pressure diesel hydrotreating, primarily due to the presence of DBTs that fractionate into the heavier end of diesel. Due to their rigid and sterically hindered structure (Figure 2.23), DBT has a difficult time reacting at the active sites of the HDS catalysts. Additionally, if the second and sixth positions of the DBT have substitutions that surround the sulfur atom, the effect becomes even more pronounced, with reaction rates 100 times slower than a molecule such as benzothiophene. These molecules are known as hindered DBTs or simply "hard sulfur." In these DBTs, it is the saturation of one of the aromatic rings that controls the rate of desulfurization. Once a ring is saturated with a naphthenic ring, the molecule can more easily bend and undergo sulfur removal at the active site of the catalyst. As saturation of the aromatic ring takes four to six hydrogen atoms, it is sensitive to hydrogen pressure effects and results in high-pressure units being built, typically operating around 1200 psi.

The removal of nitrogen is another key function of hydrotreating. The nitrogen compounds, including amines, pyridines, pyrroles, quinolines, and other related moieties, are performed for two reasons. First, these molecules can exhibit reactivity in the final products, which can cause stability and color problems during storage. Second, these molecules have a basic character and interfere with the acid catalysis of other units in the refinery, such as hydrocrackers and FCCs, by titrating the active sites. These molecules occupy the catalyst sites and reduce the effective amount of catalyst available for reaction, leading to larger reactors and higher operating temperatures in order to compensate.

Nitrogen is removed by the combined chemistries of aromatic hydrogenation, hydrogenolysis, and denitrogenation. The set of reactions are collectively known as hydrodenitrogenation (HDN), and the end result is the generation of ammonia as the final disposition of the nitrogen molecule. Ammonia is removed from the gas products through a water wash step.

Although olefins are not found in unprocessed crude oil, many processed streams, especially those coming from the FCC or coking units where significant thermal processing has occurred, contain olefins or olefin-substituted molecules. These may be linear, branched, or cycloalkenes. Most product specs cap the amount of olefins present as they can remain unstable (reactive) during

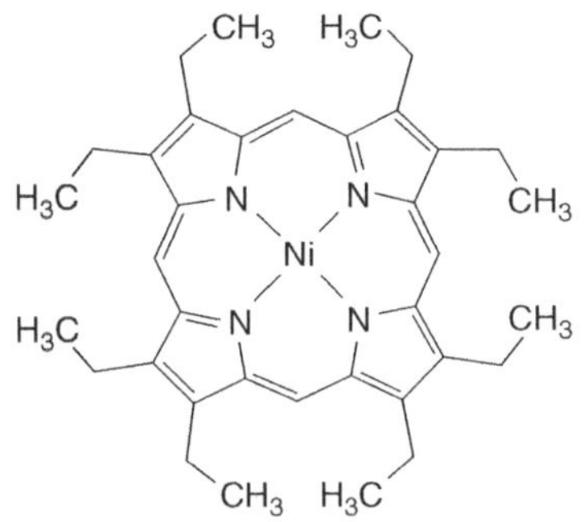

Figure 2.24 Nickel porphyrin ring structure.

transport and storage, leading to the generation of gums, foulant, haze, and color. Hydrotreating is effective at saturating double bonds, and at hydrotreating conditions, olefin saturation is rapid and exothermic.

For heavier petroleum fractions, hydrotreating is also used to remove metals from organic structures, primarily from nickel and vanadyl porphyrin rings (Figure 2.24). Metals are found in ppm levels in most crudes, and the metal is popped from the porphyrin ring rapidly in hydrodemetallization (HDM) operation. Upon removal, the metals plate out on the catalyst, slowly decreasing its activity. To alleviate this in hydrotreating, a guard bed is typically employed when metals are present, which consists of a small, low-temperature reactor before the main reactor. Metals react quickly, and the catalyst in the guard bed is sacrificed to protect the larger bed that follows. The guard bed is often easily switched out of service to be regenerated or refilled with fresh catalyst.

The last major function of hydrotreating operations is aromatic saturation. A number of properties are affected by the aromaticity, and hydrotreating is a primary method of converting aromatics into saturated products ("AroSat"). AroSat consists of adding hydrogen to aromatic rings, resulting in naphthenic rings. In polyaromatic compounds, this saturation proceeds in a stepwise fashion, saturating one ring at a time.

In aromatic saturation, the effects of hydrogen pressure and temperature are particularly important. Some AroSat units operate very close to the thermodynamic limit of aromatic saturation, which can limit conversion. In units where complete elimination of aromatics is necessary (e.g. production of white oils), the reactors operate at moderate temperatures and very high pressures, near 2000 psi, to push equilibrium in a favorable direction. Residual aromatics in these systems can lead to color and haze issues. At lower pressures and higher temperatures, the formation of PNAs may occur as equilibrium prefers their creation.

Due to the interactions of the catalyst with nitrogen compounds, metals, and coke from PNA production, the activity of a hydrotreating catalyst will slowly decrease over time. Activity is managed by control of the reactor inlet temperature to compensate for this deactivation. Once heat transfer or product specs limitations are reached due to the increased operating temperature, the reactor is taken offline, and its catalyst is replaced or regenerated. Lifetimes of catalyst beds are measured in months to years.

2.6.2.4.2 Hydrocracking

Hydrocracking is another variant of hydroprocessing that has much in common with hydrotreating mentioned earlier. The primary difference is that the sites on the catalyst also begin to promote cracking reactions as well as the HDS, HDN, HDM, and saturation reactions.

Like the FCC and coking processes, hydrocracking accomplishes molecular weight reduction of the feed. However, since it is using chemistries similar to hydrotreating simultaneously, the products are often much cleaner and more saturated. A schematic of a typical hydrocracker is given in Figure 2.25.

From a process perspective, an additional feature of hydrocrackers is the inclusion of a recycle loop, where a post-reactor fractionator sends a large portion of its

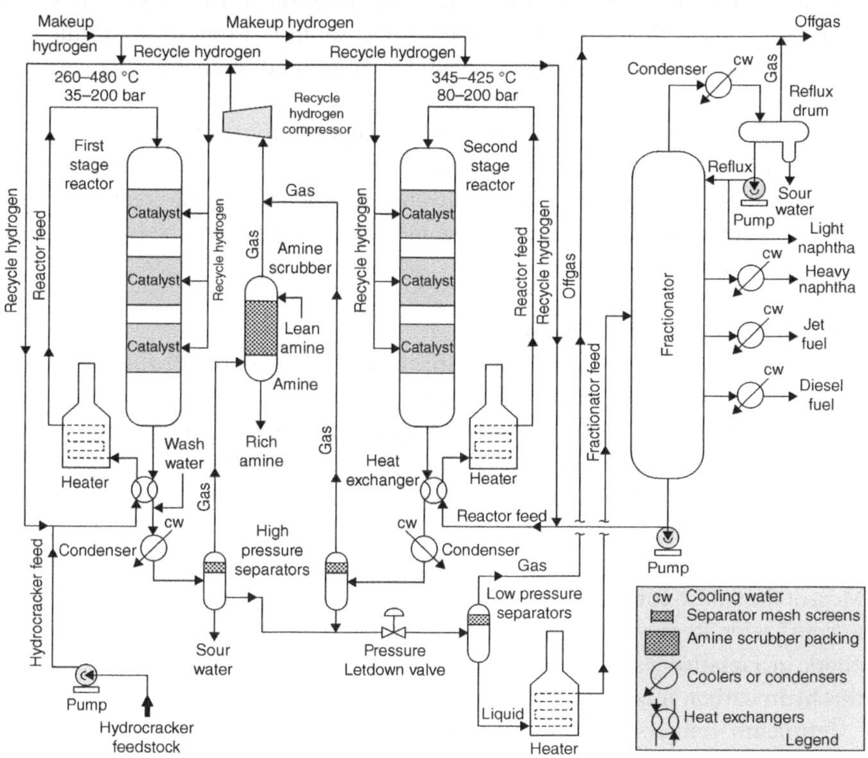

Figure 2.25 Hydrocracking unit. Source: Beychok (2009). Licenced Under CC BY SA 3.0.

bottoms stream back to the reactor feed. A small purge stream in the recycle loop prevents the accumulation of uncrackable compounds, typically PNAs, if the operating pressure is not high enough. Additionally, hydrocrackers usually involve a hydrotreating reactor and a hydrocracking reactor. These may be separate reactors or simply beds of differing catalyst within the same reactor. In some cases, additional reactor stages are added, where sulfur and nitrogen are nearly completely removed before going to the cracking reactor. This sweet feed enables much higher catalyst activity and higher per-pass conversions.

2.6.2.4.3 Hydroisomerization

The last class of hydroprocessing commonly used in refining is hydroisomerization. As with hydrocracking, hydroisomerization has much in common with hydrotreating. The primary difference is again with the catalyst, where it is specifically designed to isomerize alkyl chains and add small amounts of methyl branching. A number of low-temperature properties, such as pour point and freeze point in lube basestocks, diesel, and jet fuel, are very sensitive to branching. The addition of branches frustrates the crystallization of the alkyl chains and enables liquids to flow at lower temperatures. Naphtha isomerization units are also employed for motor gasoline, where the addition of branches increases the octane number of the fuel.

In all hydroprocessing operations, but most importantly, in hydroisomerization, some cracking to light gases takes place, as cracking reactions can never be completely reduced to zero in real systems. Even in hydrocracking, the creation of light gases is avoided when possible, as these molecules have a lower value than their liquid counterparts.

2.6.2.5 Sweetening and Treating

Sweetening and treating are not as critical as conversion and distillation in the downstream refinery processes; however, they are essential to the performance and safety of finished products. Sweetening and treating are means of removing molecules containing heteroatoms (sulfur, nitrogen, and oxygen) as well as other contaminants in the hydrocarbon mixture (e.g. dissolved metals, inorganic salts, and soluble salts dissolved in emulsified water).

The term “sweetening” means reducing “sour” sulfur-containing molecules. Most of the sulfur-containing molecules in the crude oil hydrocarbon mixture have a “sour” odor. Removing these sulfur-containing molecules removes these “sour” odors; in a relative way, this means to reduce “sour” flavor or means “sweetening” this hydrocarbon mixture.

Petroleum refiners have a choice of several different sweetening and treating processes, but the majority of them are the elimination of unwanted sulfur-containing molecules. A variety of intermediate and finished products, including middle distillates, gasoline, kerosene, jet fuel, and sour gases are water removed and sweetened. Sweetening and treating processes can be accomplished at an intermediate stage in the refining process, or just before sending the finished product to market or storage. Choices of sweetening and treating methods depend on the nature of the petroleum fractions, amount and type of impurities in the fractions to be treated, the extent to which the process removes the impurities, and end-product specifications. Sweetening and treating materials include acids, solvents, alkalis, oxidizing, and adsorption agents.

In acid treating, the sulfuric acid treating process is commonly used in the refinery. With the sulfuric acid treating process, a partial or complete removal of unsaturated hydrocarbons, sulfur, nitrogen, and oxygen-containing molecules occurs. This acid process is very effective to improve the odor, color, and stability.

The caustic treating process is a caustic wash with sodium (or potassium) hydroxide to improve odor and color by removing organic acids (naphthenic acids, phenols) and sulfur-containing molecules (mercaptans, H_2S). By combining the caustic solution with various solubility promoters (e.g. methyl alcohol and cresols), mercaptans, as well as oxygen and nitrogen compounds, can be effectively removed from petroleum fractions. The clay/lime treating process is applied to acid-treated oil to remove any residual traces of asphaltene molecules in the stream and products.

2.6.2.5.1 *Waste Treatment*

Pollutants are a consequence of petroleum refining and must be contained and remediated. Compliance with government regulation for the emission of gases and discharge of water requires considerable resources for proper treatment and monitoring. Unplanned discharges of crude and refined petroleum must be reported immediately, followed by remediation.

2.6.2.5.2 *Wastewater Treatment*

Petroleum refining consumes large quantities of water and steam in processing crude – one of the main reasons many refineries are located next to lakes or rivers. Groundwater or seawater is used in areas where surface freshwater is in short supply. Most of the water consumed is used in the desalting process, which generates wastewater containing suspended and dissolved solids, emulsified whole crude, dissolved aromatic hydrocarbons (BTEX), and other water-soluble species. Steam is used in many of the refining processes and condenses into "sour water" that can contain hydrocarbons, heteroatomic compounds (e.g. phenols, mercaptans, cyanides, and amines), dissolved gases such as H_2S and NH_3, and acids (H_2SO_4 and HF). The sour water is typically recycled for use in the desalters. Water separated out from feedstock oil tanks is an additional source of aqueous waste.

The contaminated water is sent to a wastewater treatment plant (WTP) that is usually located onsite. The treatment may involve one or more additional settling tanks for oil/water separation. The release of volatile hydrocarbons, such as benzene, may require additional controls for capture. The primary treatment begins with a gravity separation (American Petroleum Institute [API] separator), where oil that floats to the surface is scraped off. This aqueous effluent still contains dispersed fine solids and the residual oil is sent to for secondary separation. This is either a dissolved air floatation unit, which uses electrostatics and chemicals to coagulate the fines into particles that will settle with the addition of air bubbles, or an induced air floatation system based on a rotor–disperser mechanism. The secondary aqueous effluent is then sent for further treatment, usually a biological system similar to that used in municipal sewage treatment plants or a membrane bioreactor. Final tertiary treatments may involve filtration through sand or activated carbon, and chemical oxidation.

A refinery may also be required to treat water used in cooling systems, rainwater runoff, and sanitary sewage water before discharging into the environment. This may be treated by a separate WTP, and special provisions may be needed to collect and treat water used to put out fires.

2.6.2.5.3 Hydrogen Sulfide Scrubbing

The hydrogen sulfide scrubbing process is a common method used to remove or reduce the hydrogen sulfide concentration in the feed gas stream. The hydrogen sulfide scrubbing process can be performed at ambient temperature via an activated charcoal absorption approach. Hydrogen sulfide scrubbing processes from high-temperature catalytic hydrogenation are followed by zinc oxide treating. The selection of the hydrogen sulfide scrubbing process will be based on the feed molecules and the nature of the contaminants.

2.6.2.5.4 Sulfur Recovery

The sulfur recovery process converts hydrogen sulfide in sour gases and hydrocarbon streams to elemental sulfur. Known as the Claus sulfur recovery unit, the process uses both thermal and catalytic-conversion reactions, producing elemental sulfur by burning H_2S under controlled conditions. Sulfur vapors are condensed and recovered in a series of condensers, each subsequently operated at a lower temperature. Claus sulfur recovery units also operate at oil fields to remove H_2S prior to transport.

2.6.3 Fuel Products

Petroleum refining creates a wide array of fuel, lubricant, and chemical products. However, by volume, most products fall into one of a few large categories. For fuels, the largest component of refinery output, products are primarily distinguished by their boiling point range. Although other properties are extremely important for performance and regulatory needs, distillation towers are often the last processing step, and as they separate based on boiling point (or volatility), the products are inherently linked to the boiling point as well.

Each product has many specifications that must be met in order to qualify as a given product and grade of that product. These specifications can come from government mandates, pipeline or transportation limitations, accepted industry standards, or performance decisions set by the refiner. Often, it may be the case that on-spec products are not produced all the time, but rather that various blendstocks are produced and then stored onsite. These blendstocks can then be later reblended to meet specifications and optimize the product slate out of the refinery while also giving operating flexibility to the refinery.

2.6.3.1 Mogas (Motor Gasoline)

Mogas, short for motor gasoline, is the gasoline product familiar for most passenger vehicle use. Mogas streams are often referred to as CBOB (conventional blendstock for oxygenate blending) or RBOB (reformulated blendstock for oxygenate blending). A third blendstock, CARBOB (California reformulated blendstock for oxygenate blending), is specific to the state of California. The "oxygenate blending" in these refining products refers to the need to blend in an oxygenated material, typically Environmental Protection Agency (EPA)-mandated ethanol into the gasoline. Given the potential corrosion concerns that ethanol can bring to pipelines, the ethanol is blended to the gasoline after pipeline transportation. The conventional and reformulated aspects of these blendstocks refer to certain specifications, notably volatility. The addition of ethanol, a high vapor-pressure component, to the gasoline increases its overall volatility, so the blendstock must have some of the lighter components removed to compensate. Note that other oxygenates have also been used in the past but are mostly phased out, especially in North America. These include methyl *tert*-butyl ether (MTBE) and ethyl *tert*-butyl ether (ETBE).

Consumers are most familiar with the octane number of the gasoline they purchase. This test, which comes in two varieties, research octane number (RON) and motor octane number (MON), measures a fuel propensity to autoignite when compressed. In a traditional internal combustion engine, it is desirable for ignition to occur via the spark plug to allow for precise timing and a clean burn, and therefore, autoignition is an undesirable property. Octane number is defined on a 0–100 scale (although values can exceed 100), with 0 defined as the autoignition characteristics of *n*-heptane (poor, easily autoignited), and 100 as 2,2,4-trimethylpentane, or more commonly known as isooctane (excellent). As may be inferred, the more branches on the mogas molecules, its octane number tends to improve. Aromatics also have very high-octane numbers, but may be limited by other specifications.

Mogas specs can vary by grade, but many have a common set of import inspections. Testing of these properties include, among others, its boiling point distribution or volatility (ASTM-D86 2019), sulfur content (ASTM-D2622 2016), aromatics (ASTM-D5769 2015), benzene content (ASTM-D3606 2017), octane number (ASTM-D2700 2019), olefin content (ASTM-D1319 2019), and Reid vapor pressure (RVP) (ASTM-D5191 2019).

Many specifications on mogas are driven by environmental or toxicity regulations. The table from the EPA below describes some of the various types of controls on mogas and their method of addressing (Table 2.1).

2.6.3.2 Diesel

Diesel fuel is the primary fuel used for heavy-duty commercial vehicles, as well as some marine use. As with mogas, numerous grades are available depending on the intended end-use, season, and local regulations. Diesel has a higher boiling point range than motor gasoline.

As with mogas, diesel fuel is also characterized by its combustion properties – cetane number in the case of diesel – which also measures a fuel's propensity to autoignite. Unlike mogas, autoignition is a desirable property in a diesel engine, as high compression is used to ignite the fuel–air mixture. Cetane number is also defined on a 0–100 scale, with 1-methylnaphthalene defined as 0, and *n*-hexadecane defined as 100. In diesel fuels, straight chains are preferred from a cetane number standpoint.

The presence of high-cetane molecules with long straight chains introduces another problem. The carbon number of diesel fuel molecules is high enough that waxy crystals can begin to form as the temperature drops. These crystals can present a serious problem in many applications as they can cause a sharp rise in viscosity and eventually gel the fuel in very cold temperatures. In many cases, cetane number and low temperatures properties such as cloud point and pour point need to be balanced for a given application. It is also common to introduce additives, such as cetane improvers or pour point improvers, to break the interdependence of these two key properties.

Post-refinery, some diesel streams may be blended with biodiesel streams, which are chemically quite distinct from refined diesel. These streams come from various vegetable oils or recycled greases and fats. As triglycerides, these biochemicals are too viscous to be used without dilution or modification. Enzymatic or catalytic transesterification converts triglycerides to glycerol and fatty acid methyl esters.

Diesel specs include, among others, boiling point distribution (ASTM-D86 2019), cetane number (ASTM-D976 2016), cloud point (ASTM-D2500 2017), pour point (ASTM-D97 2017), sulfur, oxidative stability (ASTM-D2274 2019), and viscosity (ASTM-D445 2019).

Table 2.1 Selected EPA controls on mogas and associated methods.

Regulatory program	Overview	Pollutants addressed
Sulfur content in gasoline	Limits the sulfur content in gasoline, which allows for cleaner-burning fuel and the use of advanced emissions control technologies in cars and trucks This program is implemented at the federal level	Smog (ground-level ozone) and particulate matter
Air toxics standards for gasoline	Limits the content of toxic chemicals in gasoline This program is implemented at the federal level	Benzene and other hydrocarbons such as 1,3-butadiene, formaldehyde, acetaldehyde, acrolein, and naphthalene
Reformulated gasoline (RFG)	This standard requires the use of gasoline that meets more stringent specifications resulting in cleaner burning fuel which reduces emissions of the pollutants that contribute to ground-level ozone and particulate matter pollution This program is implemented at the state and local level with federal oversight by EPA	Smog (ground-level ozone) and particulate matter
Reid vapor pressure (RVP)	This gasoline standard requires the use of specially formulated gasoline that evaporates less at higher temperatures than regular gasoline. This reduces emissions of the volatile organic compounds and hydrocarbons that result in higher levels of ground-level ozone in the hot summer months This program is implemented at the state and local level with federal oversight by EPA	Smog (ground-level ozone)
Winter oxygenates	This program mandates the addition of oxygenates to gasoline in the winter, which reduces inefficient combustion of fuel during cold weather which can result in higher emissions of carbon monoxide This program is implemented at the state and local level with federal oversight by EPA.	Carbon monoxide (CO)

Source: EPA (2016).

As with mogas, additional specifications on diesel originate from pollution concerns during combustion in engines. The US EPA concerns itself with soot and particulate matter, NO_x compounds that contribute to smog, hydrocarbons, carbon monoxide, and any other hazardous air pollutants.

2.6.3.3 Jet Fuels/Kerosene

In between the mogas and diesel fuel boiling point distributions are jet fuel and kerosenes. Jet fuel comes in numerous grades, especially when including high-density military fuels, some of which are closer to pure-component chemical streams than the traditional complex mixture of hydrocarbons in other fuels.

Jet fuels tend to have many more specifications that other fuels, due to the potential consequences of engine trouble while in flight as opposed to land or sea-based transportation. In addition, jet fuels need to have a high energy density for efficiency. This leads to a number of specifications that center around reliability, both in the engine and in the fuel tank – the fuel needs always to flow, even at extremely cold temperatures, so low-temperature properties such as freeze point are crucial. While sitting in a tank, reactive materials in the fuel can create gums that can impede flow or cause corrosion problems. Finally, the static buildup can cause problems in flight, so conductive additives are added to jet fuel as static dissipators.

Jet specifications include, among others, boiling point distribution (ASTM-D86 2019), color (ASTM-D156 2015), smoke point (ASTM-D1322 2019), specific energy (ASTM-D4809 2018), copper corrosion (ASTM D130), freeze point (ASTM-D2386 2019), sulfur (IP-336 2004), thermal stability (ASTM-D3241 2019), particle count (IP-564 2013), electrical conductivity (ASTM-D2624 2015), and density (ASTM-D4052 2018).

2.6.3.4 Fuel Oil

The heaviest (by boiling point) of common refinery fuels is fuel oil. It is typically used for heating applications and by large marine vessel engines. Fuel oils can extend to the heaviest components of petroleum and can be quite viscous, depending on the ultimate use.

Due to the heavy, viscous nature of fuel oils, these fuels can contain measurable amounts of non-hydrocarbon compounds, which are often limited via specification. These include specific elements such as aluminum and vanadium and also generic "ash" that remain after combustion. As an example, marine fuel specifications include, among others, acid content (ASTM-D664 2018), sulfur content (ISO-8754 2003), density (ISO-3675 1998), flash point (ISO-2719 2016), distillation (ISO-3405 2019), ash content (ISO-6245 2001), water content (ISO-3733 1999), and viscosity at 50 °C (ISO-3104 1994). As mentioned earlier, marine fuels are in a current period of transition as IMO regulations start to limit the sulfur content in an effort to reduce SO_x pollution from marine sources.

2.6.3.5 Liquefied Petroleum Gas (LPG)

The lightest of the refined fuels, LPG is primarily C_3 components and used for heating and some industrial uses. It is chemically far simpler than most complex fuel streams, and as such, has fewer specifications, most of which focus on impurities. These include sulfur (ASTM-D2784 2011), moisture (ASTM-D2713 2015), and butane and heavier components (ASTM-D2163 2019).

2.7 Petrochemicals

Not all of the crude feedstock is converted into fuel. Early kerosene refineries disposed of the unwanted naphtha and residuum in nearby pits or streams. The desire to utilize the byproducts gave birth to the petrochemical industry that developed in parallel with refining (Figure 2.26). A petrochemical is any substance other than a fuel that is made from oil or natural gas (Figure 2.27) (Honeywell 2011).

2.7.1 Olefins: Prime and Higher Olefins

The major portion of the petrochemical industry manufactures intermediate products used to make plastics and synthetic rubbers. These may be polymers of hydrocarbons (e.g. ethylene, propylene, butadiene, and styrene) or intermediates containing oxygen (e.g. acetal, acrylic, ethylene glycol, and polycarbonate), nitrogen (nylon, acrylonitrile butadiene styrene), or chloride (polyvinyl chloride).

Figure 2.26 ExxonMobil's Singapore Chemical Plant is fully integrated with the adjacent refinery and can process a broad range of feedstock for the production of olefins, performance polymers, aromatics, and oxo alcohol. Source: Photo courtesy of ExxonMobil.

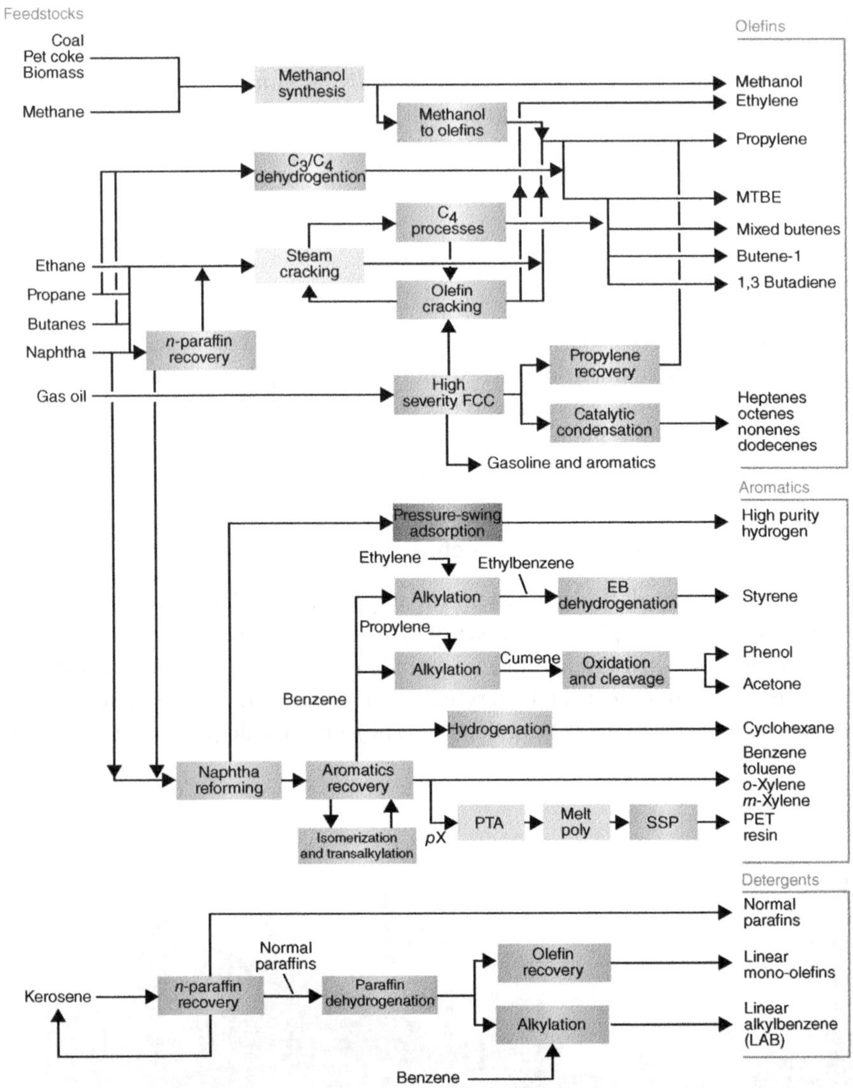

Figure 2.27 Petrochemicals flow scheme. PTA, purified terephthalic acid; SSP, solid state polycondensation. Source: Image courtesy of Honeywell UOP (Honeywell 2011). Copyright 2019, UOP LLC.

Olefins, once a discarded byproduct, now undergo extensive purification via cryogenic distillation to achieve the extremely high purities needed to synthesize high molecular weight polymers. Besides the C_2–C_4 olefins, which come from FCC, cokers, and steam cracking, many integrated refineries also oligomerize these light olefins to produce higher olefins, which enable additional derivatives such as surfactants, plasticizers, rubbers, and lubricants.

2.7.2 Aromatics

During the reforming process, naphtha streams are converted to aromatic-rich streams. While these aromatics contribute to the mogas pool of molecules, especially with their high-octane number, they are often separated out for their much higher value as a chemical feedstock than as a burnable fuel. In particular, benzene, toluene, xylene isomers, and ethylbenzene are extracted to high purity for petrochemical derivatives. These are purified via solvent extraction and distillation, and the difficult xylene isomers are additionally purified via simulated moving bed (SMB) adsorption processes or by crystallization. Additionally, xylenes are often isomerized in a loop in this process to achieve higher yields of the more desirable *para*-xylene, which is a key building block for many derivatives, including polyethylene terephthalate (PET), also known as polyester.

2.7.3 Lubes

The rise of the automobile also pushed the need for better lubricants. Specialized refineries for the production of lubricants were established in China by the Northern Wei Dynasty as early as 512 BCE. But petroleum-based lubricants were mostly inferior to animal fats and beeswax, and only highly paraffinic crudes were suitable for making lubes. Lubricant basestocks come from the same boiling point range as fuel oils. Starting in the 1920s, various methods were developed to refine petroleum feedstock into higher quality lubes by removing the worst heteroatomic compounds via clay absorption, acid treating, and selective solvent extraction. In the 1950s, hydrotreating and hydrocracking catalysis methods were developed, where hydrogen was added to the basestock. The development of additional refining processes, such as catalytic dewaxing, wax hydroisomerization, and hydrofinishing, allows greater flexibility in feedstock composition yielding high-performance products.

Synthetic lubes composed of polymers such as poly-α-olefins (PAOs) were first marketed by Gulf Oil in 1951 and improved upon by Mobil in the 1970s. Since then, a variety of synthetic chemical compounds, such as esters and polyalkylene glycols (PAGs), have been formulated as synthetic lubricants. Synthetic lubricants have many performance advantages over conventional motor oils; however, they are much more expensive and still need to be changed regularly.

2.7.4 Other Products

The petrochemical industry produces hydrocarbon and halogenated solvents and precursor compounds, such as naphthenic acids, used in the manufacture of soaps and detergents, adhesives, agricultural fertilizers and pest/weed controls, medicines, and adhesives.

2.8 The Future of Petroleum

Modern society depends on energy and products derived from fossil fuels. The twentieth century marks a period of unprecedented growth in the exploration and production of crude oil and natural gas, displacing coal as the main source of energy. The global demand for energy is expected to increase by 25% by 2040 as the standard of living improves in developing nations. With an increasing awareness of the necessity to cut greenhouse gas emissions and to foster recycling, reuse, and improved efficiencies, a significant portion of this energy demand will be offset by wind, solar, and nuclear power. Natural gas and oil consumption are expected to increase but remain close to their current levels, while coal will continue to decline (Figure 2.28). The promotion of the increased use of natural gas over coal for power generation is viewed as a transitional state that will cut CO_2 emissions.

Each year the major oil companies, BP, ExxonMobil, Shell, Total, and others, release to the public their outlook on energy that lists their projections of future demands and supplies as well as potential impacts of new regulations and geopolitical change. The current projections, made out to 2040 (or 2050), generally agree on the key driver, an increasing global population with increasing energy demand, and the key challenges of energy conservation and the development of energy sources with lower or zero carbon emission. The absolute values on the future demand for fossil fuels differ between the companies depending on how the weight the various influences are factored, but all agree that the demand will rise and that there are adequate resources to meet the demand. CO_2 emissions are also projected to increase, though at a slower rate than in past decades, until turning downward around 2040.

The International Energy Agency (IEA) Stated Policies Scenario (STEP) agrees fairly well with the industry projections with a small growth in oil and significant growth in natural gas over the next 20 years (Figure 2.29). Under STEP, CO_2 emission will continue to grow after 2020 though at a much reduced rate. In terms of petroleum production, the IEA STEP aligns with the industry outlooks with a small growth in oil and significant growth in natural gas. These outlooks contrast with IEA models that are needed for meeting the 2 °C limit on global warming, the goal of the Paris Agreement. Under the IEA Sustainable Development Scenario (SDS), CO_2 emissions would peak around 2020 and then decline through the increased efficiency in fossil fuel and energy use, a greater reliance on renewables

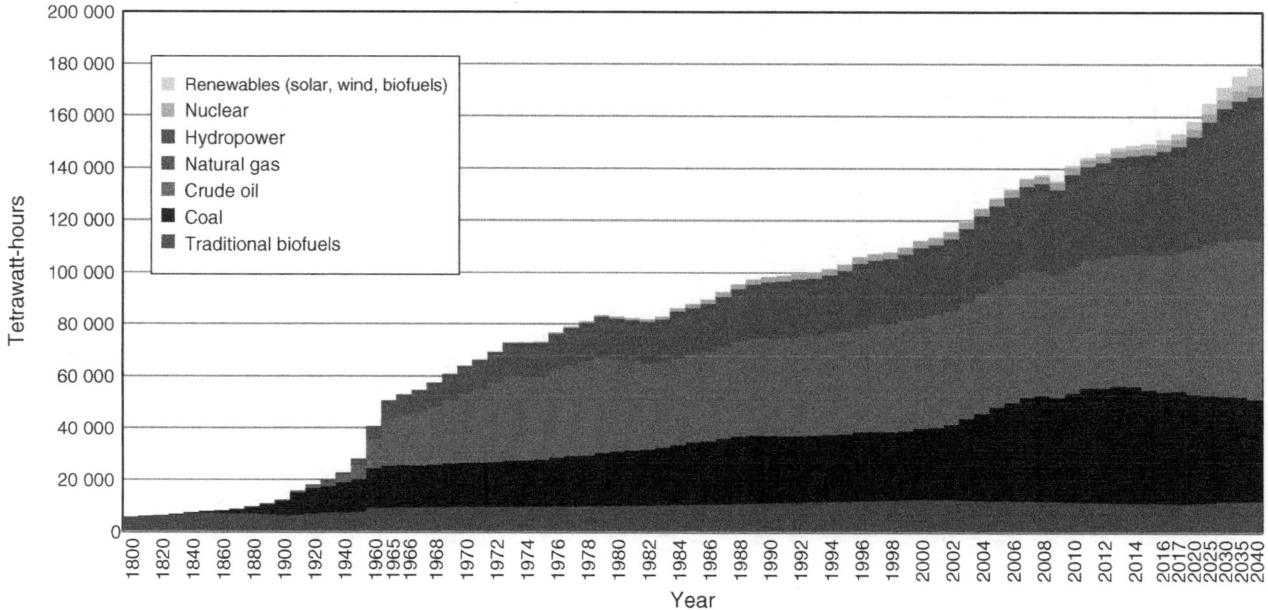

Figure 2.28 World energy consumption through time. Source: Data from Ritchie and Roser (2018). Reprinted with permission of ExxonMobil (2019). Image courtesy of ExxonMobil.

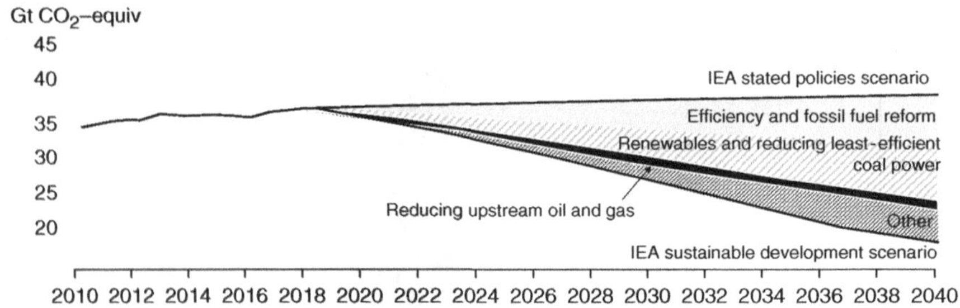

Figure 2.29 International Energy Agency Sustainable Development Scenario proposed decreases in CO_2 emissions to meet the target of a 1.5 °C increase in global temperature by 2040. Source: Reproduced with permission from IEA (2018), World Energy Outlook. All rights reserved.

and nuclear, and a major commitment to carbon capture. The SDS would reduce CO_2 emissions to zero by ~2050, and even then the chance of achieving only a 1.5 °C increase in global temperature is only 50%. New technological advances are needed to obtain the SDS goals without a major disruption of the global economy.

The reader is directed toward these books for additional information and details on the topics discussed in this chapter.

- *Refinery Engineering: Integrated Process Modeling and Optimization* (Chang et al. 2012).
- *Practical Petroleum Geochemistry for Exploration and Production* (Dembicki 2016).
- *Oil 101* (Downey 2009).
- *Springer Handbook of Petroleum Technology* (2nd edition) (Hsu and Robinson 2017).
- *Nontechnical Guide to Petroleum Geology, Exploration, Drilling and Production* (4th edition) (Hyne 2019)
- *Introduction to Petroleum Exploration and Engineering* (Palmer 2017).
- The *Chemistry and Technology of Petroleum* (5th edition) (Speight 2014).
- *Handbook of Petroleum Refining* (Speight 2016).
- *The Prize: The Epic Quest for Oil, Money and Power* (Yergin 2008).

References

Ablard, P., Bell, C., Cook, D. et al. (2012). The expanding role of mud logging. *Oilfield Review* 24: 24–41.

AHDictionary (2020). *Mummy*, 5ee. The American Heritage® Dictionary of the English Language. Retrieved from https://www.ahdictionary.com.

Al-Sibahy, A. and Edwards, R. (2017). Characterization of the clay masonry units and construction technique at the ancient city of Nippur. *Engineering Structures* 147: 517–529. https://doi.org/10.1016/j.engstruct.2017.06.017.

API (2019). *API 1509 Annex E-API Base Oil Interchangeability Guidelines for Passenger Car Motor Oils and Diesel Engine Oils (Revised September 2019)*. Online. American Petroleum Institute (API). Retrieved from https://www.api.org.

Argáez, C., Batta, E., Mansilla, J. et al. (2011). The origin of black pigmentation in a sample of Mexican prehispanic human bones. *Journal of Archaeological Science* 38: 2979–2988. https://doi.org/10.1016/j.jas.2011.06.014.

ARPA (2016). *Innovative Drilling Technology: Low-Contact Drilling Technology to Enable Economical Enhanced Geothermal System (EGS) Wells*. Online. US Department of Energy. Retrieved from https://arpa-e.energy.gov.

ASTM-D1319. (2019). *Standard test method for hydrocarbon types in liquid petroleum products by fluorescent indicator adsorption*. West Conshohocken, PA: ASTM International. Retrieved from https://www.astm.org.

ASTM-D1322. (2019). *Standard test method for smoke point of kerosene and aviation turbine fuel*. West Conshohocken, PA: ASTM International. Retrieved from https://www.astm.org.

ASTM-D156. (2015). *Standard test method for Saybolt color of petroleum products (Saybolt Chromometer Method)*. West Conshohocken, PA: ASTM International. Retrieved from https://www.astm.org.

ASTM-D189. (2019). *Standard test method for Conradson carbon residue of petroleum products*. West Conshohocken, PA: ASTM International. Retrieved from https://www.astm.org.

ASTM-D2163. (2019). *Standard test method for determination of hydrocarbons in liquefied petroleum (LP) gases and propane/propene mixtures by gas chromatography*. West Conshohocken, PA: ASTM International. Retrieved from https://www.astm.org.

ASTM-D2270. (2016). *Standard practice for calculating viscosity index from kinematic viscosity at 40 °C and 100 °C*. West Conshohocken, PA: ASTM International. Retrieved from https://www.astm.org.

ASTM-D2274. (2019). *Standard test method for oxidation stability of distillate fuel oil (Accelerated Method)*. West Conshohocken, PA: ASTM International. Retrieved from https://www.astm.org.

ASTM-D2386. (2019). *Standard test method for freezing point of aviation fuels*. West Conshohocken, PA: ASTM International. Retrieved from https://www.astm.org.

ASTM-D2500. (2017). *Standard test method for cloud point of petroleum products and liquid fuels*. West Conshohocken, PA: ASTM International. Retrieved from https://www.astm.org.

ASTM-D2624. (2015). *Standard test methods for electrical conductivity of aviation and distillate fuels*. West Conshohocken, PA: ASTM International. Retrieved from https://www.astm.org.

ASTM-D2700. (2019). *Standard test method for motor octane number of spark-ignition engine fuel*. West Conshohocken, PA: ASTM International. Retrieved from https://www.astm.org.

ASTM-D2713. (2015). *Standard test method for dryness of propane (Valve Freeze Method)*. West Conshohocken, PA: ASTM International. Retrieved from https://www.astm.org.

ASTM-D2784 (2011). *Standard test method for sulfur in liquefied petroleum gases (oxy-hydrogen burner or lamp) (withdrawn 2016)*. West Conshohocken, PA: ASTM International. Retrieved from https://www.astm.org.

ASTM-D3241. (2019). *Standard test method for thermal oxidation stability of aviation turbine fuels*. West Conshohocken, PA: ASTM International. Retrieved from https://www.astm.org.

ASTM-D3606. (2017). *Standard test method for determination of benzene and toluene in spark ignition fuels by gas chromatography*. West Conshohocken, PA: ASTM International. Retrieved from https://www.astm.org.

ASTM-D4052. (2018). *Standard test method for density, relative density, and API gravity of liquids by digital density meter*. West Conshohocken, PA: ASTM International. Retrieved from https://www.astm.org.

ASTM-D445. (2019). *Standard test method for kinematic viscosity of transparent and opaque liquids (and calculation of dynamic viscosity)*. West Conshohocken, PA: ASTM International. Retrieved from https://www.astm.org.

ASTM-D4530. (2015). *Standard test method for determination of carbon residue (Micro Method)*. West Conshohocken, PA: ASTM International. Retrieved from https://www.astm.org.

ASTM-D4809. (2018). *Standard test method for heat of combustion of liquid hydrocarbon fuels by bomb calorimeter (Precision Method)*. West Conshohocken, PA: ASTM International. Retrieved from https://www.astm.org.

ASTM-D5191. (2019). *Standard test method for vapor pressure of petroleum products and liquid fuels (Mini Method)*. West Conshohocken, PA: ASTM International. Retrieved from https://www.astm.org.

ASTM-D5769. (2015). *Standard test method for determination of benzene, toluene, and total aromatics in finished gasolines by gas chromatography/mass spectrometry*. West Conshohocken, PA: ASTM International. Retrieved from https://www.astm.org.

ASTM-D5950. (2014). *Standard test method for pour point of petroleum products (Automatic Tilt Method)*. West Conshohocken, PA: ASTM International. Retrieved from https://www.astm.org.

ASTM-D664. (2018). *Standard test method for acid number of petroleum products by potentiometric titration*. West Conshohocken, PA: ASTM International. Retrieved from https://www.astm.org.

ASTM-D86. (2019). *Standard test method for distillation of petroleum products and liquid fuels at atmospheric pressure*. West Conshohocken, PA: ASTM International. Retrieved from https://www.astm.org.

ASTM-D97. (2017). *Standard test method for pour point of petroleum products.* West Conshohocken, PA: ASTM International. Retrieved from https://www.astm.org.

ASTM-D976. (2016). *Standard test method for calculated cetane index of distillate fuels.* West Conshohocken, PA: ASTM International. Retrieved from https://www.astm .org.

ASTM-D2622. (2016). *Standard test method for sulfur in petroleum products by wavelength dispersive X-ray fluorescence spectrometry.* West Conshohocken, PA: ASTM International. Retrieved from https://www.astm.org.

Azais, M. (1845). *Explanation and History of the Artesian Well of Grenelle.* Boston, MA: J.N. Bang.

Beicip. (2019). Field development: integrated field studies. Retrieved from http:// www.beicip.com/integrated-field-studies (accessed 18 March 2020).

Beychok, M. (2009). Schematic flow diagram of a typical hydrocracker. Retrieved from http://en.citizendium.org/wiki/Hydrocracking (accessed 18 March 2020).

Boëda, E., Bonilauri, S., Connan, J. et al. (2008). Middle Palaeolithic bitumen use at Umm el Tlel around 70 000 BP. *Antiquity* 82: 853–861. https://doi.org/10.1017/ S0003598X00097623.

Bohacs, K.M., Grabowski, G.J., Carroll, A.R. et al. (2005). Production, destruction, and dilution – the many pathways to source-rock development. In: *The Deposition of Organic-Carbon-Rich Sediments: Models, Mechanisms, and Consequences* (ed. N.B. Harris), 61–101. Tulsa, OK: SEPM (Society for Sedimentary Geology).

Brown, K.M. (2016). Asphaltum (bitumen) production in everyday life on the California Channel Islands. *Journal of Anthropological Archaeology* 41: 74–87. https://doi.org/10.1016/j.jaa.2015.11.003.

Cârciumaru, M., Ion, R.M., Niţu, E.C., and Ştefănescu, R. (2012). New evidence of adhesive as hafting material on Middle and Upper Palaeolithic artefacts from Gura Cheii-Râşnov Cave (Romania). *Journal of Archaeological Science* 39: 1942–1950. https://doi.org/10.1016/j.jas.2012.02.016.

Chang, A.F., Pashikanti, K., and Liu, Y.A. (2012). *Refinery Engineering: Integrated Process Modeling and Optimization*, 1e. Weinheim, Germany: Wiley-VCH.

Clark, K.A., Ikram, S., and Evershed, R.P. (2016). The significance of petroleum bitumen in ancient Egyptian mummies. *Philosophical Transactions of the Royal Society A: Mathematical, Physical and Engineering Sciences* 347: 20160229. https:// doi.org/10.1098/rsta.2016.0229.

Connan, J. (1999). Use and trade of bitumen in antiquity and prehistory: Molecular archaeology reveals secrets of past civilizations. *Philisophical Transactions of the Royal Society B* 354: 33–50. https://doi.org/10.1098/rstb.1999.0358.

Connan, J., Carter, R., Crawford, H. et al. (2005). A comparative geochemical study of bituminous boat remains from H_3, As-Sabiyah (Kuwait), and RJ-2, Ra's al-Jinz (Oman). *Arabian Archaeology and Epigraphy* 16: 21–66. https://doi.org/10.1111/j .1600-0471.2005.00041.

Corma, A., Sauvanaud, L., Mathieu, Y. et al. (2018). Direct crude oil cracking for producing chemicals: Thermal cracking modeling. *Fuel* 211: 726–736. https://doi.org/10.1016/j.fuel.2017.09.099.

Dembicki, H. (2016). *Practical Petroleum Geochemistry for Exploration and Production*, 1e. Cambridge, MA: Elsevier.

Downey, M.P. (2009). *Oil 101*. New York, NY: Wooden Table Press LLC.

Eberle, A.P.R., King, H.E., Ravikovitch, P.I. et al. (2016). Direct measure of the dense methane phase in gas shale organic porosity by neutron scattering. *Energy and Fuels* 30: 9022–9027. https://doi.org/10.1021/acs.energyfuels.6b01548.

EIA (2012). *Crude Oil Distillation and the Definition of Refinery Capacity*. Online. US Energy Information Administration (EIA). Retrieved from https://www.eia.gov.

EPA (2016). *Learn About Gasoline*. Online. US Environmental Protection Agency (EPA). Retrieved from https://www.epa.gov.

Etiope, G. (2015). Seeps in the ancient world: Myths, religions, and social development. In: *Natural Gas Seepage: The Earth's Hydrocarbon Degassing* (ed. G. Etiope), 183–194. New York, NY: Springer.

ExxonMobil. (2018). Discovering hidden hydrocarbons: using seismic-imaging technology to map formations far below the Earth's surface. Retrieved from https://corporate.exxonmobil.com (accessed 18 March 2020).

ExxonMobil. (2019). 2019 outlook for energy: A perspective to 2040. Retrieved from https://corporate.exxonmobil.com (accessed 18 March 2020).

Feng, L., Hu, Y., Hall, C.A.S., and Wang, J. (2013). *The Chinese Oil Industry: History and Future*. Beijing, China: Springer Nature.

Gray, G. R., and Tschirley, N. K. (1975). Drilling fluids programme for ultra-deep wells in the United States. *Paper Presented at the World Petroleum Congress*, Tokyo, Japan (11–16 May 1975).

Green, M. (2014a). Fault trap. Retrieved from https://en.wikipedia.org (accessed 18 March 2020).

Green, M. (2014b). Salt dome trap. Retrieved from https://en.wikipedia.org (accessed 18 March 2020).

Green, M. (2014c). Stratigraphic trap. Wikipedia. Retrieved from https://en.wikipedia.org (accessed 18 March 2020).

Green, M. (2017). Anticline trap. Retrieved from https://en.wikipedia.org (accessed 18 March 2020).

Honeywell. (2011). Petrochemicals flow scheme. Retrieved from https://www.uop.com/petrochemicals-flow-scheme (accessed 18 March 2020).

Honeywell. (2013). UOP CCR Platforming™ process. Retrieved from https://www.uop.com/reforming-ccr-platforming (accessed 18 March 2020).

Hsu, C.S. and Robinson, P.R. (2017). *Springer Handbook of Petroleum Technology*, 2e. New York, NY: Springer International Publishing.

Hunt, T.S. (1861). *Notes on the History of Petroleum or Rock Oil*, The Canadian Naturalist, vol. VI, 1–15. Geological Survey of Canada.

Hunt, T.S. (1863). Contributions to the chemical and geological history of bitumens and of pyroschists or bituminous shales. *American Journal of Science and Arts* 35: 157–171. Retrieved from https://archive.org/details/mobot31753002152624.

Hutton, J. (1788). Theory of the Earth; An investigation of the laws observable in composition, dissolution, and restoration of land upon the globe. In: *Transactions of the Royal Society of Edinburgh* (ed. The Royal Society of Edinburgh), vol. 1, 209–304. Scotland: Royal Society of Edinburgh.

Hyne, N.J. (2019). *Nontechnical Guide to Petroleum Geology, Exploration, Drilling and Production*, 4e. Tulsa, OK: PennWell.

IEA (2018). *World Energy Outlook 2018*. Paris: International Energy Agency (IEA). Retrieved from https://www.iea.org/reports/world-energy-outlook-2018.

IP-336 (2004). *Petroleum Products – Determination of Sulfur Content – Energy-Dispersive-X-ray Fluorescence Method. (IP336-2937259)*. London: Energy Institute.

IP-564 (2013). *Determination of the Level of Cleanliness of Aviation Turbine Fuel – Laboratory Automatic Particle Counter Method (IP564-2936450)*. London: Energy Institute.

ISO-2719. (2016). *Determination of flash point – Pensky-Martens closed cup method*. Online. International Organization for Standardization. Retrieved from https://www.iso.org/standard/62263.html.

ISO-3104. (1994). *Petroleum products – transparent and opaque liquids – determination of kinematic viscosity and calculation of dynamic viscosity*. Online. International Organization for Standardization (ISO). Retrieved from https://www.iso.org/standard/8252.html.

ISO-3405. (2019). *Petroleum and related products from natural or synthetic sources – determination of distillation characteristics at atmospheric pressure*. Online. International Organization for Standardization. Retrieved from https://www.iso.org/standard/67956.html.

ISO-3675. (1998). *Crude petroleum and liquid petroleum products – laboratory determination of density – hydrometer method*. Online. International Organization for Standardization (ISO). Retrieved from https://www.iso.org/standard/26326.html.

ISO-3733. (1999). *Petroleum products and bituminous materials – determination of water – distillation method*. Online. International Organization for Standardization (ISO). Retrieved from https://www.iso.org/standard/9219.html.

ISO-6245. (2001). *Petroleum products – determination of ash*. Online. International Organization for Standardization (ISO). Retrieved from https://www.iso.org/standard/31156.html.

ISO-8754. (2003). *Petroleum products – determination of sulfur content – energy-dispersive X-ray fluorescence spectrometry*. Online. International Organization for Standardization (ISO). Retrieved from https://www.iso.org/standard/30062.html.

Jacques Girard. Batillus tanker in Saint-Nazaire (1976). Retrieved from https://en.wikipedia.org (accessed 18 March 2020).

Khun, O. (2004). Ancient Chinese drilling. *Recorder* 29 (6). Retrieved from https://csegrecorder.com/articles/view/ancient-chinese-drilling.

Logan, W.E. (1842). *Preliminary Report, 1842*. Montreal, Canada: Geological Survey of Canada.

Logan, W.E. (1843). *Report of Progress for the Year 1843*. Montreal, Canada: Geological Survey of Canada.

Lyell, C. (1830–1833). *Principles of Geology: Being an Attempt to Explain the Former Changes of the Earth's Surface, by Reference to Causes Now in Operation*, vol. 1–3. London: John Murray.

Mendeleev, D. (1877). L'origine du Petrole. *La Revue Scientifique* 18 (2): 409–416. Retrieved from https://gallica.bnf.fr/ark:/12148/bpt6k2150895/f412.image.

Mullin, J.H. (1885). The advantages in the use of the diamond drill (read at the Jackson meeting, June 1885). *Ohio Mining Journal* 3: 43–49.

NOAA (2010). *Oil and Gas Exploration*. Online. National Oceanic and Atmospheric Administration (NOAA). Retrieved from https://oceanexplorer.noaa.gov.

OSHA (1999). *OSHA Technical Manual*. Online. OSHA. Retrieved from https://www.osha.gov.

Palmer, A.C. (2017). *Introduction to Petroleum Exploration and Engineering*. Hackensack, NJ: World Scientific Publishing Co. Pte. Ltd.

Pathak, M., Kweon, H., Deo, M., and Huang, H. (2017). Kerogen swelling and confinement: Its implication on fluid thermodynamic properties in shales. *Scientific Reports* 7: 12530. https://doi.org/10.1038/s41598-017-12982-4.

Pettijohn, F.J. (1957). *Sedimentary Rocks*, 3e. New York, NY: Harper and Row.

Pollastro, R.M. (2003). *Total Petroleum Systems of the Paleozoic and Jurassic, Greater Ghawar Uplift and Adjoining Provinces of Central Saudi Arabia and Northern Arabian-Persian Gulf*. Online, U.S. Geological Survey Bulletin 2202-H. US Geological Survey (USGS). Retrieved from http://pubs.usgs.gov/bul/b2202-h.

Popov, Y.A., Pevzner, S.L., Pimenov, V.P., and Romushkevich, R.A. (1999). New geothermal data from the Kola superdeep well SG-3. *Tectonophysics* 306: 345–366. https://doi.org/10.1016/S0040-1951(99)00065-7.

PSU. (2018). FSC 432: Petroleum processing – alkylation. Retrieved from https://www.e-education.psu.edu/fsc432/content/alkylation (accessed 18 March 2020).

Raux, F. (2010). *The Spinner*. Online. RMN-Grand Palais (Musée du Louvre). Retrieved from https://www.louvre.fr/en/oeuvre-notices/spinner.

Reed, R. and Loucks, R. (2007). Imaging nanoscale pores in the Mississippian Barnett Shale of the northern Fort Worth Basin. *Paper presented at the AAPG Annual Convention*, Long Beach, CA (1–4 April 2007).

Ritchie, H., & Roser, M. (2018). Energy. University of Oxford. Retrieved from https://ourworldindata.org/energy (accessed 18 March 2020).

Sadeghbeigi, R. (2012). *Fluid Catalytic Cracking Handbook*, 3e. New York, NY: Elsevier.

Speight, J.G. (2014). *The Chemistry and Technology of Petroleum (Chemical Industries)*, 5e. Boca Raton, FL: CRC Press.

Speight, J.G. (2016). *Handbook of Petroleum Refining*, 1e. Boca Raton, FL: CRC Press.

Sterling, G. H., Krebs, J. E., and Dunn, F. P. (1989). The Bullwinkle project: An overview. *Paper presented at the Offshore Technology Conference*, Houston, Texas (1–4 May 1989).

Sternbach, C.A., Shoup, R.C., Wilhelm, C.T., and Reynolds, A.S. (2017). *Heritage of the Petroleum Geologists*, vol. 2. USA: American Association of Petroleum Geologists (AAPG).

Steward, D.B. (2007). *The Barnett Shale Play: Phoenix for the Fort Worth Basin, a History*. Fort Worth, TX: Fort Worth Geological Society.

Treibs, A. (1936). Chlorophyll- und Häminderivate in organischen Mineralstoffen. *Angewandte Chemie* 49: 682–686. https://doi.org/10.1002/ange.19360493803.

UOP. (2020). Alkylation. Retrieved from https://www.uop.com/processing-solutions/refining/gasoline/alkylation (accessed 18 March 2020).

Walker, M. W. (2012). Pushing the extended-reach envelope at Sakhalin: An operator. *Paper presented at the IADC/SPE Drilling Conference and Exhibition*, San Diego, California (6–8 March 2012).

Yergin, D. (2008). *The Prize: The Epic Quest for Oil, Money and Power*. New York, NY: Free Press.

Yule, H. and Oldman, T. (1855). *Reports of the Mission to Ava in 1855*. Calcutta: Baptist Mission Press.

Zhang, J., Shan, H., Chen, X. et al. (2014). Multifunctional two-stage riser fluid catalytic cracking process. *Applied Petrochemical Research* 4: 395–400. https://doi.org/10.1007/s13203-014-0079-5.

Zoroaster (1908). Modellmix. Retrieved from https://commons.wikimedia.org (accessed 18 March 2020).

Part III

Analytical Techniques Utilized in the Petroleum Industry

3

Petroleum Analysis Through Conventional Analytical Techniques

Melissa N. Dunkle[1] *and William L. Winniford*[2]

[1]*Analytical Science, Dow Benelux, BV, Herbert H. Dowweg 5, Building 446, Hoek, The Netherlands*
[2]*Analytical Science, The Dow Chemical Company, 230 Abner Jackson Parkway, Lake Jackson, TX, USA*

3.1 Introduction to Petroleum Analysis

Characterization methods are of utmost importance for the evaluation of petroleum, as the composition and nature of the carbon present determine how the sourced petroleum is routed through the refining process, affecting not only process design, but also operational control for refining facilities. As has been indicated in Chapter 2, sourced petroleum has a wide array of diversity, ranging from light gases to heavy oils and even heavier tars. To complicate matters even further, these materials are made up of different compound classes, namely, *n*-paraffins, *iso*-paraffins, olefins, naphthenes, and aromatics, or PIONA for short, in addition to the presence of heteroatoms. This means that there is no "one size fits all" approach for the analysis of sourced petroleum. This chapter will give an overview of the importance of petroleum characterization and how the analytical techniques evolved throughout the years, providing us with the conventional analyses still performed routinely today in the petroleum (oil and gas) and petrochemical industry.

3.2 Brief History on Petroleum Analysis

After the discovery of petroleum, the need to analyze it was quickly realized. In 1927, the American Petroleum Institute (API) established Research Project No. 6, where the main objective was to identify and determine the chemical constituents of commercial petroleum fractions (Ettre and Horvath 1975). This was an enormous undertaking, where initially, only physical separation mechanisms, such as

Analytical Techniques in the Oil and Gas Industry for Environmental Monitoring, First Edition.
Edited by Melissa N. Dunkle and William L. Winniford.

distillation, were possible. However, from 1945 on, liquid chromatography (LC) using very long columns was implemented, which eventually, by 1951, nearly 25 years after Project No. 6 started, resulted in the isolation and identification of approximately 100 compounds in petroleum (Ettre 2004). It was this work that established that all hydrocarbons in naturally sourced petroleum are composed of only three chemical groups, paraffins, naphthenes, and aromatics. After another eight years of research, only 46% of the representative petroleum product being analyzed was characterized, in which 169 individual hydrocarbons were isolated and identified (Rossini and Mair 1959). It should be noted that the pure standards utilized were also developed through API Research Project No. 6 (Streiff et al. 1948), where the API standard and API Research hydrocarbons became the gold standards in hydrocarbons analysis. Even though API Research Project No. 6 is no longer active, ending in 1967, the API still exists and offers high-quality analytical standards and other services for all aspects of the petroleum industry (https://www.api.org). It is believed that petroleum may very well contain over 100,000 different components (Bartle and Myers 2002).

3.2.1 How Petroleum Analysis Influenced Developments in Gas Chromatography

Since the invention of gas chromatography (GC) in 1952 by James and Martin (1952), its promise for hydrocarbons analysis was quickly exploited by the petroleum industry, as at that time, petroleum was quickly replacing coal as the predominant source of liquid fuels and chemical feedstocks (Bartle and Myers 2002). The theory described by James and Martin dealt with a mobile gas phase flowing over an immobilized liquid phase. Once the sample is introduced, it is transported through the system by the mobile gas phase, where partitioning with the non-volatile liquid phase immobilized in the separation column occurs. The components in the sample are separated by the large number of partitions that occur, and the mobile gas phase transports the separated components through the system to the detector. This was the first example of a chromatographic separation based on volatility, where James and Martin demonstrated the separation and quantification of the 12 components of a C_1–C_5 fatty acid mixture (Figure 3.1).

It should be noted that the first examples of GC equipment were all in-house constructions, including the system by James and Martin. Be that as it may, it was quickly realized that GC, in all of its potential, could not be fully exploited without the availability of proper instrumentation (Ettre 2004). In the mid-1950s, the first commercially available systems were offered by Perkin Elmer, Fisher/Gulf, Barber Coleman, Podbielniak (all US based), Pye Unicam (UK based), and Shimadzu (Asia based) (McNair 2010). Initial systems were isothermal; it wasn't until the late 1950s to early 1960s that temperature-programmed systems became commercially

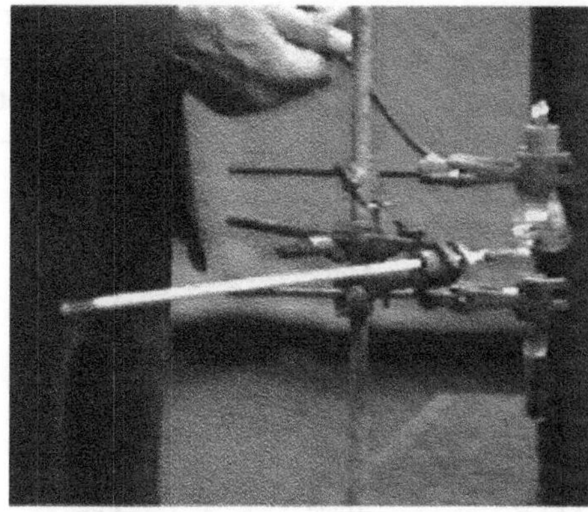

Figure 3.1 Video photograph of Martin and James's GC. Source: Bartle and Myers (2002). Reproduced with permission, Copyright (2002), Elsevier.

available. Other limitations of early GC systems included detector options and compatibility as well as a lack of column versatility. Continuing advancements were being made in the field of GC, including column and detector technology; these topics will be discussed, as well as their applicability toward the petroleum and petrochemical industry.

3.2.1.1 Detector Technology

Detectors are classified into two main categories, concentration sensitive and mass sensitive. Concentration sensitive detectors are not dependent on the flow rate exiting the GC column, while mass sensitive detectors are. For example, if the flow rate were stopped, the signal of a mass sensitive detector will drop to zero; however, the concentration sensitive detector will continue to measure a constant signal. In addition, detectors can be either selective or universal. With a universal detector, theoretically, all compounds should afford a detector response; however, with a selective detector, only specific compound classes will elicit a detector response. While advancements in detector technology are discussed, it should be noted that the discussion herein is not exhaustive and is limited to some of the more precedent advancements; all detectors discussed are also summarized in Table 3.1.

3.2.1.1.1 Gas-Density Detector

The detector utilized by James and Martin in the first example of a GC separation consisted of manual titration of the eluent exiting the separation column. While this was a sufficient detection technique to prove the concept of the GC separation, such a detection technique would limit the possibilities of GC. Therefore, in 1956, Martin and James developed the gas-density meter as a sensitive detector for measuring the effluent exiting the GC column (Martin and James 1956). The detector measured small differences in pressure between the effluent gas and a reference gas. Calibration was not necessary, as the detector response was proportional to the difference in molecular weights of the components eluting and the

Table 3.1 Comparison of selected GC detectors.

GC detector	Response	Typical LOD	Selectivity
Gas density	Mass	0.5–10 ng	Universal
TCD	Concentration	1–20 ng	Universal
FTD	Mass	1–10 ng	Selective for hydrocarbons (C–H)
FID	Mass	0.1–10 pg	Selective for hydrocarbons (C–H)
NPD	Mass	1–10 pg	Selective for nitrogen and phosphorus-containing compounds
FPD	Mass	1–100 pg	Selective for sulfur and phosphorus
PFPD	Mass	1 pg	Selective for sulfur and phosphorus
ECD	Concentration	0.1–1000 pg	Selective to halogens, nitrates, and conjugated carbonyls
SCD	Mass	1–100 pg	Selective for sulfur-containing compounds
NCD	Mass	1–100 pg	Selective for nitrogen-containing compounds
Polyarc	Mass	0.1–10 pg	Selective for hydrocarbons (C–H), universal response
MSD	Mass	1–10 pg	Universal or selective
AED	Mass	1 pg	Selective
IR	Concentration	<200 pg	Universal or selective
PID	Concentration	25–200 pg	Selective for aromatics and olefins
ELCD	Mass	1–10 pg	Selective for nitrogen, sulfur, and chlorine-containing compounds
PDD	Mass	0.1–1000 pg	Universal or selective
VUV	Mass	20–200 pg	Universal or selective

carrier gas. The gas-density meter had a linear range over 3 orders of magnitude concentration range, and the sensitivity was around 5×10^{-7} g/ml for *n*-heptane. The complexity of the gas-density meter was a technical feat at that point in time, which was also its pitfall. The design and construction of the detector were so complex, that widespread use never occurred. Eventually in 1963, a simplified version of the gas-density detector was developed (Nerheim 1963); additionally, sophisticated models were developed and commercialized by Gow-Mac Instrument Co., which is a company still in existence today (https://www.gow-mac.com/); however, this detector technology is not widely used in today's instrumentation.

3.2.1.1.2 Thermal-Conductivity Detector

At the same time, another approach for GC detection was being investigated, specifically, thermal conductivity. Thermal-conductivity detection was originally developed around 1915 to measure the purity of air, and the developed detector was called a katharometer (Daynes et al. 1920). The katharometer was limited to the measurement of gases, and in the 1940s, use was extended to measuring carbon dioxide from flue gasses. It was in 1954, in an extension of James and Martin's first GC separation, that the katharometer was exploited as a method for measuring the thermal-conductivity changes of various compounds, including hydrocarbons, eluting from the GC separation column (Ray 1954). In this approach, two heated filaments were situated in the arms of a Wheatstone bridge, where one filament was situated in the eluent exiting the column and the other in a reference gas stream. The changes in resistance were recorded using a pen/chart recorder. Unlike the gas-density meter, the thermal-conductivity detector (TCD) requires calibration curves for the quantitative measurement of the samples analyzed. The TCD is sensitive to a wide range of compounds; however, it has a limited linear dynamic concentration range of around 500, and the sensitivity is roughly 10^{-6} g/ml, or an limit of detection (LOD) around 1–20 ng on column. Despite these shortcomings, the TCD has been used extensively throughout the years as a GC detector for packed column chromatography.

3.2.1.1.3 Flame Thermocouple Detector

The flame thermocouple detector (FTD), reported in 1955, was to marry the excellent sensitivity of the gas-density detector with the simple construction of the TCD (Scott 1955). This was the first example of introducing a small burning jet into the GC detector design. A thermocouple was placed directly above the jet, and when the eluent exiting the GC column contained a solute, the temperature of the flame increased, which in turn increased the output from the thermocouple. The response was linear over 3 orders of magnitude concentration range and was sensitive to around 1×10^{-6} g/ml for *n*-heptane, or an LOD around 1–10 ng on column. This detector was the forerunner to the flame ionization detector (FID), and as such had a short-lived lifespan. That being said, the design of the FTD was revisited in 1975, where a dual flame version was developed for enhanced sensitivity compared with the single flame version (Desai and Sahgal 1975); however, again, due to the popularity of the FID detector, the dual FTD also did not take off.

3.2.1.1.4 Flame Ionization Detector

In the late 1950s, major advances in detector technology were taking place. In 1958, Harley and Pretorius and McWilliam and Dewar made significant advancements on the flame-based detectors of time, developing nearly simultaneously the FID, which could be coupled to a GC separation (Harley et al. 1958; McWilliam and Dewar 1958). The FID is an extension of the FTD, where the main difference is that the ions produced in the flame are measured, as opposed to the heat generated, respectively. The FID has a linear dynamic range of 4–5 orders of magnitude and a sensitivity of 5×10^{-12} g/ml for *n*-heptane, or an LOD around 0.1–10 pg on column. Additionally, the FID is insensitive to pressure, flow rate, and temperature changes. The FID is the most commonly used detector in GC and is responsive to all compounds containing carbon, making it especially useful in the petroleum and petrochemical industry.

3.2.1.1.5 Nitrogen Phosphorus Detector

The nitrogen phosphorus detector (NPD) was developed in 1977, having a similar design to the FID, but the principle of operation is quite different (Burgett et al. 1977). A small rubidium or cesium bead is located near the hydrogen jet of the NPD, which allows partially combusted nitrogen and phosphorus compounds eluting from the column to absorb on the bead. The absorbed material interferes with the function of the bead, causing electron emission to increase, and thus a rising current at the anode is detected. The NPD is a very sensitive detector, with limits for phosphorus compounds at 1×10^{-12} g/ml and limits for nitrogen compounds at 1×10^{-11} g/ml, corresponding to an on column LOD around 1–10 pg, and has a linear dynamic range of 3–4 orders of magnitude. Like the FID, the NPD is also insensitive to pressure, flow rate, and temperature changes.

3.2.1.1.6 Flame Photometric Detector

The flame photometric detector (FPD) was first described by Grant (1958). This detector is not as sensitive as other detectors of the time; however, it makes up for this with its selectivity for sulfur and phosphorus-containing hydrocarbons. With the FPD, the emitted light from sulfur and phosphorus containing compounds is unique compared with other hydrocarbons, which enables selective filtering of the emitted light. Upon the introduction of the GC effluent into the hydrogen flame, green light is emitted from phosphorus compounds due to the presence of the HPO molecule, and blue light is emitted from sulfur compounds due to the presence of the S_2 molecule. While selective, the FPD is not as sensitive as other detectors, such as the FID, having an LOD approximately 1–100 pg on column. The linear dynamic range in phosphorus mode is around 3 orders of magnitude and only 2 orders of magnitude in sulfur mode. Another limitation of the FPD is signal quenching when hydrocarbons are present and flame-out when large quantities of solvent are introduced into the detector.

Several of these limitations were overcome with a dual-flame design, where one flame was used to decompose the sample, while the second flame was used to produce the desired optical emission (Patterson et al. 1978). The dual-flame design significantly decreased the observed noise level compared with the single flame

FPD design. A more recent development is the pulsed flame photometric detector (PFPD) in 1991 (Atar et al. 1991), and further improved in subsequent years (Amirav and Jing 1995). The PFPD has a similar design to the FPD, but due to the pulse nature of the PFPD, it offers higher sensitivity due to lower noise signal (LOD ~1 pg on column) and improved selectivity due to the reduction of chemical background.

3.2.1.1.7 Electron Capture Detector

The electron capture detector (ECD) developed by Lovelock was not the first detector to employ a radioactive source (Lovelock and Lipsky 1960); that honor goes to the β-ray ionization detector developed by Boer (1957). The β-ray ionization detector had a similar sensitivity to the TCD or around 10^{-6} g/ml. Like the FTD, the β-ray ionization detector had a short-lived life, as it was quickly supplanted by the FID and the argon ionization detector. The argon ionization detector, also developed by Lovelock, was somewhat unpopular, due to the high noise level, which meant the sensitivity was worse than an FID at around 4×10^{-11} g/ml, or around 0.2 ng on column LOD, and only 2 orders of magnitude linear dynamic range were realized (Lovelock 1958). The ECD utilizes a low energy β-ray source, ^{63}Ni, to produce electrons and ions, which are detected and converted into a chromatographic signal. The detector has two modes of operation, DC mode, where a constant potential is applied, and pulsed mode, whereas the name indicates, the potential is pulsed. ECD sensitivity is an order of magnitude better than the FID, at around 10^{-13} g/ml, and the linear dynamic range is around 2 orders of magnitude in DC mode and an impressive 5 orders of magnitude in pulsed mode.

3.2.1.1.8 Chemiluminescence Detectors

Using the principle of the FPD, where sulfur species eluting from the GC column into the flame are converted into S_2 and emit light, the sulfur chemiluminescence detector (SCD) uses the reaction of sulfur species with ozone to form SO_2 and emitted light (Kelly et al. 1983). At high temperatures, oxidative combustion occurs, which converts all sulfur-containing species into SO. The SO then reacts with ozone to form the radical SO_2^*, which then forms SO_2 and emitted light. The sensitivity of the SCD is around 4×10^{-10} g/ml, with an on column LOD around 1–100 pg, and the linear dynamic range is roughly 3 orders of magnitude.

The nitrogen chemiluminescence detector (NCD) utilizes nearly the same principle as the SCD, where nitrogen species present in the GC effluent react with ozone to form NO_2 and emitted light. Oxidative combustion occurs at high temperatures to convert nitrogen species eluting from the GC effluent into NO. The NO reacts with ozone to form the radical NO_2^*, which then forms NO_2 and emitted light. The sensitivity and linearity of the NCD are similar to those reported for the SCD.

3.2.1.1.9 Polyarc Reactor

In a recent advancement for hydrocarbons analysis, the Polyarc reactor was introduced in 2015, which can be installed on any gas chromatograph-flame ionization detector (GC-FID) instrument between the separation column and the FID detector (Beach et al. 2016). Using hydrogen and air, the Polyarc converts all organic compounds into methane, which when detected by the FID, yields a universal response, making this a quantitative carbon detector (QCD) (Maduskar et al. 2015). As is well known with FID signals, the response factors for different compound types differ, meaning that for area percent quantification, the response factors must be taken into consideration, or a substantial error will be introduced into the quantification. With a universal response, all compound classes have the same response factor.

3.2.1.1.10 Mass Spectrometry

Mass spectrometry (MS) was first developed in the early 1900s to measure the masses of atoms and to demonstrate the existence of isotopes. Sir Joseph John Thomson designed the first mass spectrometer, but his protégé, Francis W. Aston improved upon the design and increased resolution by an order of magnitude. Around the same time, A. J. Dempster in 1918 designed the first electron impact source. By the 1940s, mass spectrometers were commercially available, and MS had found its way into the petroleum industry, where it was being used to monitor the abundances of small hydrocarbons in process streams. In 1959, Gohlke described the direct introduction of the gas chromatography effluent into a mass spectrometer (GC-MS) (Gohlke 1959). However, at this point in time, due to the cost, simplicity of construction, and operation, more emphasis was placed on GC-FID compared with GC-MS in the petroleum industry. Over the years, improvements in computer technology, spectra interpretation and identification, and the development of user-friendly lower cost GC-MS instruments eventually meant that GC-MS was implemented in many modern laboratories (Gohlke and McLafferty 1993). Modern GC-MS instruments offer universal and selective detection of compounds over a wide linear dynamic range and on column LODs of around 1 pg in scan mode.

3.2.1.1.11 Other Detectors

The GC detectors discussed thus far is by no means an exhaustive list; however, it is important to briefly mention some of the other options available for GC. The atomic emission detector (AED) is a highly sensitive and selective detector; however, it has a complicated design that has prohibited wide-spread implementation when coupling to GC. The AED is an element selective detector able to excite and detect C, H, O, N, S, Cl, Br, F, P, D, and Hg using a plasma source and collecting emission spectra at different select wavelengths, respectively. An early example of an AED being coupled to a chromatographic separation was presented in 1979 (Robbins and Caruso 1979), but it wasn't until the 1990s that gas chromatography–atomic emission detector (GC-AED) began to reach the same limits of detection as other GC detectors, around 1 pg on column (Quimby and Sullivan 1990).

Another element selective detector for GC is the electrolytic conductivity detector (ELCD), where only nitrogen, sulfur, and halogen-containing compounds are detected. The ELCD operates by first thermally decomposing the GC effluent, then the gas phase reaction products are combined with a stream of deionized liquid in a gas–liquid contactor, and finally, the electrical conductivity of the liquid is measured continuously, such that any changes in conductivity are recorded as a change in signal that corresponds to elution of a compound. The first coupling of an ELCD detector to a GC separation was reported in 1962 (Piringer and Pascalau 1962), and while the ELCD has good performance with on column LODs in the range of 1–10 pg, it has never been widely used.

Fourier transform-infrared (FT-IR) detection was coupled to GC separations in the 1960s and 1970s using lab-built interfaces to couple the GC separation to stand-alone FT-IR spectrometers, where at that time, limits of detection were in the 50–100 ng range, partially due to the slow scan rates of the FT-IR spectrometers. In 1973, an interface to allow online gas chromatography–infrared (GC-IR) analysis was introduced; however, detection limits were in the 20–150 μg range (Penzias 1973). Thanks to continuing advancements in the interface between the GC separation and the FT-IR detector, as well as improved scan rates of the detectors, detection limits of 100 pg on a fully optimized system became possible; however, when considering real-world samples, typical on column LODs are 0.5–25 ng (Visser 2002).

The ionization of compounds by light radiation is also a possible detection mechanism for GC. Early models of photoionization detectors (PIDs) used radioactive sources to induce ionization; however, in the 1970s, PIDs using ultraviolet (UV) sources were developed (Sevcik and Krysl 1973) and commercialized. The PID is selective to aromatics, olefins, and other select compounds and is around 10–50 times less sensitive than the FID (Verner 1984).

A more recent advancement in GC detector technology is the pulsed discharge detector (PDD) (Vasnin et al. 1992; Wentworth et al. 1992). The PDD is both a universal and selective detector. It has two modes of operation, pulsed discharge helium ionization detector (PDHID) and pulsed discharge electron capture detector (PDECD). When operated in the PDHID mode, the detector is universal and sensitive. Additionally, the PDHID is an attractive alternative to the FID, as the number of make-up gases required for the operation of the detector are reduced to one, helium, compared with the FID, which requires hydrogen, air, and nitrogen. When operated in PDECD mode, a similar sensitivity to the ECD is achieved, but without the radioactive source.

In 2014, the vacuum ultraviolet (VUV) detector was reported as a new GC detector capable of using the UV absorbance from 120 to 240 nm to identify and quantify the compounds present in the GC effluent (Schug et al. 2014). The VUV has a linear range of 3–4 orders of magnitude and offers on column LODs from 20 to 200 pg. GC-VUV will be discussed in more detail in Chapter 4.

3.2.1.2 Column Technology

3.2.1.2.1 Packed Columns

One of the first reported separations of a liquid petroleum fraction, using the setup described by James and Martin, was performed by Desty and Whyman (1957). It should be noted that it was common in early GC separations to utilize packed columns, where inner diameters of between 1 and 5 mm and total lengths from 1 to 5 m were typical (Bartle and Myers 2002). Oftentimes, diatomaceous earth was used as the solid support. Diatomaceous earth is naturally occurring amorphous silica (SiO_2) that is soft enough to be easily ground into a fine powder (Rabel 2000). Column packing material usually consisted of 50–80 mesh (50–100 μm particles), which afforded typical plate counts around 2000 (Blomberg et al. 2002). Celite, a brand of commercially available diatomaceous earth still available today, was used by James and Martin in their early experiments (Adlard and Poole 2009). To the solid support, a liquid stationary phase would be applied. To be effective, the liquid stationary phase must be chemically inert, non-volatile, and thermally stable, as well as allow for rapid diffusion of the analytes. For these reasons, silicones are the preferred stationary phases for today's GC columns (Blomberg 2001; Rahman et al. 2015); however, until the commercialization of reproducible GC columns, many in-house prepared columns employing exotic stationary phases were evaluated. As can be imagined, the efficiency of packed columns from this time cannot be compared with what is now the standard GC column. To give an example, in an early publication by Desty and Whyman, liquid petroleum fractions in the range of 30–150 °C were separated using two different packed columns, specifically an *n*-hexatriacontane (n-C_{36}) stationary phase and a benzyldiphenyl stationary phase both coated onto a silica support; the theoretical plates for these columns were 1200 and 320, respectively (Desty and Whyman 1957). In 1960, a device was invented to improve the packing of GC columns, making it more reproducible (Ven Horst et al. 1960); however, this was limited to the simultaneous packing of a maximum of two columns (Figure 3.2). Further improvements to packed columns were made by using longer columns with smaller and more uniform particles; however, packed-column GC would remain limited in the number of compounds able to be resolved in a single separation. As was reported by many at the time, it was believed to be impossible to fully separate the C_2–C_7, or up to the 100 °C range, of petroleum fractions on a single packed column.

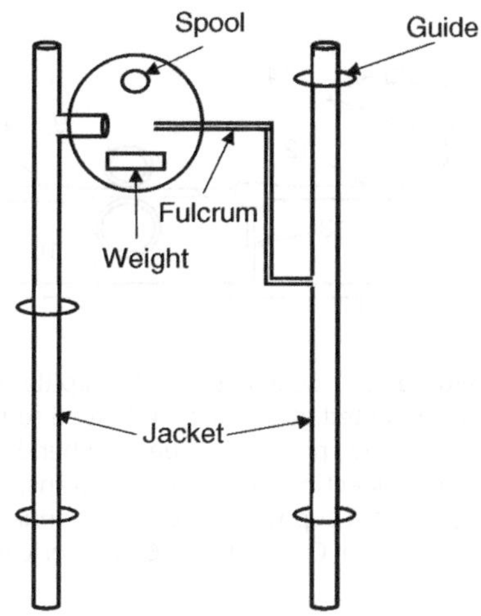

Figure 3.2 Schematic of the apparatus designed for filling two columns simultaneously. Source: Adapted with permission from Ven Horst et al. (1960). Copyright 1960, American Chemical Society.

3.2.1.2.2 Open Tubular Columns

It was for this reason that many researchers in the late 1950s began focusing on open tubular columns coated with a liquid stationary phase for use in GC. Golay reported one of the first GC separations on an open tubular coated column, as well as publishing his work on the kinetics and theory of the separation achieved on such columns (Golay 1957a,b). He theorized high-efficiency separations on coated open tubular capillary columns, especially when compared with packed columns. Several early examples of separations made using open tubular columns were in agreement with the theory. One example describes the separation of C_5–C_8 paraffins and naphthenes on a 30 m long by 0.25 mm ID column coated with squalane, which yielded 50 000 theoretical plates for *n*-heptane (Zlatkis and Lovelock 1959). Another example detailed the separation of C_5–C_8 cycloparaffins (cycloalkanes) and aromatics on a 53 m long by 0.25 mm ID column coated with poly(propylene glycol); here, over 100 000 theoretical plates were calculated (Condon 1959).

With highly efficient columns offering high-resolution separations, more compounds were able to be separated from a complex mixture in a single analysis. Yet without analytical standards, identification of the eluting compounds was difficult, and to this end, the concept of retention indices was developed. Kováts published his first article in 1958 (in German) on the relationship between the retention index of a compound on a non-polar stationary phase and its boiling point using packed columns to separate an *n*-paraffin mixture (Kováts 1958). Several benefits to using retention indices were given, including the linear dependence of the retention index on temperature, that the retention index is independent of the column constants and instrumentation used, and that it provides information on the chemical nature of the compound. Over the years, much work was performed

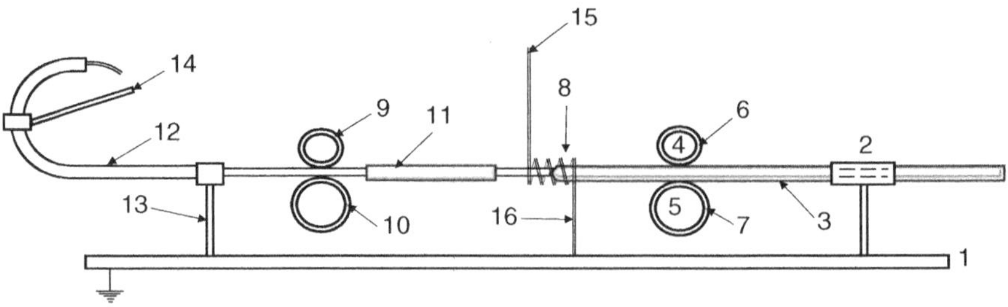

Figure 3.3 Schematic of the capillary apparatus: (1) brass base plate, (2) support, (3) thick-walled glass tube, (4, 5) feed rollers, (6, 7) rubber tires, (8) furnace, (9, 10) draw rollers, (11) porcelain tube, (12) bending tube, (13) bending tube support and ground, (14) connection to low voltage transformer 1, (15) connection to low voltage transformer 2, and (16) support and ground connection. Source: Desty et al. (1960). Reproduced with permission, Copyright (1960), American Chemical Society.

to improve the retention index system, making it more uniform and eliminating errors (Kováts and Keulemans 1964), as well as calculating the retention indices of additional compounds and ensuring that it was also applicable to open tubular capillary columns. Today, Kováts retention indices are still used and applied to the analysis of hydrocarbons; while a retention index will not provide chromatographers with identification, it is an extremely useful piece of information to help confirm an identification. This work also led to the extensive use of non-polar stationary phases; at the time, squalane, a truly non-polar stationary phase, became quite popular in the petroleum and petrochemical industry.

With a focus on improved separation by means of developments in column technology, Desty et al. reported the first apparatus (Figure 3.3) to draw long lengths of glass capillary from millimeter (mm) inner diameters to micrometer (μm) inner diameters (Desty et al. 1960).

Due to the relative inertness of glass capillaries, the advantages over stainless steel capillaries were quickly realized. In the same year, Halász and Schreyer published their design for an FID that could be coupled to capillary GC (Halász and Schreyer 1960) (in German). In the years to follow, several reports on quantitative GC-FID were published, where the first article using this detector design was published in 1961, just one year after their FID was described in the literature (Halász and Schneider 1961). Here, C_3–C_7 hydrocarbons analysis was performed on a 60 mm by 0.25 mm glass capillary column coated with squalane, and area normalization factors for quantification were verified, which opened the door for total area percent quantification used throughout the industry today. A few years later Bruderreck et al. published a detailed study on quantitative GC-FID, where linearity, repeatability, response, and normalization factors were studied for different FID detector designs (Bruderreck et al. 1964). An extremely important finding from this work was that the general applicability of using area normalization factors for quantification is dependent upon an entirely standardized FID. Sanders and Maynard from Shell confirmed in 1968 the area percent quantification of hydrocarbons in a gasoline sample with 95% confidence limits. Here, the gasoline was separated using a squalane coated stainless steel capillary column by GC-FID; an internal standard was used to confirm the results and to ensure that no heavy components were lost during the analysis. While squalane as a column coating seems foreign to today's chromatographers, this coating was the primary stationary phase for the separation of hydrocarbons by open tubular GC in the 1960s and 1970s. Perhaps one of the most impressive separations was obtained for a gasoline sample using a 0.5 mm ID by 300 m long capillary column coated with squalane, where 378 peaks were separated in just over four hours with only 17 unclassified peaks (Whittemore 1979).

3.2.1.2.3 Fused Silica Capillary Columns

While glass capillaries showed a major improvement in separation efficiency over packed columns, they were notoriously fragile and active toward highly polar compounds, which impeded progress in the field of capillary GC. However, with the invention of fused silica capillary columns (Dandeneau and Zerenner 1979), many of the pitfalls observed with glass capillaries were overcome while maintaining high separation efficiencies. Theoretical plates of 3000–5000 plates per meter are obtained with fused silica capillary columns (Rahman et al. 2015). Even though squalane was one of the most widely used stationary phases for hydrocarbons analysis, it had several shortcomings, including a low maximum operating temperature (~100 °C) and difficulties in preparing stable coatings with glass and fused silica columns (Johansen et al. 1983). As a result, squalane was predominately replaced by polydimethylsiloxane; though, squalane coated fused silica columns can still be purchased today. Nowadays, column manufacturers offer a wide array of capillary columns based on the coating procedures largely developed in the 1980s by the Lee group (Kong and Lee 1983a,b; Lee et al. 1984; Woolley et al. 1986). Using a 0.31 mm ID by 25 m long fused silica column coated with OV-101, four different Australian crude oils were analyzed by both GC-FID and GC-MS; in less than two hours, 263 peaks were separated and more than 370 compounds were identified (Zadro et al. 1985).

3.3 Conventional Analysis of Petroleum

Throughout the history of petroleum analysis, analytical methods were developed that are still in use today. These methods are standardized globally and delivered by organizations such as ASTM International, previously known as the American

Society for Testing and Materials. The petroleum and petrochemical industry considers these methods as the gold standards for petroleum fraction characterization. Methods are continually revised by the American Society for Testing and Materials (ASTM), and as such, only the revised method is valid; for this reason, even though a method may have been developed decades ago, the revised date is the new method reference date. As mentioned previously, there is no "one size fits all" method for petroleum analysis, as the diversity of the petroleum fractions is too complicated. This section details the conventional analyses, including the applicability for specific petroleum fractions still practiced today.

3.3.1 Distillation

Even before the introduction of GC, the group type analysis of petroleum was performed using distillation. In this manner, the material is distilled, and the boiling range distributions are used to characterize the amount of sample (wt%) per boiling range. In 1921, the first distillation methods were documented by the ASTM, many of which are still in use today. Different ASTM methods are available for different petroleum fractions. Crude oils, which are a heavier fraction of petroleum, are characterized using a high-efficiency distillation apparatus to yield the true boiling point profile via ASTM D2892 (ASTM-D2892 2018). For light and middle distillates, which include petroleum fractions up to n-C_{20}, ASTM D86 is used (ASTM-D86 2018). ASTM D86 uses a single plate distillation apparatus, making this method simpler than ASTM D2892. To extend the range of ASTM D86 above n-C_{20}, vacuum distillation is performed (ASTM-D1160 2018). While the distillation methods are simple in their design, these methods are laborious and time-consuming and lack reproducibility. The distillations take days to reach completion, and when considering separation efficiency, distillation offers only 5–10 theoretical plates (Speight 2001); however, this can be extended slightly by increasing the amount of sample and distillation time.

In order to mimic the distillation methods mentioned earlier but with higher efficiency, simulation distillation (SIMDIST) using a GC separation was developed. The first reported SIMDIST separation was performed in 1960 using packed column GC and was applicable to petroleum distillates ranging from −40 to 400 °C, or roughly n-C_3 to n-C_{25} hydrocarbons (Eggertsen et al. 1960). It was known that under a temperature programmed GC separation using a non-polar column, hydrocarbons elute in order of increasing boiling point. By using the retention times obtained from the analysis of a known standard with known boiling points, the retention time scale was converted into a boiling point scale; when combined with the quantitative GC data, the weight percentage of the sample per boiling point range was calculated. In 1973, ASTM D2887 was the first ASTM approved SIMDIST method for petroleum fractions possessing a final boiling point up to 540 °C, or up to n-C_{44} (ASTM-D2887 2018; Blomberg et al. 2002). While ASTM D2887 can provide SIMDIST data for a wide range of petroleum fractions, the heaviest fractions were neglected. By combining open tubular capillary columns with high temperature GC, hydrocarbons up to n-C_{140} were detected by high temperature–simulated distillation (HT-SIMDIST) (Trestianu et al. 1985). Unfortunately, the conventional polyimide coating present on the exterior of fused silica capillaries is not stable above 350 °C; even with high-temperature polyimide coatings, the temperature range is only extended to 400 °C. This led to very short column lifetimes when used at high temperature, too short to allow for the implementation of HT-SIMDIST in industry. However, with the re-introduction of metal capillary columns, operating temperatures well above 450 °C were achieved while maintaining column stability. Eventually, ASTM D6352 was introduced, which extended the final boiling point of ASTM D2887 to 700 °C, making this method applicable to petroleum fractions in the range of n-C_{10} to n-C_{90}, (ASTM-D6352 2015). To obtain SIMDIST separations of petroleum fractions above n-C_{90}, supercritical fluid chromatography is utilized; this technique will be described in full detail in Chapter 6.

3.3.2 PIONA Analyzer

A different way to perform a group type analysis on petroleum fractions is to quantify the weight percentage of the PIONA, rather than quantifying the weight percentage per boiling point range. The first automated paraffin, naphthene, aromatic (PNA) analyzer, which utilized multidimensional GC, afforded the quantitative separation of heavy naphtha into the paraffins, naphthenes, and aromatics (Boer and van Arkel 1971). By expanding upon the PNA analyzer, eventually, the PIONA analyzer, which also utilized multidimensional GC, was developed to provide quantification information for sourced lighter petroleum fractions. With the PIONA analyzer (Figure 3.4), a series of valves and columns, both packed and open tubular capillary columns, are utilized to provide separation and detection of the different PIONA groups (Johansen et al. 1983). ASTM D6839 and D5443 are available methods for PIONA analysis; however, D5443 is the preferred method when a reformer unit is available (ASTM-D5443 2018; ASTM-D6839 2018). A reformer unit, or a Reformulyzer™,[1] provides the same PIONA information as the PIONA analyzer and is compatible with D5443. The main difference of a Reformulyzer is that gasoline blends and finished gasoline can also be analyzed. While the PIONA analyzer provides quantitative information per carbon number for the various group types, its limitation is that the technique is only applicable to sourced petroleum fractions with boiling

1 ™ Trademark of AC Analytical Controls, B.V.

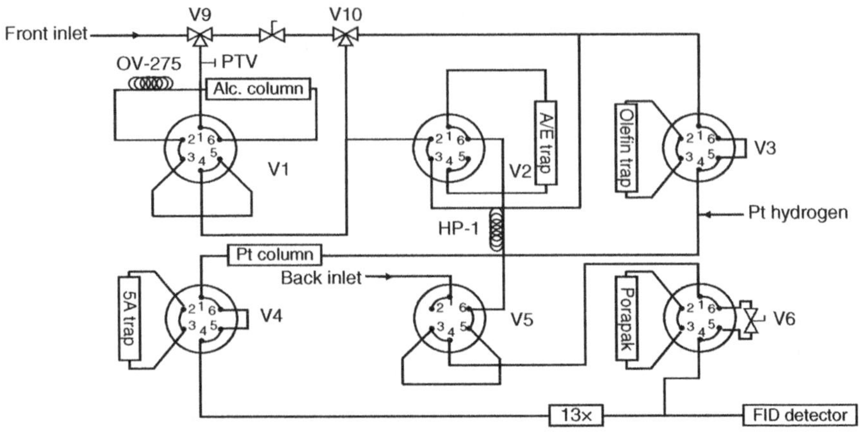

Figure 3.4 Typical PIONA instrument schematic. PTV, programmed temperature vaporizer.

ranges up to 200 °C, or roughly n-C_{11}. Despite this limitation, the PIONA analyzer is still widely used today in the petrochemical industry.

3.3.3 Detailed Hydrocarbon Analysis

The detailed hydrocarbon analysis (DHA) was developed to provide a more simplistic separation compared to the PIONA analyzer for lighter petroleum fractions. The DHA consists of a high-resolution GC separation using a single open tubular capillary column coupled to FID detection. Peak identification and retention time calibration are performed using a series of commercially available standards, and quantification, both PIONA groupings per carbon number and individual compound quantification are made using the peak area response together with known response factors and Kováts retention indices. Having the individual compounds identified and quantified provides highly detailed information over the PIONA analyzer and allows for a highly detailed characterization of the petroleum fractions being analyzed. The DHA provides accurate quantification for lighter petroleum factions with boiling ranges up to 225 °C, or roughly n-C_{13}. Several ASTM methods exist, including ASTM D5134, which only quantifies up to n-C_9 (ASTM-D5134 2017), ASTM D6729 and D6730, which utilize 100 m long capillary columns (ASTM-D6729 2014; ASTM-D6730 2016), and ASTM D6733, which utilizes a 50 m long capillary column, allowing for shorter analysis times (ASTM-D6733 2016). To give an idea of the amount of time necessary to complete an analysis, with ASTM D6730, a single analysis takes over 120 minutes to complete when helium is used as the carrier gas (Figure 3.5). In ASTM D6729 appendix X2, the carrier gas is changed to hydrogen, which

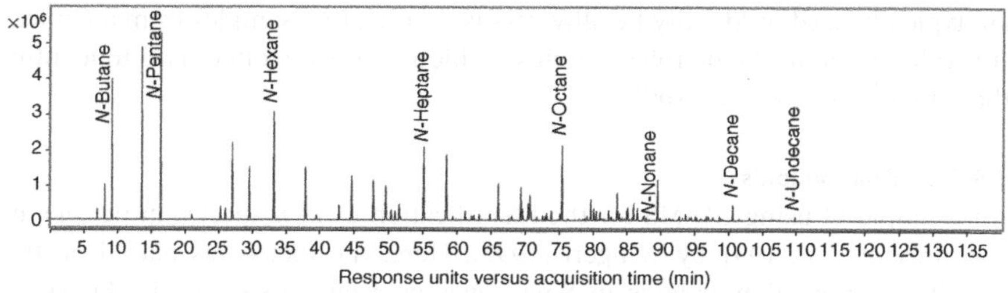

Figure 3.5 Typical chromatogram obtained from detailed hydrocarbon analysis (DHA) of a liquid hydrocarbon sample. Source: Dunkle et al. (2019). Reproduced with permission, Copyright (2019), Elsevier.

allows for shorter analysis times, around 74 minutes, while maintaining the separation of critical pairs. While hydrogen does allow for improved analysis times, not all laboratories are equipped with or prepared for the additional precautions needed when utilizing hydrogen as the carrier gas. Analysis time aside, an additional limitation of the DHA is the limited separation space afforded in a one-dimensional separation. Due to the numerous isomers present in such complicated petroleum fractions, it is possible that co-elution still exists and that misidentification is made.

While the ASTM methods listed for the DHA all utilize FID detection, the method has also been evaluated using mass spectrometric detection (Teng et al. 1994). Here, the C_4–C_{12} range of gasoline was evaluated, while another article utilizing a similar approach, focused on fuels with a higher boiling point, specifically, jet fuels with boiling ranges up to 300 °C, or roughly n-C_{17} (Lai and Song 1995). Lai and Song used GC-MS in combination with retention indices, to study the differences in jet fuels. By combining retention index information with the spectra obtained from the MS, detailed identifications were made, allowing for an accurate comparison between the coal-derived and the petroleum-derived jet fuels. Even though extremely detailed information can be discerned from the GC-MS, due to the complicated nature of quantification from MS data, the preferred detector of choice for the petroleum industry for the DHA method is still the FID. For this reason, there is no ASTM method available for the DHA with MS detection.

3.3.4 GC-MS Analysis for Unknown and Biomarker Identification

While quantification is more complicated with GC-MS data compared with GC-FID data, this does not mean that GC-MS is not utilized in the petroleum industry. GC-MS is an excellent tool for the identification of unknown compounds and biomarkers. A biomarker is a compound that can be used to trace the origin of a sample being analyzed. In the biomedical industry, biomarkers are typically used to identify healthy versus non-healthy samples from a clinical study; however, in the petroleum industry, biomarkers are often used to identify the source of a petroleum spill.

3.3.4.1 Diamondoids

An example of using GC-MS for the identification of new species in petroleum was presented in 1991 by Wingert (1992). Here, the focus was placed on the identification of diamondoids in Smackover petroleums sourced in Alabama, USA. Diamondoids are, as the name suggests, diamond-like structures consisting of only carbon and hydrogen and are present at such low concentrations that they are generally ignored. Diamondoids possess multiple fused rings that provide thermodynamic stability; however, this means that these species are resistant to thermal- and biodegradation. Adamantane is the simplest of the diamondoids, and all diamondoids are derived from this basic structure. Figure 3.6 shows the structures for adamantane and the next two diamondoid homologs. As is true with the PIONA compound classes, as diamondoids increase in carbon number, the number of structural isomers also increases.

Interestingly, when diamondoids are analyzed by GC-MS, the molecular ion is observed. This is due to their structural stability, which prevents fragmentation even when collision energy is applied. Therefore, unique spectra are obtained for diamondoids; for adamantane, a base peak of m/z 136 is obtained, and for diamantane and triamantane, base peaks of m/z 188 and 240 are obtained, respectively. Methyl- and ethyl-substituted diamondoids have also been isolated from petroleum, and their elution profiles on non-polar stationary phases, as well as their fragmentation patterns, are well known (Hála et al. 1970; Schneider et al. 1966). In the Toxey oil fraction of the Smackover petroleums analyzed by Wingert, individual concentrations of diamondoids were present in the order of 1 ppm (Wingert 1992). As diamondoids are present at such low concentrations in petroleum, single ion monitoring (SIM) mode can be utilized to enhance the sensitivity of the MS, affording detection and identification of these species. An example of the total ion chromatogram obtained for the Copeland saturates is given in Figure 3.7.

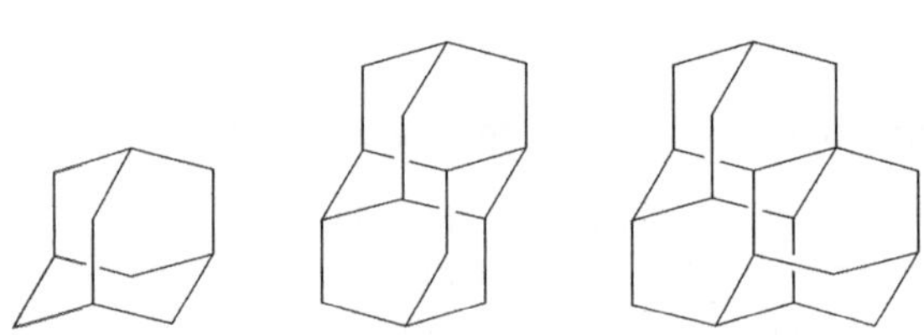

Adamantane (a) Diamantane (b) Triamantane (c)

Figure 3.6 Structures for adamantine (a), diamantine (b), and triamantane (c).

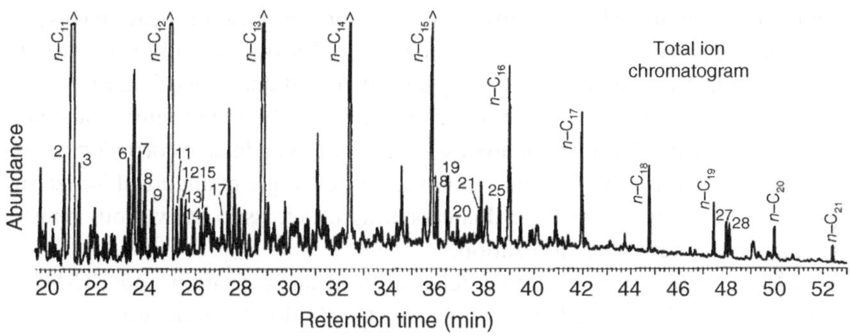

Figure 3.7 Partial total-ion chromatogram of the Copeland saturates. Source: Wingert (1992). Reproduced with permission, Copyright (1992), Elsevier.

3.3.4.2 Naphthenic Acids

Another example of the application of GC-MS is with naphthenic acids, also known as petroleum acids, which are naturally occurring components in petroleum; the naphthenic acid content in crude oils can be as high as 4%. Naphthenic acids are cycloaliphatic carboxylic acids containing cyclopentane and cyclohexane rings with the generic formula $C_nH_{2n+z}O_2$, where n is the carbon number and Z specifies the hydrogen deficiency of the ring ($Z = 0$ to -12). Naphthenic acids cover a wide carbon range, typically from $n = 7$ to 17; however, examples of higher carbon numbers also exist. Naphthenic acids have both acute and chronic toxicity to aquatic species in the environment and can be released into wastewater upon the sourcing of crude oil and bitumen. The naphthenic acid content in crude oil is also linked to the total acid number (TAN), which is determined using ASTM D664, a potentiometric titration method (ASTM-D664 2018).

In a unique example, naphthenic acids were identified and quantified in different California crude oils coming from the San Joaquin Valley using negative chemical ionization GC-MS with NF_3 as the reagent gas (Dzidic et al. 1988). With this approach, the naphthenic acids were first extracted from the crude oils; even though the extraction process was extensive, a 90% recovery (gravimetric determination) of the naphthenic acids was reported, which matched with the TAN number determined by ASTM D664. The extract was then analyzed by GC-MS, where the naphthenic acids reacted with the F^- ion formed under negative ionization, yielding a negatively charged acid and HF.

$$F^- + RCOOH \rightarrow RCOO^- + HF \tag{3.1}$$

The Kodak naphthenic acid standard was used as a reference standard in this study. Even though a high-resolution GC separation was utilized, the Kodak naphthenic standard remained chromatographically unresolved; however, due to the unique spectra obtained from negative chemical ionization, identification was possible. Upon analyzing an extract from the wastewater from a California refinery, several naphthenic acids were positively identified and quantified (Figure 3.8). To end up in the wastewater of the refinery means that the naphthenic acids were not degraded during the refining process, and as such can lead to corrosion issues. It should be noted that while this is an excellent example of chemical ionization GC-MS, not all laboratories have this capability or access to the various reagent gases required for chemical ionization.

A different study on naphthenic acids utilized derivatization and GC-MS to monitor biodegradation studies (Scott et al. 2005). In this example, water samples taken from tailings ponds were collected and analyzed. Tailings ponds are large diked ponds used to collect wastewater from the sourcing of crude oil and bitumen to avoid having hazardous chemicals enter the environment. A derivatization method (*N*-methyl-*N*-(*tert*-butyldimethylsilyl)trifluoroacetamide containing 1% *tert*-butyldimethylsilyl chloride [Clemente et al. 2004]) was used for the silylation of the naphthenic acids prior to GC-MS analysis. This derivatization step yields a stable structure showing little fragmentation while enhancing the ionization of the naphthenic acids, allowing them to be detected using electron impact GC-MS (Holowenko et al. 2002). The GC-MS results were used to study the effects of thermal degradation of the naphthenic acids from the tailings pond water samples. It was determined that while lower molecular weight naphthenic acids ($\leq C_{17}$) are easily degraded, the heavier naphthenic acids are resistant to degradation, meaning that even though the total content of naphthenic acids reduce in concentration through biodegradation, the heavy naphthenic acids can bio-accumulate in the environment.

3.3.4.3 Biomarkers

GC-MS was used to identify biomarkers to aid in source identification using freshly collected samples from the *Arrow* oil spill, which occurred 22 years prior (Wang et al. 1994). Once exposed to the environment, oil will begin to weather, meaning that changes take place, such as evaporation of the volatiles, dissolution, and degradation of components, making the characteristics of a weathered oil different from the original source. Not all components in the oil succumb to weathering effects, which makes it possible to use such compounds as biomarkers to trace back the origin of the oil source. In the article by Wang et al., 47 triterpenes and steranes were identified by GC-MS, and the ratios of the m/z 191 peaks for the C_{29}/C_{30} hopanes, which were unaffected by weathering (Figure 3.9), were used to unambiguously identify the source of the oil as originating from the tanker ship *Arrow*, which sank, spilling nearly 10,000 tons of oil into the Chedabucto Bay, Nova Scotia, Canada.

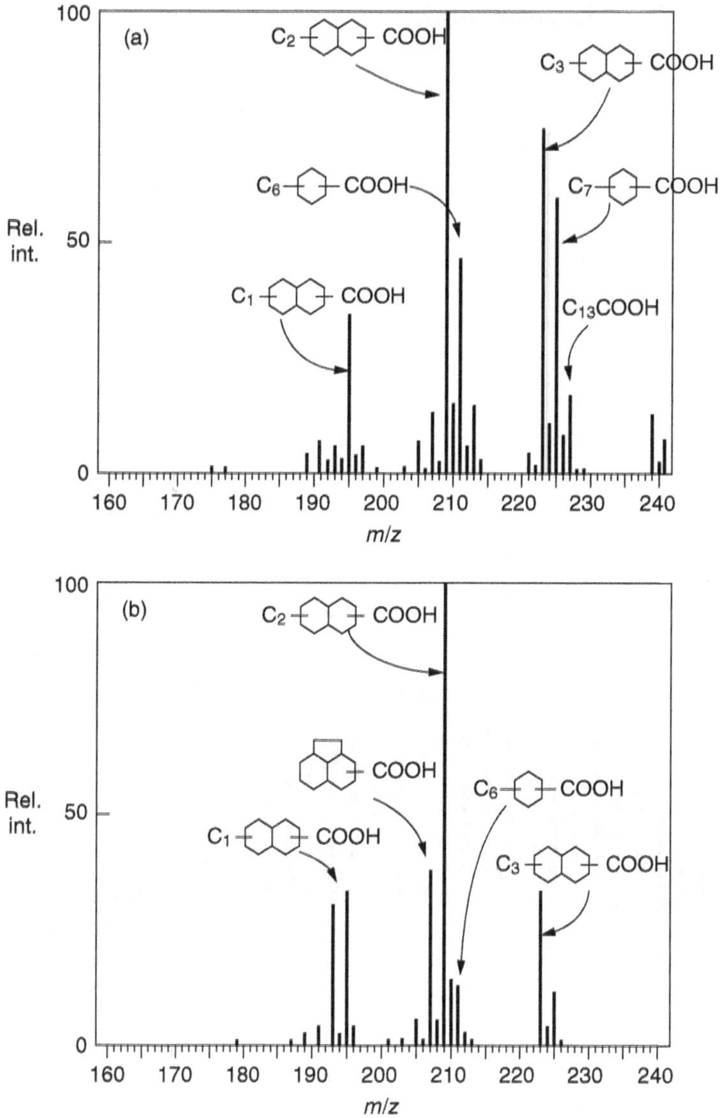

Figure 3.8 Fluoride ion mass spectra of Kodak naphthenic acid standard (a) and of wastewater acid extract from a California refinery (b) obtained at the same GC retention time. Source: Dzidic et al. (1988). Reproduced with permission, Copyright (1988), American Chemical Society.

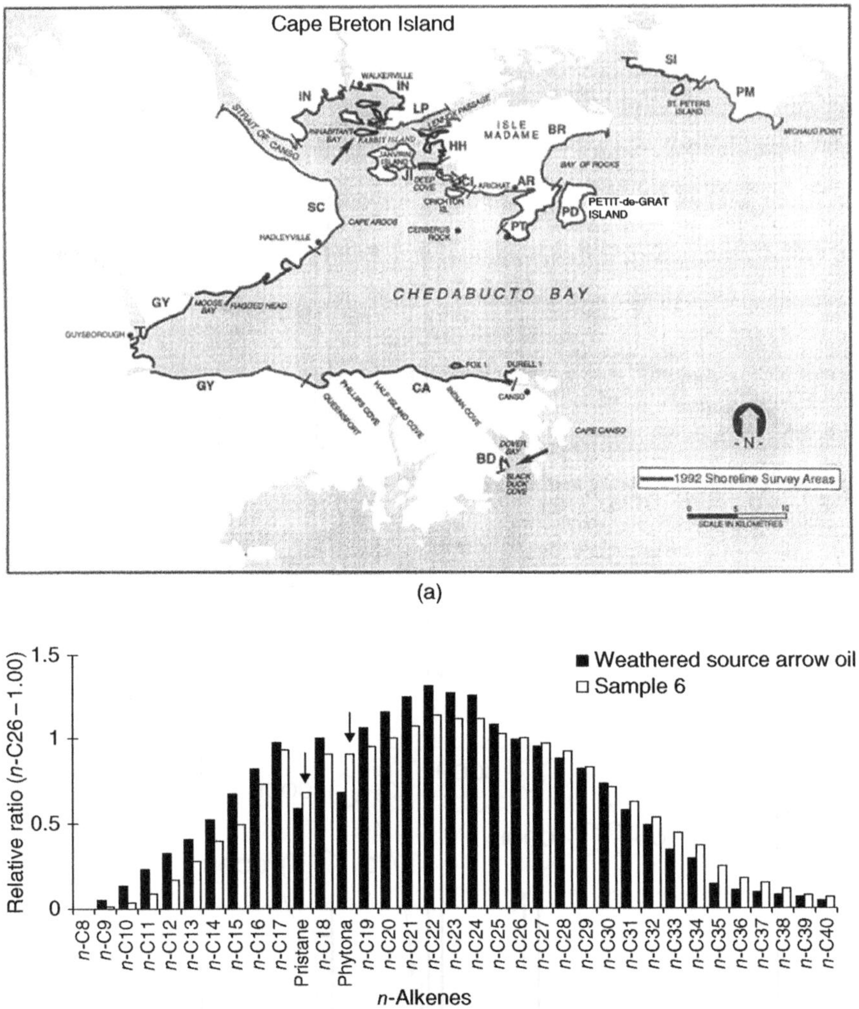

Figure 3.9 (a) Sampling locations in Chedabucto Bay and (b) comparison of the relative *n*-alkane concentrations and the *n*-alkane distribution patterns for the weather source oil and the sample S-6. The arrows indicate that before *n*-C_{26} the relative concentrations of pristine and phytane in the sample S-6 are higher than in the weathered source oil. Source: Wang et al. (1994). Reproduced with permission, Copyright (1994), American Chemical Society.

3.3.5 Total Petroleum Hydrocarbon (TPH) and Polycyclic Aromatic Hydrocarbon (PAH) and Their Environmental Impact

Total petroleum hydrocarbons (TPHs) are a generic term referring to any family or group of hydrocarbons originating from crude oil, while polycyclic aromatic hydrocarbons (PAHs) refers to a family of hydrocarbons containing two or more fused aromatic rings. PAHs occur naturally in coal, crude oil, and gasoline; however, they can also be produced when coal, oil, gasoline, and other substances are burned. While the previously mentioned techniques describe the analysis of petroleum fractions for their characterization for processing, TPH and PAH analyses focus on the environmental impact of petroleum contamination in water, soil, and air samples. Prior to remediation of an affected area, the total extent of the contamination as well as identification of the source must take place.

PAHs are an important class of environmental pollutants, as many PAHs are suspected carcinogens. PAHs are nearly ubiquitous trace contaminants of freshwater and marine sediment worldwide (Neff et al. 2005), and under the right conditions, they can be persistent in the environment. The higher molecular weight PAHs are more carcinogenic; however, these compounds tend to remain in the soil. The lower molecular weight PAHs, which are less carcinogenic, more easily find their way into groundwater due to their higher water solubility (Peters et al. 1999).

In ASTM D7678, TPHs from water and wastewater are analyzed by mid-infrared (IR) laser spectroscopy after solvent extraction from the sample (ASTM-D7678 2017). This method will quantify the amount of oil or grease present in water and wastewater resulting from contamination from a petroleum source based on the absorbance occurring between the 1370 and 1380 cm^{-1}, which corresponds to the –C–H bending of an alkane methyl unit. A separate method is available for PAH analysis from water samples, specifically, ASTM D5412, which utilizes fluorescence spectroscopy (ASTM-D5412 1993). The PAHs absorb in the 200–400 nm range and fluoresce from 280 to 500 nm. As each molecule fluoresces at a specific wavelength, in addition to quantification, detailed information on the class, or the number of aromatic rings, of the PAH can also be discerned.

Standardized methods also exist for the analysis of soil and sediment samples. ASTM D5765 allows for the quantification of TPHs from soil and sediments (ASTM-D5765 2016). Here, after TPH extraction from the solid sample, either gravimetric or GC-FID analysis is performed to quantify the amount of contamination from petroleum hydrocarbons from leaching, water quality changes, or other process conditions, to name a few. ASTM D7363 is available for the quantification of PAHs from interstitial water in sediment using a combination of solid-phase microextraction (SPME) to absorb PAHs from the sample, followed by thermal desorption into the GC-MS (ASTM D7363 2013). Using SIM mode, ions specific to PAHs can be targeted for specific identification and quantification against standards. As can be seen, for soil and sediment samples, extensive extraction protocols are necessary prior to analysis.

ASTM D6209 is available for the determination of PAHs in either gaseous or particulate form from ambient air samples (ASTM-D6209 2013). In this example, the PAHs are trapped on a sorbent and then thermally desorbed into the GC inlet for GC-MS analysis.

While standardized ASTM methods are available for water, soil, and air analysis for TPH and PAH quantification, in some cases, these analyses are not fast enough. In an example from 1999 concerning the North Cape oil spill, a GC-MS method was developed for the quantification of both TPH and PAH from seawater with a single analysis (Reddy and Quinn 1999). SIM mode was used, as from the ASTM methods mentioned earlier, PAHs have unique ions allowing for the identification and quantification of these compounds based on the number of aromatic rings present. Likewise, TPHs have a unique ion, m/z 57, which can be used for the total, rather than individual groups as with the PAHs. This was the first time that GC-MS was used to quantitatively measure TPHs, and by doing so, a single analysis was developed for the quantification of both TPHs and PAHs from seawater.

Eventually, ASTM E2143 was developed, which allows for on-site rapid screening and characterization of PAHs and the aromatic hydrocarbons from TPHs using fluorescence spectroscopy (ASTM-E2143 2013). A portable fiber-optic fluorescence spectrometer is required for the remote sensing of soil and water samples. While this method doesn't quantify total TPHs and PAHs, it is a step in the right direction, as on-site monitoring can significantly reduce the amount of time needed to obtain the results (e.g. no sample collected, shipment, sample preparation, and finally analysis).

3.3.6 Tar Analysis

Tar is a hydrocarbon-based dark brown or black viscous liquid derived from petroleum and coal, among other sources. Tar is the heaviest of the petroleum fractions and can have a molecular weight range well beyond 2000 Da.

Unfortunately, due to the high boiling ranges and high molecular weight ranges of tars, a GC method is simply out of the question. As with the lighter petroleum fractions, distillation is also possible (Herod and Stokes 1990), though there is no standardized method available from ASTM. As with the distillation methods described earlier for lighter petroleum fractions, weight percentages in the petroleum fraction based on the boiling point ranges can be estimated, but no information on the PIONA groupings can be given. In terms of the PIONA groupings, ASTM D2007 is available for the quantification of characteristic groups in petroleum-based samples with an initial boiling point above 260 °C (ASTM-D2007 2016). These characteristic groups include paraffins, aromatics, and polar compounds, which can aid in the understanding of how the oil can affect the end product. In ASTM D2007, clay-gel absorption chromatography is used; when compared with modern chromatography, this approach is dated and time-consuming. A modern take to more quickly obtain the same quantitative group type information is to apply LC followed by gravimetric quantification (Radke et al. 1980). In this approach, the polar N-, S-, and O-containing compounds were trapped on a pre-column, and the paraffins and aromatics were then separated using a short LC separation; this was then followed by gravimetric analysis, where the weight percentages of the paraffins, aromatics, and heteroatoms were determined, and good agreement with ASTM D2007 was obtained.

A new thin-layer chromatography (TLC) method for petroleum fractions was reported by Karlsen and Larter, who used the Iatroscan thin-layer chromatography–flame ionization detector (TLC-FID) for the separation of paraffins, aromatics, oil resins, and asphaltenes in petroleum fractions (Karlsen and Larter 1991). With this approach, the sample was developed on TLC rods, also called Chromarods, which were then inserted into the Iatroscan TH-10 MK-4 unit and detected and quantified by FID. The results from the TLC-FID (Figure 3.10) were compared with the LC method described earlier, and a good correlation between the two techniques was obtained; for the North Sea core extracts and oils, a correlation coefficient of 0.94 was reported. Iatron Labs, Inc. still offers the TLC-FID unit, but they have also introduced the MK-6 unit, which contains both an FID and an FPD detector.

A completely different approach to obtain paraffin and aromatic quantification from tars and other heavy petroleum fractions is to utilize nuclear magnetic resonance (NMR) spectroscopy. Paraffins and aromatics have very different resonances in ^{1}H NMR, allowing for the quantitative determination of these group types (Stompel and Bartle 1983). ^{13}C NMR in combination with various NMR acquisition parameters can also be applied to the petroleum fractions, which affords detailed information into the carbon skeletal structure of the components present (Diaz and Blanco 2003; Fischer et al. 1978). In an example from Genetti et al., correlations were made between the ^{13}C NMR results, the elemental composition obtained from CHNS (carbon, hydrogen, nitrogen, sulfur), and volatile and ash compositions obtained from ASTM D3175 and D2415, respectively, to predict the chemical structure of coal tar pitch (ASTM-D2415 2015; ASTM-D3175 2018; Genetti et al. 1999). The developed model was able to accurately predict the total volatiles and tar yields solely from the NMR data, meaning that only one analytical technique was needed to provide a plethora of information.

The analytical techniques discussed for tar have focused on providing quantitative information on the characteristic groups in order to better understand any affects there might be on the final product. Another important aspect of tar analysis is the quantification of coke formation. Coke is a carbonaceous deposit that

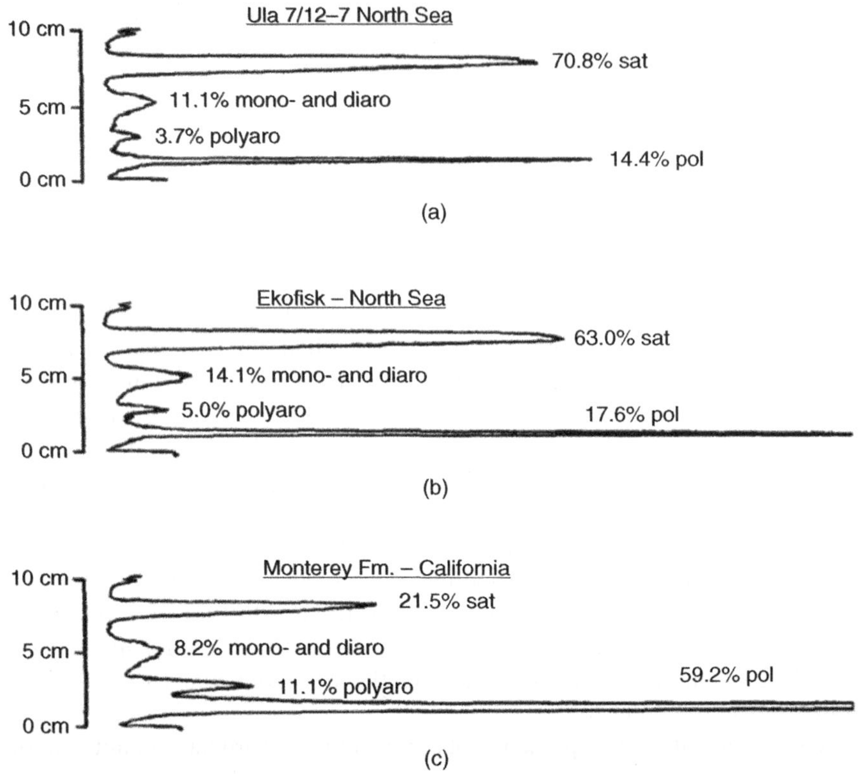

Figure 3.10 Separation and quantification of three oils on Chromarod-S III. Note the similar mono and diaromatic/polyaromatic ratio of the two North Sea oils (3.0 and 2.6 of the Ula (a) and Ekofisk (b) sample, respectively) and the drastically different Monterey Fm (c) sample, of which the off-scale polar ("asphaltene" and "resins") greatly exceed the saturated hydrocarbons. Source: Karlsen and Larter (1991). Reproduced with permission, Copyright (1991), Elsevier.

forms when a petroleum fraction is utilized as a feedstock and cracked to create other chemicals (Berry and Edgeworth-Johnstone 1944). As one can imagine, coke formation is undesired, because it not only represents a loss of product but also deposits in the cracker, which then requires an expensive de-coking step to remove it. Therefore, it is extremely useful to have information on the amount of total coke formed with the different petroleum fractions; ASTM D4715 provides quantitative data on the amount of coke formed from tar and pitch (ASTM-D4715 2017). It has long been known that asphaltenes are present in petroleum, but it has also been shown that asphaltenes are related to coke formation. Asphaltenes are the highest molecular weight compounds naturally occurring in petroleum, and they are the last molecules in petroleum fractions to combust completely. In a study on Brazilian crude oil distillation residues, it was shown how increasing the asphaltene content also increased the amount of coke formed (Goncalves et al. 2007). Therefore, quantitative information on the asphaltene content is another important parameter to consider with coke formation. ASTM D6560 utilizes distillation to isolate and then quantify the asphaltene content in various petroleum fractions (ASTM-D6560 2017).

3.3.7 Analysis of Heteroatoms and Heavy Metals

Up to this point, little attention has been placed on the analysis of the impurities in petroleum fractions; however, this information can be crucial for process control as well as for environmental hazards. Hydrocarbons containing heteroatoms, such as sulfur, nitrogen, or oxygen, can be present at varying levels, even up to percentage levels in petroleum. The presence of these heteroatoms originates from the degradation of the biological matrices (animal and vegetative) that make up petroleum. In addition to these impurities, trace levels (e.g. part per million, ppm) of inorganic heavy metals can also be present. The most common heavy metals identified in crude oil include vanadium, nickel, and iron; however, more than 40 trace elements have been detected (Nadkarni 2011).

3.3.7.1 Heteroatoms

For the analysis of heteroatoms, bulk analytical techniques are utilized to provide total sulfur, total nitrogen, and total oxygen content of the petroleum fraction. The benefit of bulk analyses is their simplicity and easy implementation in routine environments, which explains why several ASTM methods are available. As was observed with the characterization of the hydrocarbon composition, due to the wide range of petroleum fractions, several different methods are required for the analysis of the heteroatoms. Total sulfur in light hydrocarbons can be achieved by UV fluorescence (ASTM-D5453 2016) or in an online analysis for gaseous fuels using a chemical analyzer (ASTM-D7166 2015). For the heavier petroleum fractions, total sulfur measurements in coal and coke are accomplished using high-temperature tube furnace combustion with infrared absorption (ASTM-D5016 2016). The total nitrogen analysis for liquid hydrocarbons covers a much wider range than the total sulfur analysis; here, liquid hydrocarbons with a boiling range from 50 to 400 °C are analyzed by oxidative combustion followed by chemiluminescence detection (ASTM-D4629 2017). Total oxygen analysis has the most constriction in the range of applicable petroleum fractions. ASTM 5622 is limited to gasoline and methanol fuels, where total oxygen values are obtained from reductive pyrolysis (ASTM-D5622 2017). In addition to the limited total oxygen analysis, a more detailed quantification of specific oxygenates is possible through the use of ASTM 4815, which is used for the determination of methyl *tert*-butylether, ethyl *tert*-butylether, *tert*-amylmethylether, diisopropylether, methanol, ethanol, isopropanol, *n*-propanol, isobutanol, *tert*-butanol, *sec*-butanol, *n*-butanol, and *tert*-pentanol in gasoline by GC-FID (ASTM-D4815 2016).

3.3.7.2 Heavy Metals

While petroleum is mostly made up of organic compounds, inorganics are also present; heavy metals such as vanadium, nickel, iron, chromium, and molybdenum can be found at low levels in petroleum. These heavy metals reflect the geochemical conditions of petroleum source beds (Whitherspoon and Nagashima 1957). Even at low levels, these heavy metals can be detrimental to the cracking process as well as distillation of the petroleum fractions; during cracking, the heavy metals can deposit on the catalyst, and in distillation, they can be concentrated in the residue or even be transferred into the various distillation cuts, making their way into the final products (Powell 1954). As vanadium and nickel are found at the highest levels in petroleum, ranging from the ppb to ppm levels, these metals are of special significance; additionally, the vanadium/nickel ratio can be used to infer geochemical information, such as the source of the petroleum (Lewan and Maynard 1982). Elemental analyses using either X-ray fluorescence (XRF) or inductively coupled plasma (ICP) are the traditional methods to analyze heavy metals. In terms of a standardized method, ASTM D5708 is available, which utilizes ICP coupled to AED for the quantification of nickel, vanadium, and iron in crude oil (ASTM-D5708 2015).

3.3.8 Additional Analytical Applications for Petroleum

This chapter has focused on some of the more conventional analytical techniques available for the analysis of petroleum; however, this is by no means an exhaustive discussion of the topic. The advancement of analytical instrumentation is an ever-evolving process; this can especially be seen in the rapid advancement of not only the GC equipment but also the detector and column technology discussed herein.

The remaining chapters in this section are dedicated to the other analytical techniques available for the analysis of petroleum. The focus will be given on the application of advanced GC techniques, various LC possibilities, supercritical fluid chromatography, and online and process analytics for petroleum.

References

Adlard, E.R. and Poole, C.F. (2009). Gas chromatography: Historical development. In: *Handbook of Methods and Instrumentation in Separation Science*, vol. 1 (eds. I.D. Wilson and C.F. Poole), 259–265. London: Elsevier Ltd.

Amirav, A. and Jing, H. (1995). Pulsed flame photometer detector for gas chromatography. *Analytical Chemistry* 67: 3305–3318. https://doi.org/10.1021/ac00114a030.

ASTM-D1160 (2018). *Standard test method for distillation of petroleum products at reduced pressure*. West Conshohocken, PA: ASTM International. Retrieved from www.astm.org.

ASTM-D2007 (2016). *Standard test method for characteristic groups in rubber extender and processing oils and other petroleum-derived oils by the clay-gel absorption chromatographic method*. West Conshohocken, PA: ASTM International. Retrieved from www.astm.org.

ASTM-D2415 (2015). *Standard test method for ash in coal tar and pitch*. West Conshohocken, PA: ASTM International. Retrieved from www.astm.org.

ASTM-D2887 (2018). *Standard test method for boiling range distribution of petroleum fractions by gas chromatography*. West Conshohocken, PA: ASTM International. Retrieved from www.astm.org.

ASTM-D2892 (2018). *Standard test method for distillation of crude petroleum (15-theoretical plate column)*. West Conshohocken, PA: ASTM International. Retrieved from www.astm.org.

ASTM-D3175 (2018). *Standard test method for volatile matter in the analysis sample of coal and coke*. West Conshohocken, PA: ASTM International. Retrieved from www .astm.org.

ASTM-D4629 (2017). *Standard test method for trace nitrogen in liquid hydrocarbons by syringe/inlet oxidative combustion and chemiluminescence detection*. West Conshohocken, PA: ASTM International. Retrieved from www.astm.org.

ASTM-D4715 (2017). *Standard test method for coking value of tar and pitch (Alcan)*. West Conshohocken, PA: ASTM International. Retrieved from www.astm.org.

ASTM-D4815 (2016). *Standard test method for determination of MTBE, ETBE, TAME, DIPE, tertiary-amyl alcohol and C1 to C4 alcohols in gasoline by gas chromatography*. West Conshohocken, PA: ASTM International. Retrieved from www.astm.org.

ASTM-D5016 (2016). *Standard test method for total sulfur in coal and coke combustion residues using a high-temperature tube furnace combustion method with infrared absorption*. West Conshohocken, PA: ASTM International. Retrieved from www .astm.org.

ASTM-D5134 (2017). *Standard test method for detailed analysis of petroleum naphthas through n-nonane by capillary gas chromatography*. West Conshohocken, PA: ASTM International. Retrieved from www.astm.org.

ASTM-D5412 (1993). *Standard test method for quantification of complex polycyclic aromatic hydrocarbon mixtures or petroleum oils in water*. West Conshohocken, PA: ASTM International. Retrieved from www.astm.org.

ASTM-D5443 (2018). *Standard test method for paraffin, naphthene, and aromatic hydrocarbon type analysis in petroleum distillates through 200 °C by multi-dimensional gas chromatography*. West Conshohocken, PA: ASTM International. Retrieved from www.astm.org.

ASTM-D5453 (2016). *Standard test method for determination of total sulfur in light hydrocarbons, spark ignition engine fuel, diesel engine fuel, and engine oil by ultraviolet fluorescence*. West Conshohocken, PA: ASTM International. Retrieved from www.astm.org.

ASTM-D5708 (2015). *Standard test methods for determination of nickel, vanadium, and iron in crude oils and residual fuels by inductively coupled plasma (ICP) atomic emission spectrometry*. West Conshohocken, PA: ASTM International. Retrieved from www.astm.org.

ASTM-D5765 (2016). *Standard practice for solvent extraction of total petroleum hydrocarbons from soils and sediments using closed vessel microwave heating*. West Conshohocken, PA: ASTM International. Retrieved from www.astm.org.

ASTM-D6209 (2013). *Standard test method for determination of gaseous and particulate polycyclic aromatic hydrocarbons in ambient air (collection on sorbent-backed filters with gas chromatographic/mass spectrometric analysis)*. West Conshohocken, PA: ASTM International. Retrieved from www.astm.org.

ASTM-D6352 (2015). *Standard test method for boiling range distribution of petroleum distillates in boiling range from 174 °C to 700 °C by gas chromatography*. West Conshohocken, PA: ASTM International. Retrieved from www.astm.org.

ASTM-D6560 (2017). *Standard test method for determination of asphaltenes (heptane insolubles) in crude petroleum and petroleum products*. West Conshohocken, PA: ASTM International. Retrieved from www.astm.org.

ASTM-D5622 (2017). *Standard test methods for determination of total oxygen in gasoline and methanol fuels by reductive pyrolysis*. West Conshohocken, PA: ASTM International. Retrieved from www.astm.org.

ASTM-D664 (2018). *Standard test method for acid number of petroleum products by potentiometric titration*. West Conshohocken, PA: ASTM International. Retrieved from www.astm.org.

ASTM-D6729 (2014). *Standard test method for determination of individual components in spark ignition engine fuels by 100 metre capillary high resolution gas chromatography*. West Conshohocken, PA: ASTM International. Retrieved from www.astm.org.

ASTM-D6730 (2016). *Standard test method for determination of individual components in spark ignition engine fuels by 100-metre capillary (with precolumn) high-resolution gas chromatography*. West Conshohocken, PA: ASTM International. Retrieved from www.astm.org.

ASTM-D6733 (2016). *Standard test method for determination of individual components in spark ignition engine fuels by 50-metre capillary high resolution gas chromatography*. West Conshohocken, PA: ASTM International. Retrieved from www.astm.org.

ASTM-D6839 (2018). *Standard test method for hydrocarbon types, oxygenated compounds, and benzene in spark ignition engine fuels by gas chromatography*. West Conshohocken, PA: ASTM International. Retrieved from www.astm.org.

ASTM-D7166 (2015). *Standard practice for total sulfur analyzer based on-line/at-line for sulfur content of gaseous fuels*. West Conshohocken, PA: ASTM International. Retrieved from www.astm.org.

ASTM-D7363 (2013). *Standard test method for determination of parent and alkyl polycyclic aromatics in sediment pore water using solid-phase microextraction and gas chromatography/mass spectrometry in selected ion monitoring mode*. West Conshohocken, PA: ASTM International Retrieved from www.astm.org.

ASTM-D7678 (2017). *Standard test method for total oil and grease (TOG) and total petroleum hydrocarbons (TPH) in water and wastewater with solvent extraction using Mid-IR laser spectroscopy*. West Conshohocken, PA: ASTM International. Retrieved from www.astm.org.

ASTM-D86 (2018). *Standard test method for distillation of petroleum products and liquid fuels at atmospheric pressure*. West Conshohocken, PA: ASTM International. Retrieved from www.astm.org.

ASTM-E2143 (2013). *Standard test method for using field-portable fiber optics synchronous fluorescence spectrometer for quantification of field samples for aromatic and polycyclic aromatic hydrocarbons*. West Conshohocken, PA: ASTM International. Retrieved from www.astm.org.

Atar, E., Cheskis, S., and Amirav, A. (1991). Pulsed flame – a novel concept for molecular detection. *Analytical Chemistry* 63: 2061–2064. https://doi.org/10.1021/ac00018a031.

Bartle, K.D. and Myers, P. (2002). History of gas chromatography. *Trends in Analytical Chemistry* 21: 547–557. https://doi.org/10.1016/S0165-9936(02)00806-3.

Beach, C.A., Krumm, C., Spanjers, C.S. et al. (2016). Quantitative carbon detector for enhanced detection of molecules in foods, pharmaceuticals, cosmetics, flavors, and fuels. *Analyst* 141: 1627–1632. https://doi.org/10.1039/c5an02552d.

Berry, A.G.V. and Edgeworth-Johnstone, R. (1944). Petroleum coke. Formation and properties. *Analytical Chemistry* 36: 1140–1144. https://doi.org/10.1021/ie50420a015.

Blomberg, L.G. (2001). Silicone stationary phases for gas chromatography. *LCGC Europe* February: 1–6.

Blomberg, J., Schoenmakers, P.J., and Brinkman, U.A.T. (2002). Gas chromatographic methods for oil analysis. *Journal of Chromatography A* 972: 137–173. https://doi.org/10.1016/S0021-9673(02)00995-0.

Boer, H. (1957). A comparison of detection methods for gas chromatography including detection by beta ray ionization. In: *Vapor Phase Chromatography* (eds. D.H. Desty and C.L.A. Harbourn), 169–184. London: Butterworths Scientific Publications.

Boer, H. and van Arkel, P. (1971). An automatic PNA analyzer for (heavy) naphtha. *Chromatographia* 4: 300–308. https://doi.org/10.1007/BF02282905.

Bruderreck, H., Schneider, W., and Halasz, I. (1964). Quantitative gas chromatographic analysis of hydrocarbons with capillary columns and flame ionization detector. *Analytical Chemistry* 36: 461–473. https://doi.org/10.1021/ac60209a044.

Burgett, C.A., Smith, D.H., and Bente, H.B. (1977). The nitrogen-phosphorus detector and its applications in gas chromatography. *Journal of Chromatography A* 134: 57–64. https://doi.org/10.1016/S0021-9673(00)82569-8.

Clemente, J.S., MacKinnon, M.D., and Fedorak, P.M. (2004). Aerobic biodegradation of two commercial naphthenic acids preparations. *Environmental Science and Technology* 38: 1009–1016. https://doi.org/10.1021/es030543j.

Condon, R.D. (1959). Design considerations of gas chromatography system employing high efficiency Golay columns. *Analytical Chemistry* 31: 1717–1722. https://doi.org/10.1021/ac60154a054.

Dandeneau, R.D. and Zerenner, E.H. (1979). An investigation of glasses for capillary chromatography. *Journal of High Resolution Chromatography* 2: 351–356. https://doi.org/10.1002/jhrc.1240020617.

Daynes, H.A., Shakespear, G.A., and Smith, S.W.J. (1920). The theory of katharometer. *Proceedings of the Royal Society of London Series A* 97: 273–286. https://doi.org/10.1098/rspa.1920.0033.

Desai, B.K. and Sahgal, P.N. (1975). Dual flame thermocouple detector for gas chromatography. *Journal of Chromatography* 104: 47–49. https://doi.org/10.1016/S0021-9673(01)85485-6.

Desty, D.H. and Whyman, H.F. (1957). Application of gas-liquid chromatography to analysis of liquid petroleum fractions. *Analytical Chemistry* 29: 320–329. https://doi.org/10.1021/ac60123a001.

Desty, D.H., Haresnape, J.N., and Whyman, B.H.F. (1960). Construction of long lengths of coiled glass capillary. *Analytical Chemistry* 32: 302–304. https://doi.org/10.1021/ac60158a057.

Diaz, C. and Blanco, C.G. (2003). NMR: A powerful tool in the characterization of coal tar pitch. *Energy and Fuels* 17: 907–913. https://doi.org/10.1021/ef020114r.

Dunkle, M.N., Pijcke, P., Winniford, B., and Bellos, G. (2019). Quantification of the composition of liquid hydrocarbon streams: Comparing the GC-VUV to DHA and GC × GC. *Journal of Chromatography A* 1587: 239–246. https://doi.org/10.1016/j.chroma.2018.12.026.

Dzidic, I., Somerville, A.C., Raia, J.C., and Hart, H.V. (1988). Determination of naphthenic acids in California crudes and refinery wastewaters by fluoride ion chemical ionization mass spectrometry. *Analytical Chemistry* 60: 1318–1323. https://doi.org/10.1021/ac00164a015.

Eggertsen, F.T., Groennings, S., and Holst, J.J. (1960). Analytical distillation by gas chromatography. *Analytical Chemistry* 32: 904–909. https://doi.org/10.1021/ac60164a002.

Ettre, L.S. (2004). The evolution of the application of gas chromatography. *LCGC Europe* 17: 292–299.

Ettre, L.S. and Horvath, C. (1975). Foundation of modern liquid chromatography. *Analytical Chemistry* 47: 422A–446A. https://doi.org/10.1021/ac60354a041.

Fischer, P., Stadelhofer, J.W., and Zander, M. (1978). Structural investigation of coal-tar pitches and coal extracts by ^{13}C NMR spectroscopy. *Fuel* 57: 345–352. https://doi.org/10.1016/0016-2361(78)90171-0.

Genetti, D., Fletcher, T.H., and Pugmire, R.J. (1999). Development and application of a correlation of ^{13}C NMR chemical structural analyses of coal based on elemental composition and volatile matter content. *Energy and Fuels* 13: 60–68.

Gohlke, R.S. (1959). Time-of-flight mass spectrometry and gas-liquid partition chromatography. *Analytical Chemistry* 31: 535–541. https://doi.org/10.1021/ac50164a024.

Gohlke, R.S. and McLafferty, F.W. (1993). Early gas chromatography/mass spectrometry. *Journal of the American Society for Mass Spectrometry* 4: 367–371. https://doi.org/10.1016/1044-0305(93)85001-E.

Golay, M.J.E. (1957a). A performance index for gas chromatographic columns. *Nature* 180: 435–436. https://doi.org/10.1038/180435b0.

Golay, M.J.E. (1957b). Vapor phase chromatography and Telegrapher's equation. *Analytical Chemistry* 29: 928–932. https://doi.org/10.1021/ac60126a019.

Goncalves, M.L.A., Ribeiro, D.A., Teixeira, A.M.R.F., and Teixeira, M.A.G. (2007). Influence of asphaltenes on coke formation during the thermal cracking of different Brazilian distillation residues. *Fuel* 86: 619–623. https://doi.org/10.1016/j.fuel.2006.08.022.

Grant, D.W. (1958). An emissivity detector for gas chromatography. In: *Gas Chromatography* (ed. D.H. Desty), 153–177. Washington, DC: Butterworths.

Hála, S., Eyem, J., Burkhard, J., and Landa, S. (1970). Retention indices of adamantanes. *Journal of Chromatographic Science* 8: 203–209. https://doi.org/10.1093/chromsci/8.4.203.

Halász, I. and Schneider, W. (1961). Quantitative gas chromatographic analysis of hydrocarbons with capillary column and flame ionization detector. *Analytical Chemistry* 33: 978–982. https://doi.org/10.1021/ac60176a034.

Halász, I. and Schreyer, G. (1960). Erfahrungen mit Kapillarkolonnen-Flammenionisationsdetektor-Systemen bei der Gas-Chromatographie. *Chemie Ingenieur Technik* 32: 675–685. https://doi.org/10.1002/cite.330321007.

Harley, J., Nel, W., and Pretorius, V. (1958). Flame ionization detector for gas chromatography. *Nature* 181: 177–178. https://doi.org/10.1038/181177a0.

Herod, A.A. and Stokes, B.J. (1990). Separation schemes for coal and tar analysis – A comparison of methods. *Fuel Processing Technology* 24: 45–51. https://doi.org/10.1016/0378-3820(90)90041-P.

Holowenko, F.M., MacKinnon, M.D., and Fedorak, P.M. (2002). Characterization of naphthenic acids in oil sands wastewaters by gas chromatography-mass spectrometry. *Water Research* 36: 2843–2855. https://doi.org/10.1016/S0043-1354(01)00492-4.

James, A.T. and Martin, A.J.P. (1952). Gas-liquid partition chromatography: The separation and micro-estimation of volatile fatty acids from formic acid to dodecanoic acid. *Biochemical Journal* 50: 679–690. https://doi.org/10.1042/bj0500679.

Johansen, N.G., Ettre, L.S., and Miller, R.L. (1983). Quantitative analysis of hydrocarbons by structural group type in gasoline and distillates: I. Gas chromatography. *Journal of Chromatography* 256: 393–417. https://doi.org/10.1016/S0021-9673(01)88258-3.

Karlsen, D.A. and Larter, S.R. (1991). Analysis of petroleum fractions by TLC-FID: Applications to petroleum reservoir description. *Organic Geochemistry* 17: 603–617. https://doi.org/10.1016/0146-6380(91)90004-4.

Kelly, T.J., Gaffney, J.S., Phillips, M.F., and Tanner, R.L. (1983). Chemiluminescent detection of reduced sulfur compounds with ozone. *Analytical Chemistry* 55: 135–138. https://doi.org/10.1021/ac00252a034.

Kong, R.C. and Lee, M.L. (1983a). Static coating of capillary columns using mixed solvent. *Chromatographia* 17: 451–453. https://doi.org/10.1007/BF02262929.

Kong, R.C. and Lee, M.L. (1983b). Static coating of small diameter capillary columns at elevated temperatures. *Journal of High Resolution Chromatography* 6: 319–321. https://doi.org/10.1002/jhrc.1240060607.

Kováts, E. and Keulemans, A.I.M. (1964). The Kováts retention index system. *Analytical Chemistry* 36: 31A–41A. https://doi.org/10.1021/ac60214a727.

Kováts, E. (1958). Gas-chromatographische Charakterisierung organischer Verbindung en Teil 1: Retentionsindices aliphatischer Halogenide, Alkohole, Aldehyde und Ketone. *Helvetica Chimica Acta* 41: 1915–1932. https://doi.org/10.1002/hlca.19580410703.

Lai, W.C. and Song, C. (1995). Temperature-programmed retention indices for G.C. and G.C.-M.S. analysis of coal- and petroleum-derived liquid fuels. *Fuel* 74: 1436–1451. https://doi.org/10.1016/0016-2361(95)00108-H.

Lee, M.L., Kong, R.C., Woolley, C.L., and Bradshaw, J.S. (1984). Fused silica capillary column technology for gas chromatography. *Journal of Chromatographic Science* 22: 136–142. https://doi.org/10.1093/chromsci/22.4.136.

Lewan, M.D. and Maynard, J.B. (1982). Factors controlling enrichment of vanadium and nickel in the bitumen of organic sedimentary rocks. *Geochimica et Cosmochimica Acta* 46: 2546–2560. https://doi.org/10.1016/0016-7037(82)90377-5.

Lovelock, J.E. (1958). A sensitive detector for gas chromatography. *Journal of Chromatography A* 1: 35–46. https://doi.org/10.1016/S0021-9673(00)93398-3.

Lovelock, J.E. and Lipsky, S.R. (1960). Electron affinity spectroscopy – A new method for the identification of functional groups in chemical compounds separated by gas chromatography. *Journal of the American Chemical Society* 83: 431–433. https://doi.org/10.1021/ja01487a045.

Maduskar, S., Teixeira, A.R., Paulsen, A.D. et al. (2015). Quantitative carbon detector (QCD) for calibration-free, high-resolution characterization of complex mixtures. *Lab on a Chip* 15: 440–447. https://doi.org/10.1039/C4LC01180E.

Martin, A.J.P. and James, A.T. (1956). Gas-liquid chromatography: The gas-density meter, a new apparatus for the detection of vapours in flowing gas streams. *Biochemical Journal* 63: 138–143. https://doi.org/10.1042/bj0630138.

McNair, H. (2010). A history of gas chromatography: My early experiences. *LCGC Europe* 28: 138–144.

McWilliam, I.G. and Dewar, R.A. (1958). Flame ionization detector for gas chromatography. *Nature* 181: 760. https://doi.org/10.1038/187060a0.

Nadkarni, R.A.K. (2011). Elemental analysis of crude oils using spectroscopic methods. In: *Spectroscopic Analysis of Petroleum Products and Lubricants* (ed. R.A.K. Nadkarni), 605–625. West Conshohocken, PA: ASTM International.

Neff, J.M., Stout, S.A., and Gunstert, D.G. (2005). Ecological risk assessment of polycyclic aromatic hydrocarbons in sediments: identifying sources and ecological hazard. *Integrated Environmental Assessment and Management* 1: 22–33. https://doi.org/10.1897/IEAM_2004a-016.1.

Nerheim, A.G. (1963). A gas-density detector for gas chromatography. *Analytical Chemistry* 35: 1640–1644. https://doi.org/10.1021/ac60204a028.

Patterson, P.L., Howe, R.L., and Abu-Shumays, A. (1978). A dual-flame photometric detector for sulfur and phosphorus compounds in gas chromatograph effluents. *Analytical Chemistry* 50: 339–344. https://doi.org/10.1021/ac50024a042.

Penzias, G.J. (1973). Gas chromatograph peaks identified on-line by a new grating infrared spectrophotometer. *Analytical Chemistry* 45: 890–895. https://doi.org/10.1021/ac60328a002.

Peters, C.A., Knightes, C.D., and Brown, D.G. (1999). Long-term composition dynamics of PAH-containing NAPLs and implications for risk assessment. *Environmental Science and Technology* 33: 1499–1507. https://doi.org/10.1021/es981203e.

Piringer, O. and Pascalau, M. (1962). Ein neuer Detektor für die Gaschromatographie. *Journal of Chromatography* 8: 410–412. https://doi.org/10.1016/S0021-9673(01)99277-5.

Powell, R. (1954). Removal of metal components from petroleum oils. US Patent No. US2778777A, filed 16 February 1954.

Quimby, B.D. and Sullivan, J.J. (1990). Evaluation of a microwave cavity, discharge tube, and gas flow system for combined gas chromatography-atomic emission detection. *Analytical Chemistry* 62: 1027–1034. https://doi.org/10.1021/ac00209a013.

Rabel, F. (2000). Chromatography thin-layer (planar) layers. In: *Encyclopedia of Separation Science* (eds. I. Wilson, C. Poole and M. Cooke), 853–860. Salt Lake City, UT: Academic Press.

Radke, M., Willsch, H., and Welte, D.H. (1980). Preparative hydrocarbon group type determination by automated medium pressure liquid chromatography. *Analytical Chemistry* 52: 406–411. https://doi.org/10.1021/ac50053a009.

Rahman, M.M., El-Aty, A.M.A., Choi, J.H. et al. (2015). Basic overview on gas chromatography columns. In: *Analytical Separation Science* (eds. J.L. Anderson, A. Berthod, V.P. Estevez and A.P. Stalcup), 823–834. Weinheim, Germany: Wiley-VCH.

Ray, N.H. (1954). Gas chromatography. I. The separation and estimation of volatile organic compounds by gas-liquid partition chromatography. *Journal of Applied Chemistry* 4: 21–25. https://doi.org/10.1002/jctb.5010040106.

Reddy, C.R. and Quinn, J.G. (1999). GC-MS analysis of total petroleum hydrocarbons and polycyclic aromatic hydrocarbons in seawater samples after the North Cape oil spill. *Marine Pollution Bulletin* 38: 126–135. https://doi.org/10.1016/S0025-326X(98)00106-4.

Robbins, W.B. and Caruso, J.A. (1979). Development of hydride generation methods for atomic spectroscopic analysis. *Analytical Chemistry* 51: 889A–898A. https://doi.org/10.1021/ac50044a002.

Rossini, F.D. and Mair, B.J. (1959). The work of the API research project 6 on the composition of petroleum. Retrieved from OnePetro: https://www.onepetro.org (accessed 19 March 2020).

Schneider, A., Warren, R.W., and Janoski, E.J. (1966). Formation of perhydrophenalenes and polyalkyladamantanes by isomerization of tricyclic perhydroaromatics. *Journal of Organic Chemistry* 35: 1617–1625. https://doi.org/10.1021/jo01343a070.

Schug, K.A., Sawicki, I., Carlton, D.D. Jr. et al. (2014). Vacuum ultraviolet detector for gas chromatography. *Analytical Chemistry* 86: 8329–8335. https://doi.org/10.1021/ac5018343.

Scott, R.P.W. (1955). A new detector for vapour-phase partition chromatography. *Nature* 176: 793. https://doi.org/10.1038/176793a0.

Scott, A.C., Mackinnon, M.D., and Fedorak, P.M. (2005). Naphthenic acids in athabasca oil sands tailings waters are less biodegradable than commercial naphthenic acids. *Environmental Science and Technology* 39: 8388–8394. https://doi.org/10.1021/es051003k.

Sevcik, J. and Krysl, S. (1973). A photoionization detector. *Chromatographia* 6: 375–380. https://doi.org/10.1007/BF02270573.

Speight, J.G. (2001). *Handbook of Petroleum Analysis*. Hoboken, NJ: Wiley.

Stompel, Z.J. and Bartle, K.D. (1983). Structural analysis of tars from fluidized bed pyrolysis of coal: 2. ^{1}H NMR and IR spectroscopy. *Fuel* 63: 900–904. https://doi.org/10.1016/0016-2361(83)90156-4.

Streiff, A.J., Zimmerman, J.C., Soule, L.F. et al. (1948). Purification, purity, and freezing points of 30 hydrocarbons of the API-standard and API-NBS series. *Journal of Research of the National Bureau of Standards* 41: 323–357.

Teng, S.T., Williams, A.D., and Urdal, K. (1994). Detailed hydrocarbon analysis for gasoline by GC-MS (SI-PIONA). *Journal of Separation Science* 17: 469–475. https://doi.org/10.1002/jhrc.1240170601.

Trestianu, S., Zilioli, G., Sironi, A. et al. (1985). Automatic simulated distillation of heavy petroleum fractions up to 800 °C TBP by capillary gas chromatography. Part I: Possibilities and limits of the method. *Journal of High Resolution Chromatography* 8: 771–781. https://doi.org/10.1002/jhrc.1240081113.

Vasnin, S.V., Wentworth, W.E., Stearns, S.D., and Meyer, C.J. (1992). Pulsed discharge emission detector – Application to analytical spectroscopy of permanent gases. *Chromatographia* 34: 226–234. https://doi.org/10.1007/BF02268350.

Ven Horst, S.H., Ven Horst, H., and O'Connor, K. (1960). A technique for packing columns for gas chromatography. *Journal of Chemical Education* 37: 593. https://doi.org/10.1021/ed037p593.

Verner, P. (1984). Photoionization detection and its application in gas chromatography. *Journal of Chromatography* 300: 249–264. https://doi.org/10.1016/S0021-9673(01)87587-7.

Visser, T. (2002). FT-IR detection in gas chromatography. *TrAC - Trends in Analytical Chemistry* 21: 627–636. https://doi.org/10.1016/S0165-9936(02)00812-9.

Wang, Z., Fingas, M., and Sergy, G. (1994). Study of 22-year-old Arrow oil samples using biomarker compounds by GC/MS. *Environmental Science and Technology* 28: 1733–1746. https://doi.org/10.1021/es00058a027.

Wentworth, W.E., Vasnin, S.V., Stearns, S.D., and Meyer, C.J. (1992). Pulsed discharge helium ionization detector. *Chromatographia* 34: 219–225. https://doi.org/10.1007/BF02268349.

Whitherspoon, P.A. and Nagashima, K. (1957). Use of trace metals to identify Illinois crude oils. In: Illinois State Geological Survey Circular 239. Retrieved from https://core.ac.uk/download/pdf/17355269.pdf.

Whittemore, I.M. (1979). High resolution gas chromatography of the gasolines and naphthas. In: *Chromatography in Petroleum Analysis* (eds. K.H. Altgelt and T.H. Gouw), 50–70. New York, NY: Marcel Dekker, Inc.

Wingert, W.S. (1992). GC-MS analysis of diamondoid hydrocarbons in Smackover petroleums. *Fuel* 71: 37–43. https://doi.org/10.1016/0016-2361(92)90190-Y.

Woolley, C.L., Markides, K.E., Lee, M.L., and Bartle, K.D. (1986). Deactivation of small diameter fused silica capillary columns with organosilicon hydrides. *Journal of High Resolution Chromatography* 9: 506–514. https://doi.org/10.1002/jhrc.1240090904.

Zadro, S., Haken, J.K., and Pinczewski, W.V. (1985). Analysis of Australian crude oils by high-resolution capillary gas chromatography–mass spectrometry. *Journal of Chromatography* 323: 305–322. https://doi.org/10.1016/S0021-9673(01)90393-0.

Zlatkis, A. and Lovelock, J.E. (1959). Gas chromatography of hydrocarbons using capillary columns and ionization detectors. *Analytical Chemistry* 31: 620–621. https://doi.org/10.1021/ac50164a051.

4

Advanced Analytics for the Evaluation of Oil, Natural Gas, and Shale Oil/Gas

Emmie Dumont[1], Pat Sandra[1], Kyra A. Murrell[2], Frank L. Dorman[3], Allegra Leghissa[4], and Kevin A. Schug[5]

[1] *Research Institute for Chromatography, President Kennedypark 26, Kortrijk, Belgium*
[2] *Department of Chemistry, The Pennsylvania State University, 77 Pollock Road, University Park, PA, USA*
[3] *Department of Biochemistry and Molecular Biology, The Pennsylvania State University, 107 Althouse Lab, University Park, PA, USA*
[4] *Application Development, Shimadzu France, Le Luzard II, Bat A, Bd Salvador Allende Noisiel, Marne-la-Vallée, France*
[5] *Department of Chemistry and Biochemistry, The University of Texas at Arlington, 700 Planetarium Pl., Arlington TX, USA*

4.1 IRMS in the Oil and Gas Industry

4.1.1 IRMS: General

4.1.1.1 Introduction

The determination of the isotopic ratio of a given element can provide information on geographic, chemical, or biological origin. Isotope ratios of an element can become depleted or enriched through several kinetic and thermodynamic processes (isotopic fractionation), and the measurement of isotopic ratios enables one to distinguish samples with identical chemical composition. The subtle differences in isotope values differentiate between samples/compounds (isotopic fingerprint) (Kelly et al. 2018). Conventional mass spectrometers are neither sensitive nor precise enough to be able to detect the differences in natural isotopic ratios; the technique capable of tackling this is isotope ratio mass spectrometry (IRMS).

Relative differences in isotopic ratios can be measured far more precisely than absolute differences. The delta notation (δ) was introduced in order to report data on stable isotopes. The δ value is given by Eq. (4.1):

$$\delta = \frac{(R_x - R_{\text{std}}) \times 1000}{R_{\text{std}}} \tag{4.1}$$

Analytical Techniques in the Oil and Gas Industry for Environmental Monitoring, First Edition. Edited by Melissa N. Dunkle and William L. Winniford.

where R is the ratio of the heavier isotope to the lighter isotope (i.e. $^{13}C/^{12}C$, D/H, $^{18}O/^{16}O$, and $^{15}N/^{14}N$) and R_x is the ratio for the sample, whereas R_{std} is the ratio of the standard. Delta (δ) values are reported in per mil (‰) and can be either positive or negative. A positive δ value means that the sample is enriched in the heavier isotope compared with the standard; a negative δ value means that the sample is depleted in the heavier isotope, compared with the standard. Isotopic compositions of samples are measured relative to the isotopic composition of a reference gas (CO_2, H_2, N_2) that is introduced in the mass spectrometer (Muccio and Jackson 2009; Sharp 2006).

The precise comparison of the isotopic compositions is necessary and is straightforward within a laboratory. Comparing results between different laboratories is more complicated as the isotopic composition of the reference gas differs. Therefore, international standards, referred to as primary reference materials, standard mean ocean water (SMOW), atmospheric air, and PeeDeeBelemnite (PDB), are used worldwide in order to compare δ values.

4.1.1.2 Isotopic Fingerprint

Carbon flows between each reservoir in an exchange called the carbon cycle, which has slow and fast components. Any change in the cycle that shifts carbon out of one reservoir puts more carbon into the other reservoirs. The carbon from a certain reservoir is characterized by a certain isotope ratio. Shifts from one reservoir to the other implement shifts in the isotope ratio, and hence a shift in the δ value. These differences in isotope ratios observed through either anthropogenic activity (e.g. fossil fuel burning) or natural processes (e.g. volcanic eruptions) can be monitored through the measurement of the isotope ratios. Figure 4.1 shows a schematic overview of the carbon cycle and an indication of $\delta^{13}C$ values (Schoemaker 2010).

Natural processes result in broad variations in the D/H ratio found in different pools of hydrogen. These variations originate from kinetic isotope effects and physical changes, such as precipitation and evaporation. Ocean water varies slightly, between 0‰ and −10‰, while atmospheric water can be found to vary between approximately −200‰ to +100‰. Biomolecules synthesized by organisms will retain some of the D/H signature of the water which they were grown in, plus a large fractionation factor, which can be as great as several hundred per mil. The large variation of δD makes the measurement of isotope ratios of D/H a valuable extra tool in isotopic fingerprinting (Fogel and Hoering 1980).

Plants can be classified into three categories according to the photosynthetic pathway. The most common photosynthetic pathway is the Calvin cycle metabolism (C_3 pathway), typical for trees (C_3 plants). The second photosynthetic pathway is the Hatch–Slack pathway (C_4 pathway), typical for grasses (C_4 plants). The third photosynthetic pathway is the Crassulacean acid metabolism (CAM).

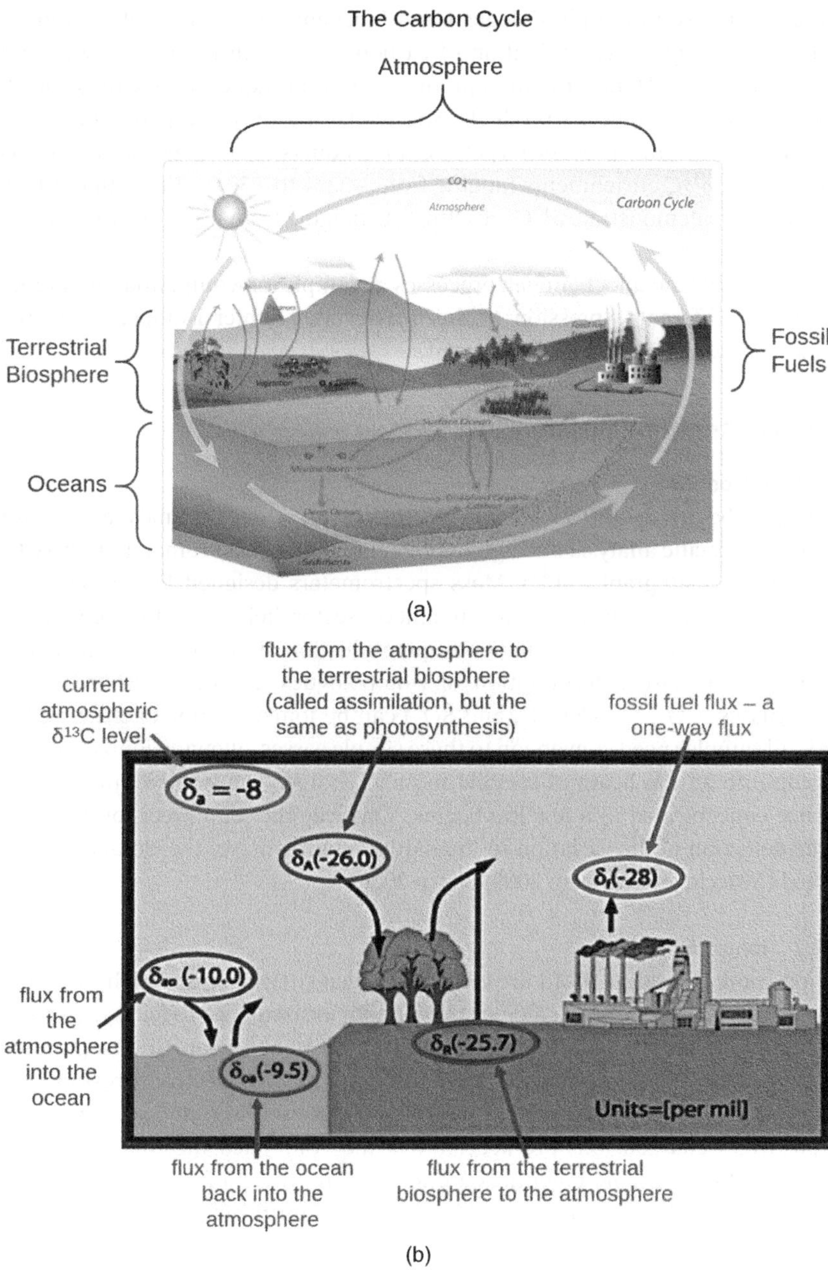

Figure 4.1 Carbon cycle (a) and an indication of $\delta^{13}C$ values (b). Source: Reproduced from Schoemaker (2010); Images provided by the National Oceanic and Atmospheric Administration (NOAA) ESRL Global Monitoring Division as part of NOAA's Ernest F. Hollings Scholarship Program.

Differences in the initial physiological and biochemical reactions of C_3 and C_4 plants result in large discrimination of carbon isotopes in the plant tissue and derivatives thereof. Hence, the final products carry a unique isotopic fingerprint, whether it be plant-derived products, species taking up these plants or plants being used in further processes (petroleum products). For example, C_3 plants typically have $\delta^{13}C$ enrichments ranging from −22‰ to −33‰. Plants that utilize C_4 metabolism demonstrate $\delta^{13}C$ enrichments ranging from −8‰ to −16‰ (Kelly et al. 2018).

All chemical and biochemical processes a sample or compound undergoes results in a different expression of the heavier and lighter isotope, giving the sample/compound its isotope fingerprint.

4.1.2 IRMS: The Technique

4.1.2.1 Introduction

IRMS can be used for bulk analysis (using an element analyzer) or for compound-specific analysis through hyphenation with gas chromatography (GC) or liquid chromatography (LC). Mass spectrometers designed for isotope-ratio measurements use dedicated static magnetic sector fields and multicollection with Faraday cups for the simultaneous quantification of the different ion beams, allowing the precise determination of isotopologues. In the ion source, only simple gasses (CO_2, N_2, CO, H_2, and SO_2) can be ionized. Prior to entering the IRMS, all samples need conversion to these simple gasses, allowing ionization and focusing into an ion beam of specific mass to charge. The isotopic information is hence concentrated in a few ion beams. This way the high precision required for the detection of the variation in the natural abundance of the isotopes can be reached (Muccio and Jackson 2009; Sharp 2006).

4.1.2.2 Ionization

The ion sources used in IRMS are electron impact (EI) sources, which provide a high ion production yield. Molecules that are not ionized need to leave the source fast enough in order to maintain the performance when coupled to separation techniques. Molecules are ionized by bombardment with electrons, resulting in the formation of positively charged ions. After extraction from the source region, the positively charged ions are accelerated and collimated through a series of lenses, to which variable potentials are applied in order to focus the ion beam. The ion beam then enters the magnetic sector analyzer through the entrance slit.

4.1.2.3 Mass Analyzer

For isotope ratio measurements, magnetic sectors are typically used. Besides the ability to quickly switch the magnetic field for the analysis of H_2 to CO, of N_2 to

CO_2, or CO_2 to SO_2, background ions (e.g. CH_4 and H_2O) functioning as quality indicators of conversion and performance of the chemical traps can be monitored. The analyzer is designed to provide low-mass resolution of 1 amu. Ions with the same kinetic energy and charge are deflected in relation to their masses. Hence, an ion with a small mass is deflected more, meaning that its trajectory has a smaller radius compared to an ion with a higher mass.

4.1.2.4 Detection

The most commonly used collector array is the universal triple collector, which is a combination of two wide Faraday cups and one smaller Faraday cup positioned in the middle. Each of the Faraday cups has its own amplifier. This combination allows the measurement of all isotope ratios on CO_2, N_2, CO, and SO_2. In case hydrogen measurement is mandatory, two additional Faraday cups are added, one on each side. The ions pick up electrons at the collectors, creating electric currents. The generated voltages are proportional to the amount of gas that enters the cup, and the differences in the relative voltages are recorded as per mil and expressed through the δ value.

4.1.2.5 Referencing

The measurement of isotope ratios requires that sample gasses need to be measured relative to a reference gas, of which the isotope ratio is known. A cylinder of calibrated reference gas can be used for long periods, and typically at the beginning and at the end of an analysis, fixed amounts of the gas are entered in the IRMS. Delta (δ) values from samples are then calculated relative to the reference gas. Standard material is available to compare interlaboratory results (SMOW, PDB).

4.1.2.6 Bulk Analysis

An elemental analyzer–isotope ratio mass spectrometry (EA-IRMS) allows the determination of the isotopic ratios of an entire sample. Samples as such, which are organic or inorganic compounds or complex mixtures thereof, first need to be converted into simple gases: CO_2, N_2, SO_2, CO, and H_2, prior to being introduced in the IRMS. In the case of bulk analysis, a defined amount of sample is weighed and placed into a combustion furnace. Here, the sample is combusted in the presence of oxygen (for N_2, CO_2, or SO_2) or pyrolyzed in a reductive environment (for H_2 and CO); nevertheless, the final products might be further treated to eliminate interferences. The combusted products are directed toward a reduction chamber in order to convert nitrous oxide to N_2 and to remove excess O_2, after which the effluent is carried through a stream of He toward the IRMS.

4.1.3 Compound Specific IRMS

4.1.3.1 Introduction

By hyphenating a separation technique (either GC or LC) to an IRMS instrument, the isotopic composition of individual compounds can be examined in a complex mixture, allowing for unique additional isotope information to be obtained. Continuous flow IRMS allows the transfer of separated compounds to the IRMS; the use of an open-split interface with an He carrier gas stream continuously transfers the sample gas from the GC or LC toward the IRMS. However, the samples from the separation cannot be directly introduced in the ion source and need conversion to simple gases. The necessary interface must be capable of not only quantitatively converting every eluting compound, but the chromatographic resolution also needs to be maintained. A comparison between the instrumental set-up in bulk analyses (elemental analyzer [EA]-interface) and compound-specific analyses (LC- and GC-interfaces) is presented in Figure 4.2 (Georgiou and Danezis 2015).

4.1.3.2 GC-IRMS

In order to determine isotope ratios of compounds eluting from a GC column, an interface is needed. Although compounds are already in the gas phase, they

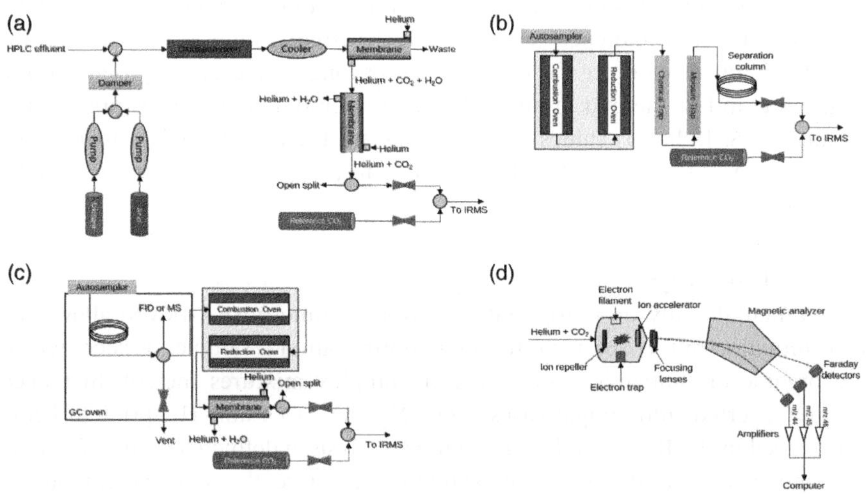

Figure 4.2 Comparison of the instrumental set-up in bulk analyses (EA-interface) and compound-specific analyses (LC- and GC-interface) (a) LC-interface. (b) EA-interface. (c) GC-interface. (d) isotope ratio mass spectrometry (IRMS). Source: Adapted from Georgiou and Danezis (2015).

need conversion to CO_2, N_2, H_2, or O_2 prior to entering the ion source, while maintaining chromatographic resolution. For GC coupling to IRMS, an oven is used, housing a reactor for the combustion process (C) and a reactor for high-temperature conversion (HTC). Samples can be introduced using a variety of injectors, including liquid autosampler, headspace injection, heated headspace injection, solid phase microextraction, thermal desorption, purge, and trap, to name a few. After sample introduction, compounds are separated on the GC column and directed toward the reactor of interest. Additionally, prior to entering the reactors, the flow can be split in order to send part of the sample toward an additional detector (e.g. flame ionization detector [FID], quadrupole mass spectrometric detection, etc.). The use of an additional detector can be beneficial when the parallel acquisition of structural information is required, as the use of combustion or HTC erases all structural information.

For $\delta^{13}C$ and $\delta^{15}N$ determination, the compounds eluting from the GC column are oxidized in a reactor at temperatures of typically 940–1000 °C, converting the compounds into CO_2, NO_x, and H_2O, where the latter is usually removed on-line through a water-permeable polymeric capillary. Any formed NO_x is immediately reduced in the capillary reactor. For the determination of $^{15}N/^{14}N$, all CO_2 is retained in a liquid nitrogen trap, to avoid isobaric interferences of CO from cracked CO_2, before being transferred to the isotope ratio mass spectrometer.

For the determination of δ^2H and $\delta^{18}O$, a quantitative conversion by pyrolysis in an HTC oven is needed to convert the organic O and H into CO and H_2. This requires an inert and reductive environment at very high temperatures. For $^{18}O/^{16}O$ determination, the reaction takes place in an inert platinum inlay placed inside a ceramic tube, which is used to protect against air and for stability. The reaction can be performed at 1280 °C due to the catalytic properties of platinum; however, for the measurement of $^2H/^1H$, an empty ceramic tube is used. The use of a catalyst inside the tube must be avoided to eliminate the risk of adsorption of H_2, which would lead to temperature-dependent fractionation. The operational temperature for conversion to H_2 is typically 1420 °C. In Figure 4.3, a schematic representation of the hyphenation of GC to IRMS is given, indicating the two interfaces used for conversion of the organic compounds into simple gases (Dumont et al. 2013).

4.1.3.3 LC-IRMS

In GC combustion, the carrier gas is helium, which does not interfere with isotope ratio measurements. The typical liquid mobile phases in LC have prevented a similar conversion when hyphenating LC to IRMS. The aqueous phase as a carrier allows using LC in a micro-EA-mode and in a compound-specific mode

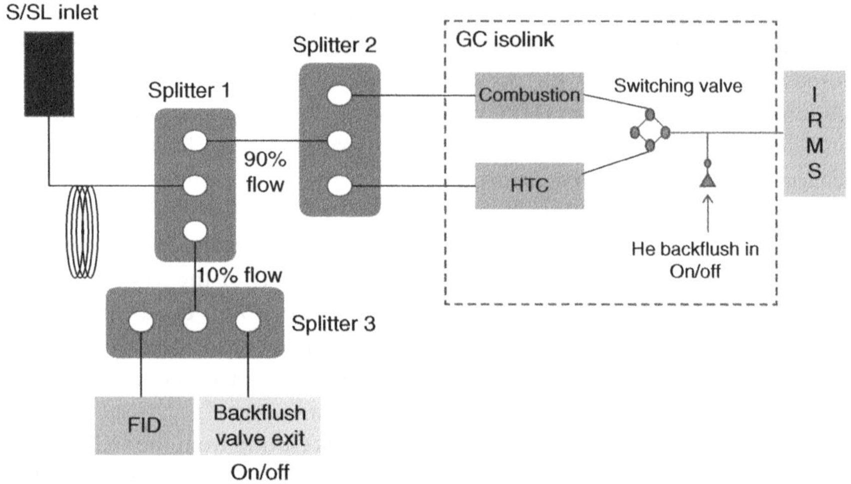

Figure 4.3 Coupling of GC to IRMS. Source: Dumont et al. (2013). Reproduced with permission, Copyright (2013), Elsevier.

(LC-mode). In micro-EA-mode, the sample can be injected via the LC system and be directed toward the interface of the IRMS (without any separation). In LC-mode, the sample is injected, compounds are separated on the LC column, and the compounds are introduced one-by-one into the interface, as a constant flow of mobile phase carries the compounds. A constant mixture of oxidation reagent and phosphoric acid is added to the mobile phase, prior to entering the oxidation reactor operating at 100 °C. All organic compounds are converted individually and quantitatively into CO_2, and the CO_2 is separated from the liquid phase by a membrane exchanger. As all carbon-containing matter is converted, mobile phases used in liquid chromatography–isotope ratio mass spectrometry (LC-IRMS) cannot contain any carbon-containing solvents or additives, as this would result in a continuous flow of organic matter into the interface. Hence, typical reversed-phase conditions cannot be combined with IRMS, severely limiting the applications in this field to the analysis of water-soluble molecules such as sugars, amino acids, short-chain organic acids, and alcohols.

4.1.3.4 Two-Dimensional GC-IRMS

Chromatographic resolution is extremely important in isotope ratio determination and for the analysis of more complex samples, such as extracts of plant material, where one-dimensional (1D) GC is often insufficient to obtain pure isolated peaks. The isotope ratio determination of coeluting peaks results in δ values that are a combination of the compounds that are not resolved. In order to obtain better separation and isolation of individual solutes from a complex profile, multidimensional chromatography offers interesting possibilities. There have been a few research groups that have focused on the use of two-dimensional (2D)-GC in combination with IRMS. Off-line 2D GC with fraction collection was recently applied by Ball et al. for the analysis of lignin phenols (Ball et al. 2012). On-line 2D GC hyphenated to IRMS was pioneered by the research group of Mosandl in the late 1990s. A commercially available 2D GC instrument using a "live switching" device was coupled to IRMS (Juchelka et al. 1998; Reichert et al. 2000); this approach has been reported for the analysis of steroids in urine (Brailsford et al. 2012) and wax compounds in tobacco leaves and smoke (Dumont et al. 2013). A 2D gas chromatography–isotope ratio mass spectrometry (GC-IRMS) approach, using a moving capillary stream switch, was used to analyze technical polychlorobiphenyl (PCB) and polychloronaphthalene (PCN) mixtures by Horii et al. (2005). For the analysis of C_2–C_5 hydrocarbons, a 2D GC set-up based on gas switching valves was used (Nara et al. 2006), and more recently, comprehensive 2D GC was coupled to IRMS (GC × GC-IRMS) (Tobias et al. 2008, 2011). In environmental analysis, heart-cutting 2D GC-IRMS was used in the analysis of monoaromatic hydrocarbons (e.g. benzene, toluene, ethylbenzene, xylenes [BTEX]) in groundwater and gas-phase samples (Ponsin et al. 2017). At a former refinery and crude oil pumping station, soil vapor extraction was implemented to remediate a historical BTEX plume, and gas-phase samples taken from the soil vapor extraction wells were subjected to heart-cutting 2D GC-IRMS. As can be seen in Figure 4.4, the heart-cuts for benzene (B) and toluene (T) were sent to the second-dimension column and the IRMS, reducing matrix effects observed in the 1D GC-IRMS separation of the same sample.

4.1.4 IRMS Applications in the Oil and Gas Industry

4.1.4.1 Introduction

A broad range of applications are described in the literature for IRMS, but only a fraction of which deal with petroleum-related topics. Most of the petroleum-related studies focus on environmental forensics associated with oil spills, contamination, or geochemical studies that address oil source and maturity.

Compound specific isotope analysis is of particular interest in the assessment of the sources of organic contaminants and their transformation reactions in the environment. A prerequisite for environmental forensics is that the isotopic composition of the sample or compound originating from different sources remains stable. Stable isotope information in environmental studies may help to differentiate among potential polluters on a local scale, to attribute specific emission sources on a regional scale and on a global scale, to quantify volatile organic compounds

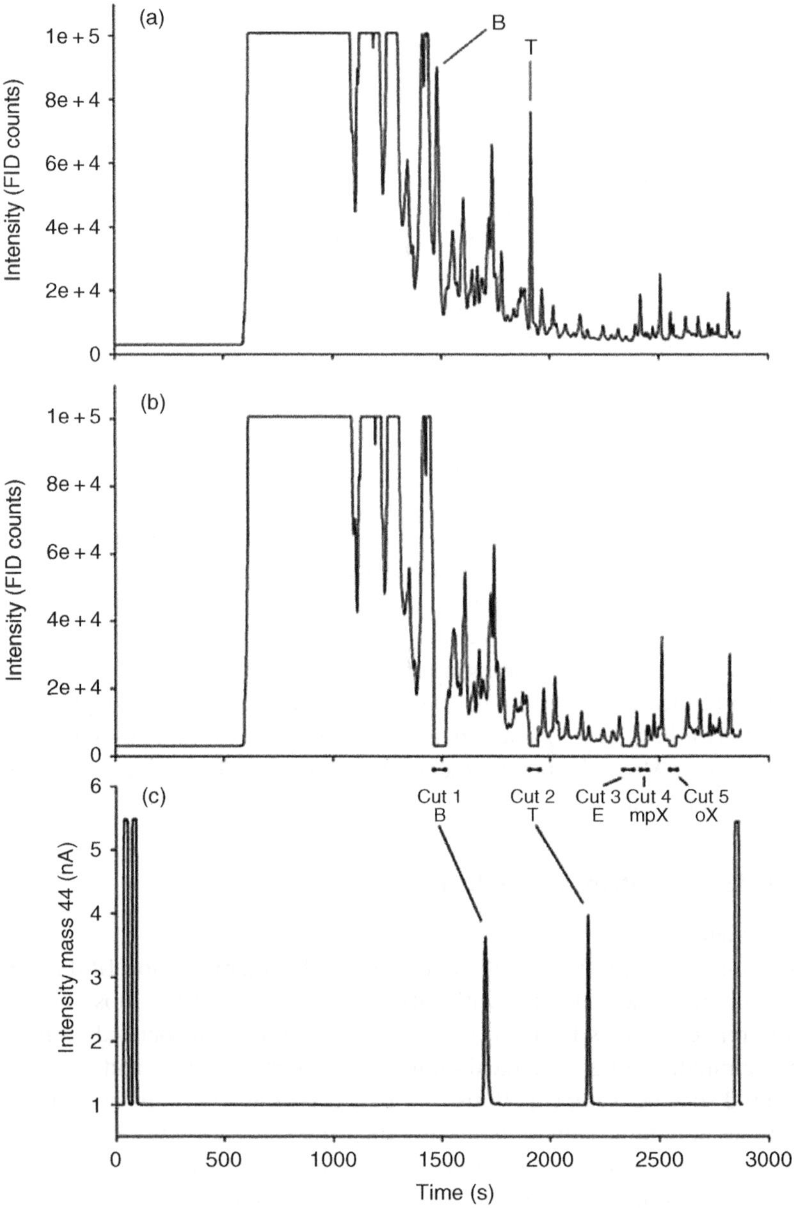

Figure 4.4 2D GC-IRMS of a sample using a CP-Wax 52 CB column (first-dimension) and a DB-5 column (second-dimension): (a) FID chromatogram without cuts (sample diluted 700 times), (b) FID chromatogram with cuts for BTEX (sample diluted 700 times), and (c) the corresponding IRMS chromatogram. Only benzene (cut 1650–1745 seconds) and toluene (cut 2130–2220 seconds) were sent to the IRMS. Source: Ponsin et al. (2017). Reproduced with permission, Copyright (2017), Elsevier.

(VOCs) and greenhouse gases (Elsner et al. 2012; Goldstein and Shaw 2003; Rockmann et al. 2003).

Apart from source identification, compound-specific isotope analysis gives the possibility to identify and quantify transformation processes. During many abiotic and biotic reactions, light isotopes are transferred, preferably from the reactant to the product pool. The residual reactant then becomes increasingly enriched in the heavier isotope.

4.1.4.2 Oil Fingerprinting

Oil fingerprinting is widely used to determine the source of an oil spill in the environment, by analyzing matching biomarkers of both spilled oil and its candidate resource (mostly by gas chromatography–mass spectrometry [GC-MS]). The unambiguous identification of oil sources and their characteristics is of utmost importance, not only for assessing the fate and environmental damage of the spill but also to provide an appropriate response to the spill (Bayona et al. 2015). Tracing back the oil spill is also important to assign responsibilities and liabilities. Recent developments in chromatography and mass spectrometry (GC-MS/MS, GC × GC-TOF) have enhanced sensitivity, selectivity, and resolution in the analysis of biomarkers for oil spill fingerprinting. The use of GC-IRMS in oil spill investigations already traces back to the previous century. GC-IRMS gained interest as being a complementary technique to GC and GC-MS for samples that have undergone extensive weathering. The use of GC-IRMS in this scientific discipline gains interest as the addition of isotopic information on biomarkers is an extra tool in comparing the oil spill to its possible source.

Hydrocarbon pollutants in aquatic environments have also been studied extensively. Sources of these contaminants include natural seepages of crude oils, ship traffic, supertankers spilling crude oil, leaking of storage tanks, and pipelines. The identification, quantification, and monitoring of these spilled pollutants are important; for example, studies from Quebec, Canada, on *n*-alkanes in sediments sampled along a river revealed the origin of a spill (Rogers and Savard 1999).

The correlation of crude oils and refined hydrocarbon products spilled in the environment with their respective resources is commonly examined through the use of GC and GC-MS. Under certain circumstances, the chemical composition of the samples can alter through extensive weathering (evaporation, water-washing, biodegradation), refer to the carbon cycle in Figure 4.1, resulting in the loss of some crucial compounds, amongst which some biomarkers necessary to link the oil spill to its source. Typically, these samples lose the more volatile *n*-alkanes, and in some cases, even all of them. As these compounds are used as biomarkers for correlation purposes, all relevant data on samples is lost, making it impossible to trace back the source of the oil spill through conventional techniques. In these cases,

GC-IRMS can aid in correlating spills to the source. Isotopic ratios of individual pyrolysis products can be used for correlation purposes with their unaltered counterparts, where some individual compounds show good isotopic correlation. Such compounds are not affected by the strong weathering conditions, which, when combined with data from the original product, make GC-IRMS a valuable tool in the comparison of oils (Mansuy et al. 1997). Asphaltenes, for example, have a similar but less condensed structure than the kerogen from which they originate; moreover they are not extensively altered by biodegradation. Pyrolyzing asphaltenes results in hydrocarbons, whose fingerprint is similar to the original oil, and hence, can be used to compare biodegraded and non-biodegraded samples.

In crude oils, there are a large number of potential biomarkers. However, for refined products, the choice is rather limited. GC-IRMS has proven to be particularly useful for refined products (gasoline), produced from varying feedstocks that could be difficult to examine through more conventional techniques (Richnow et al. 2003). The incorporation of isotopic data of specific compounds and the biomarker data provides a powerful and complementary tool for the purposes of making these correlations. Weathering processes lead to altered chromatographic profiles in refined products, and due to biodegradation, many of the *n*-alkanes in diesel samples are also removed, making a correlation of the samples challenging. Comparing the carbon isotope values of the isoprenoids in degraded and non-degraded samples allows for the determination of the relationship between the samples (Philp et al. 2002; Figure 4.5).

4.1.4.3 Air Pollution

Polycyclic aromatic hydrocarbons (PAHs) are persistent pollutants in air, soils, and sediments. The major sources to the environment are either through pyrogenic or petrogenic processes. Pyrogenic PAHs are formed by incomplete combustion of organic material, for example, cracking processes, combustion of motor fuels, or wood in forest fires. Petrogenic PAHs, on the other hand, are formed during crude oil maturation, or they are released due to fuel transportation spills. The possible sources of PAHs are well-known, and investigating whether the PAHs originate from oil, wood, or coal is important in terms of health risks and source identification. Their carbon isotopic fingerprint should reflect those of the original matter (i.e. prior to combustion). Typical $\delta^{13}C$ values for coal conversion are in the range of −23‰ to −25‰, whereas $\delta^{13}C$ values from diesel particulate typically have values of −28‰ to −31‰, allowing one to identify whether PAHs are of petrogenic or pyrogenic nature. The isotopic fingerprinting of PAHs allows one to distinguish whether the compounds originate from coal or wood-burning, or whether they

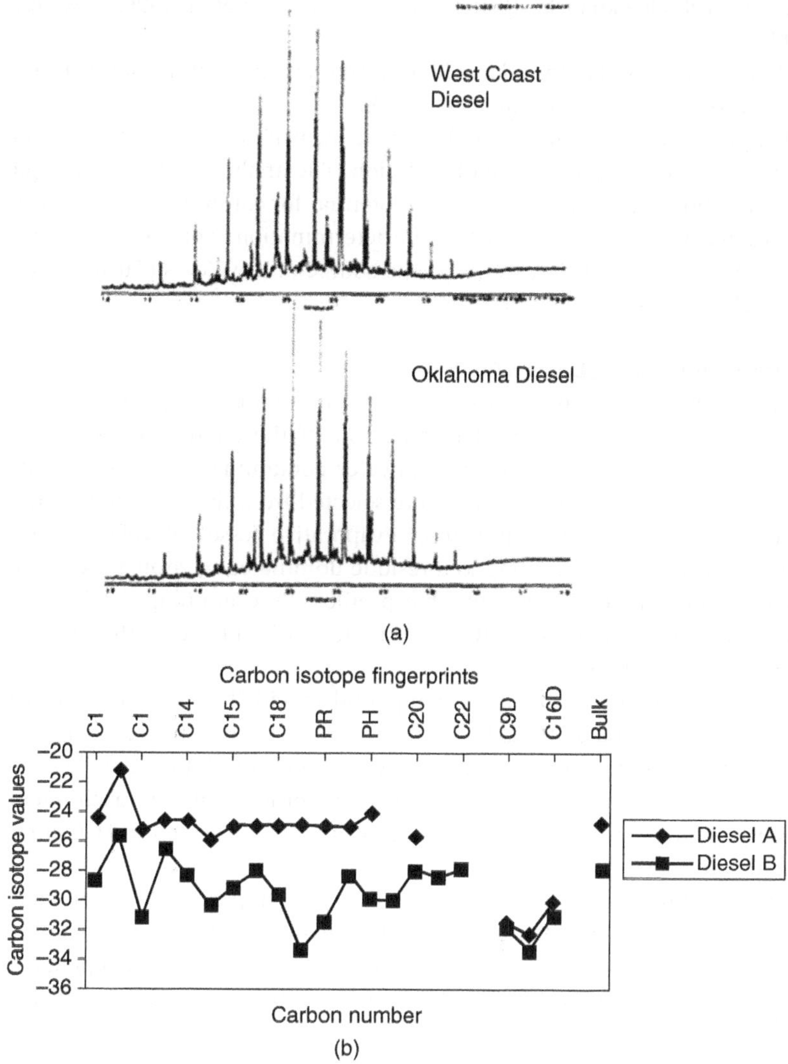

Figure 4.5 (a) Two diesel samples with similar chromatographic fingerprints. (b) Carbon and hydrogen isotope compositions for the individual components shown in (b) clearly differentiate these products as being derived from different crude oil feedstocks. Source: Philp et al. (2002). Reproduced with permission, Copyright (2002), Taylor & Francis Ltd.

originate from vehicle exhausts (Elsner et al. 2012; McRae et al. 1996; O'Malley et al. 1997).

GC-IRMS was also used in the source identification study of atmospheric PAHs in the proximity of forest fires (Okuda et al. 2002).

A dual approach where both $\delta^{13}C$ and δD were measured has gained more and more interest in revealing the origin of pollution. The analysis of VOCs through thermal desorption-GC-IRMS provides the potential for future research to trace fate and to determine the origin of VOCs. The determination of only $\delta^{13}C$ of VOCs in air is often not sufficient to determine the source of the VOCs (Turner et al. 2006; von Eckstaedt et al. 2011).

4.1.4.4 Differentiating Oil Derived Products

Differentiating gasoline supplies based on their chromatographic profile is not straightforward. Nevertheless, the determination of the source of gasoline is important, as this is often a major groundwater contaminant. In the isotopic fingerprint, several well-resolved peaks were selected from the gasoline profiles, making GC-IRMS a complementary tool. Evaporative losses of volatile compounds from fresh gasoline, leading to a residue dominated by aromatics, make it impossible to compare chromatographic profiles. By examining the isotopic composition of the compounds in the residue, it can be linked to the original gasoline (Philp et al. 2002).

The combination of measuring using compound-specific IRMS for both the δD and $\delta^{13}C$ of *n*-alkanes and plotting these values appears to be a unique tool in distinguishing diesel samples. These studies provide a high-resolution, detailed fingerprint with an enormous potential of determining origin. This approach might be extremely useful to distinguish between samples with almost identical GC chromatographic profiles.

The ability to differentiate between different diesel samples could be valuable for discovering fuel tax evasion schemes or for environmental forensic studies. Examples of tax evasion are when tax-free fuel is aimed for export but kept at the origin, and then adding tax-free fuel or lower value products to diesel fuel (Harvey et al. 2012).

4.1.4.5 Inherent Tracers for Carbon Capture and Storage (CCS)

In the oil and gas industry, CO_2 emissions are of concern. Carbon capture and storage (CCS) is a technique utilized to capture and store CO_2 in order to reduce industrial emissions and to prevent CO_2 from entering the atmosphere. When CCS is implemented, governments require monitoring of the CO_2 injected for storage, meaning that tracers are required. A complete description of tracers will be provided in Chapter 8. However, it should be briefly mentioned here that IRMS can be utilized for inherent tracer analysis.

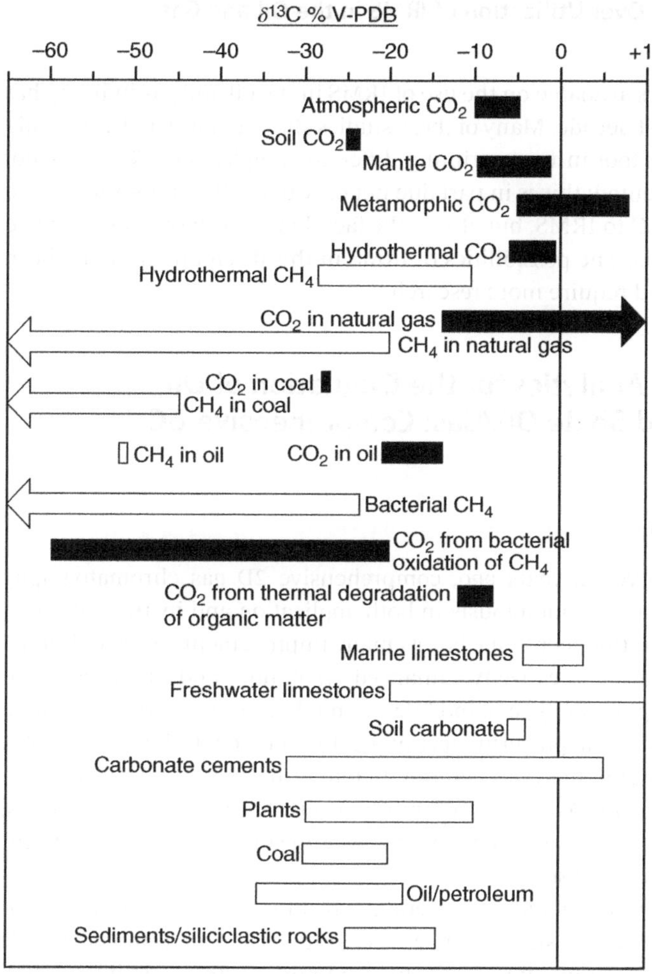

Figure 4.6 δ^{13}C values for a range of naturally occurring materials. Black boxes indicate CO_2, white boxes indicate other substances, and arrows represent values off the scale of the diagram. V-PDB, Vienne PeeDee Belemnite. Source: Flude et al. (2016). Reproduced with permission, Copyright (2016), American Chemical Society.

Inherent tracers benefit from the natural isotopes present in the molecule; in the case of CCS, the C-isotopes and O-isotopes from the captured CO_2 can be used. Flude et al. (2016) summarized the δ^{13}C values for CO_2 originating from various sources. As can be seen in Figure 4.6, there is a significant difference between CO_2 originating from natural gas, oil, and coal compared with atmospheric CO_2, indicating the potential for utilizing IRMS for inherent tracer monitoring in CCS.

4.1.5 Conclusions Over Utilization of IRMS in the Oil and Gas Industry

The number of papers available on the use of IRMS in the oil and gas industry has increased over the last decade. Many of these studies showed how GC-IRMS could be a complementary tool in fingerprinting different samples. LC-IRMS has not been widely implemented; this is in part due not only to the limitations discussed earlier on coupling LC to IRMS, but also to the fact that GC is more often applied in oil and gas analysis. The possibilities of IRMS in this discipline have not been fully exploited yet and require more research.

4.2 Advanced Analytics for the Evaluation of Oil, Natural Gas, and Shale Oil/Gas: Comprehensive GC (GC × GC)

4.2.1 Background

Since its inception over 20 years ago, comprehensive 2D gas chromatography (GC × GC) has advanced tremendously in both application and instrumentation. This innovative separation technique boasts many improvements over traditional 1D GC: much greater peak capacity, enhanced resolution, and often improved sensitivity, depending on the type of modulator and detector that are used. These features make GC × GC a powerful technique for the separation of complex mixtures and challenging matrices commonly found in samples associated with oil and gas development. Among all currently available separation techniques, GC × GC is capable of the greatest separation power for compounds with sufficient vapor pressure for GC analysis.

J.C. Giddings first wrote about the potential advantages of using multidimensional chromatography. He stated that 2D separations are determined by 1D displacement processes of two types: selective and nonselective (Giddings 1984). Selective displacements are specific to each component, an example being the interaction of a component with the column stationary phase. Nonselective displacements are the same for all components and occur from primary equilibria bulk transport properties, like vapor pressure. For the greatest peak capacity in a 2D separation, the 1D displacement techniques should be both selective and independent of each other. A combination of other 1D displacement processes, such as selective paired with nonselective, would not allow for complete separation in each dimension and thus not achieve the best peak capacity possible. Giddings also ruled that in multidimensional separations, the components resolved in the first dimension must remain separated after moving through the second dimension (Giddings 1987).

Since multidimensional chromatography requires two independent separation steps, where the separation achieved in the first dimension is preserved in the second dimension, simply having two columns connected in tandem with a continuous flow of sample does not qualify as multidimensional. Differences in migration rates between the two columns can cause separations to cancel each other out, leading to peak coelution (Giddings 1987). Continuous sample injection from the first column to the second would produce what appears to be a 1D chromatogram of a multi-phase column. Thus, multidimensional separations must have discrete, not continuous, sample transfer from the first-dimension column onto the second. There are many techniques that can accomplish this by employing either heart-cutting (MDGC, multidimensional gas chromatography) operations or the use of a modulator, which allows for comprehensive separation (GC × GC).

Heart-cutting techniques were the pioneering methods for MDGC. This form of multidimensional separation commonly occurs using a Dean's switch, in which only a small fraction of the sample from the first column is transferred to the second column for further separation. The advantage of using heart-cutting is that it allows a second, independent separation for certain sections of the 1D chromatogram where target compounds coelute. Heart-cutting does allow for the enhanced peak capacity of the targeted regions, such as for the analysis of paraffins, isoparaffins, olefins, naphthenes, and aromatics (PIONA) compounds in gasoline (Boer and van Arkel 1971). If the heart-cutting is done with two sufficiently independent columns, the peak capacity increase is additive and can be expressed as $(n_y + n_z)$, where n_y and n_z are the peak capacities of the first and second columns. MDGC is not ideal for whole sample analysis because the number of heart-cuts is limited by the availability of the second dimension column as it is separating previous heart-cuts. This technique is also limited to the application of samples with known targets. Because the timing of the heart-cuts must be precise, MDGC is not a suitable method to search for unknown compounds.

Liu and Phillips were the first to report comprehensive multidimensional GC to analyze hydrocarbon mixtures and oil samples (Liu and Philips 1991). Having developed an on-column thermal modulator interface between the first and second columns, they achieved a true comprehensive separation, where the entire sample was separated in both dimensions. The benefits of comprehensive GC × GC can be seen in Figure 4.7, where the aromatic compounds were separated in the second dimension from the alkanes of a white gas sample (Bruckner et al. 1998); as can be seen, if only the first dimension separation was utilized, coelution of the aromatics and alkanes would have occurred. In GC × GC, the separation is discrete, and the column selectivities are independent. Unlike in MDGC, the peak capacities for GC × GC from the first and second dimensions are multiplicative $(n_z \times n_y)$, as shown in Figure 4.8 (Giddings 1987). This increased peak capacity is one of the most substantial improvements achieved using GC × GC, especially in

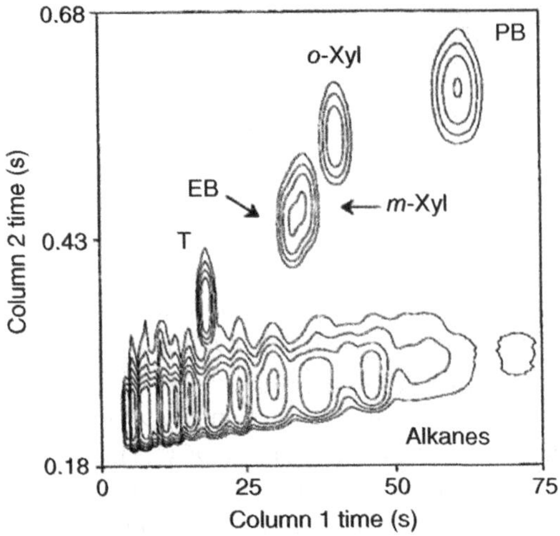

Figure 4.7 GC × GC chromatogram of a white gas sample, where the aromatic components have been separated from the alkanes. Source: Bruckner et al. (1998). Reproduced with permission, Copyright (1998), American Chemical Society.

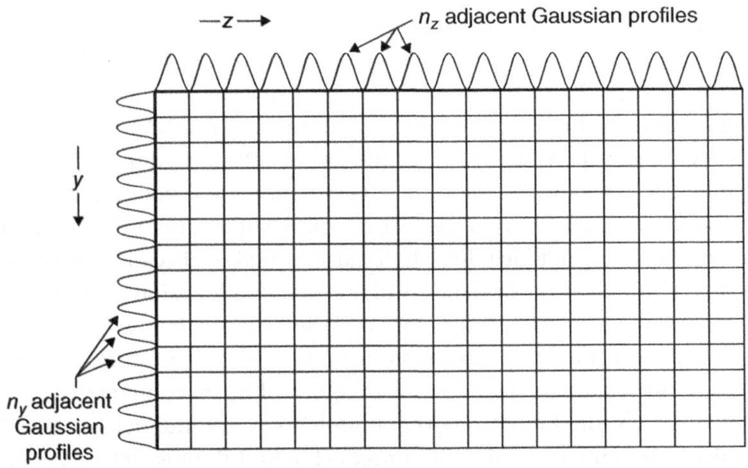

Figure 4.8 The peak capacity for GC × GC is the product of the peak capacity of the z-axis (n_z) and the y-axis (n_y), as represented by the number of boxes. Source: Giddings (1987). Reproduced with permission, Copyright (1987), John Wiley & Sons.

applications of extremely complex mixtures, such as those related to oil and gas development.

4.2.2 Basic Principles of GC × GC: Instrumentation

Despite the advantages of GC × GC over a traditional 1D separation, this 2D technique is not as widely utilized due to the increased cost and complexity in

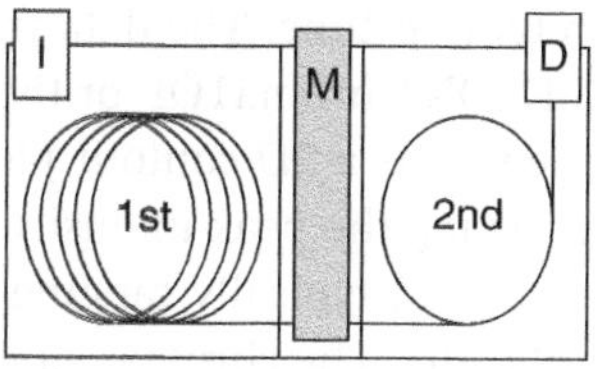

Figure 4.9 Diagram of a GC × GC system where I is the injector, M is the modulator, and D is the detector. Source: Dallüge et al. (2003). Reproduced with permission, Copyright (2003), Elsevier.

method development and data analysis. Additionally, while traditional GC has been practiced for more than a century, the technique of GC × GC is relatively new. As this field has grown in recent years, a variety of commercially available GC × GC systems have been developed. Two examples of commonly used systems are the LECO GC × GC coupled to an FID, electron capture detector (ECD), or a time-of-flight mass spectrometer (TOF-MS) and the Zoex Corporation GC × GC with the JEOL AccuTOF high-resolution TOF-MS. Historically, the most popular separations employing GC × GC are the analysis of petroleum products, though many other applications have been reported in the literature.

The main components of a GC × GC system can be broken into three parts: the dual column ensemble, the column-to-column interface termed the modulator, and the detector. A diagram of a GC × GC setup can be seen in Figure 4.9 (Dallüge et al. 2003). In GC × GC, samples are injected into the instrument in the same way as traditional 1D GC. Instead of entering the detector after separation on the primary column, the eluate enters the modulator, which is placed in between the two columns. During the entire analysis, the modulator continuously traps and reinjects packets of eluate from the first-dimension column onto the second-dimension column. Ideally, this second separation is fast because the modulator will continue to inject the first-dimension eluate throughout the analysis. The second-dimension column is typically very short (0.5–2 m) and is sometimes housed in a separate oven held at a higher temperature. To achieve the best separation, the two columns used in GC × GC should be of different stationary phases to create an orthogonal separation with the greatest peak capacity possible.

The fast, second dimension separation leads to very small peak widths ranging from 100 to 600 ms at the base (Dallüge et al. 2003). The narrow second dimension peaks require a detector that offers high sampling rates and a small internal volume. In order for these peaks to be appropriately sampled, the detector acquisition rates should be in the range of 20–200 Hz (Dallüge et al. 2003). Commonly, an FID or a TOF-MS detector is used depending on the application. Other detectors less commonly utilized are the micro electron-capture detector (μ-ECD), atomic emission detector (AED), sulfur chemiluminescence detector (SCD), and quadrupole mass spectrometers (qMSs).

The result of the GC × GC separation is a 2D chromatogram. The x-axis is the retention time of the first column separation, and the y-axis is the second column retention time. The data transformation is done by laboratory software, such as LECO's ChromaTOF or the widely applicable GC Image. The data can either be viewed as a 2D contour plot or a three-dimensional surface plot with the *z*-axis showing the peak intensity. In both of the plot types, colors are used to represent intensity, typically ranging from a cool blue to bright red in order of increasing intensity. The chromatograms generated from GC × GC contain greater data density than 1D GC, making the technique less appealing to the occasional practitioners of GC, or those with relatively simplistic separation needs. The increased cost and complexity of GC × GC may be well worth the investment for the analysis of complex samples where the greater peak capacity, improved resolution, and potentially lower sensitivity of the technique are vital to the separation and characterization of the sample.

4.2.3 Basic Principles of GC × GC: Columns

In GC × GC, both columns are heated using temperature programs inside a GC oven. Some instruments contain a smaller secondary oven that allows for independent control of the second-dimension temperature range. Typically, this oven is programmed to a slightly higher temperature than the primary oven. The addition of a secondary oven complicates the installation of the columns, but the added control of the second-dimension temperature is important for some separations. For the best separation, there should be a high degree of independence between the two columns. This is achieved by using columns of different stationary phase chemistries and, therefore, different modes of separation. Orthogonality should be maximized, meaning that the 2D chromatographic space is used effectively. Effective use of the 2D chromatographic space is characterized by an even spread of analyte peaks and compound type clustering (Jáčová et al. 2019). This discussion leads to the idea that proper orthogonal column selection is highly application-specific and should be optimized for each set of samples.

The most commonly used column set for GC × GC consists of a non-polar (non-selective) first-dimension column coupled to a polar or semi-polar (selective) secondary column. This is referred to as a normal phase column setup. The first column separates analytes based largely on their vapor pressure, which can be related to volatility. The second-dimension separation is so fast that it occurs essentially under isothermal conditions. In this way, vapor pressure contributions to separation are minimal, and the separation occurs based on column selectivity alone. Possible interactions of analytes with the column include polarity, hydrogen bonding forces, shape selectivity, and π–π interactions. The second-dimension column should be carefully selected, accounting for the chemistry of the analytes and how they can be successfully retained and separated. Considering these differences between the first- and second-dimension separations, one can see how column orthogonality plays a large role in realizing the true benefit of GC × GC separations.

The reverse phase or inverse column ensemble, utilizing a polar first-dimension column followed by a non-polar second-dimension column, is occasionally applied for petrochemical type separations. Reverse phase GC × GC is not to be confused with reverse phase LC, and the term inverse will henceforth be used to avoid confusion. The advantage of using an inverse column set is that it may provide enhanced peak shape for polar compounds and better separation of non-polar compounds. It has been shown that the inverse column configuration allows for improved resolution and separation of *n*-alkanes and branched alkanes in crude oil when compared with the normal column set (Tran et al. 2006). Using a normal phase column configuration, the mono-, di-, and tri-aromatic compounds were well retained in the second-dimension and showed good overall 2D separation. However, the low polarity hydrocarbons were poorly retained on the second dimension and did not appear separated on the 2D chromatogram, instead staying clustered at the bottom. After switching to an inverse phase column ensemble, a much greater use of chromatographic space was achieved. The alkanes now eluted earlier on the first-dimension column but were retained on the second-dimension, allowing for separation. Figure 4.10 shows both the normal and inverse phase chromatograms for the crude oil samples. A similar study by Hu et al. confirmed these results and summarized the differences between the two column sets quite nicely. The normal phase setup separates the higher molecular weight polycyclic hydrocarbons and PAHs more efficiently, but the inverse phase is more effective for the separation of medium to low molecular weight *iso*-alkanes and cycloalkanes (Hu et al. 2014).

Omais et al. observed an interesting phenomenon when analyzing a coal extract middle distillate liquid by both normal and inverse phase GC × GC. The column ensembles tested were non-polar × semi-polar and polar × semi-polar. Keeping the second-dimension column the same between both sets, they observed a reversal of the group type elution orders of paraffins, aromatics, and phenols between the two column sets. Their data shows that second-dimension separation is indeed based on the compound's volatility in inverse phase GC × GC, just as the first dimension separation is in normal phase GC × GC (Omais et al. 2012). Whether using normal or reverse phase GC × GC, the selection of a column set will involve some trial and error following the educated selection based on column phase chemistry and analyte interactions.

Common non-polar columns used in the first-dimension for normal phase separations include the 100% polydimethylsiloxane (PDMS) or the 95/5% dimethyl/diphenyl siloxane column. They typically range from 30 to 60 m in length, with some columns as long as 100 m. The film thickness is usually 0.25–1.50 μm, with an internal diameter of 0.18–0.53 mm. A common second dimension column is

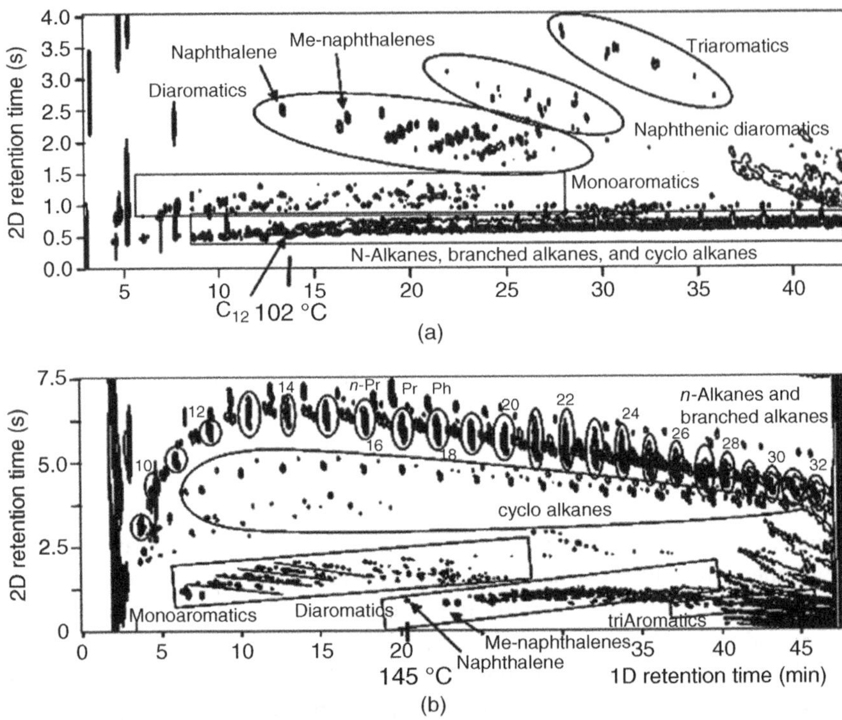

Figure 4.10 (a) Crude oil separation using a normal phase column configuration (BPX5 in 1D and BP20 in 2D). (b) Crude oil separation using a reverse phase column configuration (SolgelWax in 1D and BP1 in 2D). Source: Tran et al. (2006). Reproduced with permission, Copyright (2006), Elsevier.

a 35–50% diphenyl/dimethylpolysiloxane phase. For better separation, the analytes need to be sufficiently separated by the second dimension before the next modulation period is over and another packet of the eluate is injected. For this reason, the second-dimension columns are typically 1–2 m and 0.18–0.25 mm in diameter. Thin-film stationary phase columns (0.1–0.25 μm) are typically used for their higher efficiency and lower analyte retention than the thicker film columns. Table 4.1 details the column sets used in recent oil and gas related GC × GC separation studies.

Ionic liquid (IL) stationary phases have recently been used in comprehensive separations of petrochemical samples. ILs are organic salts with melting points at or below 100 °C, typically composed of an organic cation with an organic or inorganic counterion (Pollo et al. 2018). Some advantages of using IL stationary phase columns are their high thermal stability and increased

Table 4.1 Examples of normal and reverse phase column combinations for four petrochemical applications.

Sample	Types	1D column	2D column	Citations
Estonian pyrolysis shale oil	Reverse	Stabilwax (polyethylene glycol) 60 m × 0.25 mm × 0.25 μm	Rxi-5ms (5% diphenyl–95% dimethyl polysiloxane) 2 m × 0.15 mm × 0.15 μm	Ristic et al. (2017)
Light diesel oils	Reverse	ZB-50 (50% diphenyl–50% dimethyl polysiloxane) 60 m × 0.25 mm × 0.25 μm	ZB-5 (5% diphenyl –95% dimethyl polysiloxane) 1.5 m × 0.15 mm × 0.15 μm	Alexandrino et al. (2019b)
Hydraulic fracturing fluids and flowback	Normal	Rtx-DHA100 (100% dimethyl polysiloxane) 100 m × 0.25 mm × 0.5 μm	Rxi-17SilMS (50% diphenyl–50% dimethyl polysiloxane) 2.0 m × 0.25 mm × 0.25 μm	Piotrowski et al. (2018a)
Extra heavy gas oil	Normal	HP-5ms (5% diphenyl–95% dimethyl polysiloxane) 30 m × 0.25 mm × 0.25 μm	BPX-50 (50% diphenyl–50% dimethyl polysiloxane) 1.5 m × 0.1 mm × 0.1 μm	Avila et al. (2011)

maximum operating temperatures compared with traditional polar stationary phases. These properties make IL columns especially useful for the separation of high boiling point compounds that elute under temperatures too high for most polar wax columns. For the separation of organic sulfur compounds (OSCs) and nitrogen-containing compounds in Brazilian diesel and naphtha samples, Fontanive et al. compared column sets containing an IL stationary phase with more traditional GC × GC column sets (Fontanive et al. 2016). A DB-5MS column was coupled to an IL-59 column (1,12-di(tripropylphosphonium)dodecane bis(trifluoromethylsulfonyl)imide) in both normal and inverse configurations for comparison with the DB-5MS/DB-17 combination. The DB-5MS/IL-59 column set was found to be more effective for the separation of OSCs, while the IL-59/DB-5MS set provided better results for the nitrogen-containing compounds. Overall, having the IL-59 column in either dimension exhibited better use of the 2D separation space for both the diesel and naphtha samples. Two other studies have analyzed the use of IL stationary phase columns for the separation of aliphatic hydrocarbons in kerosene using GC × GC (Hantao et al. 2014; Nan et al. 2017). Many different types of IL stationary phases were synthesized and examined for their tunable physical and chemical properties. These studies demonstrate the value of IL columns to be tuned for the selectivity of hard to separate compounds.

4.2.4 Basic Principles of GC × GC: Modulators

The modulator is the part of the GC × GC that allows for the separation to be comprehensive and not merely heart cutting. An effective modulator has a few requirements to achieve comprehensive separation. First, it must trap all of the eluates from the first-dimension column during the pre-determined modulation period while the separation continues. Second, the modulator must refocus the packets of analytes in time or space. Lastly, it must rapidly inject the packet onto the second-dimension column for further separation.

GC × GC separation performance is significantly impacted by the modulation period or the time that the analyte packet is trapped right before re-injection into the second-dimension column. Choosing the right modulation period can be difficult, as a period that is either too long or too short can have detrimental effects on the separation. If the timing is too long, the first-dimension separation will be sacrificed because the analytes will be mixed in the modulator during the refocusing step. This will then be injected into the second column, and poorer overall separation will result because the primary separation was negatively impacted. Shorter modulation periods are preferred in order to minimize any recombination that may occur in the first dimension. If the modulation period is too short, the result will be analyte wraparound. This occurs when the modulation period is shorter than the retention time of the longest retained analytes in the second dimension. Consider a modulation period of six seconds. If all analytes elute from the second column in less than six seconds, then no wraparound occurs. If an analyte elutes in nine seconds, it will then elute with the next modulation period analytes in the second dimension with a three second retention time. This leads to the incorrect assignment of the compound's retention time that could lead to misidentification of unknown analytes. Wraparound can be corrected in some software, but it is better to optimize the modulation period in the method before data acquisition to avoid it.

The modulators used in GC × GC can be separated into two designs: thermal and valve based. Thermal modulation is more commonly applied in the practice of multidimensional chromatography. There are three types of thermal modulators – resistively heated trap, heated sweeper, and cryogenic focus (Bahaghighat et al. 2018). The most common cryogenic modulator is the quad jet design, commercialized by LECO. Cryogenic jet systems use liquid nitrogen cooled to −190 °C and can trap analytes as low as C_5 without any moving parts. The dual jet system design contains a jet that blows cryogenically cooled nitrogen gas and another jet that pulses hot air. The analytes are trapped when the cooled nitrogen gas jet is on; then, they are released when the hot jet remobilizes them. This design comes as a single stage or dual stage (quad jet) modulator. The quad jet design prevents the breakthrough of the analytes because they are focused twice. This traps the more volatile analytes and has the widest range of applications. Cryogenic focusing modulators are a proven workhorse in GC × GC, but a recent shift toward cryogen-free modulators has led to the development of more commercially available options of this kind. Both Zoex and J&X Technologies have introduced cryogen-free thermal modulators recently. The Zoex ZX2 uses a two-stage loop modulation system with a continuous cold jet as low as −90 °C and a pulsed hot jet used to re-inject the analytes. The modulation period can be set from 1 to 32 seconds and analytes as low as C_7 can be trapped and modulated. The J&X Technologies Solid State Modulator (SSM1800) utilizes a thermoelectric cooler and mica-thermic heating as well as a specific modulation column. The modulation column moves longitudinally in the hot and cold zones that can get as cold as −50 °C. J&X offers many different modulation columns that can be used for different applications and a range of analyte boiling points.

Another commonly found type of modulator is based on switching valves and is referred to as flow modulator. This type of modulation requires no cryogens and can modulate compounds ranging from C_1 to C_{40}+. The differential flow modulator design features a multi-port valve and a sample collection loop that fills with the 1D eluate and is then flushed with carrier gas to inject the eluate onto the second-dimension column. During the analysis, the first-dimension column can be operated with a typical flow rate or a slow flow rate. The sample loop would fill before a small amount of the first-dimension eluate would vent out. The second-dimension column would have a very high flow rate (20–25 ml/min) so that when the analytes in the sample loop were then injected into column 2, they are focused in time rather than focused in space as with thermal modulators. In some design variations, less than 100% of the analyte is being sampled. As long as the percentage is a sufficient representation of the whole sample, the technique can still be considered comprehensive. The main advantage is the lack of volatiles breakthrough with this type of modulation. The original differential flow modulators had second-dimension column flows that were too high to be coupled to mass spectrometry (MS). However, advances by the Mondello group combined a longer re-injection period with different restrictor column lengths to lower the flow to ~4 ml/min, making this modulator MS friendly (Franchina et al. 2016).

Flow diversion modulation has grown in use for GC × GC and is available commercially. This system is most commonly based on a Dean's switch to control the flows and pressures for eluate transfer from the first-dimension column to the second. The Agilent capillary flow technology (CFT), LECO Flux flow, and SepSolve INSIGHT modulators are commercially available flow diversion modulators that are proven to be reliable and reproducible. High flow from the second-dimension column is still a challenge for this type of flow modulation when connecting to an MS. One method to solve this is to split the flow leaving column 2 to an MS and another detector. Flow-based modulation has the advantage of instrumental simplicity versus thermal modulation but typically has less resolution in the second dimension, hence less peak capacity. When trying to determine what modulator to use for a separation, no single answer works for every analysis. Each modulator has its advantages and limitations, so it is up to the user to decide the best modulator for the application.

4.2.5 Basic Principles of GC × GC: Detectors

The last component of GC × GC is the detector. Because the analyte packets from the modulator are focused and the second-dimension separation is fast, there is very little band broadening, and peaks are small, typically 50–600 ms at the base. GC × GC detectors must be fast, able to sample at least 6–10 points per peak. Subsequently, they must collect data at a rate of 20 Hz or higher. If the data are collected at a slower rate, the peaks will not be sampled a sufficient number of times for adequate peak shape. Detectors for GC × GC must also have a small internal volume. These qualifications limit the detectors that can be coupled with GC × GC.

Historically, the FID has been the detector of choice for GC × GC due to its high sampling rates, small internal volume, and short rise time. The FID has been especially useful when applied to petrochemical samples because of their high carbon content. A study by Reddy et al. is an excellent example of the advantages of GC × GC-FID used to determine the contamination of crude oil samples with drilling fluids (Reddy et al. 2007). Miscible drilling fluids may impact the pressure/volume/temperature properties of crude oil; therefore, their presence and concentration in the oil need to be determined. The drilling fluids are identified based on their internal olefins (IOs) of C_{14} to C_{24} as they are not highly concentrated in crude oil. Using GC × GC-FID, the alkenes from the IOs drilling fluids separated into a specific area of the 2D chromatographic space not occupied by components in crude oil samples. In gas chromatography–flame ionization detector (GC-FID), these alkenes would typically coelute with other compounds. However, due to the resolution achieved with GC × GC-FID, the IOs were able to be quantified. Assuming that the IO alkenes have the same FID response factor as the crude oil hydrocarbons, the alkene group percentage and total alkene content can be calculated, confirming the presence of drilling fluids in the crude oil samples. Element-selective detectors have also been used for GC × GC applications of petroleum samples. The sulfur content in different diesel oils was characterized and quantified with GC × GC-SCD (Hua et al. 2003). The results were consistent with the American Society for Testing and Materials (ASTM) D-4294 method for the analysis of sulfur compounds in petroleum with energy-dispersive X-ray fluorescence spectrometry. GC × GC coupled to a nitrogen phosphorus detector (NPD) was used to quantify indole, alkyl carbazoles, alkyl benzocarbazoles, quinoline, alkyl quinolines, and alkylbenzoquinolines in Brazilian heavy gas oil samples (von Mühlen et al. 2007). The results showed better separation of the nitrogen compounds and improved information than their previous work on similar samples using a GC-MS instrument.

One of the most powerful detectors for GC × GC is the mass spectrometer, with the TOF-MS being the most commonly used followed by the qMS, high resolution (HR) TOF-MS, and less frequently the quadrupole time of flight (qTOF) and triple quadrupole (QqQ) MS (Tranchida et al. 2018). The mass spectral information from these detectors can be used for the identification of targets and unknowns in complex petroleum samples. The TOF-MS is best suited for GC × GC applications because of its fast acquisition rate, allowing for even the 50 ms wide peaks to be fully sampled. Even coelutions in the 2D chromatogram can be separated using mass spectral deconvolution, as long as their mass spectral fragmentation patterns differ slightly. A major benefit of using an MS is the compound library to aid in the identification of unknowns based on mass spectral similarity. Depending on the MS detector chosen, sensitive targeted information for specific compounds or accurate mass non-target information can be gained. Much like with the modulator, the choice of which MS detector to use is application dependent. As MS detectors are commonly used, a discussion of specific examples is not included in this section but is discussed later in the chapter.

Sometimes it is advantageous to separate complex mixtures with GC × GC followed by the simultaneous detection of the sample analytes using two different detectors. When using flow modulated GC × GC, the second-dimension flow may be too high for an MS detector to handle so the flow can be split to another detector, commonly an FID. This allows for accurate quantification of the sample using the FID and analyte identification with the MS. In one study, gasoline and kerosene samples were separated by GC × GC and simultaneously detected using FID and qMS (Krupčík et al. 2013). The second-dimension flow was 23 ml/min using a flow modulator, much too high for an MS turbopump to handle. Since the FID is a fast detector, the acquisition rate was 100 Hz, but the qMS only sampled at 18 Hz. Because of this, modulated peaks in the GC × GC-FID were sampled about 30 times, but the peaks in the GC × GC-qMS were sampled six to eight times (Figure 4.11). Reliable quantification was achieved with the FID because of the high sampling rate and the qMS data allowed for identification. Both quantification and unequivocal identifications could not have been achieved without coupling the two detectors.

4.2.6 Basic Principles of GC × GC: Data Processing

At the end of a GC × GC run, these short chromatograms are stacked side-by-side to produce a three-dimensional image, with the *x*-axis representing the first-dimension retention time and the *y*-axis representing the second-dimension

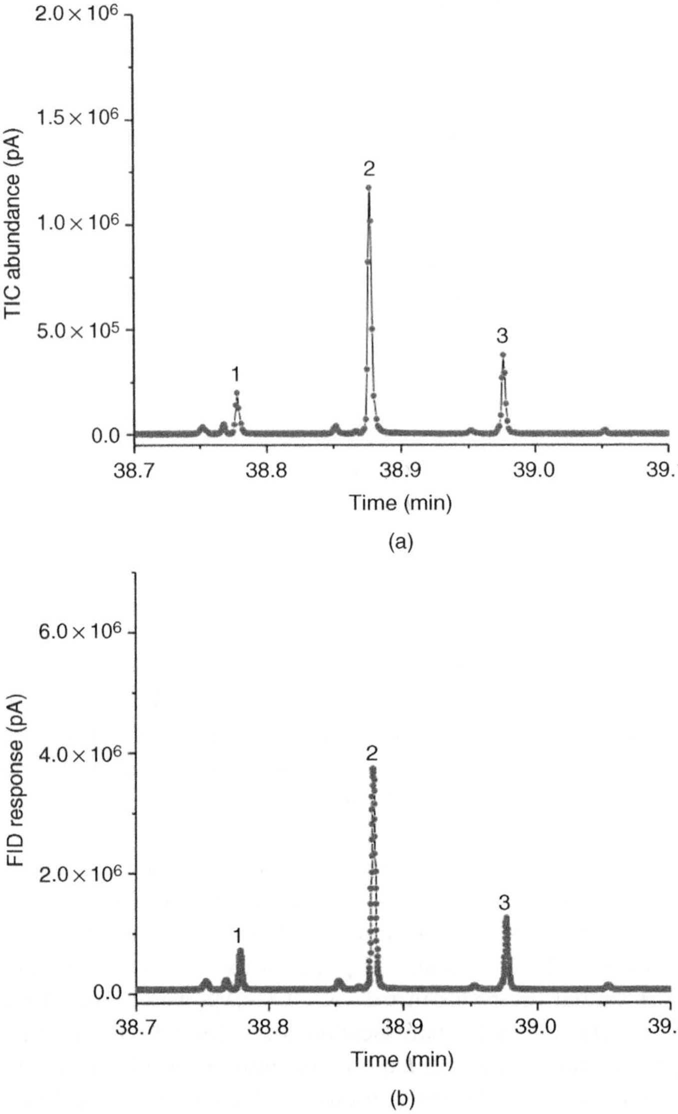

Figure 4.11 GC × GC-qMS (a) and GC × GC-FID (b) chromatogram of indane showing the sampling of each peak with the respective detectors. Source: Krupčík et al. (2013). Reproduced with permission, Copyright (2013), Elsevier.

retention times. This transformation, enabling meaningful visualization of the data, is performed by software that is either included in the instrument package (LECO's ChromaTOF or Zoex's GC Image) or a laboratory written program using tools such as MATLAB or Visual Basic. These programs collapse the image into a series of contour lines and induce false-coloration related to signal intensity for better visualization of major and minor peaks. Three-dimensional surface plots represent the same information as contour plots and can be used interchangeably.

It is important to note that the first- and second-dimension separations are not independent of one another. Although the two columns may differ in the mode of separation, the chromatography in the second dimension is largely influenced by the separation that occurs in the first dimension. Looking at a 2D chromatogram, one might notice the structured arrangement of peaks along a horizontal plane. This ordered structure is one of the biggest advantages of utilizing GC × GC for separations of complex mixtures that may contain a variety of isomeric species. In the chromatogram, homologous or isomeric groups of compounds will be placed in a diagonal configuration. This grouping is a result of column orthogonality; however, orthogonality does not always guarantee structured chromatograms. Structured chromatograms are dependent on the presence of a large number of related compounds and the correct selection of a second-dimension column that is optimized to best separate the analytes in the sample.

In order to perform a quantitative analysis of GC × GC chromatograms, the individual peaks must be integrated by combining the modulated slices. GC × GC peak integration is more complicated than 1D GC. One peak may have been modulated multiple times and is thus spread over multiple second-dimension chromatograms. Individual analyte peaks in GC × GC can be quantified with high accuracy and precision through integration in one of two ways: either by determining the peak areas of the individual second dimension chromatograms and summing them or by determining three-dimensional peak volumes through image processing techniques.

Chemometrics and other computational tools are necessary to get useful non-targeted information out of complex petroleum 2D chromatograms. GC × GC chromatograms must be pre-processed before chemometric analysis to ensure that artifacts do not interfere with the chromatographic information. Pre-processing involves baseline correction, noise reduction, normalization, and retention time shifting. This is followed by multivariate data analysis where information is gathered systematically and mathematically from the chromatogram. Pixel-based data analysis extracts information from the overall chromatographic signal instead of using individually selected peaks. Data treatment is then performed by applying a mathematical model based on multivariate statistics. More specific information on the steps for multivariate data analysis can be found in two recent review articles related to chemometrics and GC × GC (Pollo et al. 2018; Zeng et al. 2014).

4.2.7 Petrochemical Applications: Group-Type Analysis

A common application of GC × GC is for group-type analysis of petroleum samples. Petrochemical samples are unique in that the separation and characterization of the complex matrix is the main objective of the analysis. It is no surprise that petrochemical samples were the first subject of GC × GC separations. Group-type quantifications are useful to show the compositional change of a sample over time, based on specific groups of molecules. The change can be used to detail a range of important details, from the conversion of feeds in a refinery to the amount of environmental weathering after an oil spill. PIONA analysis is a commonly used example of a petroleum sample group-type analysis. Here, the hydrocarbon analysis of the sample is based on the concentrations of PIONA. Another important use of group-type analysis is for the determination and quantification of heteroatom-containing species, specifically those containing oxygen, sulfur, and nitrogen. These compounds are often monitored for product quality in petroleum samples.

In recent years, a large number of studies on group-type petrochemical analysis by GC × GC have been published. Prata et al. demonstrated the utility of GC × GC-qMS combined with multiway principal component analysis (MPCA) for the geochemical characterization of 28 Brazilian crude oils from across the country (Prata et al. 2016). A large number of crude oil samples ranging in maturity, origin, and degree of biodegradation were used to form a model to predict these geochemical properties in crude oil. Using this data, they found that biodegradation led to the greatest range of chemical variability between samples. Another study analyzed 11 Brazilian crude oil samples, utilizing principle component analysis (PCA) to determine the three most mature oils (Laakia et al. 2017). Using the group-type clustering seen in the GC × GC chromatogram, they were able to make a retention time plot showing the locations of hydrocarbons in the 2D chromatographic space, shown in Figure 4.12. The ring structures of eight unusual biomarkers were then predicted based on their location in the saturated fraction of the chromatogram. Brazilian crude oils have also been analyzed by group-type semi-quantification by GC × GC-TOFMS using both extracted ion chromatograms (EICs) and total ion chromatograms (TICs) (Coutinho et al. 2018). The findings showed that branched alkanes and *n*-alkanes were the most commonly semi-quantified components in the samples, but the branched alkane concentrations were greater.

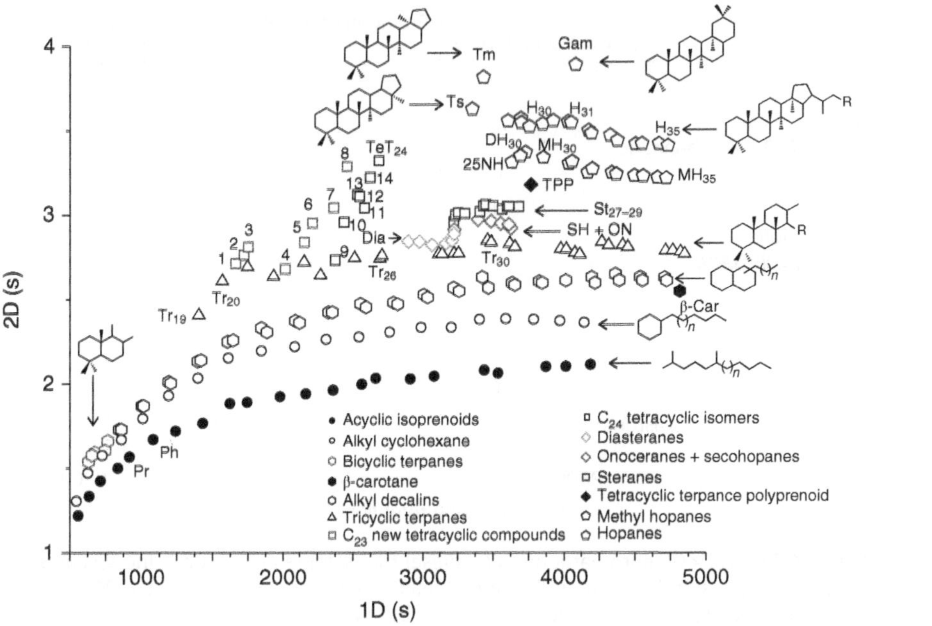

Figure 4.12 GC × GC retention time plot showing the locations of different petroleum hydrocarbons as a result of group-type clustering. Source: Laakia et al. (2017). Reproduced with permission, Copyright (2017), Elsevier.

Middle distillates, such as diesel and jet fuel, encompass a complex range of compounds and are often characterized by group-type analysis due to their complexity. Because this range of petroleum products is used for modern fuels, they are highly regulated, and GC × GC methods for analysis should give comparable results to accepted regulatory methods. The Zimmerman group has successfully applied GC × GC-TOFMS analysis to group-type quantification and classification of petroleum middle distillates. The group analyzed jet fuel, diesel fuel blended with biodiesel, and light heating oil spiked with an internal standard mix for quantification (Jennerwein et al. 2014). Data analysis was done using the Visual Basic Scripting application in the ChromaTOF software and led to the classification of 98% of the sample components, as seen in Figure 4.13. The GC × GC quantifications and characterizations demonstrated good agreement with EN 12916, high performance liquid chromatography (HPLC) method for aromatics, EN 14078, fatty acid methyl esters (FAMEs) in diesel fuel, and EN 14103, for single FAMEs distribution. In a more recent application, the same sample types were again analyzed by GC × GC-TOFMS, but now with an inverse column combination and an offline coupling to HPLC to evaluate DIN EN 12916 (Jennerwein et al. 2017). This

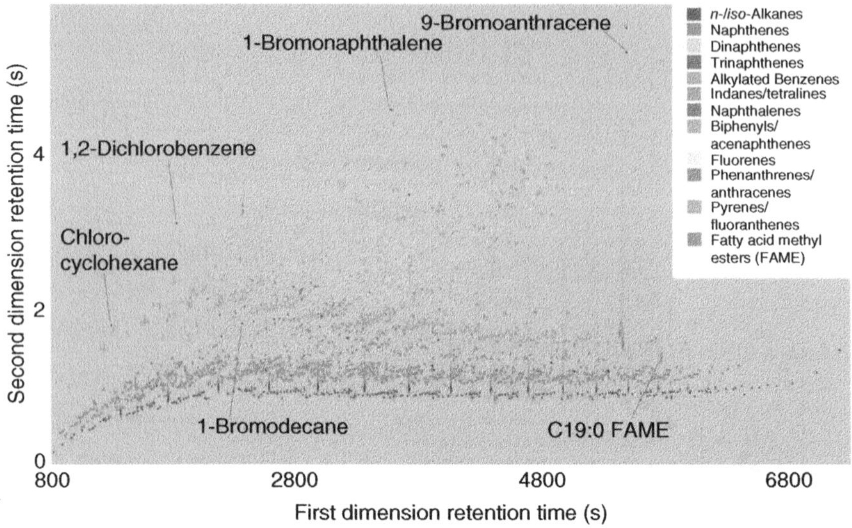

Figure 4.13 GC × GC chromatogram of a diesel fuel sample processed using Visual Basic Scripts. Internal standard compounds are labeled. Source: Jennerwein et al. (2014). Reproduced with permission, Copyright (2014), Elsevier.

work shows that GC × GC-TOFMS gives a more detailed and non-biased quantification and classification of middle distillates than the accepted regulatory method based on HPLC alone.

Thermal oxidation of jet fuels leads to the formation of hydroperoxides, which interfere with the performance of the fuel system parts. Kulsing et al. combined GC × GC-QTOFMS with GC × GC-FPD and 1D gas chromatography–flame photometric detection (GC-FPD) for the hydrocarbon and sulfur compound characterization of Merox refined jet fuels (Kulsing et al. 2017). PIONA analysis of neat and oxidized Merox jet fuels was based on the EIC exact mass values for analytes in each compound group. It did not show great differences between the neat and oxidized fuels, except in the reduction of some saturated hydrocarbons after thermal oxidation. The sulfur content was evaluated with GC × GC-FPD and 1D GC-FPD. It was found that higher polarity sulfur compounds were more common after thermal oxidation. This study gave a more comprehensive group-type analysis of jet fuel samples using a combination of chromatographic methods.

Recently, GC × GC-HRToFMS, combined with Kendrick mass defect (KMD) and knowledge-based rules (KBR), was used to create a screening approach for group-type analysis of complex petroleum samples (Weggler et al. 2019). The Kendrick mass scale defines CH_2 as exactly 14.0000 and can be used to identify classes of compounds as they differ by one alkyl group. The KMD can be calculated by the difference between the nominal Kendrick mass and the exact

Kendrick mass, which can be known when HRToFMS is utilized. In this study, KMD and KBR were combined for the binary classification of accurate mass data, into a specific compound class. With this method, analytes could be sorted into compound classes and their classifications displayed on a pseudo chromatogram, seen in Figure 4.14. Weathered diesel samples were also tested to demonstrate the classifications on real samples. Applying PCA to the samples showed clear sample clustering based on their weathering grades. This study presents a rapid screening method for petroleum samples based on accurate mass GC × GC data with KMD and KBR sample processing methods.

4.2.8 Petrochemical Applications: Contaminated Soil and Sediments

A typical problem with 1D GC analysis of petroleum samples is the ever-present hump of unresolved compounds in the chromatogram called the unresolved complex mixtures (UCMs). Little information can be gleaned from the UCM in 1D GC, but GC × GC allows for better separation and chemical class characterization of UCMs. GC × GC-FID was used to characterize the hydrocarbon UCM in diesel contaminated marine sediments (Frysinger et al. 2003). In this study, two different column combinations were used to separate different extraction fractions of the UCM and achieve the best separation of the UCM fraction. A 100% PDMS column was coupled to a 14% cyanopropylphenyl polysiloxane column in the second-dimension for the separation of the polar aromatic compounds, including monoaromatics, naphthalenes, and the multi-ring PAHs. A chiral γ-cyclodextrin phase was used in the second dimension for the shape-selective separation of branched alkanes and cycloalkanes. This was one of the first applications of GC × GC for petroleum hydrocarbon UCMs in sediments.

An important aspect of remediation of oil-contaminated soils is the complete characterization of the contamination. The more complete the characterization is, the higher the quality and specificity of the remediation for the contaminated soils can be. Typically, total petroleum hydrocarbon (TPH) analysis is done with 1D GC-FID, but Van De Weghe et al. employed normal phase GC × GC-FID for the chemical characterization of soil polluted with industrial oil (Van De Weghe et al. 2006). Comparing the typical TPH method to GC × GC-FID gave better reproducibility and higher recovery of petroleum hydrocarbons with the GC × GC method. One major advantage of the 2D chromatogram was the group-type characterization that is missing with the traditional TPH method. Another crucial aspect of petroleum-contaminated soil remediation is the leaching potential of the soils. The leaching potential corresponds to the threat of the contamination reaching groundwater. A method was presented for the aqueous solubility calculation of TPH in soils using HPLC, followed by GC × GC-FID and Raoult's law (Mao et al. 2009). The hydrocarbon fractions were divided into 79 groups

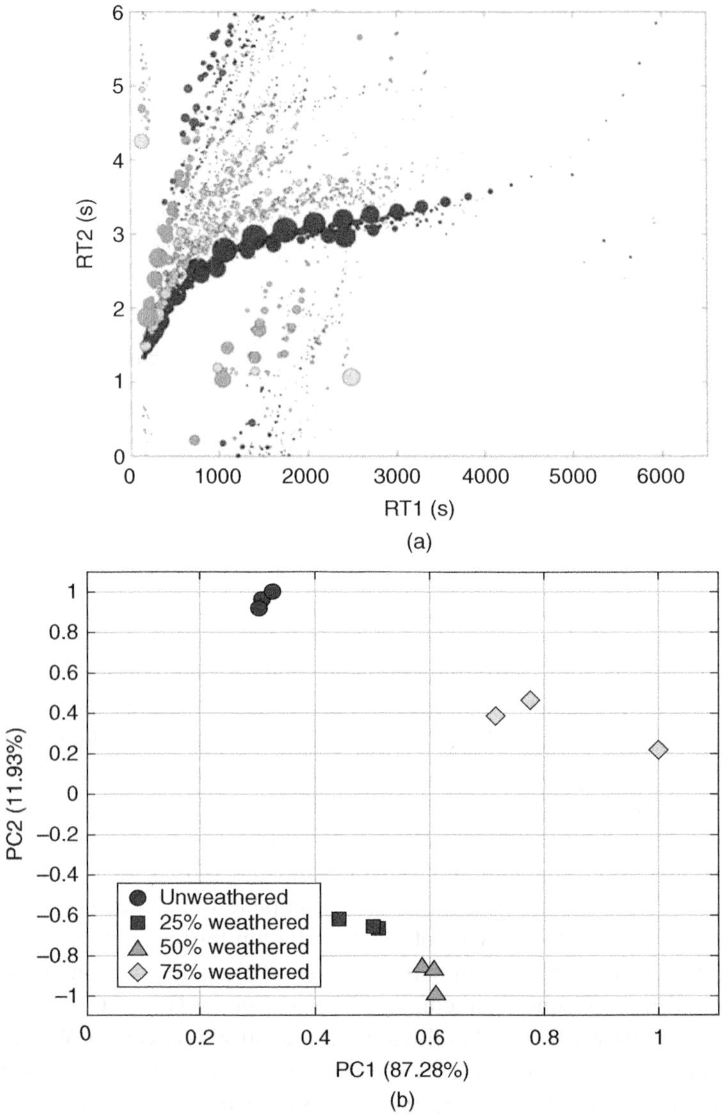

Figure 4.14 Pseudochromatographic figure of diesel sample after classifications applied (a). PCA plot of weathered diesel sample (b). Source: Weggler et al. (2019). Reproduced with permission, Copyright (2019), American Chemical Society.

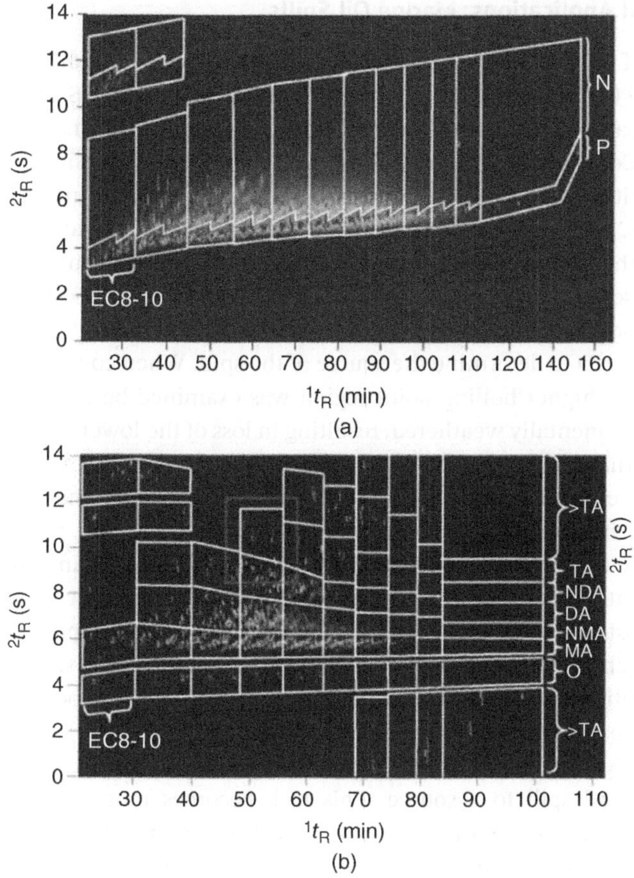

Figure 4.15 GC × GC chromatograms displaying the 79 hydrocarbon fractions, saturated fraction (a), and unsaturated fraction (b). TA, triaromatics; DA, diaromatics; MA, monoaromatics; NMA, naphthenic monoaromatics; NDA, naphthenic diaromatics Source: Mao et al. (2009). Reproduced with permission, Copyright (2009), Elsevier.

based on two GC × GC chromatograms, one for unsaturated compounds and one for saturated compounds; see Figure 4.15. Each of the group type quantifications was used to calculate the water solubility of the fraction, which is related to the leaching potential for that class. A field sample of oil-contaminated soil was tested to demonstrate the applicability of the method. The calculated and measured aqueous solubility values were similar for most of the defined hydrocarbon classes, showing good agreement between the method and the field experiment.

4.2.9 Petrochemical Applications: Marine Oil Spills

The first application of GC × GC for the source identification of a marine oil spill was published in 1999 (Gaines et al. 1999). This project was led by the US Coast Guard Marine Safety Labs, the entity in charge of forensic marine oil spill analysis, and the University of Connecticut. Prior to GC × GC, oil spill source identification was done using a multimethod approach combining fluorescence and infrared spectroscopy with GC to fingerprint regions and visually compare samples. GC/MS later became the primary method used, because more information about specific chemical classes and biomarkers could be elucidated. In the 1999 study, oil spill and possible source samples were both qualitatively and quantitatively compared by GC × GC-FID to determine the source of the spill. When comparing the chromatograms, the higher boiling point region was examined because the spill sample was environmentally weathered, resulting in loss of the lower boiling point region analytes, those below C_{12}. For the comparison of the samples, regions of the chromatograms corresponding to specific chemical classes were analyzed qualitatively (visually inspected) and quantitatively. For example, one region contained naphthalene components, while another region contained alkane and cycloalkane components, although both chromatographic regions showed that the oil sample was most similar to the source oil sample. The source sample that was selected as a match for the spill using GC × GC-FID was the same sample selected using the traditional methods. Following this study, GC × GC has become an invaluable tool for marine oil spill source identification.

GC × GC has been shown to effectively fingerprint and compare samples in order to match an oil spill to a source. This task becomes more difficult as the oil spill experiences weathering as a result of dissolution, evaporation, biodegradation, water washing, and photo-oxidation. The sample collection should be done as soon as possible to avoid the extensive weathering of the sample. Oil weathering has been studied, and GC × GC has been used to elucidate the differences between weathered and non-weathered samples. One study analyzed samples taken from oil-coated rocks as a result of the April 2003 *Bouchard 120* oil spill in Buzzards Bay, MA (Nelson et al. 2006). The weathering profiles of samples collected 12 days and 179 days after the spill were compared using GC × GC-FID and data-visualization techniques. Before comparison of the 12-day and 179-day chromatograms, they were both normalized and scaled to the conserved biomarker 17α(H)-21β(H)-hopane. Three types of advanced image processing techniques were used to qualitatively compare the samples, including difference, ratio, and addition chromatograms. In these methods, the GC × GC chromatograms of the 12-day and 179-day samples were point-by-point subtracted, divided, or added, respectively. The difference chromatograms show prominent peaks from the *n*-alkanes and low molecular weight PAHs, meaning that they were removed due to weathering. The ratio chromatograms show that the alkanes and aromatics in the n-C_{19} to n-C_{25} likely experience more weathering than one- and two-ring cycloalkanes and alkylbenzenes. The addition chromatograms could be applied to determine the sources of the weathering based on what compound classes were conserved or lost. Evaporation, water washing, and biodegradation were found to be the main causes of weathering for these samples.

In oil spill analysis of heavy fuels or crude oils, weathering resistant biomarkers, such as steranes and hopanes, can be found in the persistent heavier fractions. Diesel oil spills are less resistant to long-term weathering and do not display the same recalcitrant biomarkers. The lack of biomarkers makes the spill-source matching difficult by traditional 1D GC methods. Chemometric techniques have been combined with GC × GC data to determine diesel oil weathering biomarkers and aid in spill-source matching. In a more recent study, weathering processes were investigated for light diesel oils (n-C_9 to n-C_{20}) using GC × GC-(HR)QTOFMS and pixel-based analysis (Alexandrino et al. 2019a). Advantages of GC × GC-(HR)QTOFMS include the separation of naphthenic compounds that typically coelute in 1D GC, better detection of high molecular weight hydrocarbons ($>n$-C_{18}), and accurate mass identification of diagnostic hydrocarbon compound classes. The pixel-based analysis using PCA is a more time-efficient and accurate way of comparing samples of large data sets than the time-consuming visual inspection of 2D chromatograms. In the pixel-based analysis, mathematical methods are used to extract, pixel-by-pixel, the most relevant chromatographic information to distinguish samples in a large dataset. Two light diesel oils were weathered by evaporation, photo-oxidation, and biodegradation to determine the weathering-resistance petroleum biomarkers that could be used to accurately identify and differentiate very similar diesel oils. In addition, weathered samples were able to be correctly matched back to their diesel oil sources after the removal of chromatogram sections characteristic of evaporative weathering. In a similar study, diesel oil spill samples were matched to their sources using GC × GC-(HR)QTOFMS and 2D CHEMometric analysis of Selection Ion Chromatograms (2D-CHEMSIC) (Alexandrino et al. 2019b). 2D-CHEMSIC is a pixel-based analysis using chemometrics of 2D EICs. This method was used to determine the chemical variability of the samples in order to distinguish similar sources of diesel oil and match oil spills to their source.

Recently, Nelson et al. compared source oils from two iconic large scale oil spills that occurred in the Gulf of Mexico (Nelson et al. 2019). They combined GC × GC-FID and HRToFMS to explore the chemical complexity of the source oil from the 1979 to 1980 Ixtoc I blowout and the Macondo well oil from the 2010 Deepwater Horizon (DWH) explosion. In comparing GC × GC-HRToFMS chromatograms, the greatest difference between the source oils was in the

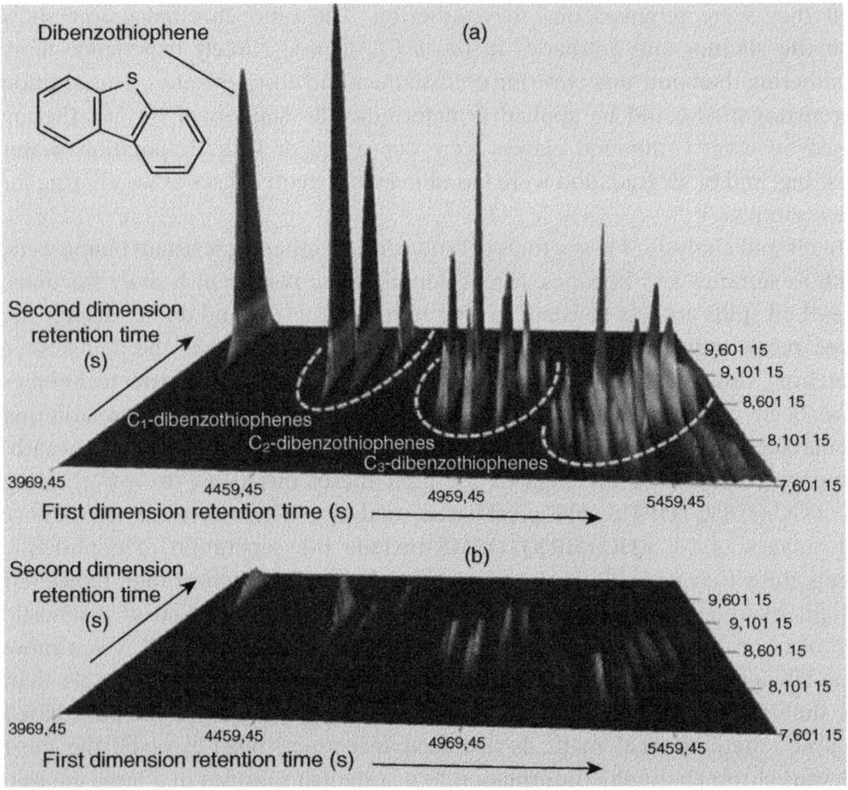

Figure 4.16 A comparison of the GC × GC chromatogram dibenzothiophene region to discriminate between the (a) Ixtoc I crude oil and the (b) DWH crude oil. Source: Nelson et al. (2019). Reproduced with permission, Copyright (2019), American Chemical Society.

dibenzothiophene and alkylated dibenzothiophene regions. As can be seen in Figure 4.16, the Ixtoc I crude oil contains more dibenzothiophenes than the DWH oil, and these compounds can be used to discriminate between the two oil types. Using the accurate mass information from the HRToFMS detector, a plot of ring double-bond equivalents (RDBE) versus carbon number was produced and allowed for the discovery of three other sulfur-containing families including phenanthrothiophenes, benzonaphthothiophenes, and chrysenothiophenes. These compound classes are useful for the petroleum fingerprinting as they vary greatly between source oils. Alkylated carbazoles and benzocarbazoles were also found extensively in the Ixtoc I crude but not in the DWH crude samples. Researchers can better characterize oil spill samples and are more likely to match the sample to the spill source using these advanced analytical techniques. GC × GC, combined with FID and MS detectors, has become a critical tool for petroleum fingerprinting and will continue to answer questions in the oil spill forensics field.

4.2.10 Petrochemical Applications: Hydraulic Fracturing

Unconventional natural gas development (UNGD), including horizontal drilling and high-volume hydraulic fracturing, requires large volumes of water with additives to create pressures high enough to fracture the shale and release the natural gas. Fracturing fluids are mostly water (90%) but also contain sand (9%) and chemical modifiers (up to 2%). Once the wastewater fluids are pumped back to the surface after use, they are called produced waters (PWs) or flowback. These PWs are difficult to treat because of their chemical complexity; therefore, they are typically disposed of in Class II injection wells. A growing concern within the public is the contamination of potable water by PWs, which can be a result of spills during transport, or leaks during storage and disposal. A 2015 study attributed a groundwater contamination event in Pennsylvania to Marcellus Shale gas development, aided by the use of GC × GC-TOFMS (Llewellyn et al. 2015). Stray natural gas had contaminated an aquifer and well waters were observed to produce a white foam when in use, although the cause of this was not immediately known. Water samples were taken from the impacted household wells and from roughly 30 PWs from surrounding UNGD locations. The 2D chromatograms for both the PWs and impacted well waters showed similar UCMs of hydrocarbons. In addition, the compound 2-*n*-butoxyethanol was positively identified in PW samples and one of the foaming drinking water wells. The evidence from this study confirmed that natural gas leaks and PW contamination are plausible environmental and human health concerns.

In order to properly treat and dispose of flowback waters, a complete chemical characterization of the waters is needed, including information about how flowback differs from the fracturing water mixtures. Using non-targeted GC × GC-TOFMS and advanced scripting algorithms, Piotrowski et al. compared fracturing fluids and the resulting flowback over a 10-day period (Piotrowski et al. 2018b). This study found that the hydrocarbon concentrations increased with time, but the relative chemical compositions of the flowback samples remained relatively constant. Data reduction methods using advanced scripting algorithms were used to examine class-specific fragmentation mass-to-charge ratios. Figure 4.17 displays the 2D chromatograms of the scripted hydrocarbons displayed as bubbles where the size indicates relative ion abundance. Additionally, they identified three common organic modifiers in the fracturing water using HRToFMS and determined their fates. Ethylene glycol was removed during the first four days in the well while glutaraldehyde and cinnamaldehyde were transformed from polymerization reactions.

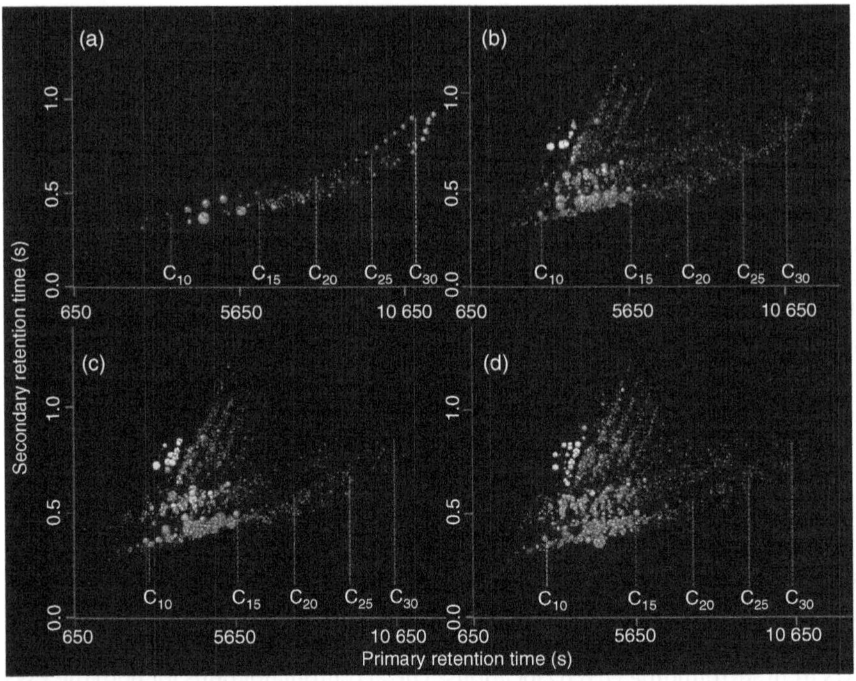

Figure 4.17 The GC × GC bubble chromatograms of flowback samples: (a) day 4, (b) day 6, (c) day 8, and (d) day 10. The lowest bands (a) show the *n*-alkanes (very dark gray) and the branched alkanes (dark gray). The remaining bands from bottom to top are gray-scaled from dark to light and show the alkenes, cyclopentanes, cyclohexanes, para-substituted cyclohexanes, fluorenes, 3-ring aromatics, and hydronaphthalenes. Source: Piotrowski et al. (2018b). Reproduced with permission, Copyright (2018), Elsevier.

As was briefly mentioned in the last study, organic modifiers used in fracturing waters may be transformed while in the well, although this area of research is limited. In an investigation of the Fayetteville Shale (Arkansas), six PW samples were analyzed with GC × GC-FID and TOFMS (Hoelzer et al. 2016). Strict compound-identification-confidence criteria were used to classify compounds according to their origin. The hydrocarbon differences seen in the 2D chromatograms allowed them to be used for chemical fingerprinting, as well as give information about the geological formation. Hopanes were found to be biomarkers from geological processes and can be used to fingerprint the individual shale formation and estimate the thermal maturity of the shale hydrocarbons. Furthermore, many compounds were found that could be attributed to disclosed fracturing water mixtures, as well as undisclosed compounds such as halogenated hydrocarbons and chloromethyl alkanoates. Halogenated methanes and acetones were determined to be unintended byproducts of the fracturing water additives. This study shows the chemical complexity of fracturing waters and PWs and underscores the need for advanced analytical techniques like GC × GC for their analysis.

PWs and flowback from shale gas wells have unique hydrocarbon patterns that are specific to the well and may be used for point-source identification of a well following a contamination event. To fully understand the chemistry of the flowback waters, a detailed hydrocarbon analysis (DHA) should also be performed on the shale rock itself. In one study, thermal desorption was used to directly introduce shale rock samples into a GC × GC-TOFMS to characterize the hydrocarbon signatures of differently sourced shale samples (Piotrowski et al. 2018a). Two shale formations, Marcellus and Utica/Point Pleasant, were sampled and compared to see if their hydrocarbon fingerprints could be used to differentiate between the two locations. The shale samples were heated to 350 °C, and the thermally desorbed analytes were directly introduced into the GC × GC for analysis. Soxhlet extraction of the shale core samples was also conducted to compare to thermal desorption. The two extraction methods produced similar results overall, but the thermal desorption performed better for more volatile compounds, while Soxhlet extraction preserved the higher molecular weight compounds better. Thermal desorption was also shown to be quicker and presents less contamination due to the lack of organic solvents used. For a comparison of the two shale rock samples, their hydrocarbon fingerprints were used. The Utica shale has a higher abundance of mono-aromatic and naphthalene compounds, while the Marcellus shale contained more branched alkanes and linear alkanes. The method used was also able to determine the presence of sulfur-containing compounds. Their identification in shale rock is important because large amounts of sulfur-containing compounds may lead to soured gas.

4.2.11 Conclusions of Utilizing GC × GC in the Oil and Gas Industry

Comprehensive GC (GC × GC) is a well-established multidimensional analytical technique where several commercial offerings with respect to instrumentation and data processing software are available. Due to the complexity of GC × GC, this technique is typically reserved for advanced users; however, due to the higher degree of detail obtained for complex samples, GC × GC is gaining in use and popularity.

Petroleum and petroleum distillates are highly complicated samples, where the enhanced separation space provided by GC × GC (compared with 1D GC) can be exploited to provide detailed sample information. As was shown, GC × GC can be coupled to different detector technologies, with FID and TOFMS being the most commonly applied detectors. Several examples were presented, ranging from the analysis of petroleum and distillates to their identification in environmental samples (soil, sediment, and water). Additional examples where GC × GC was exploited for environmental monitoring are presented in Chapter 10.

4.3 Petroleum and Hydrocarbon Analysis by Gas Chromatography: Vacuum Ultraviolet Spectroscopy

4.3.1 Introduction to GC-VUV

Even though GC has been widely recognized as the most efficient technique for the separation and speciation of volatile components, different detectors and different methods are currently used. In the past few years, the use of a vacuum ultraviolet (VUV) broadband absorption spectroscopy detector has increased significantly in importance in the petrochemical analysis field (Schug et al. 2014). Analytes separated by GC are directed through a transfer line to a VUV flow cell. The VUV detector is applicable for the detection of all molecules that absorb strongly in the 125–240 nm wavelength range (a second-generation detector measures absorption from 125 to 430 nm). Detection occurs for molecules able to absorb light in this low wavelength deep ultraviolet region, specifically undergoing $\sigma \rightarrow \sigma^*$, $n \rightarrow \sigma^*$, $\pi \rightarrow \pi^*$, and $n \rightarrow \pi^*$ electronic transitions, among others (Lelevic et al. 2019; Santos and Schug 2017; Schug et al. 2014). This accounts for virtually every molecule, and every molecule has a unique absorption spectrum. Absorption follows standard Beer–Lambert law principles, and thus, the VUV detector is one of the few GC detectors which provides both quantitative and qualitative information. Some examples of VUV absorption spectra for different compound classes are shown in Figure 4.18; these spectra provide some indication regarding how different compound classes exhibit fairly similar (e.g. for saturated hydrocarbons) or drastically different (aromatic compounds) absorption spectra.

The VUV is interfaced with the GC through a heated transfer line (usually thermostatted at 300 °C to avoid condensation of separated analytes), which contains an uncoated deactivated glass capillary to connect to the GC column in the GC oven. At the end of the transfer line, a makeup flow of carrier gas can be introduced to minimize the time spent by the analytes in the detector flow cell. The standard VUV flow cell has a path length of 10 cm and a volume of 80 μl (Schug et al. 2014). The light source and detection modules, which book-end the flow cell, are specially treated and sealed in an inert environment, but are not held under vacuum. Indeed, the VUV detector requires no vacuum to operate; VUV is simply a moniker referring to how measurements in this wavelength range were made previously in bright-source synchrotron facilities. A vacuum was needed to remove strongly absorbing species in air, such as water, carbon dioxide, and oxygen, which can interfere with VUV measurements. Luckily, GC carrier gases,

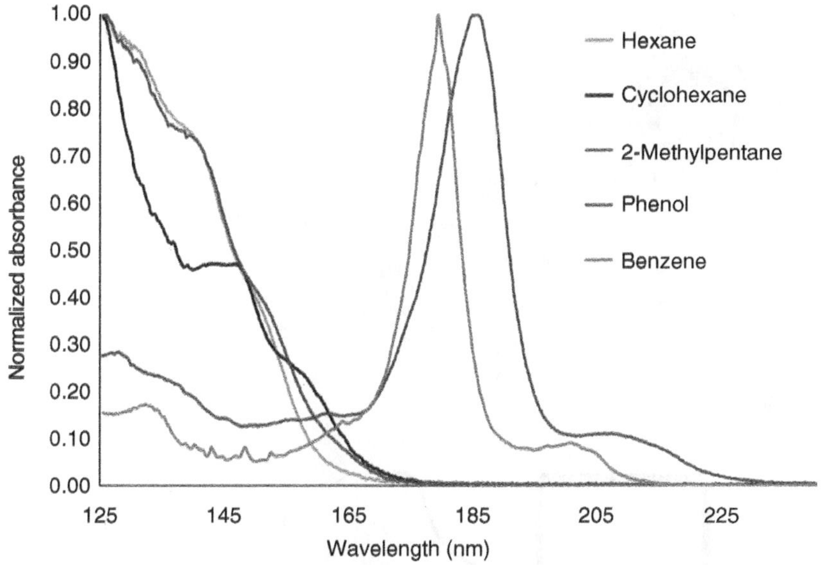

Figure 4.18 Representative VUV absorption spectra for a series of molecules representing different compound classes (*n*-hexane, 2-methyl pentane, cyclohexane, phenol, and benzene). Source: Image courtesy of VUV Analytics.

such as nitrogen, hydrogen, and helium (as well as argon), do not absorb appreciably and yield negligible and easily subtractable background signals during normal operation of the detector. The source and detector modules interface with the flow cell through magnesium fluoride windows; these ultimately control the lowest wavelength bandpass for the detector.

A modified deuterium lamp is the source, and reflective optics are used to direct the light through the flow cell and to the detector. Full range absorption spectra are recorded simultaneously (at up to 100 Hz acquisition speed) with the presence of a back-thinned charge coupled device (CCD) detector. A dark noise reading is taken at the beginning of the run, for background subtraction. To reduce background absorption, the path is purged with nitrogen or a noble gas (Santos and Schug 2017). Figure 4.19 schematically represents the VUV detector and the arrangement of its components (Santos and Schug 2017; Schug et al. 2014).

The VUV is formally a concentration sensitive detector (Liu et al. 2017), but it appears to exhibit some mass sensitive properties, in that modest increases in flow rate do not have a large effect on detector response. For this reason, the use of a makeup flow carrier gas can aid in the separation and detection of certain analytes. Increasing make-up gas flow can mitigate band broadening; however, the residence time of the analyte in the flow cell should be balanced by the acquisition rate of the detector. While the detector can acquire at up to 100 Hz, such a fast

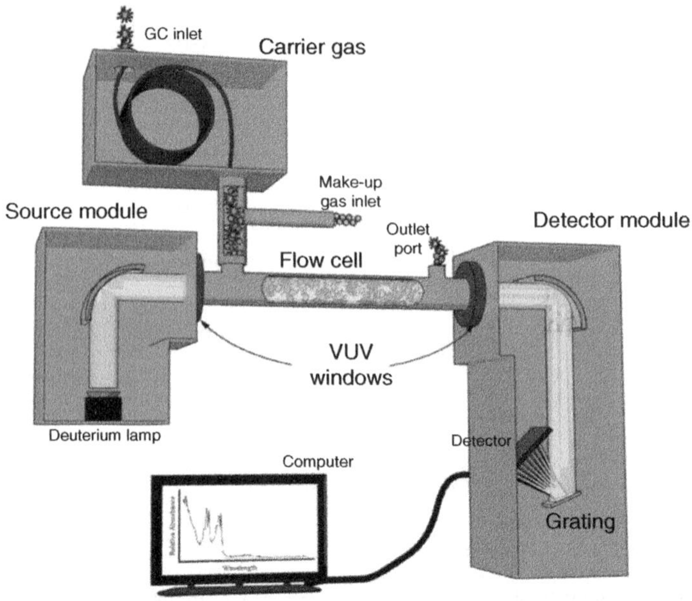

Figure 4.19 Schematic of the components of the VUV detector for GC. Source: Schug et al. (2014). Reproduced with permission, Copyright (2014), American Chemical Society.

acquisition is rarely utilized, because it will increase detector noise. The make-up flow is adjusted as needed to limit band broadening, and the acquisition speed is set just fast enough to ensure sufficient data points are acquired across each chromatographic peak. Only very short peak widths observed in fast GC and comprehensive 2D GC (GC × GC) require the use of higher acquisition speeds (Zoccali et al. 2017).

4.3.2 GC-VUV Data Processing

As full wavelength-range spectra are captured with each data point, there are a myriad of post-run data processing treatments, which can be applied using the software. One of the most powerful tools is the deconvolution of chromatographically unresolved analytes. Because virtually all molecules have unique absorption spectra, individual analyte contributions to an overlapped signal can be determined based on spectroscopic resolution. A signal associated with coelution will be comprised of an additive absorption from each species, weighted by the abundance of each. Thus, an abundance-weighted linear combination of reference library spectra can be used to replot a chromatogram, highlighting the presence of individual signals. Requirements for this process to work well include the presence of reference library spectra for each of the overlapping species and sufficient spectral dissimilarity among the species to accommodate differentiation over a wide range of relative abundances present. The more dissimilar the spectra of two coeluting species, the wider the range of relative abundances of each that can be deconvolved (Schenk et al. 2016; Skultety et al. 2017). Recent research has explored alternate deconvolution algorithms, which could accommodate the coelution of unknown species, but this development is not yet commercially available.

This universal detector is capable of providing both qualitative and quantitative information. The ability to provide qualitative information is an important feature, because most common GC detectors, such as FID, ECD, and TCD, only provide quantitative information. Qualitative assignment of individual standards can only be made based on chromatographic indicators, such as retention index (Santos and Schug 2017; Schug et al. 2014). MS is the most common detection technique, which also provides qualitative information for eluted compounds. In many ways, VUV detection is highly complementary to MS; additionally, VUV can be used to analyze labile molecules that are not suitable for MS analysis, for example, formaldehyde (Schenk et al. 2019) and some pesticides (Fan et al. 2015). Overall, isomer differentiation and facile deconvolution of coeluting species based on additive spectroscopic signals are areas where VUV can be superior to MS detection.

Initial work investigating the complementarity of VUV and MS detection was performed by Gröger et al. in the first coupling of the VUV detector to a GC × GC system (Gröger et al. 2016). The GC × GC was performed using cryo modulation. One of the most important aspects of this work was evaluating the acquisition speed of the VUV; 100 Hz acquisition speed was shown to be more than adequate to keep up with the narrow GC × GC peak widths. One initial drawback of the detector highlighted by this research was the large internal volume of the detector flow cell (80 μl). For the very small volume peaks generated by GC × GC, this large volume caused undesirable peak broadening. This drawback has been since addressed in newer generations of the detector, where the current commercially available VUV detectors are outfitted with a 37 μl flow cell. Worthy of note, in another report of GC × GC-VUV where biodiesel mixtures were investigated among other samples in a flow-modulated system (Zoccali et al. 2017), the VUV detector was able to accept separation flow rates as high as 12 ml/min; additionally, a VUV acquisition speed of 33 Hz was adequate to obtain enough data points across each peak.

The complementarity of VUV and MS was examined by Solouki and coworkers (Anthony et al. 2018a,b), where the confirmation of the identification of species in complex mixtures can be aided significantly by additional information. VUV detection, in conjunction with MS detection, improves identification through library searching strategies, while also providing a significant increase in the confidence of assignments, especially for isomeric species, which can be difficult to differentiate by GC-MS.

4.3.2.1 Time Interval Deconvolution (TID) Algorithm

A time interval deconvolution (TID) algorithm was created to provide a faster and more advanced tool for deconvolving analytes, which provides the automatic deconvolution of analytes, using approximate retention indices, the VUV spectral library, and the known additivity of absorption through the use of the Beer–Lambert law (Santos and Schug 2017; Schug et al. 2014). TID was developed and used by Walsh to characterize different gasoline samples (Walsh et al. 2016). Figure 4.20 illustrates the automatic deconvolution obtained between 1,3,5-trimethylbenzene and 2,3-dimethyloctane, which partially overlaps between 76 and 77 minutes. Figure 4.20b shows the deconvolution obtained by summing the absorbance scans within the time region and fitting the resulting total absorbance spectrum.

The TID algorithm divides the chromatogram into different time slice intervals and analyzes it based on the summed absorbance measured in each interval to identify the individual components, as well as their contribution to the total absorbance spectrum. TID can be used to target individual components or to collate the total content of different compound classes, such as in a PIONA analysis.

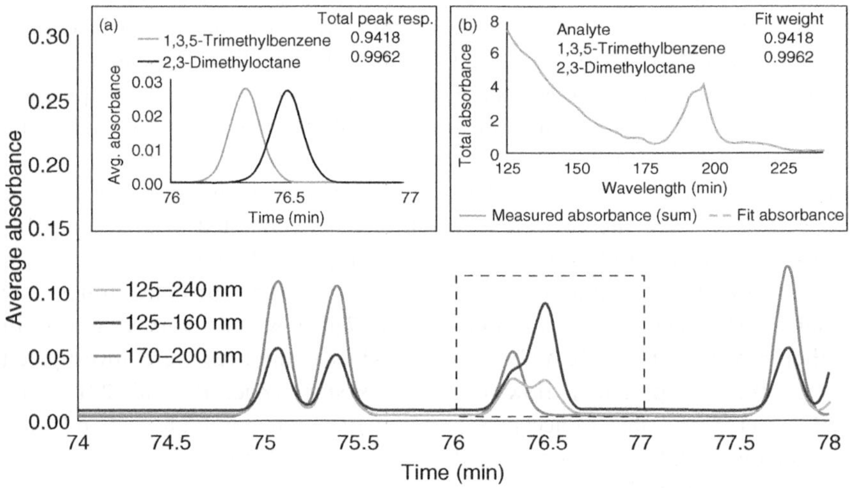

Figure 4.20 Example of deconvolution of coeluting analytes in a gasoline sample. The bottom overlay shows the same GC-VUV separation at different wavelength filters with the co-elution highlighted between 76 and 77 minutes. The deconvoluted chromatogram between 76 and 77 minutes and the corresponding spectra for 1,3,5-trimethylbenzene and 2,3-dimethyloctane are shown above. Source: Walsh et al. (2016). Reprinted with permission, Copyright (2016), American Chemical Society.

Table 4.2 Measured VUV average relative response factor (RRF) values for different classes.

Hydrocarbon class/compound	RRF
Paraffin	0.769
Isoparaffin	0.781
Olefin	0.465
Naphthene	0.786
C9+ aromatic	0.296
Methanol	1.211
Ethanol	1.029
Benzene	0.258
Iso-octane	0.674
Toluene	0.267
Ethylbenzene	0.284
Xylenes (*m*-, *p*-, *o*-)	0.284
Naphthalene	0.207
1,2-Methylnaphthalene	0.250

Values are measured relative to methane, defined at RRF = 1.000.
Source: Walsh et al. (2016). Reproduced with permission, Copyright (2016), American Chemical Society.

When the TID is used for PIONA determination, at the end of the analysis, the responses of each PIONA class and selected single compounds are calculated and converted into a percentage of the total mass contributions using relative response factors as shown in Table 4.2 (Walsh et al. 2016).

The TID algorithm underlies the recently posted ASTM D8071-19 methodology (ASTM-D8071 2019), which uses gas chromatography–vacuum ultraviolet (GC-VUV) for the determination of select hydrocarbons and oxygenates in spark-engine fuels. Table 4.3 indicates the scope of this method; compounds and classes of compounds, which can be speciated with specified precision, are listed. Theoretically and practically speaking, this method can potentially replace or subsume a large number of other ASTM methods for similar PIONA or PIONA+ determinations, which feature much more complex instrumentation, lower precision, and/or worse throughput. Current efforts are also being made to extend applications to heavier fuels; ASTM D8267-19a was developed for the determination of the total aromatic content in aviation fuels using GC-VUV.

Table 4.3 Scope of ASTM D8071-19 compound and compound class determinations with specified precision (ASTM-D8071 2019).

Property	Units	Applicable range
Paraffins	vol%	3.572–23.105
Isoparaffins	vol%	22.697–71.993
Olefins	vol%	0.011–44.002
Naphthenes	vol%	0.606–18.416
Aromatics	vol%	14.743–58.124
Methanol	vol%	0.063–3.426
Ethanol	mass%	0.042–15.991
Benzene	vol%	0.09–1.091
Toluene	vol%	0.698–31.377
Ethylbenzene	vol%	0.5–3.175
Xylenes	vol%	3.037–18.955
Naphthalene	vol%	0.019–0.779
Methylnaphthalenes	vol%	0.21–1.484

The method may be applied outside of this scope, but precision has not been determined for such determinations.

Figure 4.21 shows the potential for VUV absorption spectroscopy to differentiate a myriad of mono- and diaromatic species.

The spectral resolution is on the order of 5 nm, and gas-phase absorption spectra are not subject to solvent effects or various spectral broadening processes; thus, the spectra for individual species shown in Figure 4.21 are unique and well differentiable from one another. Ultimately, ASTM methods are likely to favor collective quantitation of a class of aromatics over individual speciation; however, individual speciation is possible.

4.3.2.2 Pseudo-absolute Quantitation

Another interesting feature of the VUV detector is the potential to quantitate an analyte observed during a GC-VUV analysis without the need for calibration. This concept is referred to as pseudo-absolute quantitation (Bai et al. 2017). Ultimately, the absorption cross-section (or the molar absorptivity of a compound across a range of wavelengths) is a physical property of a molecule, which can be determined and recorded. When an absorption cross-section is known for a compound, and that compound reaches the VUV detector, then the number of molecules present in the flow cell that create an absorption event of a specific

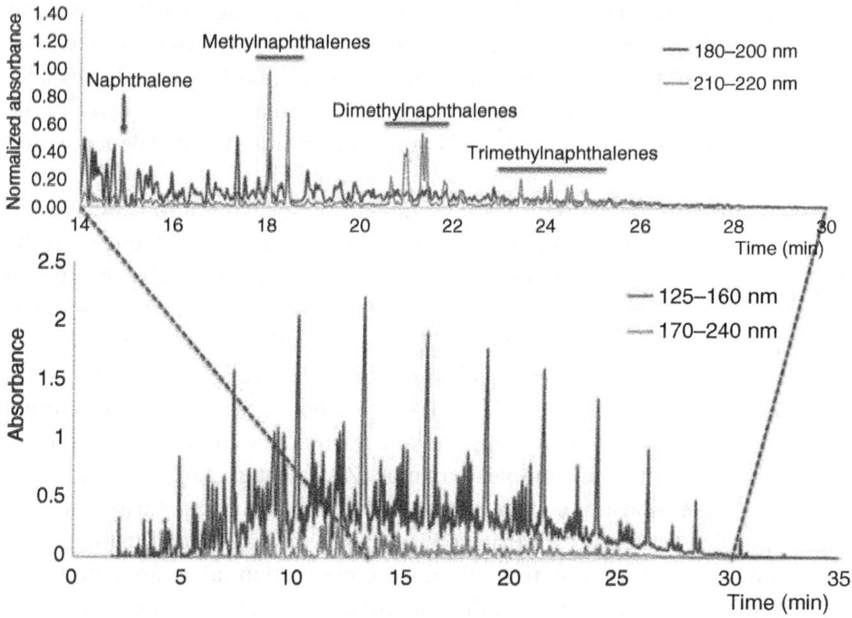

Figure 4.21 Chromatographic analysis of jet fuel using different spectral filters. Filters of 125–160 nm and 170–240 nm highlight saturates and aromatics, respectively. Aromatics absorption can be further parsed using filters 180–200 nm and 210–220 nm for mono- and diaromatics, respectively. Source: Schenk et al. (2016). Reproduced with permission, Copyright (2016), Elsevier.

magnitude can be determined. Of course, this is dependent on knowing the detector scan rate, the flow cell volume, the absorption path length, and the carrier gas flow rate, but these are all parameters set by the user.

If one were to be able to inject a sample directly into the VUV spectrometer, without loss, then an absolute determination could be made of the number of molecules present for analytes with registered cross-sections. The methodology used in GC-VUV for calibration-less quantitation is called pseudo-absolute, because, in order to relate the number of molecules observed in the flow cell back to the original sample concentration, it is necessary to determine the systematic effects of the injection process on analyte quantity. For example, using split injection results in a systematic loss of analyte between injection and detection. Overall, there can be a variety of systematic losses and gains, which can occur during GC injection. For example, it is not enough to know that the split ratio is set to 100 : 1 because the accuracy of these settings can be different from instrument to instrument. The accuracy of such settings (injection volume is another example) is not a high priority for manufacturers, because the vast majority of quantitative analyses involve calibration. As long as settings, such as the split ratio and injection volume, can be controlled precisely, then performance will be suitable for a calibration-based quantitation. For calibration-less, pseudo-absolute quantitation, if an internal standard is not used, then the systematic error in the accuracy of each GC setting that effects on-column analyte abundance would need to be determined.

Bai et al. introduced GC-VUV-based pseudo-absolute quantitation and showed how this concept could be used to determine such systematic influences on the amount of analyte injected on-column (Bai et al. 2017). Ultimately, an internal standard-based method for pseudo-absolute quantitation makes more practical sense, because the internal standard will experience the same system parameters as the analyte, and its signal can be used to normalize any systematic errors experienced by the analyte. Zoccali et al. showed how both direct- and internal standard-based pseudo-absolute quantitation approaches could be used for the determination of FAMEs in complex mixtures (Zoccali et al. 2017). Beyond these studies, this calibration-less concept for quantitation has not been further utilized or developed; however, one can surmise a large time saving for a methodology that could avoid the need for calibration. Of course, cross-section values are experimentally determined and will have a finite amount of error. These values need to be continually refined and made as precise as possible in order to provide the best accuracy for pseudo-absolute determinations.

4.3.3 GC-VUV Applications

Due to all these characteristics, the VUV detector is widely increasing its importance in the field of petrochemical analysis, with results comparable with those associated with the standardized methodologies. Bai et al. used this technology to analyze permanent gases (Bai et al. 2015). The characterized gases originated from sources, such as water wells located less than a mile away from an unconventional gas extraction well. In addition, off-gassing components from thermal runaway events exhibited by lithium–ion batteries were also analyzed. Generally, this type of analysis is performed using a gas chromatography–thermal conductivity detector (GC-TCD) or FID, the former for the analysis of nitrogen, carbon dioxide, and light hydrocarbons, and the latter for higher molecular weight hydrocarbons. With the VUV, all these components can be analyzed simultaneously, as shown in Figure 4.22 (Bai et al. 2015). The GC-VUV also provides the ability to differentiate coeluting gases, an application that would have been very difficult with GC-MS.

This detector has also been used for the analysis of diesel fuels (Weber et al. 2016). The developed method separated and speciated mono-, di-, and polyaromatic compounds through the use of an IL-based GC column. Deconvolution was used to demonstrate the ability of the GC-VUV to spectroscopically discriminate

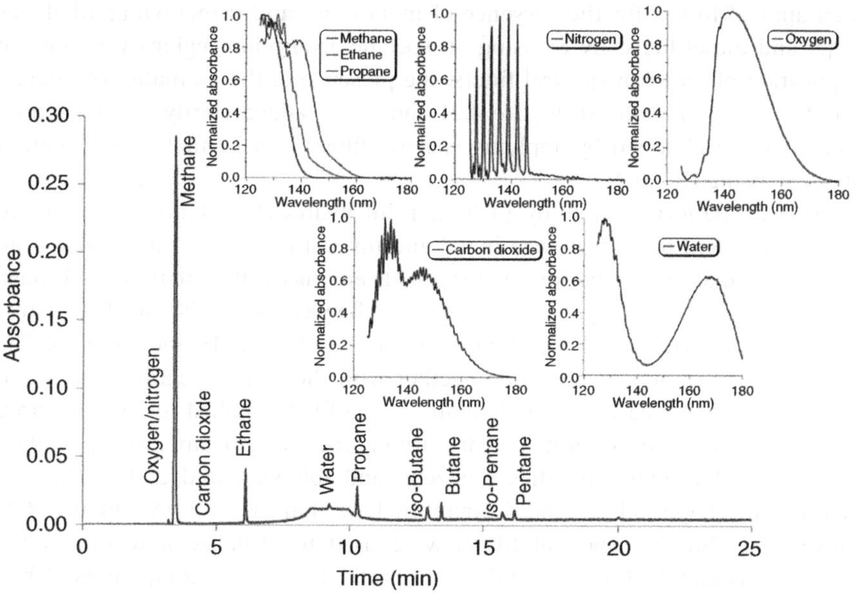

Figure 4.22 GC-VUV analysis of natural gas, including examples of common absorption spectra encountered. Source: Bai et al. (2015). Reproduced with permission, Copyright (2015), Elsevier.

between saturates and monoaromatics. The singularly novel aspect of this work was the development and use of a new dynamic filter. In this approach, the device identified the wavelength of maximum absorbance within a search window and tracks its average response in a smaller sub-window. Thus, all components can be tracked and plotted at their maximum absorption wavelength, where they will exhibit the best sensitivity. The results obtained from this method were comparable with those obtained using supercritical fluid chromatography with flame ionization detection. An absolute error of 0.8% for light diesel samples was reported. When comparing light gas oils, on the other hand, discrepancies were attributed to the incomplete resolution among saturate and monoaromatic compounds, and among monoaromatic and diaromatic groups when analyzing samples with a high fraction of species characterized by higher molecular weight (Weber et al. 2016).

Schenk et al. used the deconvolving power of the VUV detector to speciate dimethylnaphthalenes (Schenk et al. 2016). This class of compounds is challenging to analyze via GC-MS due to their chromatographic coelution and similar mass spectra. The different investigations performed using the GC-VUV showed the ability of the system to differentiate the isomers present at different ratios, up to nearly 2 orders of magnitude (1 : 99) in relative abundance. The GC-VUV was then applied to test for the presence of mono-, di-, and trimethylnaphthalenes in jet and diesel fuels. By focusing on specific absorption regions through the application of post-run spectral filters, the presence of the aromatics of interest could be better visualized. Figure 4.21 shows the added clarity for speciation, which can be obtained by applying spectral filters to a jet fuel chromatogram obtained using GC-VUV.

Bai et al. performed a study to fingerprint hydrocarbon biomarkers for age and source identification in weathered and unweathered diesel fuels (Bai et al. 2018). This analysis was performed to speciate amongst different diesel fuel spills, especially for their dating, based on the ratio between C_{17}/pristane. The results were then compared to those obtained via GC × GC-TOFMS and GC-FID. The biomarkers pristane and phytane, together with their ratios relative to C_{17} and C_{18}, were determined in the fuel samples. The filters applied for this research were 125–155 nm for saturated hydrocarbons (octane, pristane, and phytane), 170–195 nm for monoaromatics (benzene and toluene), and 200–230 nm for diaromatics (naphthalene and 1,5-naphthalene; Bai et al. 2018). Figure 4.23 shows how different spectral filters were used to visualize a region in the VUV spectrum, which corresponded to a particular class of compounds. When compared with the results obtained using GC × GC-TOFMS, it was clear that the review of the data of the GC-VUV was easier to perform, but it was also unquestionable that the presence of these different classes and the differences among sample compositions may be better visualized through the fingerprint associated with the GC × GC-TOFMS analysis (Bai et al. 2018).

The analysis of diesel spill biomarkers is often performed using GC-FID, primarily because it is instrumentation familiar to judges and lawyers who are involved with environmental forensics cases. Of course, GC-FID of diesel fuel is characterized by many overlapping signals, which can complicate accurate assignment and quantitation of individual components. The use of the GC-VUV can solve the issue of coelution and the relative possible peak overestimation through the deconvolution of the peaks and by calculating their exact contribution. Bai compared the ratios obtained through triplicate analyses of weathered diesel using different techniques and detectors, as seen in Table 4.4. The results obtained with the GC-VUV are comparable with those obtained with the GC-FID but differ more significantly from the GC × GC-TOFMS (Bai et al. 2018). GC × GC-TOFMS provides an increased resolution of species in these complex mixtures, which can be assumed to aid in the accurate assignment of abundance ratios; however, the instrumentation is more complicated, and it requires more training in order to operate effectively.

Dunkle et al. compared GC-VUV with DHA and GC × GC-FID for the analysis of liquid hydrocarbon streams (Dunkle et al. 2019). PIONA determinations by the different methods were very comparable, where all three techniques were highly

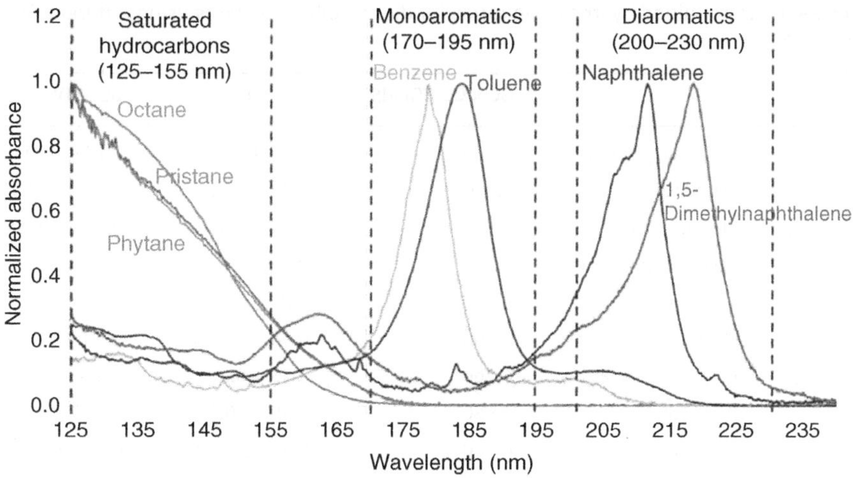

Figure 4.23 VUV absorption spectra demonstrating the potential for different filters applied to accentuate the presence of different classes of compounds in a GC-VUV chromatogram. Source: Bai et al. (2018). Reproduced with permission, Copyright (2018), Elsevier.

precise, yielding results with less than 1.3% RSD (relative standard deviation) on five injections a day for three days. Overall, the GC-VUV approach provided a high degree of accuracy for PIONA analysis (all levels within 0.6% error); GC-VUV was also overall simpler in instrumental complexity than the other techniques.

Recently, the power of theoretical computations were evaluated for hydrocarbon and PIONA compound spectral absorption prediction and developing a better understanding of hydrocarbon absorption properties. Previously, theoretical computations have been used to support GC-VUV experimental research. It can be a useful aid in the identification or discrimination of different chemical compounds, where reasonable matches between experimental theoretically computed spectra have been reported (Schenk et al. 2016; Schug et al. 2014; Skultety et al. 2017). It has also been used to discern the identity of different polyaromatic hydrocarbon oxidation products, an area of research where few pure standards exist to confirm compound identification (Ponduru et al. 2018). Time-dependent density functional theory (TD-DFT) is the preferred computational method, as it provides a good balance of accuracy and computational cost. Mao et al. used TD-DFT and experimental VUV absorption measurements to study the absorption properties of alkanes (Mao et al. 2019a). The availability of both theoretical and experimental data for a variety of linear and branched alkanes allowed conclusions to be drawn about the nature of the electronic transitions, specifically that the first strong transitions for alkanes do not always correspond to the highest occupied molecular orbital (HOMO)–lowest unoccupied molecular

Table 4.4 Determined hydrocarbon ratios for diesel spill biomarkers using different GC technologies.

		GC × GC-TOFMS	GC-FID	GC-VUV
0%	Pristane/phytane	1.360	1.2 ± 0.2	1.15 ± 0.01
	Pristane/C_{17}	0.860	0.47 ± 0.01	0.463 ± 0.005
	Phytane/C_{18}	0.710	0.39 ± 0.03	0.408 ± 0.004
25%	Pristane/phytane	2.280	2.4 ± 0.1	2.3 ± 0.4
	Pristane/C_{17}	0.960	0.89 ± 0.01	0.98 ± 0.04
	Phytane/C_{18}	0.500	0.48 ± 0.01	0.50 ± 0.09
50%	Pristane/phytane	2.260	2.23 ± 0.03	2.47 ± 0.04
	Pristane/C_{17}	1.080	0.790 ± 0.004	0.840 ± 0.005
	Phytane/C_{18}	0.480	0.380 ± 0.002	0.381 ± 0.005
75%	Pristane/phytane	1.360	2.107 ± 0.09	2.54 ± 0.08
	Pristane/C_{17}	0.890	0.777 ± 0.006	0.87 ± 0.05
	Phytane/C_{18}	0.650	0.368 ± 0.008	0.317 ± 0.006

Source: Bai et al. (2018). Reproduced with permission, Copyright (2018), Elsevier.

orbital (LUMO) transition; therefore, HOMO–LUMO energy gaps should not be used solely to understand observed experimental absorbance edges. In another heavily computational effort, Mao et al. broadened their view to evaluate different TD-DFT basis sets and functionals for accurate calculation of PIONA compound absorption spectra (Mao et al. 2019b). The chosen functional/basis set B3LYP/6-31+G** was judged to be best for locating peak positions. Theoretical line spectra, which indicate oscillator strength (absorptivity) and energy (wavelength), still need to be artificially broadened to a Gaussian shape in order to resemble experimental spectra. In many cases, the availability of a new GC-VUV detector now provides an exceptional route for obtaining experimental spectra in order to refine the accuracy of theoretical methods.

4.3.4 GC-VUV Conclusions

In summary, GC, coupled with VUV detection, offers a wide variety of different methods and beneficial attributes for the analysis of petroleum and hydrocarbons. The qualitative nature of electronic absorption spectra provides highly complementary data to MS for compound discrimination and identification; it also provides compound classification based on similar structural attributes, which yield similar absorption features in the VUV/UV absorption range. As the power of theoretical computations grows, as well as the limits of the VUV absorption technologies, it will be conceivable to continue to apply these technologies to heavier fuels and other petrochemical products. GC-VUV has quickly risen to become a potentially very valuable tool in the petrochemical analysis toolbox.

References

Alexandrino, G.L., Malmborg, J., Augusto, F., and Christensen, J.H. (2019a). Investigating weathering in light diesel oils using comprehensive two-dimensional gas chromatography–high resolution mass spectrometry and pixel-based analysis: Possibilities and limitations. *Journal of Chromatography A* 1591: 155–161. https://doi.org/10.1016/j.chroma.2019.01.042.

Alexandrino, G.L., Tomasi, G., Kienhuis, P.G.M. et al. (2019b). Forensic investigations of diesel oil spills in the environment using comprehensive two-dimensional gas chromatography-high resolution mass spectrometry and chemometrics: New perspectives in the absence of recalcitrant biomarkers. *Environmental Science and Technology* 53: 550–559. https://doi.org/10.1021/acs.est.8b05238.

Anthony, I.G.M., Brantley, M.R., Floyd, A.R. et al. (2018a). Improving accuracy and confidence of chemical identification by gas chromatography/vacuum ultraviolet spectroscopy-mass spectrometry: Parallel gas chromatography, vacuum ultraviolet, and mass spectrometry library searches. *Analytical Chemistry* 90: 12307–12313. https://doi.org/10.1021/acs.analchem.8b04028.

Anthony, I.G.M., Brantley, M.R., Gaw, C.A. et al. (2018b). Vacuum ultraviolet spectroscopy and mass spectrometry: A tandem detection approach for improved identification of gas chromatography-eluting compounds. *Analytical Chemistry* 90: 4878–4885. https://doi.org/10.1021/acs.analchem.8b00531.

ASTM-D8071 (2019). *Standard test method for determination of hydrocarbon group types and select hydrocarbon and oxygenate compounds in automotive spark-ignition engine fuel using gas chromatography with vacuum ultraviolet absorption spectroscopy detection (GC-VUV)*. West Conshohocken, PA: ASTM International.

Avila, B.M., Pereira, R., Gomes, A.O., and Azevedo, D.A. (2011). Chemical characterization of aromatic compounds in extra heavy gas oil by comprehensive two-dimensional gas chromatography coupled to time-of-flight mass spectrometry. *Journal of Chromatography A* 1218: 3208–3216. https://doi.org/10.1016/j.chroma.2010.09.051.

Bahaghighat, H.D., Freye, C.E., and Synovec, R.E. (2018). Recent advances in modulator technology for comprehensive two dimensional gas chromatography. *TrAC - Trends in Analytical Chemistry* 113: 379–391. https://doi.org/10.1016/J.TRAC.2018.04.016.

Bai, L., Smuts, J., Walsh, P. et al. (2015). Permanent gas analysis using gas chromatography with vacuum ultraviolet detection. *Journal of Chromatography A* 1388: 244–250. https://doi.org/10.1016/j.chroma.2015.02.007.

Bai, L., Smuts, J., Walsh, P. et al. (2017). Pseudo-absolute quantitative analysis using gas chromatography – vacuum ultraviolet spectroscopy – a tutorial. *Analytica Chimica Acta* 953: 10–22. https://doi.org/10.1016/j.aca.2016.11.039.

Bai, L., Smuts, J., Schenk, J. et al. (2018). Comparison of GC-VUV, GC-FID, and comprehensive two-dimensional GC–MS for the characterization of weathered and unweathered diesel fuels. *Fuel* 214: 521–527. https://doi.org/10.1016/j.fuel.2017.11.053.

Ball, G.I., Xu, L., McNichol, A.P., and Aluwihare, L.I. (2012). A two-dimensional, heart-cutting preparative gas chromatograph facilitates highly resolved single-compound isolations with utility towards compound-specific natural abundance radiocarbon (^{14}C) analyses. *Journal of Chromatography A* 1220: 122–131. https://doi.org/10.1016/j.chroma.2011.11.021.

Bayona, J.M., Dominguez, C., and Albaiges, J. (2015). Analytical developments for oil spill fingerprinting. *Trends in Environmental Analytical Chemistry* 5: 26–34. https://doi.org/10.1016/j.teac.2015.01.004.

Boer, H. and van Arkel, P. (1971). An automatic PNA analyser for (heavy) naphtha. *Chromatographia* 4: 300–308. https://doi.org/10.1007/bf02282905.

Brailsford, A., Gavrilovic, I., Ansell, R. et al. (2012). Two-dimensional gas chromatography with heart-cutting for isotope ratio mass spectrometry analysis of steroids in doping control. *Drug Testing and Analysis* 4: 962–969. https://doi.org/10.1002/dta.1379.

Bruckner, C.A., Prazen, B.J., and Synovec, R.E. (1998). Comprehensive two-dimensional high-speed gas chromatography with chemometric analysis. *Analytical Chemistry* 70: 2796–2804. https://doi.org/10.1021/ac980164m.

Coutinho, D.M., França, D., Vanini, G. et al. (2018). Rapid hydrocarbon group-type semi-quantification in crude oils by comprehensive two-dimensional gas chromatography. *Fuel* 220: 379–388. https://doi.org/10.1016/j.fuel.2018.02.009.

Dallüge, J., Beens, J., and Brinkman, U.A.T. (2003). Comprehensive two-dimensional gas chromatography: A powerful and versatile analytical tool. *Journal of Chromatography A* 1000: 69–108. https://doi.org/10.1016/S0021-9673(03)00242-5.

Dumont, E., Tienpont, B., Higashi, N. et al. (2013). Heart-cutting two-dimensional gas chromatography in combination with isotope ratio mass spectrometry for the characterization of the wax fraction in plant material. *Journal of Chromatography A* 1317: 230–238. https://doi.org/10.1016/j.chroma.2013.04.033.

Dunkle, M.N., Pijcke, P., Winniford, W.L., and Bellos, G. (2019). Quantification of the composition of liquid hydrocarbon streams: Comparing the GC-VUV to DHA and GC × GC. *Journal of Chromatography A* 1587: 239–246. https://doi.org/10.1016/j.chroma.2018.12.026.

von Eckstaedt, C.V., Grice, K., Ioppolo-Armanios, M. et al. (2011). δD and $\delta^{13}C$ analyses of atmospheric volatile organic compounds by thermal desorption gas chromatography isotope ratio mass spectrometry. *Journal of Chromatography A* 1218: 6511–6517. https://doi.org/10.1016/j.chroma.2011.06.098.

Elsner, M., Jochmann, M.A., Hofstetter, T.B. et al. (2012). Current challenges in compound-specific stable isotope analysis of environmental organic contaminants. *Analytical and Bioanalytical Chemistry* 403: 2471–2491. https://doi.org/10.1007/s00216-011-5683-y.

Fan, H., Smuts, J., Walsh, P. et al. (2015). Gas chromatography–vacuum ultraviolet spectroscopy for multiclass pesticide identification. *Journal of Chromatography A* 1389: 120–127. https://doi.org/10.1016/j.chroma.2015.02.035.

Flude, S., Johnson, G., Gilfillan, S.M.V., and Haszeldine, R.S. (2016). Inherent tracers for carbon capture and storage in sedimentary formations: Composition and applications. *Environmental Science and Technology* 50: 7939–7955. https://doi.org/10.1021/acs.est.6b01548.

Fogel, M. and Hoering, T. (1980). Biogeochemistry of the stable hydrogen isotopes. *Geochimica et Cosmochimica Acta* 44: 1197–1206. https://doi.org/10.1016/0016-7037(80)90073-3.

Fontanive, F.C., Souza-Silva, É.A., da Silva, J.M. et al. (2016). Characterization of sulfur and nitrogen compounds in Brazilian petroleum derivatives using ionic liquid capillary columns in comprehensive two-dimensional gas chromatography with time-of-flight mass spectrometric detection. *Journal of Chromatography A* 1461: 131–143. https://doi.org/10.1016/j.chroma.2016.07.025.

Franchina, F.A., Maimone, M., Tranchida, P.Q., and Mondello, L. (2016). Flow modulation comprehensive two-dimensional gas chromatography–mass spectrometry using ≈4 mL min^{-1} gas flows. *Journal of Chromatography A* 1441: 134–139. https://doi.org/10.1016/j.chroma.2016.02.041.

Frysinger, G.S., Gaines, R.B., Xu, L., and Reddy, C.M. (2003). Resolving the unresolved complex mixture in petroleum-contaminated sediments. *Environmental Science and Technology* 37: 1653–1662. https://doi.org/10.1021/es020742n.

Gaines, R.B., Frysinger, G.S., Hendrick-Smith, M.S., and Stuart, J.D. (1999). Oil spill source identification by comprehensive two-dimensional gas chromatography. *Environmental Science and Technology* 33: 2106–2112. https://doi.org/10.1021/es9810484.

Georgiou, C.A. and Danezis, G.P. (2015). Elemental and isotopic mass spectrometry. In: *Advanced Mass Spectrometry for Food Safety and Quality*, vol. 68 (ed. Y. Picó), 131–243. Amsterdam: Elsevier.

Giddings, J.C. (1984). Two-dimensional separations: Concept and promise. *Analytical Chemistry* 56: 1258A–1270A. https://doi.org/10.1021/ac00276a003.

Giddings, J.C. (1987). Concepts and comparisons in multidimensional separation. *Journal of High Resolution Chromatography* 10: 319–323. https://doi.org/10.1002/jhrc.1240100517.

Goldstein, A.H. and Shaw, S.L. (2003). Isotopes of volatile organic compounds: An emerging approach for studying atmospheric budgets and chemistry. *Chemical Reviews* 103: 5025–5048. https://doi.org/10.1021/cr0206566.

Gröger, T., Gruber, B., Harrison, D. et al. (2016). A vacuum ultraviolet absorption array spectrometer as a selective detector for comprehensive two-dimensional gas chromatography: Concept and first results. *Analytical Chemistry* 88: 3031–3039. https://doi.org/10.1021/acs.analchem.5b02472.

Hantao, L.W., Najafi, A., Zhang, C. et al. (2014). Tuning the selectivity of ionic liquid stationary phases for enhanced separation of nonpolar analytes in kerosene using multidimensional gas chromatography. *Analytical Chemistry* 86: 3717–3721. https://doi.org/10.1021/ac5004129.

Harvey, S.D., Jarman, K.H., Moran, J.J. et al. (2012). Characterization of diesel fuel by chemical separation combined with capillary gas chromatography (GC) isotope ratio mass spectrometry (IRMS). *Talanta* 99: 262–269. https://doi.org/10.1016/j.talanta.2012.05.049.

Hoelzer, K., Sumner, A.J., Karatum, O. et al. (2016). Indications of transformation products from hydraulic fracturing additives in shale-gas wastewater. *Environmental Science and Technology* 50: 8036–8048. https://doi.org/10.1021/acs.est.6b00430.

Horii, Y., Kannan, K., Petrick, G. et al. (2005). Congener-specific carbon isotopic analysis of technical PCB and PCN mixtures using two-dimensional gas chromatography–isotope ratio mass spectrometry. *Environmental Science and Technology* 39: 4206–4212. https://doi.org/10.1021/es050133q.

Hu, S.Z., Li, S.F., Cao, J. et al. (2014). Indications of transformation products from hydraulic fracturing additives in shale-gas wastewater. *Petroleum Science and Technology* 32: 565–574. https://doi.org/10.1080/10916466.2011.598896.

Hua, R., Li, Y., Liu, W. et al. (2003). Determination of sulfur-containing compounds in diesel oils by comprehensive two-dimensional gas chromatography with a sulfur chemiluminescence detector. *Journal of Chromatography A* 1019: 101–109. https://doi.org/10.1016/j.chroma.2003.08.048.

Jáčová, J., Gardlo, A., Dimandja, J.-M.D. et al. (2019). Impact of sample dimensionality on orthogonality metrics in comprehensive two-dimensional separations. *Analytica Chimica Acta* 1064: 138–149. https://doi.org/10.1016/j.aca.2019.03.018.

Jennerwein, M.K., Eschner, M., Gröger, T. et al. (2014). Complete group-type quantification of petroleum middle distillates based on comprehensive two-dimensional gas chromatography time-of-flight mass spectrometry (GC × GC-TOFMS) and visual basic scripting. *Energy and Fuels* 28: 5670–5681. https://doi.org/10.1021/ef501247h.

Jennerwein, M.K., Sutherland, A.C., Eschner, M. et al. (2017). Quantitative analysis of modern fuels derived from middle distillates – the impact of diverse compositions on standard methods evaluated by an offline hyphenation of HPLC-refractive index detection with GC × GC-TOFMS. *Fuel* 187: 16–25. https://doi.org/10.1016/j.fuel.2016.09.033.

Juchelka, D., Beck, T., Hener, U. et al. (1998). Multidimensional gas chromatography coupled on-line with isotope ratio mass spectrometry (MDGC-IRMS): Progress in the analytical authentication of genuine flavor components. *Journal of High Resolution Chromatography* 21: 145–151. https://doi.org/10.1002/(SICI)1521-4168(19980301)21:3<145::AID-JHRC145>3.0.CO;2-Z.

Kelly, S., Brodie, C., and Hilkert, A. (2018). Isotopic-spectroscopic technique: Stable isotope-ratio mass spectrometry (IRMS). In: *Modern Techniques for Food Authentication* (ed. D.-W. Sun), 349–413. Elsevier.

Krupčík, J., Gorovenko, R., Špánik, I. et al. (2013). Flow-modulated comprehensive two-dimensional gas chromatography with simultaneous flame ionization and quadrupole mass spectrometric detection. *Journal of Chromatography A* 1280: 104–111. https://doi.org/10.1016/j.chroma.2013.01.015.

Kulsing, C., Rawson, P., Webster, R.L. et al. (2017). Group-type analysis of hydrocarbons and sulfur compounds in thermally stressed Merox jet fuel samples. *Energy and Fuels* 31: 8978–8984. https://doi.org/10.1021/acs.energyfuels.7b01119.

Laakia, J., Casilli, A., Araújo, B.Q. et al. (2017). Characterization of unusual tetracyclic compounds and possible novel maturity parameters for Brazilian crude oils using comprehensive two-dimensional gas chromatography-time of flight mass spectrometry. *Organic Geochemistry* 106: 93–104. https://doi.org/10.1016/j.orggeochem.2016.10.012.

Lelevic, A., Souchon, V., Moreaud, M. et al. (2019). Gas chromatography vacuum ultraviolet spectroscopy: A review. *Journal of Separation Science* https://doi.org/10.1002/jssc.201900770.

Liu, Z. and Philips, J.B. (1991). Comprehensive two-dimensional gas chromatography using an on-column thermal modulator interface. *Journal of Chromatographic Science* 29: 227–231. https://doi.org/10.1093/chromsci/29.6.227.

Liu, H., Raffin, G., Trutt, G., and Randon, J. (2017). Is vacuum ultraviolet detector a concentration or a mass dependent detector? *Journal of Chromatography A* 1530: 171–175. https://doi.org/10.1016/j.chroma.2017.11.028.

Llewellyn, G.T., Dorman, F., Westland, J.L. et al. (2015). Evaluating a groundwater supply contamination incident attributed to Marcellus Shale gas development. *Proceedings of the National Academy of Sciences of the United States of America* 11: 6325–6330. https://doi.org/10.1073/pnas.1420279112.

Mansuy, L., Philp, R.P., and Allen, J. (1997). Source identification of oil spills based on the isotopic composition of individual components in weathered oil samples. *Environmental Science and Technology* 31: 3417–3425. https://doi.org/10.1021/es970068n.

Mao, D., Lookman, R., Van De Weghe, H. et al. (2009). Aqueous solubility calculation for petroleum mixtures in soil using comprehensive two-dimensional gas chromatography analysis data. *Journal of Chromatography A* 1216: 2873–2880. https://doi.org/10.1016/j.chroma.2008.08.072.

Mao, J.X., Kroll, P., and Schug, K.A. (2019a). Vacuum ultraviolet absorbance of alkanes: an experimental and theoretical investigation. *Structural Chemistry* 30: 2217–2224. https://doi.org/10.1007/s11224-019-01412-y.

Mao, J.X., Walsh, P., Kroll, P., and Schug, K.A. (2019b). Simulation of vacuum ultraviolet absorption spectra: Paraffin, isoparaffin, olefin, naphthene, and aromatic hydrocarbon class compounds. *Applied Spectroscopy* 74: 72–80. https://doi.org/10.1177/0003702819875132.

McRae, C., Love, G.D., Murray, I.P. et al. (1996). Potential of gas chromatography isotope ratio mass spectrometry to source polycyclic aromatic hydrocarbon emissions. *Analytical Communications* 33: 331–333. https://doi.org/10.1039/AC9963300331.

Muccio, Z. and Jackson, G.P. (2009). Isotope ratio mass spectrometry. *Analyst* 134: 213–222. https://doi.org/10.1039/B808232D.

von Mühlen, C., de Oliveira, E.C., Morrison, P.D. et al. (2007). Qualitative and quantitative study of nitrogen-containing compounds in heavy gas oil using comprehensive two-dimensional gas chromatography with nitrogen phosphorus detection. *Journal of Separation Science* 30: 3223–3232. https://doi.org/10.1002/jssc.200700172.

Nan, H., Zhang, C., O'Brien, R.A. et al. (2017). Lipidic ionic liquid stationary phases for the separation of aliphatic hydrocarbons by comprehensive two-dimensional gas chromatography. *Journal of Chromatography A* 1481: 127–136. https://doi.org/10.1016/j.chroma.2016.12.032.

Nara, H., Nakagawa, F., and Yoshida, N. (2006). Development of two-dimensional gas chromatography/isotope ratio mass spectrometry for the stable carbon isotopic analysis of C_2–C_5 non-methane hydrocarbons emitted from biomass burning. *Rapid Communications in Mass Spectrometry* 20: 241–247. https://doi.org/10.1002/rcm.2302.

Nelson, R.K., Kile, B.M., Plata, D.L. et al. (2006). Tracking the weathering of an oil spill with comprehensive two-dimensional gas chromatography. *Environmental Forensics* 7: 33–44. https://doi.org/10.1080/15275920500506758.

Nelson, R.K., Gosselin, K.M., Hollander, D.J. et al. (2019). Exploring the complexity of two iconic crude oil spills in the Gulf of Mexico (Ixtoc I and Deepwater Horizon) using comprehensive two-dimensional gas chromatography (GC × GC). *Energy and Fuels* 33: 3922–3925. https://doi.org/10.1021/acs.energyfuels.8b04384.

Okuda, T., Kumata, H., Naraoka, H., and Takada, H. (2002). Origin of atmospheric polycyclic aromatic hydrocarbons (PAHs) in Chinese cities solved by compound-specific stable carbon isotopic analyses. *Organic Geochemistry* 33: 1737–1745. https://doi.org/10.1016/S0146-6380(02)00180-8.

Omais, B., Courtiade, M., Charon, N. et al. (2012). Reversal of elution order in a single second dimension by changing the first column nature in comprehensive two-dimensional gas chromatography. *Journal of Chromatography A* 1255: 196–201. https://doi.org/10.1016/j.chroma.2012.07.042.

O'Malley, V.P., Burke, R.A., and Schlotzhauer, W.S. (1997). Using GC-MS/Combustion/IRMS to determine the $^{13}C/^{12}C$ ratios of individual hydrocarbons produced from the combustion of biomass materials – application to biomass burning. *Organic Geochemistry* 27: 567–581. https://doi.org/10.1016/S0146-6380(97)00087-9.

Philp, R.P., Allen, J., and Kuder, T. (2002). The use of the isotopic composition of individual compounds for correlating spilled oils and refined products in the environment with suspected sources. *Environmental Forensics* 3: 341–348. https://doi.org/10.1006/enfo.2002.0104.

Piotrowski, P.K., Tasker, T.L., Burgos, W.D., and Dorman, F.L. (2018a). Applications of thermal desorption coupled to comprehensive two-dimensional gas chromatography/time-of-flight mass spectrometry for hydrocarbon fingerprinting of hydraulically fractured shale rocks. *Journal of Chromatography A* 1579: 99–105. https://doi.org/10.1016/j.chroma.2018.10.021.

Piotrowski, P.K., Weggler, B.A., Barth-Naftilan, E. et al. (2018b). Non-targeted chemical characterization of a Marcellus shale gas well through GC × GC with scripting algorithms and high-resolution time-of-flight mass spectrometry. *Fuel* 215: 363–369. https://doi.org/10.1016/j.fuel.2017.11.026.

Pollo, B.J., Alexandrino, G.L., Augusto, F., and Hantao, L.W. (2018). The impact of comprehensive two-dimensional gas chromatography on oil & gas analysis: Recent advances and applications in petroleum industry. *TrAC - Trends in Analytical Chemistry* 105: 202–217. https://doi.org/10.1016/j.trac.2018.05.007.

Ponduru, T.T., Qiu, C., Mao, J.X. et al. (2018). Copper(I)-based oxidation of polycyclic aromatic hydrocarbons and product elucidation using vacuum ultraviolet spectroscopy and theoretical spectral calculations. *New Journal of Chemistry* 42: 19442–19449. https://doi.org/10.1039/C8NJ04740E.

Ponsin, V., Buscheck, T., and Hunkeler, D. (2017). Heart-cutting two-dimensional gas chromatography-isotope ratio mass spectrometry analysis of monoaromatic hydrocarbons in complex groundwater and gas-phase samples. *Journal of Chromatography A* 1492: 117–128. https://doi.org/10.1016/j.chroma.2017.02.060.

Prata, P.S., Alexandrino, G.L., Mogollón, N.G.S., and Augusto, F. (2016). Discriminating Brazilian crude oils using comprehensive two-dimensional gas chromatography–mass spectrometry and multiway principal component analysis. *Journal of Chromatography A* 1472: 99–106. https://doi.org/10.1016/j.chroma.2016.10.044.

Reddy, C.M., Nelson, R.K., Sylva, S.P. et al. (2007). Identification and quantification of alkene-based drilling fluids in crude oils by comprehensive two-dimensional gas chromatography with flame ionization detection. *Journal of Chromatography A* 1148: 100–107. https://doi.org/10.1016/j.chroma.2007.03.001.

Reichert, S., Fischer, D., Assche, S., and Mosandl, A. (2000). Stable isotope labelling in biosynthetic studies of dill ether, using enantioselective multidimensional gas chromatography, online coupled with isotope ratio mass spectrometry. *Flavour and Fragrance Journal* 15: 303–308. https://doi.org/10.1002/1099-1026(200009/10)15:5<303::AID-FFJ912>3.0.CO;2-2.

Richnow, H.H., Meckenstock, R.U., Reitzel, L.A. et al. (2003). In situ biodegradation determined by carbon isotope fractionation of aromatic hydrocarbons in an anaerobic landfill leachate plume (Vejen, Denmark). *Journal of Contaminant Hydrology* 64: 59–72. https://doi.org/10.1016/S0169-7722(02)00104-3.

Ristic, N., Djokic, M., Konist, A. et al. (2017). Quantitative compositional analysis of Estonian shale oil using comprehensive two dimensional gas chromatography. *Fuel Processing Technology* 167: 241–249. https://doi.org/10.1016/j.fuproc.2017.07.008.

Rockmann, T., Kaiser, J., Brenninkmeijer, C.A., and Brand, W.A. (2003). Gas chromatography/isotope-ratio mass spectrometry method for high-precision position-dependent ^{15}N and ^{18}O measurements of atmospheric nitrous oxide. *Rapid Communications in Mass Spectrometry* 17: 1897–1908. https://doi.org/10.1002/rcm.1132.

Rogers, K.M. and Savard, M.M. (1999). Detection of petroleum contamination in river sediments from Quebec City region using GC-IRMS. *Organic Geochemistry* 30: 1559–1569. https://doi.org/10.1016/S0146-6380(99)00132-1.

Santos, I.C. and Schug, K.A. (2017). Recent advances and applications of gas chromatography vacuum ultraviolet spectroscopy. *Journal of Separation Science* 40: 138–151. https://doi.org/10.1002/jssc.201601023.

Schenk, J., Mao, J.X., Smuts, J. et al. (2016). Analysis and deconvolution of dimethylnaphthalene isomers using gas chromatography vacuum ultraviolet spectroscopy and theoretical computations. *Analytica Chimica Acta* 945: 1–8. https://doi.org/10.1016/j.aca.2016.09.021.

Schenk, J., Carlton, D.D., Smuts, J.P. et al. (2019). Lab-simulated downhole leaching of formaldehyde from proppants by high performance liquid chromatography (HPLC), headspace gas chromatography–vacuum ultraviolet (HS-GC-VUV) spectroscopy, and headspace gas chromatography-mass spectrometry (HS-GC-MS). *Environmental Science: Processes & Impacts* 21: 214–223. https://doi.org/10.1039/c8em00342d.

Schoemaker, L. (2010). *The Basics: Isotope Fingerprints*. Online. National Oceanic and Atmospheric Administration (NOAA). Retrieved from http://esrl.noaa.gov/gmd.

Schug, K.A., Sawicki, I., Carlton, D.D.J. et al. (2014). Vacuum ultraviolet detector for gas chromatography. *Analytical Chemistry* 86: 8329–8335. https://doi.org/10.1021/ac5018343.

Sharp, Z. (2006). *Principles of Stable Isotope Geochemistry*. Prentice Hall.

Skultety, L., Frycak, P., Qiu, C. et al. (2017). Resolution of isomeric new designer stimulants using gas chromatography–vacuum ultraviolet spectroscopy and theoretical computations. *Analytica Chimica Acta* 971: 55–67. https://doi.org/10.1016/j.aca.2017.03.023.

Tobias, H.J., Sacks, G., Zhang, Y., and Brenna, J. (2008). Detection of synthetic testosterone use by novel comprehensive two-dimensional gas chromatography combustion-isotope ratio mass spectrometry. *Analytical Chemistry* 80: 8613–8621. https://doi.org/10.1021/ac801511d.

Tobias, H.J., Zhang, Y., Auchus, R., and Brenna, J. (2011). Detection of synthetic testosterone use by novel comprehensive two-dimensional gas chromatography combustion-isotope ratio mass spectrometry. *Analytical Chemistry* 83: 7158–7165. https://doi.org/10.1021/ac2015849.

Tran, T.C., Logan, G.A., Grosjean, E. et al. (2006). Comparison of column phase configurations for comprehensive two dimensional gas chromatographic analysis of crude oil and bitumen. *Organic Geochemistry* 37: 1190–1194. https://doi.org/10.1016/j.orggeochem.2006.05.0.

Tranchida, P.Q., Aloisi, I., Giocastro, B., and Mondello, L. (2018). Current state of comprehensive two-dimensional gas chromatography-mass spectrometry with focus on processes of ionization. *TrAC - Trends in Analytical Chemistry* 105: 360–366. https://doi.org/10.1016/j.trac.2018.05.016.

Turner, N., Jones, M., Grice, K. et al. (2006). $\delta^{13}C$ of volatile organic compounds (VOCs) in airborne samples by thermal desorption-gas chromatography-isotope ratio-mass spectrometry (TD-GC-IR-MS). *Atmospheric Environment* 40: 3381–3388. https://doi.org/10.1016/j.atmosenv.2006.01.030.

Van De Weghe, H., Vanermen, G., Gemoets, J. et al. (2006). Application of comprehensive two-dimensional gas chromatography for the assessment of oil contaminated soils. *Journal of Chromatography A* 1137: 91–100. https://doi.org/10.1016/j.chroma.2006.10.014.

Walsh, P., Garbalena, M., and Schug, K.A. (2016). Rapid analysis and time interval deconvolution for comprehensive fuel compound group classification and speciation using gas chromatography–vacuum ultraviolet spectroscopy. *Analytical Chemistry* 88: 11130–11138. https://doi.org/10.1021/acs.analchem.6b03226.

Weber, B.M., Walsh, P., and Harynuk, J.J. (2016). Determination of hydrocarbon group-type of diesel fuels by gas chromatography with vacuum ultraviolet detection. *Analytical Chemistry* 88: 5809–5817. https://doi.org/10.1021/acs.analchem.6b00383.

Weggler, B.A., Gruber, B., and Dorman, F.L. (2019). Rapid screening of complex matrices: utilizing Kendrick mass defect to enhance knowledge-based group type evaluation of GC × GC-HR(EI)ToFMS data. (Brief-report). *Analytical Chemistry* 91: 10949–10954. https://doi.org/10.1021/acs.analchem.9b01750.

Zeng, Z., Li, J., Hugel, H.M. et al. (2014). Interpretation of comprehensive two-dimensional gas chromatography data using advanced chemometrics. *TrAC - Trends in Analytical Chemistry* 53: 150–166. https://doi.org/10.1016/j.trac.2013.08.009.

Zoccali, M., Schug, K.A., Walsh, P. et al. (2017). Flow-modulated comprehensive two-dimensional gas chromatography combined with a vacuum ultraviolet detector for the analysis of complex mixtures. *Journal of Chromatography A* 1497: 135–143. https://doi.org/10.1016/j.chroma.2017.03.073.

5

Liquid Chromatography: Applications for the Oil and Gas Industry

Denice van Herwerden, Bob W. J. Pirok, and Peter J. Schoenmakers

Department of Analytical Chemistry, Van't Hoff Institute for Molecular Science, University of Amsterdam, Amsterdam, The Netherlands

5.1 Introduction

5.1.1 Petroleum Industry

The petroleum industry represents a major part (estimated at around 40%) of the chemical industry (Sharma 2018). Petrochemicals are compounds that are derived from the products of the refining process of petroleum and natural gasses. These compounds are essential in a variety of application fields, such as functional materials, detergents, pharmaceuticals, food, and agriculture. For monitoring these compounds in either an environmental or petrochemical context, analytical chromatographic techniques, such as gas chromatography (GC), liquid chromatography (LC), and supercritical fluid chromatography (SFC) are often used.

The analysis of the composition of petroleum resources can be important for refining processes. For example, heavy polyaromatic hydrocarbons (PAHs) may accumulate in the hydrocracking process, resulting in deposition or catalyst deactivation (Panda et al. 2018). Another class of analytes that may cause catalyst deactivation are silicon-containing compounds, due to their reactivity with the catalysts that are used for hydrodesulfurization and demetallization processes. Besides catalyst inhibition, compounds such as phenols also have a negative effect on the biological degradation of toxic compounds in petrochemical wastewater (Ravanchi et al. 2009).

The petroleum industry has a significant impact on environmental pollution (Wang et al. 2015). Petrochemicals have become common pollutants due to the contamination of the environment with waste from the petroleum industry (Pavlova and Ivanova 2003). This industrial waste can contain harmful

Analytical Techniques in the Oil and Gas Industry for Environmental Monitoring, First Edition.
Edited by Melissa N. Dunkle and William L. Winniford.

compounds, such as phenols, PAHs, and naphthenic acids (NAs). The latter are acutely toxic to aquatic organisms at low concentrations and are released into the environment through petrochemical wastewater (Wang and Kasperski 2010).

The analysis of these petrochemical products can be conducted using a variety of analytical methods, including GC, LC, and SFC. In this chapter, we will focus on the LC techniques used in the petroleum industry. Therefore, we will provide a brief introduction into the basic principles of LC, LC hyphenated with mass spectrometry (LC-MS), and two-dimensional liquid chromatography (2D-LC). These chromatographic techniques can be used for different purposes, such as regulation of a biodegradation process, prefractionation or clean-up prior to GC analysis, carbon group-type analysis, or for the monitoring of specific toxic compounds. The authors intend to provide an overview of contemporary LC techniques for samples related to oil, refining, and wastewater encountered in the oil and petrochemical industries. The focus will be on new or emerging methods described in public literature, rather than on established, standardized methods (American Society for Testing and Materials [ASTM], Institute of Petroleum [IP], Institut Français du Pétrole [IFP], etc.). Furthermore, the impact of recent advances in LC in the petroleum industry will be assessed. Such developments include the use of core–shell particles, ultra-high-performance liquid chromatography (UHPLC), and 2D-LC.

5.1.2 Introduction to Liquid Chromatography

GC is a routinely used chromatographic technique for the analysis of petroleum-related compounds (Blomberg et al. 2002; Pirok et al. 2019). However, GC is limited to the analysis of volatile compounds that are stable at the high temperatures applied (Castillo et al. 1998). When these requirements are not met, LC can be a suitable alternative, because it does not require high temperatures, and the analytes are already in the liquid phase (Nizio et al. 2012). Another advantage of LC is the availability of a variety of separation selectivities. Examples include reversed-phase liquid chromatography (RPLC), normal-phase liquid chromatography (NPLC), size-exclusion chromatography (SEC), and hydrophilic-interaction liquid chromatography (HILIC) (Pirok et al. 2019). When a higher sample throughput is desired, UHPLC can be used instead of conventional high-performance liquid chromatography (HPLC). The maximum pressures encountered in HPLC are 400 bar (40 MPa, 6000 psi), whereas contemporary UHPLC systems have pressure limits of 1000–1500 bar (100–150 MPa, 15 000–22 000 psi). Such higher pressures allow the use of sub-2 μm particles, provided that these are packed in short columns (Jakobsen et al. 2017; Pirok et al. 2019). UHPLC technology provides high separation efficiencies (i.e. numbers of theoretical plates) and reasonable peak capacities (i.e. numbers of equally resolved peaks that may fit in the chromatogram) in a much shorter time than HPLC. As a result, the resolution can be improved threefold in similar run times, or equal resolution can be achieved much faster (Frenich et al. 2014). Another development that can improve the separation efficiency is the re-emergence of core–shell particles. Such particles consist of a solid core and a porous outer layer (González-Ruiz et al. 2015; Hayes et al. 2014). The intraparticle porosity and the total pore volume are reduced (Pirok et al. 2019), but higher efficiencies are obtained, thanks to shorter diffusion path lengths and, perhaps more importantly, more uniform particle diameters (Gritti et al. 2010). As a result, analysis times may be reduced and/or chromatographic resolution may be enhanced.

Despite 50 years of development in LC, practical implementations of the technique offer much lower peak capacities than common GC systems. This – and the lack of sensitive, (near-) universal detectors – renders LC less suitable for the separation of complex samples, such as petroleum (Mills et al. 1999; Pirok et al. 2019). In the statistical-overlap theory (Davis and Giddings 1983), it is assumed that all sample components are randomly divided across a chromatogram. It follows from this theory that samples that contain more than approximately 20 components have a high probability of showing overlapping peaks in LC (Pirok et al. 2019). Comprehensive two-dimensional LC (LC × LC) may be used to achieve a higher peak capacity for complex samples. In this 2D-LC method, the first-dimension effluent is split into many successive fractions, each of which is subjected to a (fast) second-dimension separation. High-resolution 1D-LC, using long, narrow columns, may yield a peak capacity of 1000 in a 16-hours overnight run (Eeltink et al. 2010), which implies a peak-production rate (peak capacity per unit time) of about one peak per minute. In contrast, LC × LC may yield a peak capacity of 4800 in 80 minutes (Gargano et al. 2016), or roughly one peak per second. The actual usefulness of this increased peak capacity depends on the orthogonality between the two dimensions. High orthogonality can be achieved by selecting a retention mechanism for the two dimensions that are as different as possible. A high degree orthogonality also reduces the uncertainty of peak identification (Ochoa et al. 2018) and enhances the possibility of structure appearing in the 2D-chromatogram if the (dominant) sample dimensions match the separation dimensions (Giddings 1995).

In specific cases, only one or a few peaks of the first dimension are of interest for further analysis, the total peak capacity is less relevant, and heart-cut two-dimensional liquid chromatography (LC-LC) can be employed (Pirok et al. 2019). In this technique, only one or a few fractions of the first-dimension effluent are subjected to the second dimension. This greatly simplifies the process, and it allows for a longer, high-resolution second-dimension separation. A disadvantage of LC × LC is that the method-development process is relatively difficult and time-consuming in comparison with 1D-LC and heart-cut 2D-LC. This is thought to be one of the major reasons why LC × LC methods are not yet frequently applied in industry (Pirok et al. 2018).

Returning to the application of LC in the petrochemical industry, a final advantage that deserves mentioning is the possibility to separate and identify polar, ionic, and high-molecular-weight compounds using hyphenation of LC with mass spectrometry (MS) (Castillo and Barceló 1997). For example, the analysis of NAs with gas chromatography-mass spectrometry (GC-MS) requires a derivatization step prior to injection, which increases the method complexity and may instigate sample losses. In contrast, LC-MS allows direct analysis of NAs. A limitation of LC-MS is that only volatile salts can be used for mobile-phase buffers. Also, if multiple components enter the MS simultaneously, this may cause matrix effects or ion suppression for specific analytes (van Beek et al. 2018). Thus, a good separation is desirable prior to MS analysis. Another advantage of using LC-MS instead of GC-MS is the various ionization sources that may be used, such as electrospray ionization (ESI), atmospheric-pressure chemical ionization (APCI), and atmospheric-pressure photo-ionization (APPI) (Han et al. 2018). These are soft ionization methods that generate few fragment ions, resulting in enhanced detection and simplified MS spectra. If additional identification, verification, or selective quantification is needed, tandem-MS (MS/MS) methods may also be applied.

5.2 Group-Type Separations

Group-type separation is commonly used to analyze quantitatively or to preparatively obtain specific fractions of oil or refined products, such as fractions containing aromatics, resins, or saturates (Liu et al. 2010). For the analysis of middle distillates, which have boiling points in the range from 150 to 400 °C, standard LC methods, such as ASTM D-6379, EN-12916, or IP-391, can be used (Kartanowicz et al. 2007). In recent literature, LC separations of middle-distillate fractions have mainly been used as a prefractionation step prior to GC analysis (Adam et al. 2007; Jennerwein et al. 2017; Vendeuvre et al. 2005). This will be further discussed in Section 5.5 (and Table 5.4). In the current section, we will focus on the group-type separations of heavier petroleum fractions with higher boiling points (>400 °C) that are incompatible with GC analysis (Liu et al. 2010).

5.2.1 Group-Type Separations of Heavy Distillates

Compositional analysis of petroleum samples can provide important information for the optimization of refining and production processes, which is especially interesting due to the decreasing availability of sweet crude oil

(Bissada et al. 2016b; Matisová 1985). For instance, the effective use of heavy oils in the petroleum-refining industry strongly depends on the ring number of aromatics present. Thus, the chemical composition may influence the processing scheme (Li et al. 1998; Stevenson 1971). Group-type analysis can be used to either obtain detailed compositional information or to determine the relative amounts of the individual hydrocarbon groups (Miller 1982). An example is shown with the separation of heavy oils into saturates, aromatics, resins, and asphaltenes (SARA), of which the saturates can be further separated into paraffins, iso-paraffins, and naphthenes (PIN) (Bissada et al. 2016a). The resins and asphaltenes are non-paraffinic fractions containing hetero-elements, such as nitrogen, sulfur, and oxygen, and many polar substituents.

Traditionally, the determination of hydrocarbon group types in heavy-oil fractions has been performed with a fluorescence-indicator adsorption method, which allowed saturates, olefins, and aromatics to be determined (Matisová 1985; Pearson and Gharfeh 1986). This method was based on classical liquid-displacement chromatography with activated silica gel containing fluorescent dyes (Matisová 1985). After separation, the percentage volume of the fractions was determined with limited precision by manually measuring the band lengths under ultraviolet (UV) light. Because polar components are detected as aromatics, this method has a low selectivity. It also requires long analysis times, on the order of three to four hours. The introduction of HPLC not only improved the accuracy and precision of the quantified hydrocarbon groups, but also reduced the time required for group-type separations (Matisová 1985). Initially, the separation was performed on a single column with activated silica. However, this required an activation step, and polar analytes tended to adsorb irreversibly on the columns, resulting in poor repeatability. To avoid this, multi-column methods have been introduced, for example, incorporating a cyano column prior to the silica column to retain polar components (Pearson and Gharfeh 1986).

A large number of methods have been developed to separate SARA fractions, each using two or three columns and detectors (Carbognani and Izquierdo 1990; Pearson and Gharfeh 1986; Robbins 1998). Prior to injection into the HPLC, the sample preparation involves de-asphalting. The maltene fraction is obtained by adding an excess of hexane, leaving behind a polar, insoluble asphaltene fraction. Methods to separate the maltene fraction make use of valves, which allows columns to be temporarily taken out of the flow stream and components to be backflushed from the column (Figure 5.1). For example, Pearson and Gharfeh used a cyano column and two amino-cyano columns for separating the maltene fraction (Pearson and Gharfeh 1986). During injection, all three columns are used, and the resins and aromatics are retained on the first and second columns, respectively. After detection of the saturates, valves A and B, are switched to exclude the cyano column and to backflush the aromatics from the amino-cyano

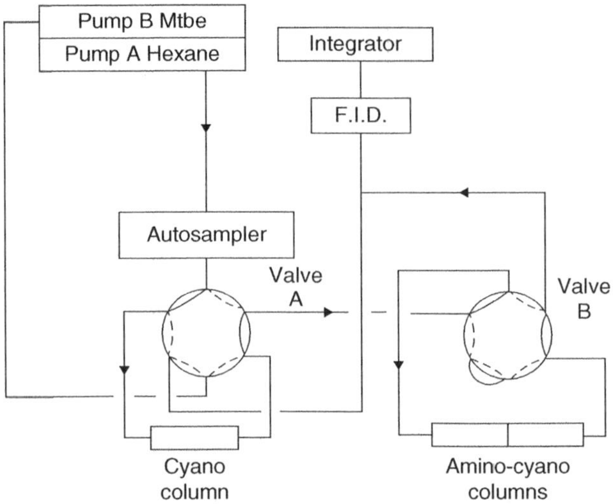

Figure 5.1 Schematic overview of the SARA method of Pearson and Gharfeh that applies valves to either use both columns or only one and allows backflushing of the first column to elute the resins (Pearson and Gharfeh 1986). The indicated valve positions are used during injection, while pump A is on. After the detection of the saturates, valve A and B are switched to leave out the cyano column and to backflush the aromatics from the amino-cyano columns. Finally, pump B is turned on to elute the resins from the cyano column. Source: Pearson and Gharfeh (1986). Reproduced with permission, Copyright (1986), American Chemical Society.

columns. Then, pump B is turned on to elute the resins from the cyano column. The obtained fractions can be further analyzed with other techniques, such as GC and MS (Robson et al. 2017).

A more recent method by Bissada et al. utilized a similar principle for the fully automated separation and fraction collection of SARA and PIN fractions (Bissada et al. 2016a,b). This method does not require de-asphalted samples (Figure 5.2a), reducing the probability of cross-contamination and enhancing the precision (Bissada et al. 2016a). However, the separation between asphaltenes and resins still requires further optimization, because both fractions are eluted from the first column with different mobile-phase compositions and are not fully resolved. The total process takes 60 minutes, including system re-conditioning, and the fractions can be used for further analysis. The saturate fraction obtained from the SARA separation can be separated into PIN fractions by another multi-column method (Bissada et al. 2016a). Instead of using a cyano and a silica column, two molecular-sieve columns (5.4 and 6.2 Å) are implemented (Bissada et al. 2016a,b). The method uses different temperatures to retain the *n*- and iso-paraffins on the molecular-sieve columns, and temperature-gradient programs to elute these fractions (Figure 5.2b).

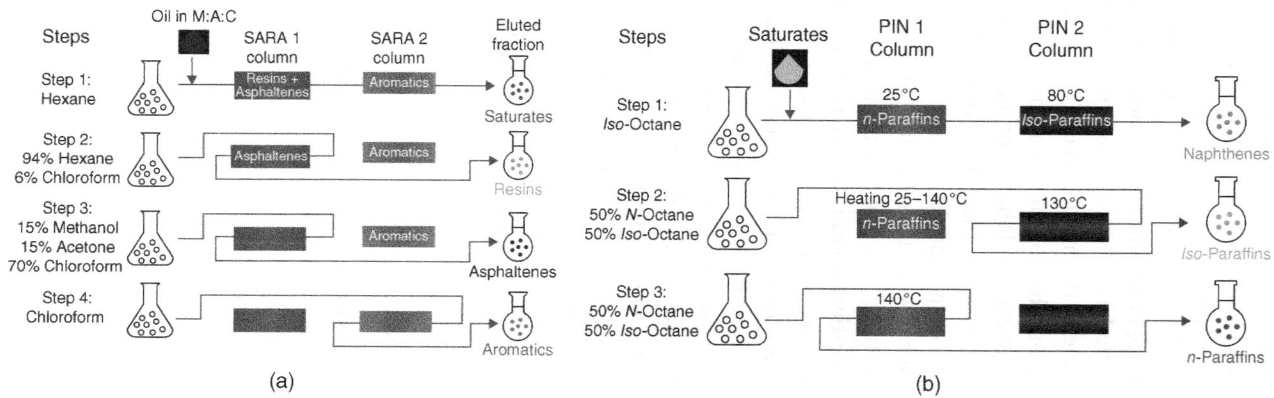

Figure 5.2 Schematic overview of the fraction-collection steps of the SARA (a) and PIN (b) methods of Bissada et al. (2016a,b). First, for the SARA method, the unretained saturates fraction is collected. Then, subsequently, the resins and asphaltenes are eluted from the first column by backflushing with different mobile phases with increasing polarity. Lastly, the aromatic fraction is collected. For the PIN method, the saturate fraction is injected, and the naphthenes are collected. Then, the mobile phase composition is changed, and the column temperatures are increased to first elute the *iso*-paraffins from the second column and then the *n*-paraffins from the first column. Source: Bissada et al. (2016a,b). Reproduced with permission, Copyright (2016), Elsevier.

Both the SARA and PIN methods of Bissada et al. are based on the selective trapping of hydrocarbon fractions rather than on a simultaneous separation, which may explain why recent developments in LC for higher separation efficiencies are not directly applicable to these SARA and PIN analyses (Pirok et al. 2019). While earlier methods required a de-asphalting step prior to LC separation, the method of Bissada et al. can be applied to untreated crude oils. Samples were dissolved in a mixture of acetone, methanol, and chloroform, and then injected into the HPLC system using *n*-hexane as the mobile phase (Bissada et al. 2016b). Alternative methods to avoid the de-asphalting step were mentioned, using, for example, a Teflon packed column that can resolubilize the asphaltenes. However, such alternative methods were not used in the SARA method developed by Bissada et al. A clear explanation as to why asphaltene-containing samples can be directly injected in an HPLC system has so far been lacking (Bissada et al. 2016b).

The use of LC × LC for separating saturates, aromatics, and resins of de-asphalted heavy oils has been investigated (van Beek et al. 2018). Biphenyl and octadecyl-silica were used as first-dimension and second-dimension columns, respectively. A methanol to tetrahydrofuran (THF) gradient was used in the first dimension and a methanol to dichloromethane gradient in the second dimension. The method was optimized using computer-aided optimization algorithms (Program for Interpretive Optimization of 2D Resolution [PIOTR]) that can optimize the separation of analytes based on a few experiments (Pirok et al. 2016). Optimization of the method was only marginally successful for the de-asphalted sample because the LC × LC chromatogram did not feature distinct peaks. Only the group of saturates were fully separated from the aromatics and resins. This could be improved by re-injecting separated fractions or by using MS for tracking specific components.

5.2.2 Other Group-Type Separations

One of the most general group-type separations is SARA, which has been discussed earlier. However, elaborate methods have also been developed. For example, Li et al. and Robbins et al. also separated the mono-, di-, and polyaromatics (Li et al. 1998; Robbins 1998). Additionally, the analysis of olefins can be of interest in optimizing the refining processes (Badoni et al. 1992; Heshka et al. 2019). Olefins are generated by thermal and catalytic cracking of heavy oils into lighter molecules in the absence of excess hydrogen. Among the lighter molecules are olefins (i.e. alkenes), which contain double bonds and which may give rise to instability and fouling issues (Heshka et al. 2019). Tomić et al. described a method to separate the olefins from the aromatics, but not from the saturates (Tomić et al. 2009); as observed with other methods, they used selective detection of olefins, which was possible with UV (diode array detector [DAD]) detection, because saturates do not absorb in the UV range (Heshka et al. 2019; Lienne et al. 1987; Tomić et al. 2009).

It is also possible to study the presence of certain elements, instead of entire molecules. Examples include the analysis of sulfur-, nitrogen-, and oxygen-containing compounds. Although such element-specific methods are mainly performed by GC, several LC-based methods for sulfur-containing compounds have been reported in the literature. Such methods are dedicated for non-volatile samples (Han et al. 2018). If LC is used for the heavier petroleum fractions, it is used in conjunction with MS. For determining nitrogen-containing compounds, LC-MS has also become the method of choice (Table 5.1).

5.3 Molecular-Weight Distribution

Besides separating petroleum samples into different hydrocarbon groups, molecular-weight distributions (MWDs) of petroleum samples or fractions can be determined. This can be achieved with SEC, which separates compounds based on the hydrodynamic volume. The separation principle is based on the exclusion of analytes from a fraction of the pores in the stationary-phase particles, resulting in elution before the column dead time (Pirok et al. 2017). Large molecules cannot enter the smaller pores, which results in them being eluted before smaller molecules. An important factor is that there should be no interaction between the analytes and the stationary phase. If this condition is not met, analyte elution times will not be solely dependent on their hydrodynamic volumes, meaning accurate MWDs cannot be determined.

SEC requires excellent solubility of the analytes in a suitable solvent (Herod et al. 2007). The mobile-phase solvents that have mainly been used for the analysis of petroleum-related samples are THF and 1-methyl-2-pyrrolidinone (NMP). Although THF has been frequently used, it is not suitable for dissolving large aromatic molecules or polar resins, which results in incomplete recoveries and inaccurate MWDs (Al-Muhareb et al. 2006; Herod et al. 2007; Herod et al. 2003). When using THF, interactions between the stationary phase and aromatic analytes can also not be completely prevented, which again will result in inaccurate MWDs and sometimes even in the elution of analytes after the solvent signal that is indicative for the total pore volume (Al-Muhareb et al. 2006). NMP has been used to overcome these limitations. However, even in NMP, asphaltenes are only dissolved around 50%. Morgan et al. utilized a 6 : 1 mixture of NMP and chloroform for the analysis of maltenes and the soluble asphaltene fraction (Morgan et al. 2010). Herod et al. examined the influence of the addition of LiBr to NMP, which resulted in a decreased solubility and increased analyte-stationary-phase interactions, resulting in longer elution times. Heptane has also been used as a mobile

Table 5.1 Overview of methods for group-type separations.

Application (hydrocarbon groups)	Mechanism (selectivity)	Detection	Remarks	References
SARA related group-type separations				
Heavy oils (saturates, aromatics, resins)	LC × LC: ^{1}D RPLC (biphenyl), ^{2}D SEC (C18)	DAD, CAD	Computer-aided optimization techniques	van Beek et al. (2018)
Separation and quantitation of saturate fraction of SARA into PIN	SARA NPLC (cyano and silica) PIN SEC (5.4 and 6.2 Å)	ELS gravimetry (off-line)	SEC columns rapidly heated and cooled	Bissada et al. (2016a)
Separation and quantitation of SARA fractions	NPLC (cyano and silica)	ELS gravimetry (off-line)	Backflushed resins asphaltenes 60 min	Bissada et al. (2016b)
Gasoline (saturates, alkanes, aromatics)	NPLC (NH_2)	DAD		Kamiński et al. (2004)
Heavy distillates (saturates, 1-4 ring aromatics, resins)	NPLC (PAC, DNAP)	DAD (204–430 nm)	Optional: columns in series	Robbins (1998)
Heavy petroleum fractions (saturates, monoaromatics, diaromatics, polyaromatics, resins)	NPLC (silica gel and alumina)	On-line: RI, UV Off-line: SFS		Liu et al. (2010)
De-asphalted vacuum residues (saturates, aromatics, resins)	NPLC (cyano and silica)	RI, UV-Vis	Three columns in series (cyano-silica-silica)	Carbognani and Izquierdo (1990)
De-asphalted crude oil (saturates, aromatics, resins)	NPLC (cyano and 2× aminocyano)	FID	LC-FID interface, columns in series	Pearson and Gharfeh (1986)
Coal	NPLC (cyano and silica)			Lanças and McNair (1985)
Standards (SARA)	NPLC (cyano)	RI	resins backflushed	Miller (1982)
Other group-type separations				
Olefins				
Cracked petroleum fractions (mono- and di- olefins)	NPLC (silver silica)	RI, UV	$\lambda = 210, 240, 254$ nm	Heshka et al. (2019)

(*continued*)

Table 5.1 (Continued)

Application (hydrocarbon groups)	Mechanism (selectivity)	Detection	Remarks	References
Fluid catalytic cracking (saturates, olefins, aromatics)	NPLC (amino silica)	UV/DAD	$\lambda = 190–367$	Tomić et al. (2009)
Light and heavy petroleum products (saturates, olefins, aromatics)	NPLC (silver silica)	RI, UV	$\lambda = 210$	Lienne et al. (1987)
Gasoline (saturates, olefins)	NPLC (silica gel)	RI		Jinno et al. (1980)
Nitro and sulfur containing compounds				
Coal (nitro compounds)	RPLC (C18)	MS	Ultrasonic extraction	Xia et al. (2016)
Petroleum (nitro compounds)	NPLC (dinitroanilinopropyl silica)	MS	FT-ICR	Oro and Lucy (2013)
Petroleum (nitro compounds)	Multiple NPLC columns	UV		Oro and Lucy (2010)
Crude oil (polyaromatic sulfur hydrocarbons)	NPLC (polystyrene divinylbenzene)	MS/MS		Rudzinski and Rai (2005)
Fossil fuels (polyaromatic sulfur hydrocarbons)	NPLC (aminopropylsilane)	UV	Semi-preparative, further GC analysis	Mössner and Wise (1999)
Monitoring of molecular transformation of hydrotreating (sulfur compounds)	NPLC (dinitroanilinopropyl silica)	MS		Qian and Hsu (1992)
Heavy distillates (nitro compounds)	NPLC (dinitroanilinopropyl silica)	MS		Hsu et al. (1994)
Characterization of heavy hydrocarbons (sulfur compounds)	NPLC (dinitroanilinopropyl silica)	MS		Hsu et al. (1991)

SFS, synchronous fluorescence spectroscopy; FID, flame ionization detector; LC-FID, liquid chromatography–flame ionization detector.

phase, but the resulting separation cannot be identified as SEC. Interactions of the aromatic components with the stationary phase were maximized, resulting in elution after the dead volume (Al-Muhareb et al. 2006). Al-Muhareb et al. used this method in combination with both UV and evaporative light scattering (ELS) detections, detecting the aliphatic components with ELS detection before the column dead volume and the aromatics with UV detection thereafter (Al-Muhareb et al. 2006).

For determining the number-average or weight-average molecular weight, it is assumed that the size (hydrodynamic volume) of the calibration standards corresponds to that of analytes of similar molecular weight (Herod et al. 2007). The standards that are used in these methods are often a combination of polymeric standards and smaller compounds such as tetraphenyl or benzopyrene. The most common calibration standard is polystyrene (PS) (Table 5.2). However, there are differences in chemical structure between the sample components and these calibration standards, which can result in inaccurate molecular-weight estimates. Boczkaj et al. utilized SEC with *n*-alkane standards for determining the boiling-point range of high-boiling petroleum fractions, which could not be determined with GC-based methods due to the temperature limitation (boiling points <400 °C) (Boczkaj et al. 2015).

Most methods either use a Mixed-D or a Mixed-E column from Polymer Laboratories or several columns in series (Table 5.2). The development of core–shell particles may result in similar performance as fully porous particles in very short analysis times when used for SEC (Pirok et al. 2017; Schure and Moran 2017). However, columns packed with core–shell particles exhibit a limited molecular-weight range. A solution may be to couple multiple core–shell columns with different porosities in series, although this would lead to an increase in analysis time (Pirok et al. 2017). Ultimately, a mix-bed column seems most attractive, as it allows to maintain the short analysis times while extending the molecular-weight range, at the expense of a (slight) decrease in resolution. According to the literature, core–shell particles have not yet been used for SEC analysis in the petroleum industry (Table 5.2).

5.4 Target Analysis

5.4.1 Polyaromatic Hydrocarbons

PAHs occur naturally in fossil fuels and end up in the environment via multiple paths, for example, through the release of wastewater from the petroleum and petrochemical industry (Chen 2004; Panda et al. 2018; Pugazhendi et al. 2017). PAHs are compounds that contain multiple fused aromatic rings that share at least

Table 5.2 Overview of SEC methods used for the determination of molecular-weight distributions in the petroleum industry.

Application	Column/pore-size (particle size), mobile phase	Detection	Temperature, calibration standards	References
High-boiling petroleum fractions	Phenogel 50 and 100 Å in series (5 μm), THF	RI	30 °C, *n*-alkanes	Boczkaj et al. (2015)
Coal tar pitch	Mixed-D[a], NMP	DAD	80 °C, PS PMMA polysaccharide	George et al. (2010)
Bitumen	500 Å (10 μm), THF			Le Guern et al. (2010)
Maltene and soluble asphaltene fraction	Mixed-E and Mixed-D[a], NMP:chloroform (6 : 1)	UV	80 °C, PS	Morgan et al. (2010)
Pentane and toluene soluble fractions of petroleum residues and coal tars	Mixed-D[a], NMP	UV-fluorescence	80 °C, PS	Karaca et al. (2008)
Coal liquids, petroleum vacuum residues	Mixed-E[a], heptane	UV, ELS	20 °C, *n*-alkanes	Al-Muhareb et al. (2006)
Average MW of high-boiling petroleum fractions	PSM 60 and 300 Å in series (5 μm), THF	RI	20 °C, PS tetraphenyl benzopyrene	Sharma et al. (2003)
Coal liquid/ petroleum residues	Mixed-E and Mixed-D[a], NMP	UV, ELS	85 and 80 °C, respectively	Herod et al. (2003)
Fingerprinting of crude oils	500 and 50 Å, chloroform	DAD	35 °C	Varotsis and Pasadakis (1998)

a) Mixed-D and Mixed-E columns are packed with 5-μm and 3-μm particles, respectively, and exhibit molecular-weight ranges of 100–200 000 and 100–20 000, respectively.

two carbon atoms with another aromatic ring (Panda et al. 2018). This chemical structure results in hydrophobic compounds with low solubility in water, and it renders PAHs to be persistent in the environment (Pugazhendi et al. 2017; Manoli and Samara 1999). This persistence, in combination with the toxicity, mutagenicity, and potential carcinogenicity of a number of PAHs, poses a risk to aquatic and human life. Consequently, 16 PAHs are listed as priority pollutants by the United States Environmental Protection Agency (EPA) and by other governments (Cai et al. 2009; Pugazhendi et al. 2017). Therefore, chromatographic techniques have been developed for monitoring these compounds.

While both GC and LC have been used for the analysis of PAHs (Cai et al. 2009), LC is generally applied for the analysis of larger PAHs due to their low volatility (Panda et al. 2018); moreover, higher molecular-weight PAHs are known to be subject to thermal degradation during GC analysis (Cai et al. 2009). Examples of LC methods that have been used to analyze PAHs in petroleum-related samples are available in the literature; one specific example developed by Witt uses RPLC with an octadecyl-functionalized silica (C18) stationary phase and fluorescence detection (Witt 1995). More recently, Cai et al. developed a high-throughput UHPLC method that also utilizes a C18-functionalized stationary phase and fluorescence detection for the analysis of wastewater effluent. This latter method had a much shorter overall analysis time (Cai et al. 2009).

Methods described in the literature utilizing 2D-LC for the analysis of PAHs are still scarce. Murahashi used comprehensive LC × LC for the separation of PAHs, where he was able to separate the analytes both based on their size and polarity (Murahashi 2003). A pentabromobenzyl (PBB) column was used in the first dimension, which separated the analytes according to the carbon number, and a C18 stationary phase was used in the second dimension, separating the PAHs based on both carbon number and polarity. A more recent example is given in Figure 5.3, where different columns were evaluated for the LC × LC of a wastewater extract.

It has become clear from multiple methods that reversed-phase LC with octadecyl (C18) stationary phases is the most generally used technique for the analysis of PAHs, which is not surprising because of the hydrophobic nature of the PAHs (Pugazhendi et al. 2017). However, due to the aromaticity of the PAHs, the use of stationary phases that can exhibit π–π interactions has also been explored. Panda et al. tested multiple columns, including stationary phases that could undergo π–π interactions, and the results showed C18 stationary phases to yield the best resolution and selectivity (Panda et al. 2018). Likewise, polymeric C18 stationary phases, containing cross-linked silanes on the silica surface, were found to provide a better selectivity for separating the PAHs than monomeric stationary phases (Wise et al. 1993; Figure 5.4).

As for the detection of the PAHs, UV or fluorescence detection is often used. Both are well-suited techniques, where fluorescence has greater selectivity than UV detection, due to the variations in excitation and emission wavelengths. This makes the technique especially suitable for the quantitation of individual PAHs. However, a disadvantage is that not all PAHs exhibit fluorescence. For the non-fluorescent PAHs, UV detection provides a more universal solution, especially when the full spectra are measured, which show characteristic maxima and minima for each PAH (Fetzer and Biggs 1987). Besides UV and fluorescence detection, Cai et al. described a method to separate and identify PAHs, with

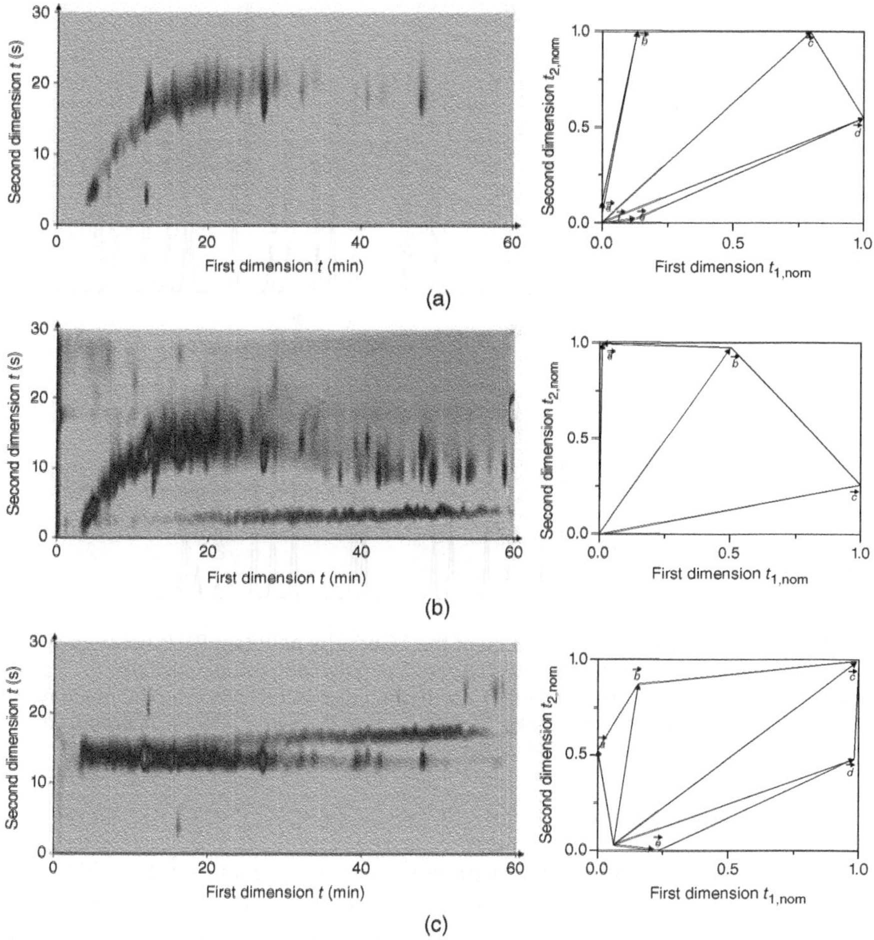

Figure 5.3 Contour plot of the LC × LC separation of a wastewater extract using different stationary phase combinations: (a) C18 × Phenyl-Hexyl, (b) C18 × PFP, and (c) C18 × HILIC and their corresponding surface coverage calculation by vectors. Detection was performed using UV at 290 nm and a sampling frequency of 5 Hz. Source: Ouyang et al. (2015). Reproduced with permission, Copyright (2015), Elsevier.

emphasis on the heavier PAHs, by using MS detection (Cai et al. 2009). Here, RPLC was coupled to tandem MS (MS/MS) with APPI. The method allowed the simultaneous characterization of both polar and non-polar PAHs. Polar compounds are less volatile and often less stable than non-polar molecules and are, thus, difficult to analyze with GC.

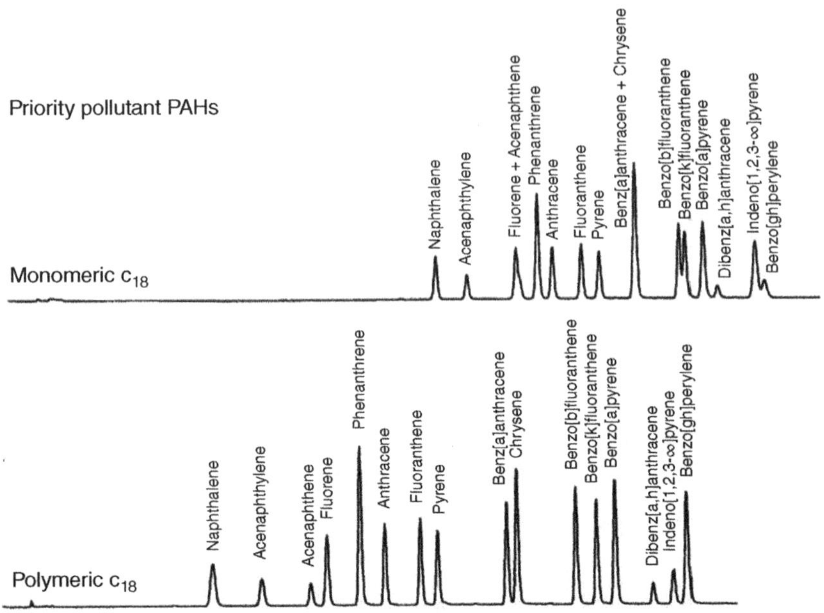

Figure 5.4 Difference in the separation of the 16 priority pollutant PAHs between a polymeric and a monomeric RPLC C18 column. The polymeric column contains cross-linked silanes on the silica surface and has a higher selectivity than the monomeric column, from which a number of PAHs coelute. Source: Wise et al. (1993). Reproduced with permission, Copyright (1993), Elsevier.

5.4.2 Naphthenic Acids

An alternative to conventional petroleum sources are oil sands, which are obtained from surface mining (Bauer et al. 2019). To obtain the heavy oils, akin to short residues or bitumen, from the oil sands, a hot-water-extraction process is utilized that generates large amounts of oil-sands-process-affected water (OSPW) (Martin et al. 2010). A fraction of the organic compounds that are present in this water are NAs, which feature many aliphatic and alicyclic carboxylic acids (El-Din et al. 2011; Martin et al. 2008). The general formula for these carboxylic acids is $C_nH_{2n+z}O_2$, where n is the number of carbon atoms, and z is the hydrogen deficiency, which is equal to zero or negative. However, more recently, it has been found that this description is insufficient to describe all NA components, because of the presence of more-oxygenated polycarboxylic acids and NAs with heteroatoms in OSPW (Brunswick et al. 2015). NAs are persistent in the aquatic environment and toxic to the aquatic biota at low concentrations. Therefore, sensitive analytical techniques are required (Wang and Kasperski 2010). Currently, OSPW is being stored until the NAs are successfully remediated,

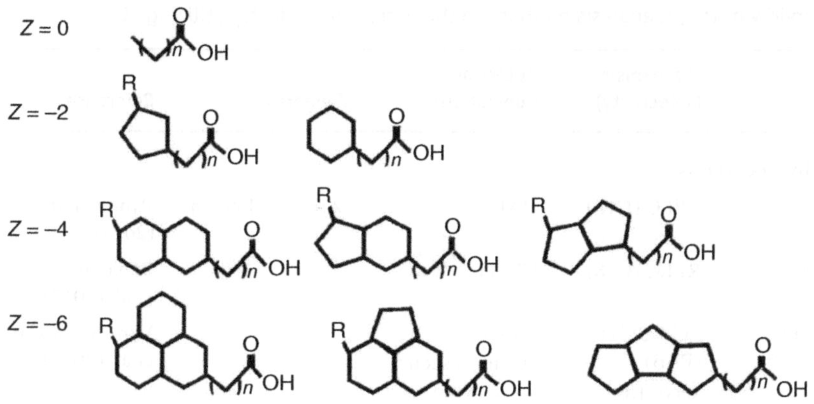

Figure 5.5 Examples from Headley et al. of classical NA structures. Source: Headley et al. (2008). Reproduced with permission, Copyright (2008), John Wiley & Sons.

because there is a no-discharge policy for these pollutants, necessitating techniques for characterization and for monitoring degradation (Martin et al. 2008; Wang et al. 2015; Figure 5.5).

LC techniques have been used to monitor degradation, such as the method of Martin et al. which allowed monitoring of NA biodegradation accelerated by ozonation (Martin et al. 2010). For this purpose, a combination of LC with high-resolution mass spectrometry (HRMS) was used. Brunswick et al. developed an LC-MS method for trace analysis of NAs (Brunswick et al. 2015). They tested multiple RPLC columns and found that the Agilent Poroshell 120 SB C18 achieved the best separation at 50 °C. Most published methods use RPLC C18 columns (Table 5.3); however, Wang et al. utilized an octyl-functionalized silica (C8) column (Wang and Kasperski 2010). This was ascribed to cationic process chemicals possibly present in OSPW, which were thought to give rise to strong interaction with the hydroxyl (silanol) groups of the C18 column. However, silanol groups on a C8 stationary phase may be equal in number (and more accessible) than those on a C18 phase. The observations of Wang and Kasperski may, therefore, have also been due to variations between brands (ZORBAX Eclipse Plus C8 and Thermo-Scientific Aquasil C18 column).

Besides the monitoring of degradation processes of NAs, characterization methods are also important for studying the toxicity of NAs present in the OSPW and for determining their contribution to the total acid number (Headley et al. 2008, 2013). When only quantitative results on chemical composition are obtained, it does not provide insight into, for example, the extent to which the NAs are corrosive (Clemente et al. 2003). Challenges for characterization are the separation of isobaric compounds, which are analytes with equal molecular

Table 5.3 Overview of target analysis methods in the petroleum industry utilizing LC.

Application	Mechanism (selectivity)	Detection (ionization)	Remarks	References
Polyaromatic hydrocarbons				
Heavy PAHs in petroleum	RPLC (C18)	DAD	$\lambda = 200$–400 nm	Panda et al. (2018)
Biodegradation monitoring	RPLC (C18)	UV-Vis		Pugazhendi et al. (2017)
PAHs in vacuum residues	LC-LC: ^{1}D (PBB) ^{2}D (C18)	^{1}D DAD ^{2}D fluorescence		Vanhoenacker et al. (2016)
PAH standards	RPLC (C18)	MS/MS (APPI)	UHPLC	Cai et al. (2009)
Water analysis	RPLC (C18)	UV-Vis		Wang et al. (2007)
Trace-level analysis in water	RPLC (C18)	Fluorescence		Chen (2004)
Gasoline analysis	LC × LC	UV-Vis	^{1}D NPLC (NH_2), ^{2}D RPLC (C18), $\lambda = 254$ nm	Murahashi (2003)
Seawater	RPLC (C18)	fluorescence		Witt (1995)
Naphthenic acids				
Biodegradation monitoring	UHPLC (BEH C18)	MS (ESI^-)	Column at 50 °C	Wang et al. (2015)
Degradation monitoring	UHPLC (BEH phenyl)	HRMS	Column at 50 °C	Martin et al. (2010)
NAs in aqueous solution	RPLC (C8)	MS/MS		Wang and Kasperski (2010)
Quantification of NAs and transformation products	RPLC (C18)	HRMS	Capillary HPLC	Bataineh et al. (2006)
Quantification and characterization	RPLC (C18)	HRMS (ESI)	Capillary HPLC	Martin et al. (2008)
Quantification	RPLC (C18)	DAD		Yen et al. (2004)
Biodegradation monitoring	RPLC (C18)	UV-Vis		Clemente et al. (2003)
Phenols				

(*continued*)

Table 5.3 (Continued)

Application	Mechanism (selectivity)	Detection (ionization)	Remarks	References
Wastewater analysis	RPLC (C18)	Photodiode array	Dispersive liquid–liquid microextraction	Salcedo et al. (2019)
Wastewater analysis of 11 priority pollutant phenols	RPLC (C18)	DAD	Different λ for all compounds, dispersive liquid–liquid microextraction	Saraji and Marzban (2010)
Aqueous samples	RPLC (C18)	UV-Vis	Liquid phase microextraction with backflush extraction	Zhao and Lee (2001)
Aqueous samples	RPLC (C18)	UV-Vis	Off-line preconcentration	Gawdzik et al. (1990)
Crude oil analysis	RPLC (C18)	Electro-chemical	Liquid–liquid extraction	MacCrehan and Brown-Thomas (1987)
Aqueous samples	RPLC (C18)	UV-Vis		Chao and Suatoni (1982)
Trace analysis of priority pollutant phenols in water	RPLC (C18)	UV		Realini (1981)

weights, but different chemical formulas, and isomeric compounds, which are analytes that have the same chemical formula (Headley et al. 2008). For this purpose, both (2D-)GC-MS with electron ionization (EI) and LC-MS with APPI and ESI have been used. However, GC analysis of NAs requires a derivatization step to stabilize the labile-carboxylic acid group prior to (hot) injection and prior to separation. The derivatization step must be monitored by duplicate analysis, and the samples cannot be exposed to water, because the ester formed may then hydrolyze back to the initial acid. The reader is referred to Section 3.3.4.2, for additional information on the GC analysis of NAs. Another advantage of LC-MS is the availability of soft ionization techniques such as ESI and APCI. No derivatization step is required, and relatively little fragmentation is observed (Wang and Kasperski 2010). The addition of an LC dimension to unit-resolution MS decreases the number of misclassifications, because isobaric components may be separated. Liquid chromatography-high-resolution mass spectrometry

(LC-HRMS) methods have been developed to decrease misassignments even further, which yield a higher mass accuracy. For example, the method of Martin et al. utilized HPLC-HRMS with ESI as ionization for accurate characterization of NAs (Martin et al. 2010).

More recently, a method for the monitoring of the biodegradation of NAs has been developed that utilizes ultra-high-performance liquid chromatography-mass spectrometry (UHPLC-MS) with ESI (Wang et al. 2015). This method allows identification of various NA species, without the requirement of an expensive HRMS. Martin et al. used a comprehensive LC × LC separation to investigate the potential of separating isomeric species (Martin et al. 2008). Besides this, little literature exists of LC × LC applications for NAs, although it may potentially be a powerful technique for source fingerprinting and for studies focusing on the fate of NAs in the environment. Due to the complexity of NA mixtures, LC analysis alone does not suffice; however, LC can be a powerful tool in combination with other techniques, such as MS, which is the most-often used detection method (Table 5.3).

5.4.3 Phenols

Phenolic compounds are used in multiple industries, including the petroleum industry, and they can be released into the environment by, for example, the disposal of wastewater (Saraji and Marzban 2010). These compounds are persistent and toxic for both aquatic life and humans (Barrios-Martinez et al. 2006). The different phenolic compound classes can each follow different metabolic pathways and toxicokinetic patterns, and 11 phenols are listed as priority pollutants by the EPA (Gawdzik et al. 1990; Zhao and Lee 2001). Phenolic compounds are used in a variety of processes in the petroleum industry, and consequently, they are present in wastewater (Barrios-Martinez et al. 2006). These compounds can inhibit or eliminate the biodegradation of other compounds by microorganisms. As was the case with NAs, GC analysis of phenols usually requires a derivatization step, which increases the sample-preparation time and the chance of errors (Saraji and Marzban 2010; Zhao and Lee 2001). However, LC analysis often requires a preconcentration step to achieve sufficiently low detection limits for trace analysis (Zhao and Lee 2001).

All methods for the analysis of phenols in wastewater use RPLC with C18 columns (Table 5.3). The main difference between the various methods originates from the sample preparation protocol being used. The most recent methods from Salcedo et al. and Saraji et al. feature dispersive liquid–liquid microextraction, which is a method wherein a high contact area between the aqueous phase and the extraction solvent is achieved by using a disperser solvent that gives rise to a cloudy solution (Salcedo et al. 2019; Saraji and Marzban 2010). The advantages of these methods are their low cost, fast extraction, and high enrichment factors.

5.5 LC as a Pre-separation Technique for GC Analysis

As mentioned previously, petroleum samples are highly complex, and with the limited peak capacity of LC, it is not possible to separate all the individual components. Although the peak capacities offered by GC are much higher than those of LC, GC by itself does not usually suffice to separate highly complex samples (Mao et al. 2008). An additional LC separation will decrease the number of components that are simultaneously introduced into the GC. Such pre-separations are used in a variety of applications in the petroleum industry (Table 5.4). The additional LC dimension is often used for separating the different hydrocarbon groups with, for example, aromatic, saturated, unsaturated, and polar functionalities (Apffel and McNair 1983; Kelly and Bartle 1994). Besides a few methods that use SEC or RPLC, NPLC is generally used for these applications. An advantage is that the mobile phases used are relatively non-polar and volatile, which allows for more efficient solvent evaporation in the subsequent GC analysis (Kelly and Bartle 1994).

The LC separations can either be performed off-line or on-line (Kelly and Bartle 1994). For the off-line LC methods, either one fraction containing an entire chromatographic peak or multiple fractions at constant intervals can be obtained for subsequent analysis by GC (Mao et al. 2008). Methods in which entire peaks are collected typically involve a concentration step of the eluent fraction to decrease the sample volume before injection in the GC system, which can be advantageous for trace analysis. For example, Dutriez et al. used an evaporation step prior to high-temperature two-dimensional gas chromatography (2D-GC) (Dutriez et al. 2010a). Mao et al. developed a comprehensive two-dimensional gas chromatography (GC × GC) method with an off-line HPLC pre-fractionation step in which multiple fractions were collected across (part of) the chromatogram at a constant time interval (Mao et al. 2008). However, these off-line methods are time-consuming and present potential risks of sample loss and contamination (Kelly and Bartle 1994).

One solution would be the on-line coupling of liquid chromatography-gas chromatography (LC-GC). However, such a coupling would require a dedicated interface, plus significant skills and careful optimization. Harborne et al. utilized a solvent splitter for reducing the eluent volume to be more compatible with GC injection volumes (Raglione et al. 1987). Another possibility for reducing the solvent volume is to couple capillary LC with GC, as demonstrated by Cortes et al. (1985). Other methods for the on-line coupling of LC with GC are based on trapping of the components at the start of the GC column to allow for the injection of higher volumes (Kelly and Bartle 1994). For this purpose, either an on-column interface or a loop-type interface can be used. In the latter method, the LC fraction of interest is stored in a loop prior to injection in the GC, and the trapping of low volatiles is achieved by instantaneous solvent evaporation. This method can be used for

Table 5.4 Overview of methods that use LC as a pre-fractionation or clean-up method with relevance to the petroleum industry.

Application	Mechanism (selectivity)	Method (on/off-line coupled, interface)	References
Aromatics in middle-distillates	NPLC (amino)	GC × GC-MS (off-line)	Jennerwein et al. (2017)
PAHs	2D-LC: ^{1}D RPLC (C18) ^{2}D RPLC (PBB)	2D-GC-MS (on-line, large volume injection in solvent vent mode)	Lim et al. (2017)
PAHs in coal tar	NPLC (NH_2)	GC-MS (off-line)	Wilson et al. (2017)
Characterization of vacuum gas oils	Preparative LC (clay gel adsorption)	High-temperature GC × GC (off-line)	Dutriez et al. (2010b)
Monitoring pf hydrocarbon attenuation	NPLC (silver-silica)	GC × GC (off-line)	Mao et al. (2009a)
Heavy petroleum products	NPLC (silver-silica)	GC × GC (off-line)	Mao et al. (2009c)
Oils and oil polluted soils	NPLC (silver-silica)	GC × GC (off-line)	Mao et al. (2009b), Mao et al. (2008)
Hydrocarbon analysis of middle distillates	preparative NPLC (mix of silica and γ-alumina)	GC × GC (off-line)	Adam et al. (2007)
Characterization of middle distillates	NPLC (mix of silica and alumina)	GC × GC (off-line)	Vendeuvre et al. (2005)
Sulfur-containing compounds in middle-distillates	NPLC (amino silane silica)	High resolution GC (off-line)	Beens and Tijssen (1997)
Chlorinated PAHs	NPLC (nitropropyl-, 2-(1-pyrenel) ethyldimethyl- silica)	GC-MS (on-line, loop)	Östman and Nilsson (1992)
Group-type separation of tar	NPLC (amino-, silver-silica)	GC (on-line, on-column)	Duquet et al. (1988)
Dibenzothiophene	NPLC (aminopropyl silane)	GC (on-line, loop)	Berthou and Dreano (1988)
Chlorinated benzenes in fuel oil	capillary NPLC (silica)	GC (on-line, loop)	Cortes et al. (1985)
Solvent refined coal	NPLC, RPLC	GC-MS (on-line, flow splitter)	Raglione et al. (1987)
Characterization of gasoline and diesel	NPLC (silica)	GC (on-line, on-column)	Apffel and McNair (1983)

large-volume injections; however, even moderately volatile analytes may (partly) evaporate with the solvent, preventing their (quantitative) analysis with this on-line interface. An advantage of the loop-type method is the possibility to add an internal standard to the LC fraction (or LC eluent) before GC analysis because of the known injection volume. This latter option was used by Berthou et al. for the analysis of dibenzothiophenes in the environment (Berthou and Dreano 1988).

On-column interfaces allow for the direct introduction of the desired fraction through the GC injector (Kelly and Bartle 1994). During injection it is important that the flow rate of the LC eluent is (slightly) higher than the solvent-evaporation speed, to ensure that the volatile components are trapped in the solvent present in the retention gap, which is a piece of tubing without stationary phase installed at the beginning of the GC column. Although the on-column method allows the analysis of volatiles, quantitation of such analytes is only possible if the internal standard co-elutes with the peak of interest. The addition of a known amount of internal standard is not possible, because the LC eluent is directly injected in the GC, causing the exact injection volume to be unknown. Since both the on-column and loop-type interfaces do not allow quantitative analysis of volatile analytes, an off-line method may be required. This may partly explain the large number of recently published off-line methods for LC pre-fractionation (Table 5.4). In general, when a suitable off-line method or on-line interface was selected, the additional LC separation improved the GC analysis.

5.6 Conclusions

In general, LC is used in situations in which GC does not suffice for analyzing the petroleum-related samples. This includes the heavier petroleum fractions, due to their high boiling points, thermally labile compounds, and, for example, acidic compounds that would require a derivatization step prior to GC analysis. Furthermore, LC has been used as a pre-separation technique to decrease the sample dimensionality prior to GC analysis. Either on- or off-line coupling may be used, with the latter being more frequently used in recent years. As for recent developments in LC, UHPLC appears to be increasingly applied to petroleum-related analyses. The use of core–shell particles may be beneficial, for example, for determining MWDs of petroleum samples, with reduced analysis times being foreseen, without a concomitant loss of resolution. However, applications of core–shell particles have not yet been reported in the literature for petroleum-related samples. Finally, only a few methods have so far been described that use 2D-LC, either in heart-cut (LC-LC) or comprehensive (LC × LC) mode. This may – at least in part – be due to the costly and time-consuming method development and optimization. One LC × LC method for analyzing heavy oil fractions was developed using computer-aided optimization algorithms. Such algorithms may greatly reduce the time required for method development and optimization, provided that suitable standards are available. The latter can be difficult considering the complexity of heavy-oil samples. As a general conclusion, recent developments in LC seem to make their way into the petroleum industry, but very slowly and, so far, not very frequently.

References

Adam, F., Bertoncini, F., Thiébaut, D. et al. (2007). Towards comprehensive hydrocarbons analysis of middle distillates by LC–GC × GC. *Journal of Chromatographic Science* 45: 643–649. https://doi.org/10.1093/chromsci/45.10.643.

Al-Muhareb, E.M., Karaca, F., Morgan, T.J. et al. (2006). Size exclusion chromatography for the unambiguous detection of aliphatics in fractions from petroleum vacuum residues, coal liquids, and standard materials, in the presence of aromatics. *Energy and Fuels* 20: 1165–1174. https://doi.org/10.1021/ef0600098.

Apffel, J.A. and McNair, H. (1983). Hydrocarbon group-type analyses by on-line multi-dimensional chromatography. II. Liquid chromatography-gas chromatography. *Journal of Chromatography A* 279: 139–144. https://doi.org/10.1016/S0021-9673(01)93611-8.

Badoni, R.P., Bhagat, S.D., and Joshi, G.C. (1992). Analysis of olefinic hydrocarbons in cracked petroleum stocks: A review. *Fuel* 71: 483–491. https://doi.org/10.1016/0016-2361(92)90144-D.

Barrios-Martinez, A., Barbot, E., Marrot, B. et al. (2006). Degradation of synthetic phenol-containing wastewaters by MBR. *Journal of Membrane Science* 281: 288–296. https://doi.org/10.1016/j.memsci.2006.03.048.

Bataineh, M., Scott, A.C., Fedorak, P.M., and Martin, J.W. (2006). Capillary HPLC/QTOF-MS for characterizing complex naphthenic acid mixtures and their microbial transformation. *Analytical Chemistry* 78: 8354–8361. https://doi.org/10.1021/ac061562p.

Bauer, A.E., Frank, R.A., Headley, J.V. et al. (2019). A preparative method for the isolation and fractionation of dissolved organic acids from bitumen-influenced waters. *Science of the Total Environment* 671: 587–597. https://doi.org/10.1016/j.scitotenv.2019.03.244.

van Beek, F.T., Edam, R., Pirok, B.W.J. et al. (2018). Comprehensive two-dimensional liquid chromatography of heavy oil. *Journal of Chromatography A* 1564: 110–119. https://doi.org/10.1016/j.chroma.2018.06.001.

Beens, J. and Tijssen, R. (1997). The characterization and quantitation of sulfur-containing compounds in (heavy) middle distillates by LC-GC-FID-SCD. *HRC Journal of High Resolution Chromatography* 20: 131–137. https://doi.org/10.1002/jhrc.1240200303.

Berthou, F. and Dreano, Y. (1988). Determination of methylated dibenzothiophenes in environmental samples by on-line HPLC-CGC. *Journal of High Resolution Chromatography* 11: 706–712. https://doi.org/10.1002/jhrc.1240111005.

Bissada, K.K.A., Tan, J., Szymezyk, E. et al. (2016a). Group-type characterization of crude oil and bitumen. Part II: Efficient separation and quantification of normal-paraffins iso-paraffins and naphthenes (PIN). *Fuel* 173: 217–221. https://doi.org/10.1016/j.fuel.2016.01.056.

Bissada, K.K.A., Tan, J., Szymezyk, E. et al. (2016b). Group-type characterization of crude oil and bitumen. Part I: Enhanced separation and quantification of saturates, aromatics, resins and asphaltenes (SARA). *Organic Geochemistry* 95: 21–28. https://doi.org/10.1016/j.orggeochem.2016.02.007.

Blomberg, J., Schoenmakers, P.J., and Brinkman, U.A.T. (2002). Gas chromatographic methods for oil analysis. *Journal of Chromatography A* 972: 137–173. https://doi.org/10.1016/S0021-9673(02)00995-0.

Boczkaj, G., Przyjazny, A., and Kamiński, M. (2015). Size-exclusion chromatography for the determination of the boiling point distribution of high-boiling petroleum fractions. *Journal of Separation Science* 38: 741–748. https://doi.org/10.1002/jssc.201401230.

Brunswick, P., Shang, D., van Aggelen, G. et al. (2015). Trace analysis of total naphthenic acids in aqueous environmental matrices by liquid chromatography/mass spectrometry-quadrupole time of flight mass spectrometry direct injection. *Journal of Chromatography A* 1405: 49–71. https://doi.org/10.1016/j.chroma.2015.05.048.

Cai, S.S., Syage, J.A., Hanold, K.A., and Balogh, M.P. (2009). Ultra performance liquid chromatography-atmospheric pressure photoionization-tandem mass spectrometry for high-sensitivity and high-throughput analysis of U.S. Environmental Protection Agency 16 priority pollutants polynuclear aromatic hydrocarbons. *Analytical Chemistry* 81: 2123–2128. https://doi.org/10.1021/ac802275e.

Carbognani, L. and Izquierdo, A. (1990). Preparative compound class separation of heavy oil vacuum residua by high performance liquid chromatography. *Fuel Science and Technology International* 8: 1–15. https://doi.org/10.1080/08843759008915909.

Castillo, M. and Barceló, D. (1997). Analysis of industrial effluents to determine endocrine-disrupting chemicals. *TrAC – Trends in Analytical Chemistry* 16: 574–583. https://doi.org/10.1016/S0165-9936(97)00090-3.

Castillo, M., Oubiña, A., and Barceló, D. (1998). Evaluation of ELISA kits followed by liquid chromatography-atmospheric pressure chemical ionization-mass spectrometry for the determination of organic pollutants in industrial effluents. *Environmental Science and Technology* 32: 2180–2184. https://doi.org/10.1021/es971042z.

Chao, G.K. and Suatoni, J.C. (1982). Determination of phenolic compounds by HPLC. *Journal of Chromatographic Science* 20: 436–440. https://doi.org/10.1093/chromsci/20.9.436.

Chen, H.-W. (2004). Determination of polycyclic aromatic hydrocarbons in water by solid-phase microextraction and liquid chromatography. *Analytical Sciences* 20: 1383–1388. https://doi/org/10.2116/analsci.20.1383.

Clemente, J.S., Yen, T.-W., and Fedorak, P.M. (2003). Development of a high performance liquid chromatography method to monitor the biodegradation of naphthenic acids. *Journal of Environmental Engineering and Science* 2: 177–186. https://doi.org/10.1139/s03-011.

Cortes, H.J., Richter, B.E., Pfeiffer, C.D., and Jensen, D.E. (1985). Determination of trace chlorinated benzenes in fuel oil by on-line multidimensional chromatography using packed-capillary liquid chromatography and capillary gas chromatography. *Journal of Chromatography* 349: 55–61. https://doi.org/10.1016/S0021-9673(00)90632-0.

Davis, J.M. and Giddings, J.C. (1983). Statistical theory of component overlap in multicomponent chromatograms. *Analytical Chemistry* 55: 418–424. https://doi.org/10.1021/ac00254a003.

Duquet, D., Dewaele, C., and Verzele, M. (1988). Coupling micro-LC and capillary GC as a powerful tool for the analysis of complex mixtures. *Journal of High Resolution Chromatography* 11: 252–256. https://doi.org/10.1002/jhrc.1240110307.

Dutriez, T., Courtiade, M., Thiébaut, D. et al. (2010a). Extended characterization of a vacuum gas oil by offline LC-high-temperature comprehensive two-dimensional gas chromatography. *Journal of Separation Science* 33: 1787–1796. https://doi.org/10.1002/jssc.201000102.

Dutriez, T., Courtiade, M., Thiébaut, D. et al. (2010b). Advances in quantitative analysis of heavy petroleum fractions by liquid chromatography-high-temperature comprehensive two-dimensional gas chromatography: Breakthrough for conversion processes. *Energy and Fuels* 24: 4430–4438. https://doi.org/10.1021/ef1002809.

Eeltink, S., Dolman, S., Vivo-Truyols, G. et al. (2010). Selection of column dimensions and gradient conditions to maximize the peak-production rate in comprehensive off-line two-dimensional liquid chromatography using monolithic columns. *Analytical Chemistry* 82 https://doi.org/10.1021/ac101514d.

El-Din, M.G., Fu, H., Wang, N. et al. (2011). Naphthenic acids speciation and removal during petroleum-coke adsorption and ozonation of oil sands process-affected water. *Science of the Total Environment* 409: 5119–5125. https://doi.org/10.1016/j.scitotenv.2011.08.033.

Fetzer, J.C. and Biggs, W.R. (1987). Retention behavior of large polycyclic aromatics in bonded-phase high-performance liquid chromatography. *Journal of Chromatography A* 386: 87–101. https://doi.org/10.1016/S0021-9673(01)94587-X.

Frenich, A.G., Romero-González, R., and del Mar Aguilera-Luiz, M. (2014). Comprehensive analysis of toxics (pesticides, veterinary drugs and mycotoxins) in food by UHPLC-MS. *TrAC - Trends in Analytical Chemistry* 63: 158–169. https://doi.org/10.1016/j.trac.2014.06.020.

Gargano, A.F.G., Duffin, M., Navarro, P., and Schoenmakers, P.J.P.J. (2016). Reducing dilution and analysis time in online comprehensive two-dimensional liquid chromatography by active modulation. *Analytical Chemistry* 88: 1785–1793. https://doi.org/10.1021/acs.analchem.5b04051.

Gawdzik, B., Gawdzik, J., and Czerwińska-Bil, U. (1990). Use of polymeric sorbents for the off-line preconcentration of priority pollutant phenols from water for high-performance liquid chromatographic analysis. *Journal of Chromatography A* 509: 135–140. https://doi.org/10.1016/S0021-9673(01)93247-9.

George, A., Morgan, T.J., Alvarez, P. et al. (2010). Fractionation of a coal tar pitch by ultra-filtration, and characterization by size exclusion chromatography, UV-fluorescence and laser desorption-mass spectroscopy. *Fuel* 89: 2953–2970. https://doi.org/10.1016/j.fuel.2010.04.011.

Giddings, J.C. (1995). Sample dimensionality: A predictor of order-disorder in component peak distribution in multidimensional separation. *Journal of Chromatography A* 703: 3–15. https://doi.org/10.1016/0021-9673(95)00249-M.

González-Ruiz, V., Olives, A.I., and Martín, M.A. (2015). Core-shell particles lead the way to renewing high-performance liquid chromatography. *TrAC – Trends in Analytical Chemistry* 64: 17–28. https://doi.org/10.1016/j.trac.2014.08.008.

Gritti, F., Leonardis, I., Abia, J., and Guiochon, G. (2010). Physical properties and structure of fine core-shell particles used as packing materials for chromatography. Relationships between particle characteristics and column performance. *Journal of Chromatography A* 1217: 3819–3843. https://doi.org/10.1016/j.chroma.2010.04.026.

Han, Y., Zhang, Y., Xu, C., and Hsu, C.S. (2018). Molecular characterization of sulfur-containing compounds in petroleum. *Fuel* 221: 144–158. https://doi.org/10.1016/j.fuel.2018.02.110.

Hayes, R., Ahmed, A., Edge, T., and Zhang, H. (2014). Core-shell particles: Preparation, fundamentals and applications in high performance liquid chromatography. *Journal of Chromatography A* 1357: 36–52. https://doi.org/10.1016/j.chroma.2014.05.010.

Headley, J.V., Peru, K.M., and Barrow, M.P. (2008). Mass spectrometric characterisation of naphthenic acids in environmental samples: A review. *Mass Spectrometry Reviews* 28: 121–134. https://doi.org/10.1002/mas.20185.

Headley, J.V., Peru, K.M., Mohamed, M.H. et al. (2013). Chemical fingerprinting of naphthenic acids and oil sands process waters-A review of analytical methods for environmental samples. *Journal of Environmental Science and Health – Part A Toxic/Hazardous Substances and Environmental Engineering* 48: 1145–1163. https://doi.org/10.1080/10934529.2013.776332.

Herod, A.A., Zhuo, Y., and Kandiyoti, R. (2003). Size-exclusion chromatography of large molecules from coal liquids, petroleum residues, soots, biomass tars and humic substances. *Journal of Biochemical and Biophysical Methods* 56: 335–361. https://doi.org/10.1016/S0165-022X(03)00070-8.

Herod, A.A., Bartle, K.D., and Kandiyoti, R. (2007). Characterization of heavy hydrocarbons by chromatographic and mass spectrometric methods: An overview. *Energy and Fuels* 21: 2176–2203. https://doi.org/10.1021/ef060642t.

Heshka, N.E., Baltazar, M., and Chen, J. (2019). Separation and quantification of olefins and diolefins in cracked petroleum fractions using silver-ion high performance liquid chromatography. *Petroleum Science and Technology* 37: 1808–1816. https://doi.org/10.1080/10916466.2019.1605378.

Hsu, C.S., Qian, K., Aczel, T. et al. (1991). On-line liquid chromatography/mass spectrometry for heavy hydrocarbon characterization. *Energy and Fuels* 5: 395–398. https://doi.org/10.1021/ef00027a007.

Hsu, C.S., Qian, K., and Robbins, W.K. (1994). Nitrogen speciation of polar petroleum compounds by compound class separation and on-line liquid chromatography – mass spectrometry (LC-MS). *Journal of High Resolution Chromatography* 17: 271–276. https://doi.org/10.1002/jhrc.1240170415.

Jakobsen, S.S., Christensen, J.H., Verdier, S. et al. (2017). Increasing flexibility in two-dimensional liquid chromatography by pulsed elution of the first dimension: A proof of concept. *Analytical Chemistry* 89: 8723–8730. https://doi.org/10.1021/acs.analchem.7b00758.

Jennerwein, M.K., Sutherland, A.C., Eschner, M. et al. (2017). Quantitative analysis of modern fuels derived from middle distillates – the impact of diverse compositions on standard methods evaluated by an offline hyphenation of HPLC-refractive index detection with GC × GC-TOFMS. *Fuel* 187: 16–25. https://doi.org/10.1016/j.fuel.2016.09.033.

Jinno, K., Nomura, H., and Hirata, Y. (1980). Group-type analysis of gasoline range materials by high performance liquid chromatography. *Journal of High Resolution Chromatography* 3: 503–506. https://doi.org/10.1002/jhrc.1240031004.

Kamiński, M., Kartanowicz, R., and Przyjazny, A. (2004). Application of high-performance liquid chromatography with ultraviolet diode array detection and refractive index detection to the determination of class composition and to the analysis of gasoline. *Journal of Chromatography A* 1029: 77–85. https://doi.org/10.1016/j.chroma.2003.12.053.

Karaca, F., Millan-Agorio, M., Morgan, T.J. et al. (2008). The pentane- and toluene-soluble fractions of a petroleum residue and three coal tars by size exclusion chromatography and UV-fluorescence spectroscopy. *Oil & Gas Science and Technology* 63: 129–137. https://doi.org/10.2516/ogst.

Kartanowicz, R., Gilgenast, E., and Kamiński, M. (2007). Group-type analysis of middle distillates by test method IP-391/EN-12916/ASTM D 6379 in terms of resolution and selectivity of chromatographic columns. *Chemia Analityczna* 52: 265–280.

Kelly, G.W. and Bartle, K.D. (1994). The use of combined LC-GC for the analysis of fuel products: A review. *Journal of High Resolution Chromatography* 17: 390–397. https://doi.org/10.1002/jhrc.1240170604.

Lanças, F.M. and McNair, H.M. (1985). Characterization of high-ash Brazilian (mina do leao) coal-derived liquids by high performance liquid chromatography group type separation. *Journal of Liquid Chromatography* 8: 239–259. https://doi.org/10.1080/01483918508067075.

Le Guern, M., Chailleux, E., Farcas, F. et al. (2010). Physico-chemical analysis of five hard bitumens: Identification of chemical species and molecular organization before and after artificial aging. *Fuel* 89: 3330–3339. https://doi.org/10.1016/j.fuel.2010.04.035.

Li, Y., Deng, X., and Yu, W. (1998). Group-type analyses of heavy petroleum fractions by preparative liquid chromatography and synchronous fluorescence spectrometry: Analyses of aromatics by ring number of Liaohe vacuum gas oil, coker gas oil and heavy cycle oil. *Fuel* 77: 277–284. https://doi.org/10.1016/S0016-2361(97)00195-6.

Lienne, M., Gareil, P., Rosset, R. et al. (1987). Synthesis of silver and copper(II) silica gels for the liquid chromatographic separation of olefinic hydrocarbons in the group analysis of light and heavy petroleum compounds. *Journal of chromatography* 395: 255–270. https://doi.org/10.1016/B978-0-444-43020-5.50013-8.

Lim, H., Ahmed, T.M., Bergvall, C., and Westerholm, R. (2017). Automated clean-up, separation and detection of polycyclic aromatic hydrocarbons in particulate matter extracts using a 2D-LC/2D-GC system: A method translation from two FIDs to two MS detectors. *Analytical and Bioanalytical Chemistry* 409: 5619–5629. https://doi.org/10.1007/s00216-017-0509-1.

Liu, P., Shi, Q., Chung, K.H. et al. (2010). Molecular characterization of sulfur compounds in Venezuela crude oil and its SARA fractions by electrospray ionization Fourier transform ion cyclotron resonance mass spectrometry. *Energy and Fuels* 24: 5089–5096. https://doi.org/10.1021/ef100904k.

MacCrehan, W.A. and Brown-Thomas, J.M. (1987). Determination of phenols in petroleum crude oils using liquid chromatography with electrochemical detection. *Analytical Chemistry* 59: 477–479. https://doi.org/10.1021/ac00130a021.

Manoli, E. and Samara, C. (1999). Polycyclic aromatic hydrocarbons in natural waters: Sources, occurrence and analysis. *TrAC - Trends in Analytical Chemistry* 18: 417–428. doi:https://doi.org/10.1016/S0165-9936(99)00111-9.

Mao, D., Van De Weghe, H., Diels, L. et al. (2008). High-performance liquid chromatography fractionation using a silver-modified column followed by two-dimensional comprehensive gas chromatography for detailed group-type characterization of oils and oil pollutions. *Journal of Chromatography A* 1179: 33–40. https://doi.org/10.1016/j.chroma.2007.09.085.

Mao, D., Lookman, R., Van De Weghe, H. et al. (2009a). Detailed analysis of petroleum hydrocarbon attenuation in biopiles by high-performance liquid chromatography followed by comprehensive two-dimensional gas chromatography. *Journal of Chromatography A* 1216: 1524–1527. https://doi.org/10.1016/j.chroma.2008.12.087.

Mao, D., Lookman, R., Van De Weghe, H. et al. (2009b). Combining HPLC-GC × GC, GC × GC/ToF-MS, and selected ecotoxicity assays for detailed monitoring of petroleum hydrocarbon degradation in soil and leaching water. *Environmental Science and Technology* 43: 7651–7657. https://doi.org/10.1021/es9015603.

Mao, D., Van De Weghe, H., Lookman, R. et al. (2009c). Resolving the unresolved complex mixture in motor oils using high-performance liquid chromatography followed by comprehensive two-dimensional gas chromatography. *Fuel* 88: 312–318. https://doi.org/10.1016/j.fuel.2008.08.021.

Martin, J.W., Han, X., Peru, K.M., and Headley, J.V. (2008). Comparison of high- and low-resolution electrospray ionization mass spectrometry for the analysis of naphthenic acid mixtures in oil sands process water. *Rapid Communications in Mass Spectrometry* 22: 1919–1924. https://doi.org/10.1002/rcm.

Martin, J.W., Barri, T., Han, X. et al. (2010). Ozonation of oil sands process-affected water accelerates microbial bioremediation. *Environmental Science and Technology* 44: 8350–8356. https://doi.org/10.1021/es101556z.

Matisová, E. (1985). Quantitative analysis of hydrocarbons in gasolines. *TrAC - Trends in Analytical Chemistry* 4: 175–178. https://doi.org/10.1016/0165-9936(85)88025-0.

Miller, R. (1982). Hydrocarbon class fractionation with bonded-phase liquid chromatography. *Analytical Chemistry* 54: 1742–1746. https://doi.org/10.1021/ac00248a021.

Mills, M.A., McDonald, T.J., Bonner, J.S. et al. (1999). Method for quantifying the fate of petroleum in the environment. *Chemosphere* 39: 2563–2582. https://doi.org/10.1016/S0045-6535(99)00163-0.

Morgan, T.J., George, A., Alvarez-Rodriguez, P. et al. (2010). Estimating molecular masses of petroleum-derived fractions: High mass (>2000u) materials in maltenes and asphaltenes from Maya crude oil. *Journal of Chromatography A* 1217: 3804–3818. https://doi.org/10.1016/j.chroma.2010.04.024.

Mössner, S.G. and Wise, S.A. (1999). Determination of polycyclic aromatic sulfur heterocycles in fossil fuel-related samples. *Analytical Chemistry* 71: 58–69. https://doi.org/10.1021/ac980664f.

Murahashi, T. (2003). Comprehensive two-dimensional high-performance liquid chromatography for the separation of polycyclic aromatic hydrocarbons. *Analyst* 128: 611–615. https://doi.org/10.1039/b212643e.

Nizio, K.D., McGinitie, T.M., and Harynuk, J.J. (2012). Comprehensive multidimensional separations for the analysis of petroleum. *Journal of Chromatography A* 1255: 12–23. https://doi.org/10.1016/j.chroma.2012.01.078.

Ochoa, C.M., Shoenmakers, P., Mallet, C.R., and Lurie, I.S. (2018). Decreasing the uncertainty of peak assignments for the analysis of synthetic cathinones using multi-dimensional ultra-high performance liquid chromatography. *Analytical Methods* 10: 3178–3187. https://doi.org/10.1039/c8ay00565f.

Oro, N.E. and Lucy, C.A. (2010). Comparison of hypercrosslinked polystyrene columns for the separation of nitrogen group-types in petroleum using High Performance Liquid Chromatography. *Journal of Chromatography A* 1217: 6178–6185. https://doi.org/10.1016/j.chroma.2010.07.078.

Oro, N.E. and Lucy, C.A. (2013). Analysis of the nitrogen content of distillate cut gas oils and treated heavy gas oils using normal phase HPLC, fraction collection and petroleomic FT-ICR MS data. *Energy and Fuels* 27: 35–45. https://doi.org/10.1021/ef301116j.

Östman, C. and Nilsson, U. (1992). Coupled LC-GC-MS for on-line clean-up, separation, and identification of chlorinated polycyclic aromatic hydrocarbons at picogram levels in urban air. *Journal of High Resolution Chromatography* 15: 745–750. https://doi.org/10.1002/jhrc.1240151109.

Ouyang, Z., Leonards, P., Legler, J. et al. (2015). Comprehensive two-dimensional liquid chromatography coupled to high resolution time of flight mass spectrometry for chemical characterization of sewage treatment plant effluents. *Journal of Chromatography A* 1380: 139–145. https://doi.org/10.1016/j.chroma.2014.12.075.

Panda, S.K., Muller, H., Al-Qunaysi, T.A., and Koseoglu, O.R. (2018). Determination of heavy polycyclic aromatic hydrocarbons by non-aqueous reversed phase liquid chromatography: Application and limitation in refining streams. *Journal of Chromatography A* 1533: 30–37. https://doi.org/10.1016/j.chroma.2017.11.061.

Pavlova, A. and Ivanova, R. (2003). Determination of petroleum hydrocarbons and polycyclic aromatic hydrocarbons in sludge from wastewater treatment basins. *Journal of Environmental Monitoring* 5: 319–323. https://doi.org/10.1039/b208157c.

Pearson, C.D. and Gharfeh, S.G. (1986). Automated high-performance liquid chromatography determination of hydrocarbon types in crude oil residues using a flame ionization detector. *Analytical Chemistry* 58: 307–311. https://doi.org/10.1021/ac00293a010.

Pirok, B.W.J., Stoll, D.R., and Schoenmakers, P.J. (2019). Recent developments in two-dimensional liquid chromatography: Fundamental improvements for practical applications. *Analytical Chemistry* 91: 240–263. https://doi.org/10.1021/acs.analchem.8b04841.

Pirok, B.W.J., Pous-Torres, S., Ortiz-Bolsico, C. et al. (2016). Program for the interpretive optimization of two-dimensional resolution. *Journal of Chromatography A* 1450: 29–37. https://doi.org/10.1016/j.chroma.2016.04.061.

Pirok, B.W.J., Breuer, P., Hoppe, S.J.M. et al. (2017). Size-exclusion chromatography using core-shell particles. *Journal of Chromatography A* 1486: 96–102. https://doi.org/10.1016/j.chroma.2016.12.015.

Pirok, B.W.J., Gargano, A.F.G., and Schoenmakers, P.J. (2018). Optimizing separations in online comprehensive two-dimensional liquid chromatography. *Journal of Separation Science* 41: 68–98. https://doi.org/10.1002/jssc.201700863.

Pugazhendi, A., Abbad Wazin, H., Qari, H. et al. (2017). Biodegradation of low and high molecular weight hydrocarbons in petroleum refinery wastewater by a thermophilic bacterial consortium. *Environmental Technology (United Kingdom)* 38: 2381–2391. https://doi.org/10.1080/09593330.2016.1262460.

Qian, K. and Hsu, C.S. (1992). Molecular transformation in hydrotreating processes studied by on-line liquid chromatography/mass spectrometry. *Analytical Chemistry* 64: 2327–2333. https://doi.org/10.1021/ac00044a005.

Raglione, T.V., Troskosky, J.A., and Hartwick, R.A. (1987). On-line microbore high-performance liquid chromatography-capillary gas chromatography-mass spectrometry II. Application to the analysis of solvent refined coal. *Journal of Chromatography* 409: 213–221. https://doi.org/10.1016/S0021-9673(01)86797-2.

Ravanchi, M.T., Kaghazchi, T., and Kargari, A. (2009). Application of membrane separation processes in petrochemical industry: a review. *Desalination* 235: 199–244. https://doi.org/10.1016/j.desal.2007.10.042.

Realini, P.A. (1981). Determination of priority pollutant phenols in water by HPLC. *Journal of Chromatographic Science* 19: 124–129. https://doi.org/10.1093/chromsci/19.3.124.

Robbins, W.K. (1998). Quantitative measurement of mass and aromaticity distributions for heavy distillates 1. Capabilities of the HPLC-2 System. *Journal of Chromatographic Science* 36: 457–466. https://doi.org/10.1093/chromsci/36.9.457.

Robson, W.J., Sutton, P.A., McCormack, P. et al. (2017). Class type separation of the polar and apolar components of petroleum. *Analytical Chemistry* 89: 2919–2927. https://doi.org/10.1021/acs.analchem.6b04202.

Rudzinski, W.E. and Rai, V. (2005). Detection of polyaromatic sulfur heterocycles in crude oil using postcolumn addition of tropylium and tandem mass spectrometry. *Energy and Fuels* 19: 1611–1618. https://doi.org/10.1021/ef0400666.

Salcedo, G.M., Kupski, L., Degang, L. et al. (2019). Determination of fifteen phenols in wastewater from petroleum refinery samples using a dispersive liquid – liquid microextraction and liquid chromatography with a photodiode array detector. *Microchemical Journal* 146: 722–728. https://doi.org/10.1016/j.microc.2019.01.075.

Saraji, M. and Marzban, M. (2010). Determination of 11 priority pollutant phenols in wastewater using dispersive liquid-liquid microextraction followed by high-performance liquid chromatography-diode-array detection. *Analytical and Bioanalytical Chemistry* 396: 2685–2693. https://doi.org/10.1007/s00216-010-3496-z.

Schure, M.R. and Moran, R.E. (2017). Size exclusion chromatography with superficially porous particles. *Journal of Chromatography A* 1480: 11–19. https://doi.org/10.1016/j.chroma.2016.12.016.

Sharma, A. (2018). Hazardous effects of petrochemical industries: A review. *Recent Advances in Petroleum Science* 3: 2–4. https://doi.org/10.19080/rapsci.2017.03.555607.

Sharma, B.K., Sarowha, S.L.S., and Bhagat, S.D. (2003). Chromatographic characterization of high-boiling petroleum fractions. *Journal of Separation Science* 26: 1657–1664. https://doi.org/10.1002/jssc.200301506.

Stevenson, R. (1971). Rapid separation of petroleum fuels by hydrocarbon type. *Journal of Chromatographic Science* 9: 257–262. https://doi.org/10.1093/chromsci/9.5.257.

Tomić, T., Babić, S., Nasipak, N.U. et al. (2009). Determination of alkenes in cracking products by normal-phase high-performance liquid chromatography-diode array detection. *Journal of Chromatography A* 1216: 3819–3824. https://doi.org/10.1016/j.chroma.2009.02.062.

Vanhoenacker, G., Steenbeke, M., David, F., Sandra, P., & Sandra, K. (2016). *Analysis of Polycyclic Aromatic Hydrocarbons in Petroleum Vacuum Residues by Multiple Heart-Cutting LC Using the Agilent 1290 Infinity 2D-LC Solution*. (5991-6549EN). Agilent Technologies

Varotsis, N. and Pasadakis, N. (1998). Determination of the composition of commingled crude oil streams by HPLC-SEC with UV and IR detection. *Journal of Liquid Chromatography and Related Technologies* 21: 657–667. https://doi.org/10.1080/10826079808005848.

Vendeuvre, C., Ruiz-Guerrero, R., Bertoncini, F. et al. (2005). Characterisation of middle-distillates by comprehensive two-dimensional gas chromatography (GC × GC): a powerful alternative for performing various standard analysis of middle-distillates. *Journal of Chromatography A* 1086: 21–28. https://doi.org/10.1016/j.chroma.2005.05.106.

Wang, X. and Kasperski, K.L. (2010). Analysis of naphthenic acids in aqueous solution using HPLC-MS/MS. *Analytical Methods* 2: 1715–1722. https://doi.org/10.1039/c0ay00204f.

Wang, W.D., Huang, Y.M., Shu, W.Q., and Cao, J. (2007). Multiwalled carbon nanotubes as adsorbents of solid-phase extraction for determination of polycyclic aromatic hydrocarbons in environmental waters coupled with high-performance liquid chromatography. *Journal of Chromatography A* 1173: 27–36. https://doi.org/10.1016/j.chroma.2007.10.027.

Wang, B., Wan, Y., Gao, Y. et al. (2015). Occurrences and behaviors of naphthenic acids in a petroleum refinery wastewater treatment plant. *Environmental Science and Technology* 49: 5796–5804. https://doi.org/10.1021/es505809g.

Wilson, W.B., Hayes, H.V., Sander, L.C. et al. (2017). Qualitative characterization of SRM 1597a coal tar for polycyclic aromatic hydrocarbons and methyl-substituted derivatives via normal-phase liquid chromatography and gas chromatography/mass spectrometry. *Analytical and Bioanalytical Chemistry* 409: 5171–5183. https://doi.org/10.1007/s00216-017-0464-x.

Wise, S.A., Sander, L.C., and May, W.E. (1993). Determination of polycyclic aromatic hydrocarbons by liquid chromatography. *Journal of Chromatography A* 642: 329–349. https://doi.org/10.1016/0021-9673(93)80097-R.

Witt, G. (1995). Polycyclic aromatic hydrocarbons in water and sediment of the Baltic sea. *Environmental Science and Technology* 31: 237–248. https://doi.org/10.1021/es00062a024.

Xia, J.L., Fan, X., You, C.Y. et al. (2016). Sequential ultrasonic extraction of a Chinese coal and characterization of nitrogen-containing compounds in the extracts using high-performance liquid chromatography with mass spectrometry. *Journal of Separation Science* 39: 2491–2498. https://doi.org/10.1002/jssc.201600190.

Yen, T.W., Marsh, W.P., MacKinnon, M.D., and Fedorak, P.M. (2004). Measuring naphthenic acids concentrations in aqueous environmental samples by liquid chromatography. *Journal of Chromatography A* 1033: 83–90. https://doi.org/10.1016/j.chroma.2004.01.030.

Zhao, L. and Lee, H.K. (2001). Determination of phenols in water using liquid phase microextraction with back extraction combined with high-performance liquid chromatography. *Journal of Chromatography A* 931: 95–105. https://doi.org/10.1016/S0021-9673(01)01199-2.

6

Supercritical Fluids in Chromatography: Applications to the Oil and Gas Industry

Didier Thiébaut[1] *and Robert M. Campbell*[2]

[1] *Laboratory of Analytical Sciences, Bioanalytics, and Miniaturization, UMR 8231 Chemistry, Biology and Innovation, CNRS-ESPCI Paris, PSL University, Paris, France*
[2] *Analytical Science, Midland – Michigan Operations, The Dow Chemical Company, Midland, MI, USA*

6.1 Introduction

In the beginning, supercritical fluid chromatography (SFC) was an extension of gas chromatography (GC). The first report related to SFC was published by Klesper et al. (1962) on the separation of porphyrins using "high pressure gas chromatography above critical pressure." It was the demonstration that a fluid operated above the critical point could allow the separation of compounds at a lower temperature than GC. Sie et al. (1966) and McLaren et al. (1968) works showed that CO_2 at high pressure could solvate heavy compounds, indicating a supercritical fluid (SF), could behave as a liquid owing to a liquid like density while having a lower viscosity than a liquid, enabling better kinetics than a liquid in all aspects of their use as a solvent.

The pioneers of SFC used GC capillary columns (Novotny et al. 1981) with a GC derived system, while attempts to use packed columns from LC also showed great potential with SFC systems derived from liquid chromatography (LC). It must be emphasized that SFC on capillary GC columns performed in the 1980s could be compared with today's nano-ultra high performance liquid chromatography (UHPLC) from the perspective of operating conditions (pressure, flow rates, etc.). One of the main advantages of SFC was the coupling of an LC-like technique to the most widely utilized GC detector, flame ionization detection (FID). Thus, one can understand that supercritical fluid chromatography-flame ionization detection (SFC-FID) could be of major interest for petroleum related samples to expand the application field of GC toward heavier cuts. Indeed, the extensive use of CO_2 as the SF was more suitable for the analysis of low polarity compounds, such as

Analytical Techniques in the Oil and Gas Industry for Environmental Monitoring, First Edition.
Edited by Melissa N. Dunkle and William L. Winniford.

hydrocarbons, compared with more polar compounds that are partly soluble or even not soluble in neat carbon dioxide.

Now, the "many faces" (Lesellier and West 2015) of SFC are mainly restricted to the use of packed columns for pharmaceutical applications, where polar compounds are of interest. For these cases, the mobile phase is a mixture of CO_2, a polar solvent (5–50%, also called the polar modifier), and additives added in a small amount, similar as in LC (c. 0.1%), including acids, bases, or water. The other applications of SFC at the analytical scale can be considered as a niche in the SFC niche, which includes the petroleum applications. To the knowledge of the authors, capillary SFC is only used for petroleum applications.

It is worth mentioning that SFC can also be used in semi and preparative scale applications (Francotte 2017; Tarafder 2017), as well as for extraction (SFE), mainly at the industrial scale. For such industrial implementation, preparative SFC or SFE are considered as green techniques, because the SF is easier to recycle compared with solvents, the process is faster owing to better kinetics, and the use of organic solvents is reduced because most of the phase is composed of carbon dioxide.

In this chapter, the application of SFC to analytical scale separations of petroleum related compounds will be reviewed. Brief general information on packed column supercritical fluid chromatography (pSFC) and capillary supercritical fluid chromatography (cSFC) will be given in the first section of this chapter before the dedicated SFC applications are presented.

6.2 Basics of SFC

In SFC, the mobile phase is a fluid, CO_2 in most applications, used above the critical pressure. If the temperature is also above the critical pressure, then the fluid is in the supercritical state. Below the critical temperature, the fluid is in the subcritical state. The point is, for chromatography, pressure (P) and temperature (T) conditions must be high enough so that there is only one phase in the mobile phase; this is of major importance when polar modifiers are added because the coordinates of the critical point are changed versus the amount of solvent added to the SF. Therefore, at low temperatures, the fluid exhibits liquid-like properties (viscosity, density, and diffusivity), while at a higher temperature, the properties will change between those of gases and liquids. The fluid will be more gas-like only at high temperatures for chromatography, above 100–150 °C, and as a result, so will be the chromatography.

The solubility of compounds increases with the density of the SF. However, the kinetics are better at low density, because the viscosity is lower. There is an antinomy between the thermodynamics and the kinetics in SFC, because the best kinetics are obtained when the fluid exhibits lower solvation properties (i.e. at low density).

The density plays a major role on retention when neat carbon dioxide is used as the mobile phase (Poe 2017). Despite the fact that the polarity of the CO_2 increases when the density increases (Giddings et al. 1968), the CO_2 polarity is controversial (Berger 1997). It is estimated that the polarity of CO_2 is within the range of that of alkanes, and optimistically, of dichloromethane. This means that neat CO_2 cannot dissolve polar compounds, and as such, is the reason why, as in LC, the polarity of the mobile phase can be modified by adding polar solvents (polar modifiers) to the CO_2.

When modifiers are added to the SF, the properties are much more dependent on the percentage of modifier used in the mobile phase. In that case, the SFC mechanism can be roughly associated to normal phase liquid chromatography (NPLC), despite the fact that there is still an effect of the density of the mobile phase on retention. As in NPLC, the higher the polarity of the modifier, the higher the effect of this percentage in the mobile phase on the retention. It must be noticed that the solubility of water in the CO_2 is low (c. 1000–2000 ppm, depending on P–T). In practice, water should only be added to a binary mobile phase such as CO_2-methanol.

Of course, the addition of liquids as modifiers to the CO_2 will also vary the density and the viscosity toward those of liquids, and as a result, the kinetics will also be varied. However, diffusion coefficients are still three times higher for situations when 20% methanol is added to the CO_2, ensuring improved kinetics in SFC compared with LC (Sassiat et al. 1987) at lower concentrations of modifiers. At higher modifier concentration, SFC can be viewed as an LC-like technique, offering enhanced fluidity (Cui and Olesik 1991; Lesellier and West 2015).

No matter what the conditions, the mobile phase always plays an active role in the solvating process of solutes, and therefore, cannot at all be considered as a carrier gas as with GC. For this reason, no matter what the column used, the density and the composition of the mobile phase, SFC separations have a mechanism similar to LC, because the same type of interactions are involved between solutes, mobile phase, and stationary phase. One difference comes from the CO_2 itself. Thus, different selectivity can be obtained in SFC compared with LC simply because the mobile phase contains CO_2, as it has been demonstrated for many years for chiral separations, for example Mourier et al. (1985) and West (2014).

In the two next sections, some general information concerning SFC performed with capillary columns (Section 6.2.2) and packed columns (Section 6.2.1) will be provided. The intent is not to provide a comprehensive course on SFC; rather, our aim is to provide some essential information so the reader will be aware of how SFC is very similar to LC and what the main differences are between the two techniques. As a rule of thumb, an NPLC user will be able to use packed column SFC with minimum training; however, conversely, in our view, only very experienced GC users will be able to handle capillary SFC.

6.2.1 Packed Column SFC

6.2.1.1 Implementation

As stated previously, by far, packed column SFC is the main mode of implementation of SFC. Modern analytical scale instrumentation is available from major LC manufacturers, such as Agilent Technologies, Jasco, Shimadzu, and Waters. These systems are very similar to LC systems, with the main difference consisting of the automatic backpressure regulator (BPR), placed after an ultraviolet (UV) detector, when available. The BPR is utilized to keep the mobile phase at the sub or supercritical state in all the compartments of the chromatograph because a dense mobile phase is required to keep the solvent strength of the mobile phase, and thus, ensuring the solubility of the analytes. In the Agilent Technologies system, a conventional LC system can be used as an SFC instrument with only minor modifications, such as some parts of the pump used to deliver the CO_2, the UV cell in order to deal with the CO_2 at high pressure, and the autosampler to accommodate the depressurization of the CO_2 in the injection loop. It should be noted that a module is added to the modified LC system in order to handle pressure control using a BPR and to feed the CO_2 pump of the LC system with pressurized CO_2 via an extra pump ("booster") connected to the CO_2 cylinder. Thus, the task of the CO_2 pump of the modified LC system is only to meter the flow rate of the CO_2, because the CO_2 is fed by the booster pump at a few bars below the column inlet working pressure. With such a system, gaseous CO_2 can be utilized. All the systems from other manufacturers use only one pump for delivering the CO_2; in doing so, the pump must ensure both compression and flow metering of the SF; thus, to ensure efficient pumping, the CO_2 pump must be fed by liquid CO_2 provided by a cylinder equipped with a dip tube.

The majority of SFC applications are mainly related to the pharmaceutical industry and the separation of enantiomers, where the mobile phase contains a polar modifier, or a mixture of modifiers, and additives in order to enable better solvation of the compounds of interest by the mobile phase, and thus, better chromatography. As in LC, there is also an effect of additives and polar modifiers on the stationary phase (SP) in pSFC, as the SP is solvated by the mobile phase, the mobile phase and the additives can deactivate the reactive sites of the SP, such as the silanol groups. A gradient of eluent strength can be applied by increasing the amount of the strongest solvent in the mobile phase, generally the polar modifier. To accomplish this, a second pump is necessary to add modifiers to the CO_2. However, the system must be capable of performing solvent composition gradients, just as in LC.

All commercially available LC columns can be used in pSFC mode, just as long as they are pressure and CO_2 resistant. Similarly to what is performed in LC, UHPLC columns packed with sub-2 μm particles, as well as fused core columns, can also be used in pSFC for increased efficiency and resolution power (Berger 2010; Lesellier 2012; Sarazin et al. 2011). However, dedicated packed SFC columns have been developed quite recently, including ethyl pyridine bonded silica, diethylamine (DEA), and Daicel columns, to name a few, to extend the selectivity offering and to allow better separation of polar/basic compounds for pharmaceutical applications. In order to have an idea on the behavior of SFC columns, the reader is referred to the Lesellier and West (2015) paper on the classification of LC and pSFC columns.

The main detectors utilized in packed column SFC are as follows:

- In dense conditions with a high-pressure cell, such as a UV spectrophotometer where dedicated cells are available, or rarely Fourier transform infra-red spectrometer (FTIR) (Thiébaut et al. 1999), where the low-pressure cell has to be modified.
- Or after decompression of the mobile phase, such as MS, light scattering detector, Corona Cad, chemiluminescence, and GC detectors such as the flame ionization detector (FID), thermionic detector (TID), and atomic emission detector (AED), for example. These cases are only applicable if the mobile phase does not contain modifiers or additives that could interfere with detection.

Detectors requiring decompression of the mobile phase prior to the detection, such as MS, which can be considered one of the most important and very active topics in pSFC (Pilařová et al. 2019), owing to the information that can be obtained on compounds, are generally achieved by diverting a portion of the flow after the column outlet and before the BPR. Using the most recent systems offering a BPR with very low dead volume (Shimadzu and Agilent technologies), the MS can also be placed after the BPR. When the MS is placed by diverting flow prior to the BPR, a special piece of tubing called a "restrictor" must be installed in order to perform both the decompression and the introduction of the mobile phase in the MS source (APCI, ESI, APPI, etc.). This tubing is generally a small inner diameter (ID) tube that creates a pressure drop according to Darcy's law, where at the operating pressure regulated by the BPR, the flow rate in the MS will depend on the ID and length of the tubing, and of the viscosity of the fluid. Compared with LC-MS hyphenation, there are many possibilities to interface pSFC to MS (Ciclet et al. 2018; Losacco et al. 2019); however, the flow rate of the mobile phase entering the MS in pSFC mode is not easily measured in a precise manner (Grand-Guillaume Perrenoud et al. 2014). The effluent flow reaching the MS will vary if a gradient of modifier is applied during the separation because the viscosity of the mobile phase will change. As in LC, the mobile phase can be modified to favor the ionization of compounds through the addition of solvents or additives before entering the source via an extra pump. Using modern mass spectrometers developed for LC, it has been reported that the SFC response is comparable with, and in some cases, better than LC (Losacco et al. 2019; Pilařová et al. 2019). All types of mass spectrometers can be used, such as quadrupole, time of flight, and ion mobility, provided the flow of the mobile phase entering the source is in a similar range to what is utilized in LC.

With GC detectors, such as FID, the restrictor must ensure the decompression of the mobile phase a few millimeters below the detection window using a low volume restrictor, such as an integral restrictor or a frit restrictor (Thiébaut et al. 1999).

6.2.1.2 Applications of Packed Column SFC

As a rule of thumb, the criterion described by Terry Berger can be used: any molecule soluble in methanol could (should!) be analyzed by SFC using CO_2 and polar modifiers. Recent trends, using a high amount of modifiers in the CO_2 in order to add water in the mobile phase, allow the elution of polar compounds having a Log *P* below zero (Losacco et al. 2019).

As in LC, the separation will need to ensure that a sufficient selectivity is obtained using an appropriate combination of a stationary phase and a mobile phase composed of, in most cases, CO_2, polar modifiers, and additives. Separations without polar modifiers in the mobile phase are restricted to low (hydrocarbons) to medium polarity compounds (such as polymer or oil additives), and commonly hyphenated to a detector that is not, or only partly, compatible with polar modifiers and additives, such as FID, FTIR, and AED.

For column selection, a similar approach to what is done for chiral separations can be followed (Maftouh et al. 2005), where a set of 5–10 dedicated columns is generally utilized to quickly screen which stationary phases give some selectivity for further optimization. Using a system offering a sophisticated multi solvent dispenser and column switching, up to 12 columns can be investigated with different mobile phases, such as methanol, ethanol, isopropanol, etc. eventually also containing additives.

For non-chiral separations, a different set of stationary phases is involved. The choice of columns can be dictated by the user's experience, column preferences based on robustness and applications, and the literature, of course. Normal phase columns (diol, aminopropyl) and dedicated SFC columns can be tested (ethyl pyridine, 2PIC, Daicel SFC A and B, etc.). A reversed phase column can also be added, despite the fact that it is not typically used in NPLC. However, the presence of residual silanol groups and alkyl chains will modulate the interactions and, consequently, the selectivity.

The packed column SFC separations of hydrocarbons and oil related compounds will be discussed in dedicated parts of this chapter.

6.2.2 Capillary SFC

To the best of our knowledge, capillary SFC is only used for Simulated Distillation (SIMDIST) applications in the petroleum industry. Thus, the reader will find some details on the system in Section 6.3 and in Figure 6.1. Only some general considerations on capillary SFC operation are mentioned here.

In capillary SFC, GC capillary columns are implemented. They should have the lowest ID possible (e.g. 100 or 50 μm). The stationary phase should be immobilized and offer good stability over time, such as MS dedicated GC columns. Injection volume/quantity should be very low (e.g. below 1 μl) in order to avoid overloading of the column, because the amount of the stationary phase is very low compared with packed columns.

Generally, neat CO_2 is used as the mobile phase. Therefore, the only way to increase the solvent strength during a separation is pressure programming of the mobile phase, such that the density of the mobile phase is increased during the separation (Jentof and Gouw 1970). Pressure programming is available on most commercial systems; the flow rate of the SF in the column is only a few μl/min or less, similar to nano-LC. Owing to this low flow rate, FID or any GC detector

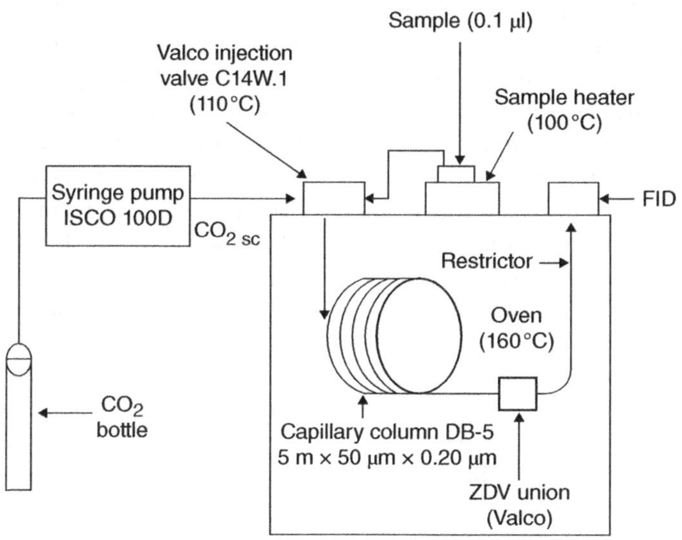

Figure 6.1 Scheme of capillary SFC apparatus used for simulated distillation. Source: Dulaurent et al. (2007). Reproduced with permission of Institut Français du Pétrole.

can be directly coupled to capillary SFC, while UV detection would require a high-pressure nano cell having a very low volume. Last but not least, such low flow rates are not compatible with the dead volumes of the BPR, and overall, the commercial BPRs are inefficient at such low flow rates. This means that a restrictor must be used prior to the FID detector, as indicated in the SIMDIST section to follow.

A cSFC system is available from Selerity Technologies (Salt Lake City, USA).

However, it is possible to implement cSFC using a conventional system by splitting the pump effluent before entering the column. By doing so, the pressure can be set up by the BPR, which is placed in a diverted flow situation where the flow rate in the column will rely on the geometry of a restrictor placed between the column and detector (MS or FID), as described for packed columns (Lavison-Bompard et al. 2012). The use of both a restrictor and flow splitting is the first weak point of capillary SFC, a second one being the fact that capillary SFC is a nano-chromatographic technique, and a third one being the duration of separations compared to packed column SFC, which due to the length of capillary column is much higher than that of packed columns while the linear velocity should be smaller.

6.3 Simulated Distillation (SIMDIST)

SIMDIST is used in the petroleum industry to provide the hydrocarbon distribution of the sample in weight percent versus the boiling range of the fraction (expressed in Atmospheric Equivalent Boiling Point, AEBP) for many types of samples, feeds, and cuts treated in refining and conversion processes. SIMDIST analysis requires a calibration curve between the boiling point of normal paraffins and their elution temperature or retention time. Adequate experimental conditions (column and conditions) enable SIMDIST results to be in agreement with true boiling point (TBP) distillation curves, as described in ASTM D2892 (ASTM-D2892 2018). In SIMDIST, the elution of compounds having the same boiling point should be the same whatever their chemical structure, which is referred to as SIMDIST selectivity, which is described in more detail in the following text. In GC SIMDIST, the final boiling point (FBP) obtained is higher than 538 °C (1000 °F, ASTM D2887 (ASTM-D2887 2019)), compared with 400 °C for ASTM D2892 and 565 °C for ASTM D5236 (ASTM-D5236 2018) involving lower pressures (c. 10 Pa). High-temperature gas chromatography (HTGC) as described in ASTM methods D6352 (ASTM-D6352 2015), D7169 (ASTM-D7169 2018), and D7500 (ASTM-D7500 2015) using oven temperatures up to 450 °C, allows AEBP around 700 °C to be determined in routine analysis using nonpolar stationary phases; however, in HTGC, cracking can occur (Schwartz et al. 1987) and the lifetime of columns is reduced (Kelemidou and Severin 1996). For additional information on GC SIMDIST, the authors refer the reader to Chapter 3.

The solvent strength of carbon dioxide toward hydrocarbons enables cSFC of high molecular weight hydrocarbons (HMHs) at lower temperatures than GC, using both capillary GC columns and packed capillary LC columns. Thus, cSFC is a complementary technique to GC in order to expand the capabilities of SIMDIST toward heavy cuts.

6.3.1 Experimental

SIMDIST in cSFC is a miniaturized separation technique performed under high pressure and quite high-temperature conditions for SFC using a typical capillary SFC system equipped with FID detection (Figure 6.1).

One pump is required to deliver carbon dioxide to the column. The maximum operating pressure of the pump should be as high as possible, higher than 60 MPa, and pressure programming is mandatory. The injector must be heated to avoid the precipitation of the samples, which is mandatory for the injection of polywax used for calibration (Thiébaut 2012). The system requires a restrictor to transfer and decompress the SF from the column outlet to the FID. The flow rate of the mobile phase depends on the column ID; typically, the flow is a few μl/min for the capillary columns generally used for this application and more for micro packed columns. If higher flow rates are required (e.g. regular ID packed columns), a conventional system can be used to provide the CO_2 to the column via proper splitting of the flow rate as indicated above (Lavison-Bompard et al. 2012). More details on the operating conditions and recommendations can be found in Thiébaut (2012) and Dulaurent et al. (2007).

6.3.2 Results

Many published results are available in the literature showing the elution of alkanes over C_{100} using capillary SFC. A summary of these results has been compiled in Table 6.1, including comments on samples, conditions, and features.

When using alkyl bonded phases in packed capillary columns, the longer the alkyl chain, the stronger the retention of hydrocarbons, and thus, the higher the chromatographic selectivity (Huynh 1998; Shariff et al. 1994). High temperature (c. 170 °C) combined with high pressure enabled the elution of hydrocarbons having more than 136 carbon atoms ($>C_{136}$). At this time, the pressure limits of the entire system (e.g. pump, injector, column, etc.) were reached. As retention increases with the length of the alkyl bonded chain, a compromise had to be found to keep the final operating pressure around 50 MPa while the "SIMDIST selectivity" (expressed here as the boiling point difference between two coeluting

Table 6.1 Summary of the major contributions of SFC SIMDIST.

References	Column/oven temperature (°C)	Carbon # Sample analyzed	Comments	Final pressure (MPa)
Ph. Sotty (1994) and Ph. Sotty et al. (1993)	C_{11}-bonded silica packed capillary 180 °C	C_{132} Polywax 1000	Injection T: 130 °C	>50
Shariff et al. (1994)	Various chain length packed capillary 120 °C	C_{130} Polywax	Better RSD on C_6 column	41.5
Huynh (1998)	C_4-bonded silica custom packed capillary 160–170 °C	C_{136} Polywax 1000	Routine between C_{80} and C_{120}, better selectivity with long alkyl bonded chains (>C_4) Retention increases versus alkyl bonded chain length	48
Raynie et al. (1991)	*n*-Octylpolysiloxane capillary 150 °C	C_{100}		32
Thiébaut and Robert (1999)	PDMS capillary 180 °C	C_{108} Polywax 1000		50
Dahan (2005) and Dulaurent et al. (2007)	5% phenyl-methylpoly-siloxane 160 °C	C_{120} Polywax 1000	Routine analysis of vacuum residues using FID or SCD	55

hydrocarbons of different structure, i.e. aromatic structure versus saturate) was less than 10 °C.

It must be highlighted that non-polar alkyl bonded silica LC stationary phases could be used at elevated temperatures (more than 150 °C) without evidence of degradation of their chromatographic performance (Huynh 1998) and suited for routine analysis. Interestingly, using a C_4 bonded silica column packed with 1.9 μm particles, a very fast separation of alkanes from C_{14} to C_{80} was obtained in less than 2.5 minutes, as shown in Figure 6.2. As neither an optimization of injection conditions nor separation at higher temperature were carried out, one could expect the elution of alkanes with more than 80 carbon atoms, similarly to the results shown below with capillary columns (Sarazin et al. 2011).

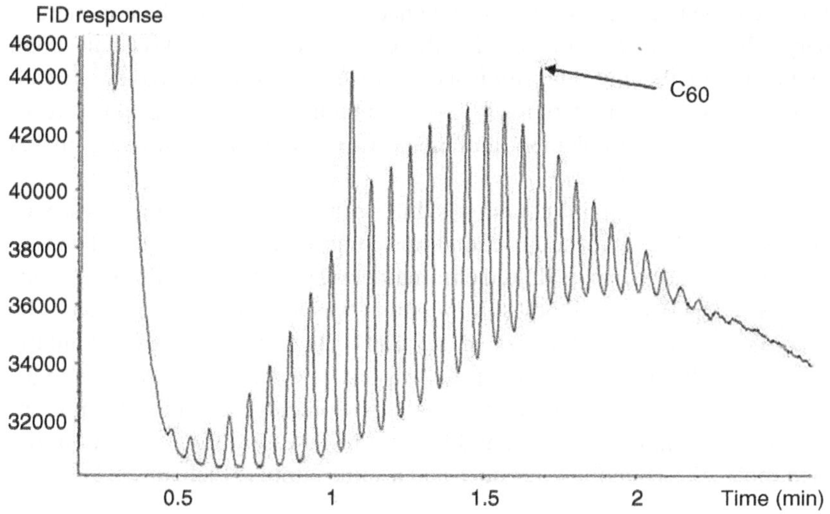

Figure 6.2 UHPSFC chromatogram of polywax 655. Column Sepax C_4 (5 cm × 0.46 cm, 1.9 μm); flow rate 2.5 ml/min, T oven = 100 °C; pressure gradient from 80 to 370 bars (99 bars/min); sample dissolved in *n*-heptane; FID. Source: Sarazin et al. (2011). Reproduced with permission, Copyright (2011), John Wiley & Sons.

In capillary SFC, very high pressure is required for eluting HMHs exceeding C_{100} at constant high temperature (>100 °C). Of course, the density of the mobile phase must be increased during the separation via pressure programming, as indicated previously.

The heaviest reported paraffin determined in cSFC SIMDIST is C_{126} (Dulaurent et al. 2007) on a 5% phenylpolydimethylsiloxane stationary phase. However, when applied to vacuum distillation residue samples, these conditions allowed the elution of compounds estimated to have more than 200 carbon atoms (BP > 900 °C). For such a BP range, the calibration is out of range, because above C_{126}, standards could not be separated and identified. Thus, the calibration curve had to be extrapolated using a logarithmic regression (Dulaurent et al. 2007). Moreover, most of the BPs of alkanes in this range of high molecular weight are not available and must also be estimated (Huynh 1998). However, it was possible to obtain previously inaccessible information on the composition of vacuum residues by comparing the SIMDIST curves of different samples (Dulaurent et al. 2007).

For samples containing a significant amount of low molecular weight compounds eluting in the solvent peak tail, the initial part of the curve (no more than two points) was corrected using GC data. Then, the SFC and GC curves could be superimposed. Beyond C_{120}, no information could be obtained from GC.

It must also be pointed out that no evidence of degradation of the commercial capillary column could be observed after three months of operation (Dahan 2005). This indicates that the immobilization of the film of the stationary phase was excellent, considering implementation with a dense mobile phase, having good solvent properties for non-polar polymers, and including those used as stationary phases in GC (Gouw and Jentoft 1972) .

Using the same setup and conditions, but with a sulfur chemiluminescence detector (SCD), SIMDIST of sulfur-containing compounds was carried out on vacuum residues. The supercritical fluid chromatography-sulfur chemiluminescence detector (SFC-SCD) showed total sulfur content was in good agreement with the information provided with the sample, 2.5–5.6%. There was no need to combine GC information because the solvent peak was not detected by SCD. Interestingly, the regular and sulfur distillation curves almost coincided, because the amount of sulfur in the samples was high, and it could be assessed that nearly all of the detected molecules from these residues had a sulfur atom (Thiébaut 2012).

6.4 Group-Type and Related Separations

Group-type analysis consists of the separation of petroleum samples into families of hydrocarbons, such as saturates, olefins, aromatics, and resins. It is referred to as SARA or SOARA (saturates, olefins, aromatics, resins, and asphaltenes). In some instances, sub-groups can be further separated, for example, the saturates can be further separated into paraffins and cycloalkanes (naphthenes), and unsaturated hydrocarbons can be further separated into mono-, di-, tri-, and polycyclic aromatic hydrocarbons (PAHs). Methods used for this purpose can be found in reference (Thiébaut and Robert 1999). In LC, standards are required in order to use refractometric detection (Thiébaut and Robert 1999); however, when using FID, no or only minor correction is needed in SFC.

SFC can be used for group type separations of middle distillates and heavy fractions. The main ASTM method is devoted to SFC separation of hydrocarbon groups in diesel fuels (ASTM D5186, released in 1991, the last revision being in 2019 (ASTM-D5186 2019)). The initial aim was the determination of total aromatic and non-aromatic content, but the application range was extended to aviation turbine fuels and blendstocks with the aim of determining non-aromatic, mono, and polyaromatic hydrocarbon groups.

The performance of the separation is evaluated by the value of resolution obtained between test compounds selected as markers of the groups to be separated; for example, the resolution between tetraline and naphthalene is used to evaluate the separation between mono aromatics and polyaromatics, which must be greater than two to fulfill the requirements of the ASTM method. This is easily obtained using a bare silica column having a high specific area and carbon dioxide as the mobile phase (M'Hamdi et al. 1998). Attempts to improve the resolution between the hydrocarbon groups involved the use of different mobile phases (Schwartz and Brownlee 1986), carbon dioxide and SF_6 mixtures (M'Hamdi et al. 1998), and of course, columns and detectors such as UV (Thiébaut and Robert 1999).

The separation of alkanes and olefins can involve a silver loaded silica column, such that π–donor interactions are developed between the unsaturated compounds and the stationary phase (Norris and Rawdon 1984). Thus, the separation of unsaturated compounds could be obtained using a multidimensional approach (Andersson et al. 1993). This leads to ASTM method D6550 (ASTM-D6550 2015) for the determination of the olefin content of gasoline by SFC.

Method D7347, revised in 2015 (ASTM-D7347 2015), also deals with the determination of olefins content in denatured ethanol. It requires a third column packed with polyvinyl alcohol to trap the alcohol before the separation of olefins can be performed, as described in ASTM D6550.

6.4.1 Heavy Samples

Multidimensional separation is required for the group type separation of heavy samples. Starting from the conditions used in separating olefins in light samples, the main difference is the replacement of the bare silica column. Instead, an NPLC column, such as cyanopropyl bonded silica, is used in order to retain polar compounds, for example, from crude oil as shown by Skaar et al. (1990) and Dutriez et al. (2012), who used this approach to couple multidimensional SFC to high-temperature GC × GC (Figure 6.3), as previously described by Adam et al. (2010) for SFC-GC × GC of gas oils. Two columns, cyanopropyl-bonded

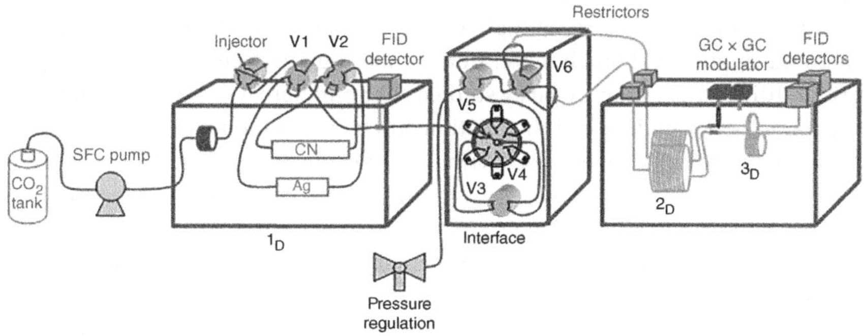

Figure 6.3 Schematic of the SFC- × 2GC × GC-FID instrument considering the set up for the heavy petroleum samples analysis with CN and Ag as SFC columns. Source: Dutriez et al. (2013). Reproduced with permission, Copyright (2013), Elsevier.

and silver-loaded silica, were used. Polar compounds were trapped on the cyano-propyl stationary phase, while hydrocarbons were transferred to the silver loaded silica. The aliphatic fraction eluted first from the silver loaded silica column. Then, the unsaturated compounds adsorbed on the silver column were backflushed before the polar compounds were backflushed from the cyanopropyl column. A baseline separation between aliphatic, unsaturated, and polar group types was reported for vacuum distillates prior to the detailed separation of the two classes of hydrocarbons by high-temperature comprehensive two-dimensional (2D) GC, leading to unprecedented results.

It must be emphasized that coupling SFC to GC or GC × GC is quite straightforward for experienced chromatographers. A small ID tubing is used as the restrictor for SFC. The outlet of the restrictor is placed in the injector of the GC, where a PTV is the best option as described by Potgieter et al. (2013); compounds transferred from the SFC are trapped (and focused) using the injector cooling facilities or in the GC oven (Adam et al. 2010) before an injection. As it could be seen on the GC × GC chromatograms of the SFC fractions, there was no significant band broadening during the transfer from SFC to GC × GC (Adam et al. 2010; Dutriez et al. 2012, 2013; Potgieter et al. 2013). However, it is obvious that SFC-HTGC × GC remains limited by the performances of HTGC × GC for the elution of heavy compounds.

6.4.2 Additives

Detection possibilities of SFC using neat CO_2 as the mobile phase were combined recently to the chromatographic features of SFC for hydrocarbons group type separation to investigate the additives of lubricating oils. Additives are added to base stocks as mixtures of selected compounds in order to improve the properties of the lubricant. Depending on their structure, additives can trap the particles formed in the engine, improve viscosity at high and low temperature, limit oxidation, limit wear of the engine, and much more. Due to their high molecular weight, the polymeric additives were not previously studied. The aim here was to understand their behavior, to better understand the relationship between their structure and activity, and to monitor the changes occurring during aging.

The separation of base stocks using the ASTM method D5186 described earlier was inadequate for efficient group type separation of mineral base stock. Therefore, a 2D column system was implemented to (i) trap polar additives on a first polar stationary phase (bare silica) while the base stock, apolar, was partially eluted, followed by (ii) backflushing the additives from the bare silica column for their separation on an apolar stationary phase (such as octadecyl bonded silica) (Lavison et al. 2007). However, coelution between less polar additives and base stock hydrocarbons still persisted, and specific detection was implemented for selective determination of most of the additives owing to the presence of heteroatoms or functions.

FTIR, UV, AED, and MS (APCI) were hyphenated to SFC to build a hyphenated SFC system (Lavison-Bompard et al. 2012). The esters added to base stocks could be monitored easily using FTIR at 1750 cm^{-1}, although CO_2 absorbs in some IR regions. As their concentration is high, the FTIR response was high enough to provide a selective signal and easy identification. For the AED, selective detection of Zn, S, P, and CN allowed the determination of the elution zones of interest prior to further investigations using FTIR Chemigrams and/or MS data. This was successfully applied to the identification of Zn-Dithiophosphate homologs, a major antioxidant family, in the additive packages added to base stocks. Unfortunately, the sensitivity of the AED when coupled to SFC was too low to provide useful information on formulated lubricants. Mass spectrometry, of course, was needed because it could provide precise structural information on the additives, as illustrated in reference (Lavison-Bompard et al. 2012).

6.5 Detailed Separations

6.5.1 Surfactant and Alkoxylate Polymer Analysis by SFC

6.5.1.1 Open Tubular Columns

Since the early days of technology development, non-ionic surfactants have been successfully characterized by both packed and capillary column SFC (Lee and Markides 1990). Pressurized carbon dioxide as a mobile phase was found to be suitable for effective elution of ethoxylated polymers to allow elution of oligomers beyond the range that could be eluted by GC. Using open tubular columns coated with a methyl silicone stationary phase and using neat CO_2 as the mobile phase at pressures up to 40 MPa, it was found that approximately double the molecular weight of ethoxylates could be eluted by SFC as compared with GC. Increasing the column temperature to 160 °C was found to improve the separation of ethoxylates, which is believed to be due to higher diffusion rates at higher temperatures. The lower densities obtained at the higher temperatures did not significantly impact the molecular weight upper limit that could be eluted. Polyethylene glycol with an average molecular weight of 400 Da (PEG 400) was evaluated by cSFC-MS using a 50 μm ID × 10 m capillary column cross-linked with 5% phenyl polymethylphenylsiloxane (SE-54 type); the mobile phase composition was carbon dioxide, isopropanol (95 : 5), and pressure programming was also applied (Kaloinoski et al. 1987); refer to Figure 6.4.

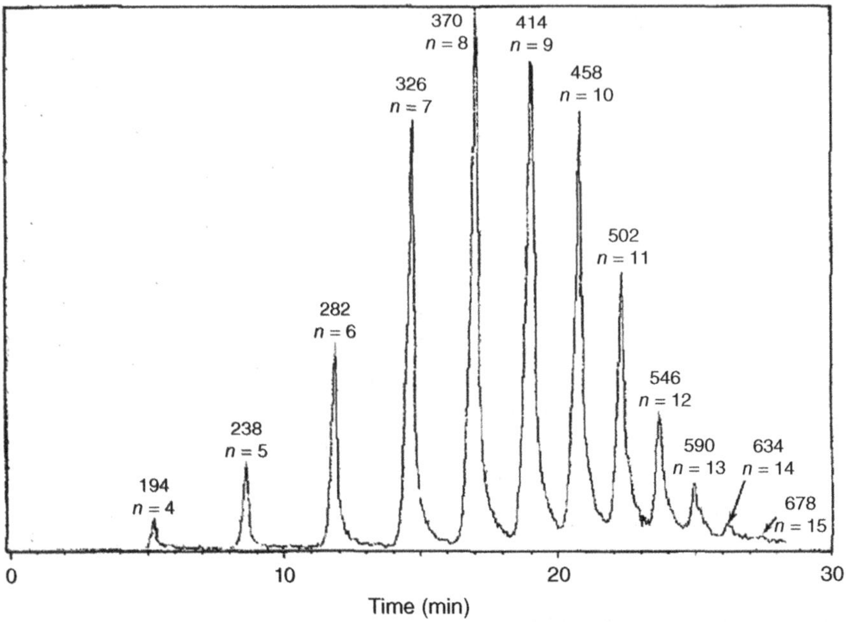

Figure 6.4 Total ion chromatogram for the cSFC-MS separation of PEG 400 using a 50 μm ID × 10 m capillary column cross-linked with 5% phenyl polymethylphenylsiloxane (SE-54 type). Ammonia was used as the CI reagent. The mobile phase was carbon dioxide-isopropanol (95 : 5), and the temperature was 125 °C. Source: Kaloinoski et al. (1987). Reproduced with permission, Copyright (1987), Elsevier.

6.5.1.2 Packed Capillary Column SFC of Surfactants

Packed capillary column technology has also been used to characterize ethoxylated surfactant materials with SFC. Figure 6.5 shows the separation of an alkylene glycol polymer, UCON™ LB-135,[1] using a packed capillary column of 250 μm diameter and FID (Anton and Berger 1998). Good peak shape was obtained for polyglycols, even with a neat CO_2 mobile phase and the bonded diol silica column with hexamethyldisilazane (HMDS) end-capping. Elevated temperatures or polar modifiers were not required for enhancing the peak shape for non-ionic surfactants with this column system.

Polar modifiers have also been used with packed capillary columns for the analysis of surfactants. Figure 6.6 shows an example where the acetonitrile composition was applied in a linear gradient to effect the separation of the oligomers of the ethoxylated C_{18} polyglycol. An excellent separation was obtained in a short analysis time. The rapid analyses obtained with SFC with packed columns were an attractive and beneficial feature for many chromatographers.

1 ™ Trademark of The Dow Chemical Company ("Dow"), or an affiliated company of Dow.

Figure 6.5 pSFC chromatogram of UCON™ LB-135 lubricant. Conditions: Lee Scientific Model 600 SFC system, 0.2 μl injection volume, 40 cm × 250 μm ID capillary column packed with diol bonded and HMDS end-capped silica particles, 10 μm, CO_2 mobile phase, programmed from 100 to 400 atm at 4 atm/min and held at 400 atm for 30 minutes; 60 °C; FID. Source: Anton and Berger (1998). Reproduced with permission, Copyright (1998), Marcel Dekker, Inc.

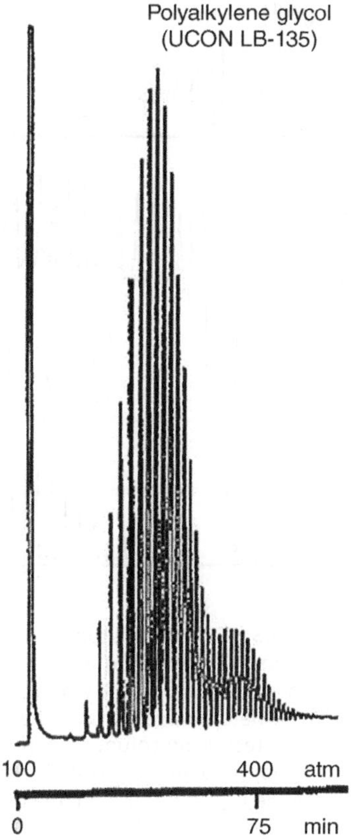

6.5.2 Packed Column SFC of Surfactants

Packed column SFC has been used extensively for the analysis of surfactants. Detection was primarily performed by evaporative light scattering (ELSD) (Anton and Berger 1998), but UV and MS detection have also been commonly used (Hoffman et al. 2004a,b,c). Typically, a silica column with bonded polar functionality, such as cyanopropyl or diol was used with either a small amount of a polar modifier, such as methanol, and either pressure programming or methanol modifier compositional programming was employed to effect the separation of the alkoxylated oligomers as shown in Figure 6.7. Rapid separation of the oligomers was obtained with good resolution. Using a pressure ramp and holding the methanol modifier composition constant, the relative response for the oligomers was very uniform. By avoiding composition gradients, which can have a large effect on detector response, the relative response across the oligomer distribution was constant, allowing for the calculation of average molecular weight and

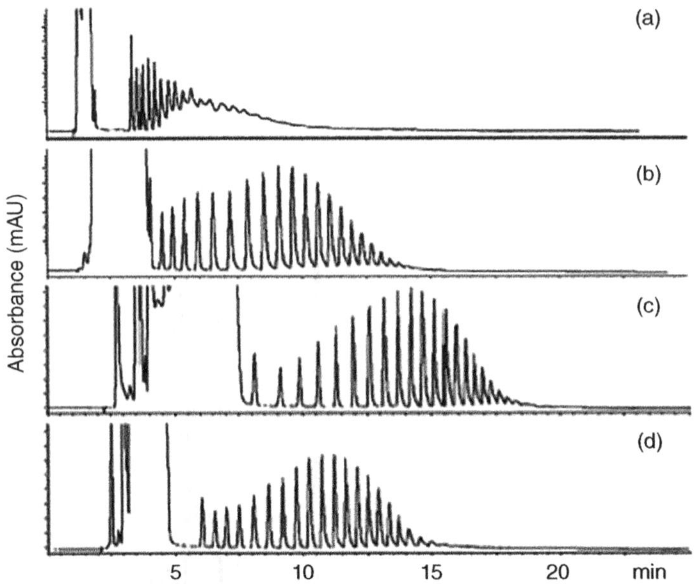

Figure 6.6 SFC chromatogram of a C_{18} polyethylene glycol derivatized with DPTMDS conditions: CO_2 modified with acetonitrile, 40 °C oven temperature, 2.4 ml/min flow rate, 120 bar outlet pressure, UV detection at 215 nm, and linear modifier gradient (1% modifier held for five minutes then increased to 20% at 1%/min, hold at 20% for five minutes) using columns of the same dimension (4.6 × 250 mm, 5 μm) but stationary phases (a) Discovery C_{18} column, (b) Discovery RP-Amide C_{16} column, (c) 2× Discovery RP-Amide C_{16} columns, and (d) Discover C_{18} column + Discovery RP-Amide C_{16} column. Source: Hoffman et al. (2004b). Reproduced with permission, Copyright (2004), Elsevier.

oligomer dispersion. A low dead-volume variable back pressure regulator was used in this work, which allowed an excellent separation with pressure gradients.

6.5.2.1 Surfactants by Sub-2 μm Particle Packed Column SFC

While surfactants have been characterized by SFC using columns with conventional particle sizes, when sub-2 μm particle columns became available, they were used successfully for the SFC analysis of surfactant materials (Campbell et al. 2013). While some early SFC practitioners predicted large resolution losses due to selectivity losses across a column with large pressure drops (Peaden and Lee 1983), more recent data suggest that these losses are only realized if the retention of the components of interest is highly dependent on the mobile phase density. For components whose retention is determined largely by the mobile phase modifier composition, the large pressure drops do not result in large selectivity losses. This

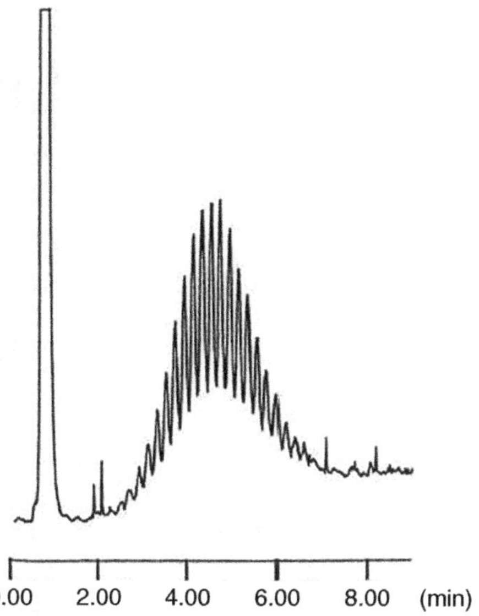

Figure 6.7 Packed column SFC Chromatogram of PEG 1500. Conditions: column: Permacoat Diol, 3 μm, 200 mm × 2 mm ID, carbon dioxide flow: 2.0 ml/min, methanol flow: 0.3 ml/min, Pressure: 250–400 bar in eight minutes with a one minute initial hold and a two-minutes final hold; detection: ELSD. Source: Anton and Berger (1998). Reproduced with permission, Copyright (1998), Marcel Dekker, Inc.

is shown in Figure 6.8, where the top chromatogram shows the SFC analysis of Triton™ X-165 polyglycol[2] using a long, 300 mm, sub-2 μm particle size column. An excellent separation was obtained despite the large pressure drop across the column. This result is compared with the chromatogram using a conventional column shown in the bottom of the figure. Both narrow peaks and low analysis times were observed with the 1.7 μm column. Approximately 100,000 theoretical plates were obtained with this 300 mm × 3.0 mm, 1.7 μm column system, whereas only 25 000 plates are typically obtained with 5 μm columns.

Pressurized gases such as carbon dioxide, while exhibiting relatively high densities, are regarded by many as anti-solvents for polymers. Thus, when combining pressurized CO_2 with a strong liquid solvent and performing a composition gradient, a size or molecular weight selectivity can be obtained. This is shown in Figure 6.9, where despite the high pressure drops across the column, the sub-2 μm particle columns exhibited size selectivity for polyethylene glycol polymers (Campbell et al. 2013). Low average molecular weight polymers eluted first, while high molecular weight materials eluted later in the carbon dioxide/methanol gradient on a cyanopropyl column. As can be seen, good separations based on molecular size were obtained with columns packed with sub-2 μm particles.

2 ™ Trademark of The Dow Chemical Company ("Dow"), or an affiliated company of Dow.

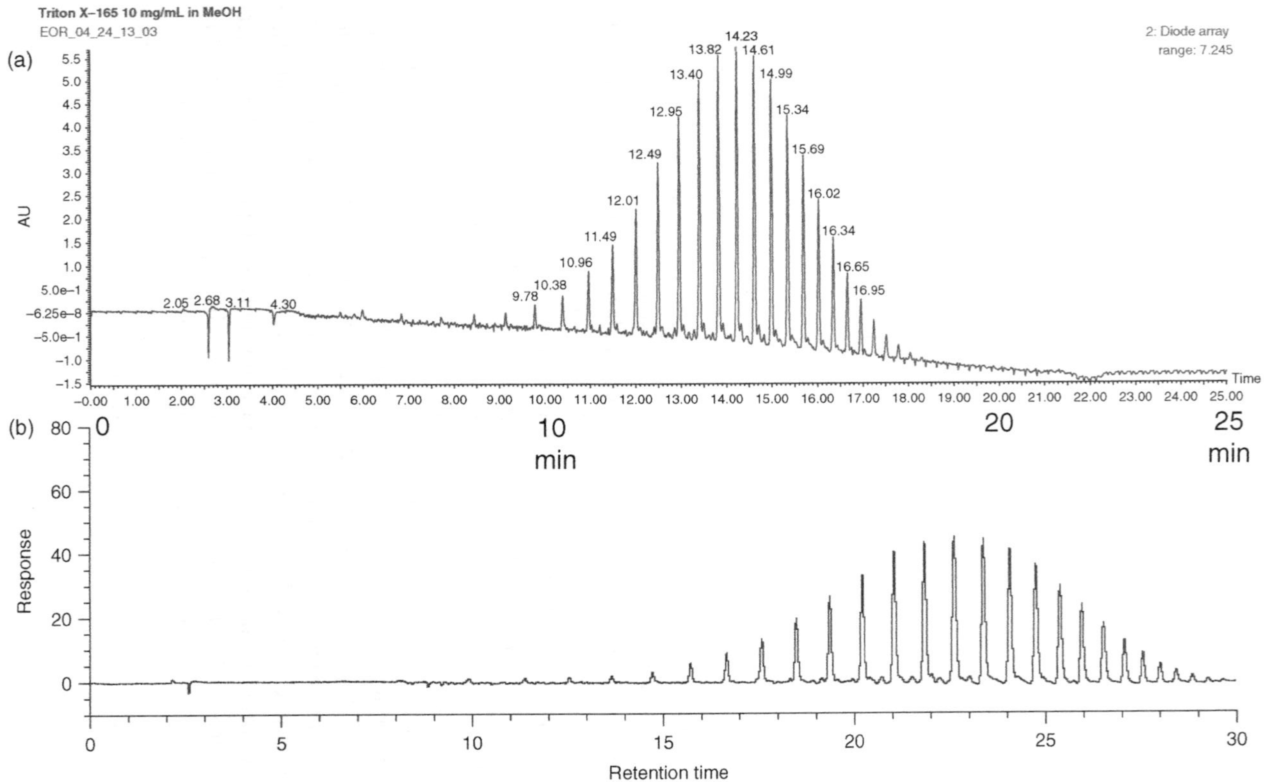

Figure 6.8 SFC chromatograms of Triton™ X-165 alkyl phenol ethoxylate. Conditions:(a) column: BEH hybrid silica, 300 × 3.0 mm, 1.7 µm, 0.8 ml/min, 2–35% methanol in CO_2 in 20 minutes, UV at 230 nm, 60 °C; (b) column: Rx-Sil (silica), 25 cm × 4.6 mm, 5 µm, 1.7 ml/min, 2–35% methanol in CO_2 in 25 minutes, 60 °C, 20 MPa backpressure, UV at 230 nm. Source: Campbell et al. (2013). Reproduced with permission of The Dow Chemical Company.

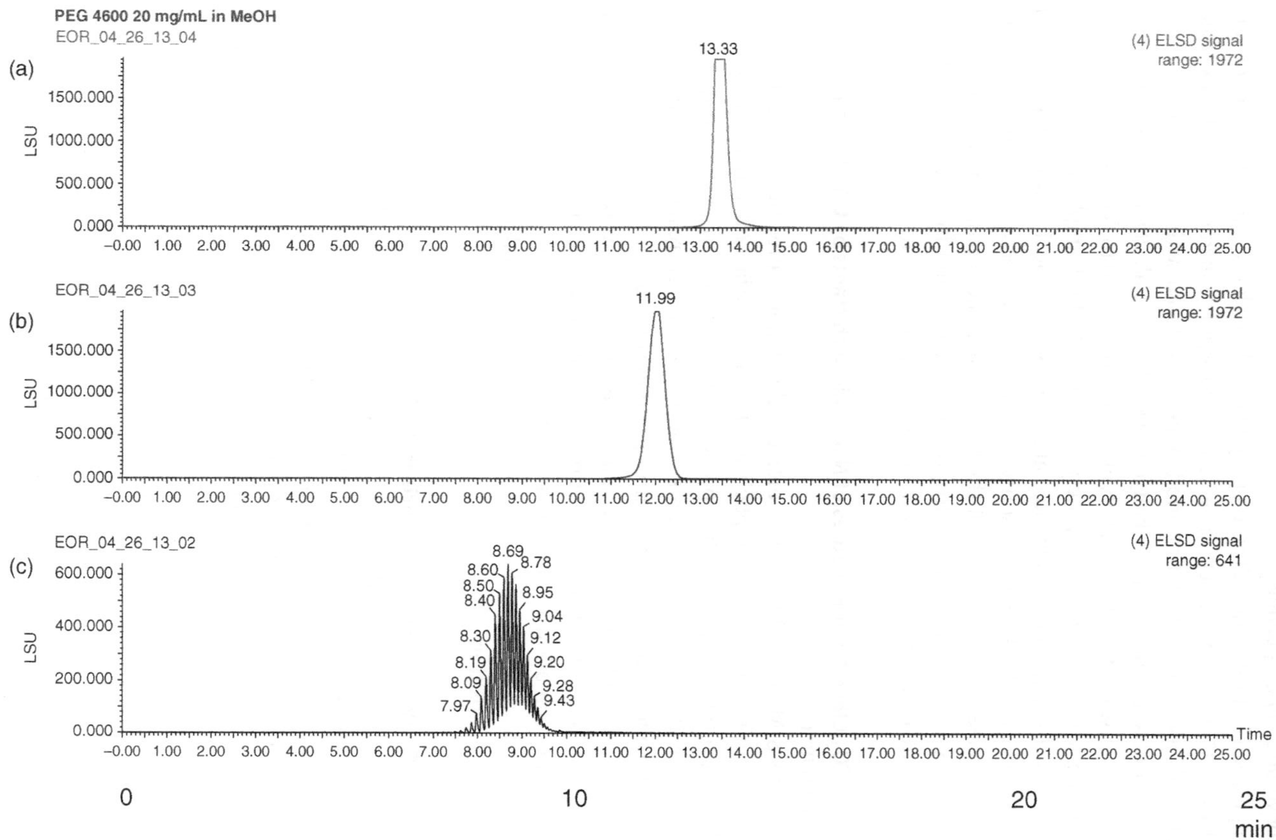

Figure 6.9 Packed column SFC of ethoxylates by size. (a) PEG 8000 MW; (b) PEG 4600 MW; (c) PEG 1450 MW. Conditions: column: Fortis CN, 150 × 3.0 mm, 1.7 μm; Mobile phase: 2–60% methanol in CO_2; 60 °C; ELSD. Source: Campbell et al. (2013). Reproduced with permission of The Dow Chemical Company.

6.5.2.2 Surfactant Characterization by SFC/MS: Software-Assisted Deconvolution of Co-polymers

Complex surfactants, including co-polymeric alkoxylates, have been characterized by SFC with MS detection, where electrospray ionization (ESI) was used extensively as the MS ionization mode when interfaced with SFC for the characterization of surfactants and polyglycols (Hoffman et al. 2004a). More recently, computer software, for example, Polymerix, has been used to deconvolute complex co-polymeric surfactants in conjunction with SFC/MS (Campbell et al. 2015). In this case, a co-polymer of ethoxylated and propoxylated alcohol was characterized by SFC/MS and the Polymerix software was used to deconvolute the multiply charged species to build a map of the EO/PO oligomer composition. This study was done with conventional quadrupole ion analysis, but high resolution MS is recommended for the best results with the Polymerix deconvolution software.

6.5.2.3 CO_2 Cloud Point Pressures of Non-ionic Surfactants by Capillary and Packed Column SFC

Similar to SIMDIST, the carbon dioxide cloud point pressures of surfactants were determined by SFC (Campbell et al. 2015). Surfactant retention time on both open tubular columns with pressure gradients and packed columns with carbon dioxide/methanol gradients was shown to correlate with measured carbon dioxide cloud point pressure at isothermal conditions. The surfactant response was divided into quartiles, and the last quartile was found to give the best correlation to cloud point pressure. Based on the collected data, a model was developed which allowed prediction of the surfactant CO_2 cloud point pressure with input descriptors, including the surfactant structural features and known oligomer composition.

6.5.2.4 CO_2/Water Partition Coefficients by SFC

More recently, SFC has been used to determine the physical properties of surfactants, such as their CO_2-water partition coefficients (Campbell et al. 2017). In this study, the retention time of non-ionic surfactants in gradient packed column SFC was found to correlate with directly measured CO_2-water partition coefficients when plotted on a log-log scale. Furthermore, a five-factor model was developed where both the retention time and partition coefficient could be predicted based on surfactant descriptors derived from the structure and oligomer composition, as shown in Figure 6.10. A silica-based column with diol functionality was found to give the best correlation, but many columns also showed a good correlation of retention time with the partition coefficient. Detection was by ELSD, and methanol was used as the polar modifier in the carbon dioxide gradient mobile phase system. The eluted oligomer distribution of each surfactant was divided into

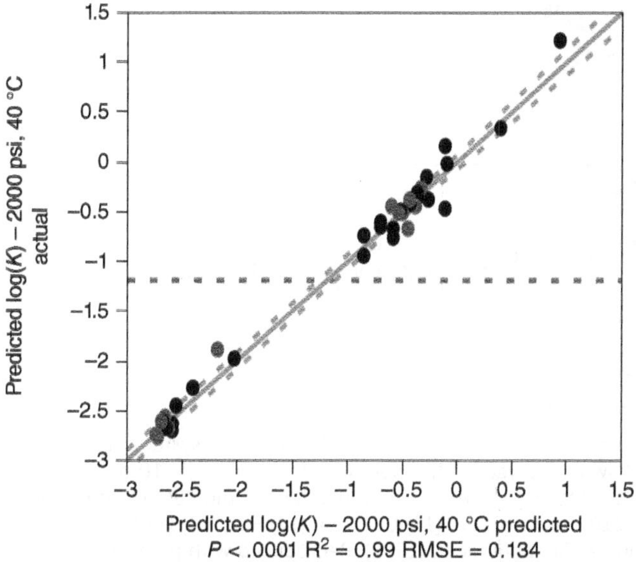

Figure 6.10 Plot of correlation of actual log *K* with predicted log *K* using the developed model (2000 psi, 40 °C). Source: Campbell et al. (2017). Reproduced with permission of The Dow Chemical Company.

quartiles by area response, whereupon the first quartile was found to give the best correlation of retention time and partition coefficient.

6.5.2.5 SFC of Ionic Surfactants

In addition to the analysis of non-ionic surfactants, SFC has also been successfully applied to the characterization and determination of ionic surfactant species (Hoffman et al. 2003; Pinkston et al. 2004; Zheng et al. 2005a,b,c,d, 2006). In this work, the SF chromatographic conditions are similar to conventional SFC, but up to 15% water can be added to the polar modifier mobile phase component such as methanol to facilitate elution. Mobile phase additives, such as ammonium hydroxide and ammonium acetate, were used to facilitate elution of any ionic species, which may exhibit ion exchange retention on the stationary phase.

6.5.3 Capillary SFC of Surfactants

6.5.3.1 Large Volume Injection in Capillary SFC

With conventional split injection, capillary SFC is limited to the analysis of only fairly concentrated solutions. For environmental applications, it is frequently preferred and sometimes required to perform analyses of very dilute solutions.

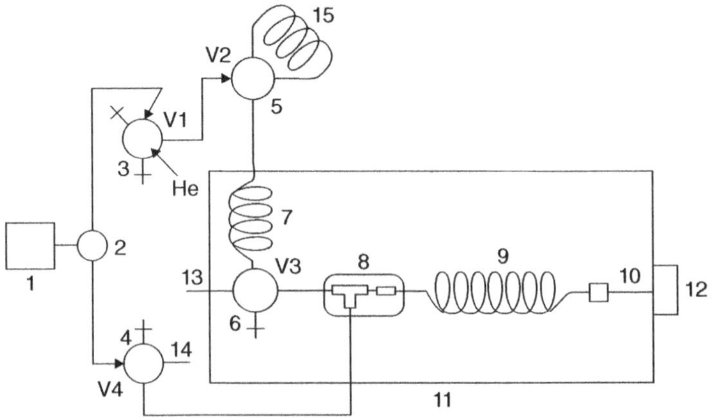

Figure 6.11 Schematic diagram of the large-volume injection cSFC system: (1) SFC pump; (2) tee; (3) 10-port switching valve; (4) 4-port switching valve; (5) 10-port switching valve; (6) 8-port switching valve; (7) capillary inlet column; (8) SFC Interface; (9) Capillary SFC column; (10) frit restrictor; (11) GC oven; (12) FID; (13) vent; (14) vent; and (15) sample loop. Source: Campbell et al. (1992). Reproduced with permission of The Dow Chemical Company.

To allow the determination of analytes at low concentration levels, a large volume injection system was developed for capillary SFC (Campbell et al. 1992; Cortes et al. 1992). This injection system provided for a pre-column where the liquid sample was deposited, as shown in Figure 6.11. It was then subjected to a flow of inert gas to evaporate and remove the solvent. Then the pre-column was pressurized with the carbon dioxide mobile phase, and the sample was swept into the column for analysis after switching the injection valve. The sample was concentrated at the head of the analytical column and analyzed via pressure programming as in conventional analyses with FID. This system was applied to the analysis of dilute solutions of polyglycol surfactants

6.5.3.2 Splitless Injection in Capillary SFC

Similar in concept to splitless injection in GC, splitless injection technology was also developed for capillary SFC (Campbell 1998). With conventional split injection, only fairly concentrated sample solutions could be analyzed. Using splitless injection, dilute solutions could be injected and analyzed, greatly improving the sensitivity of the technique. In conventional timed split technology, the sample loop was filled, and the rotor was left in the mobile phase flush for only short periods of time, on the order of 100 ms. This effected about a 1–20 split. In splitless mode, the filled loop as a groove in the valve rotor was left in the inject position for a much longer period of 20 s to one or two minutes, allowing for the full loop to be injected. After the one to two-minute injection period, the valve rotor was

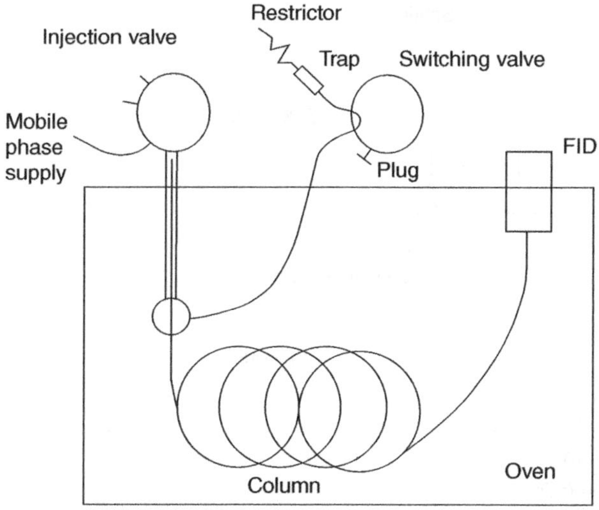

Figure 6.12 Schematic diagram of the SFC splitless injection instrumentation system. Source: Campbell (1998). Reproduced with permission of The Dow Chemical Company.

returned to the load position, see Figure 6.12, and the splitter valve was actuated to turn on the split flow, effectively removing the tail on the large solvent peak. An uncoated inlet affixed to the column was used to prevent distortion and broadening of early eluting peaks due to solvent flooding, similar to large volume injection in capillary GC. An SF chromatogram of several hydrocarbon compounds injected via splitless injection is shown in Figure 6.13. The precision of splitless injection as the raw area relative standard deviation of five measurements was reported at 3% (Felix 2000). Capillary SFC with split and splitless injection is available commercially from Selerity Technologies.

6.5.4 Separations of Polyaromatic Hydrocarbons (PAHs)

As an indication from the previous parts of this chapter, SFC is well suited for the separation of PAHs. PAHs are registered as major pollutants, and their monitoring is required in order to comply with the limits defined by the authorities such as the US Environmental Protection Agency and the European Community. For example, the European Chemical Agency has defined 1 mg/kg (0.0001% by weight) of benzo[a]pyrene in extender oil. In toys, the limit of listed PAHs is 0.5 mg/kg (0.00005% by weight) in any of their rubber or plastic components that come into direct as well as prolonged or short-term repetitive contact with the human skin or the oral cavity, under normal or reasonable conditions. The indicated techniques for their determination is LC and GC/MS (Reach 2010).

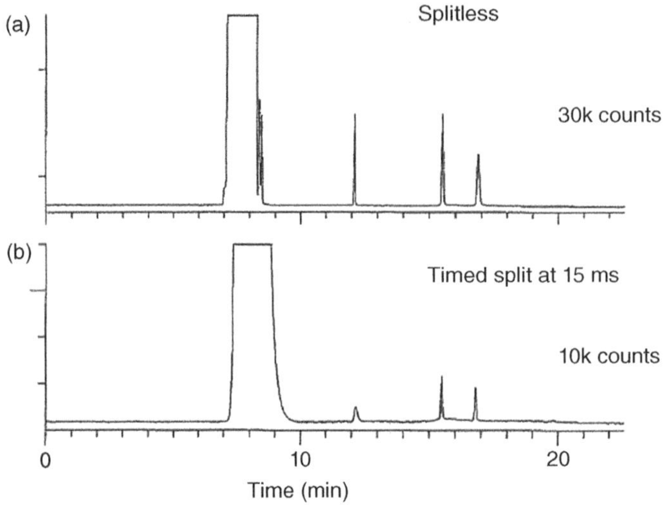

Figure 6.13 SF Chromatogram of C_7, iC_8, C_{10}, C_{20}, and C_{30} hydrocarbons dissolved in hexane at 400 ppm, (a) 23 second splitless injection, (b) timed split injection at 15 ms. Conditions: Column: methyl silicone, 3 m × 50 μm ID with integral uncoated inlet, 3 m (Sensar Larson/Davis), 0.2 μl internal loop injection, CO_2 at 70 atm for 10 minutes then ramped at 15 atm/min to 260 atm, 40 °C, FID. Source: Campbell (1998). Reproduced with permission of The Dow Chemical Company.

However, nice examples of the separation of PAHs by pSFC have been presented for many years (Sandra et al. 1998), or earlier by Berger and Wilson, where the coupling of 11 packed columns enabled 220,000 plates in the column. The efficiency was high enough to provide a GC-like chromatogram of a complex synthetic mixture of PAHs and of a chimney extract (Berger and Wilson 1993). The community is still active in this area, as it can be understood from the recent literature that despite LC and GC methods being the preferred methods because they are well established and the chromatographs are available elsewhere, some recent examples from the literature are commented below.

An SFC/APCI-MS separation of PAHs in coffee and dark beer was published in 2018 using a C_{18} bonded silica column using a gradient of acetonitrile containing 0.5% of formic acid. The authors indicate that APCI provided a better detection limit (5 ng/ml of benzo[a] anthracene) than in APPI (Yoshioka et al. 2018). However, a nicer and faster separation of 16 PAHs in rubber tires was recently reported (Tang et al. 2018). Despite the detection limit being in the range 10–50 ng/g, this is the first report on the determination of 16 PAHs in tire rubber by UHPSFC-APPI-MS/MS. The separation was performed at 14 MPa and 50 °C in less than five minutes using a Torus 2 picolylamine column with a modifier (methanol) gradient. The partial resolution observed on three pairs of compounds could certainly be improved through further investigation on the nature of stationary phases, as only ethyl pyridine and 2 picolylamine phases were investigated.

In a more recent paper (Lübeck et al. 2019), a quite similar approach was taken to the fast separation of 16 PAHs indicated earlier, where the SFC capabilities were applied to the combined analysis of PAHS and oxygenated PAHs (OPAHs) in unconventional oils. SFC was hyphenated to UV detection and quadrupole time-of-flight mass spectrometer (QTOF-MS) with negative electrospray ionization (ESI^-). Methanol, ethanol, and isopropanol were investigated as modifiers with 2 picolylamine (2-PIC), diol, and ethylene bridged bare silica (BEH) columns. Due to the polarity of the columns, the elution order was correlated with the polarity of solutes for OPAHs. The elution of OPAH families depended on the nature of the stationary phase; for example, the authors report the retention of phenolic acids gave rather broad peaks on the 2-PIC column, while the peak shape was much better on the diol column. The best separation between PAHs and OPAHs was obtained using the BEH column. The MS detection was strongly affected by the composition of the modifier, including the amount of water.

With the recent release by Shimadzu of an analytical SFE-SFC system, some expansion of on-line SFE-SFC applications at the analytical scale is expected. As an example, the application of SFE-SFC-APCI-MS allowed the determination of PAHs in soil. A quite long (25 cm) and uncommon column (Cholester) at 50 °C was implemented using a CO_2-acetonitrile gradient. The separation duration was 30 minutes, longer than previous examples, but near baseline resolution was reported. Moreover, the entire method could be validated for concentrations of 10–1500 ng of PAHs per gram of soil in Certified Reference Material sediment, clay, and sand, while the limits of detection were in the range 0.001–5 ng/g using the Multiple Reaction Monitoring mode provided by MS/MS (Wicker et al. 2018). It must be added that the use of smaller particles with optimized column lengths should allow for a reduction of the analysis time to less than five minutes.

6.5.5 SFC in Multidimensional Separations

6.5.5.1 LC × SFC

As it has been shown in Section 6.4.1, for SFC-GC × GC, SFC has an important role to play in hyphenated techniques for the detailed separation of petroleum fractions. Packed column SFC and its hyphenation with LC is also being investigated for the separation of bio-oils.

UHPSFC was investigated recently for the separation of pyrolysis oils. For this application, pSFC was hyphenated to UV (Crepier et al. 2017) and to HRMS (APCI) (Crepier et al. 2018). Using an ethyl pyridine bonded silica and an mobile phase containing acetonitrile and water as modifiers, 120 compounds could be detected.

The hyphenation of pSFC as a second dimension separation with reversed phase LC used in a first dimension was demonstrated for the analysis of an aqueous extract of bio-oil (Sarrut et al. 2015a).

LC × SFC has also been reported to allow the analysis of phenolic compounds obtained from depolymerized lignin samples (Sun et al. 2018), which was based on a previous demonstration of the separation power of SFC for these compounds; 11 major derivatives were separated by pSFC in less than six minutes without derivation (Sun et al. 2016), and in a more convincing work using UHPSFC coupled to QTOF MS, where the resolution of 32 of 40 compounds was reported (Prothmann et al. 2017). In their interesting work in LC × SFC, the authors compared the use of trapping columns as collection loops between the two dimensions; three types of trapping columns were investigated in association with a reversed phase column (C_{18} bonded silica) in the first dimension while four different columns were tested for the pSFC second dimension. Whatever the second pSFC column, the 2D separation system was found to offer a high degree of orthogonality for the standard mixture. Using a diol column in the SFC dimension, high orthogonality was also obtained on a real sample as it can be seen in Figure 6.14.

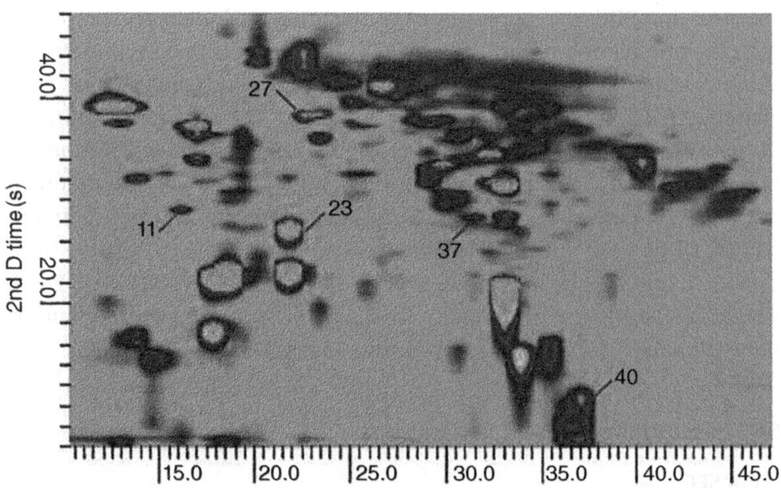

Figure 6.14 RPLC × SFC separation of a lignin depolymerized sample using two trapping columns in the interface. First dimension column: Agilent Zorbax Eclipse Plus C_{18} column (2.1 × 100 mm, 1.8 μm); $T = 50\,°C$; gradient starting from 10 to 90 vol% of acetonitrile in water in 60 minutes: flow rate 0.05 ml/min; injection volume was 5 μl. Second dimension column: Waters UPC^{2}Torus Diol (High density diol, 3.0 × 50 mm, 1.7 μm); $T = 55\,°C$; gradient starting from 7 to 32 vol% of acetonitrile-water (75/25 v/v) in 0.49 minutes: Flow rate 0.05 ml/min; 2× Agilent Eclipse Plus phenyl-hexyl, 2.1 × 5 mm, 1.8 μm trapping columns were used in the interface; the modulation time was 0.8 minutes. Source: Sun et al. (2018). Reproduced with permission, Copyright (2018), Elsevier.

6.5.5.2 Feasibility of SFC × SFC

The feasibility of comprehensive 2D SFC (SFC × SFC) was demonstrated in 2003 by Hirata et al. (2003). In this work, the transfer between the two dimensions was obtained by stop flow operation in the first dimension in order to allow the decompression of the mobile phase of the first dimension prior to its transfer to the second column via a sampling loop.

In order to avoid total decompression of the mobile phase and the risk of compound loss during transfer between the two columns (precipitation and lack of solubilization of compounds), an instrument was constructed on the basis of on-line comprehensive 2D LC. An LC × LC-like interface was placed between the two SFC dimensions (Guibal et al. 2012); it consisted of a two-loop switching valve that allowed for the collection of the first dimension column effluent and the subsequent transfer to the second dimension; during the time a loop is being filled by the mobile phase coming from the first dimension, the analysis of the content of the second loop must be performed in the second dimension.

As a column having a conventional diameter was used in the first dimension, the first-dimension column flow rate was split prior to modulation. By doing so, a UV detector could be used after both dimensions (on the split flow in the first dimension). After the second-dimension column, the flow was diverted to implement an FID for easy detection of saturated hydrocarbons. Later, the interface developed by Guibal and coworkers was improved to perform solute focusing between the two dimensions, referred to here as active modulation (Petkovic et al. 2018), in order to mimic the behavior of a GC × GC cryogenic modulator. It involved in the transfer toward the second column a lower density than the first separation, so that the compounds were focused at the head of the second-dimension column. Next, the second-dimension separation was performed at high pressure so that a pressure gradient was generated in the second dimension column during the 18-second separation. The pressure changes were obtained using three pumps: one for the first dimension, one for the second dimension, and one for the transfer between the two dimensions. The pressures applied were controlled by three independent BPRs associated to each of the respective pumps. A second switching valve was added to the initial set up so different pressures could be applied to the second-dimension column during the transfer and the separation steps. Figure 6.15 shows the improvement of the chromatogram obtained using active modulation. It is obvious from this figure that the chromatogram obtained using active modulation resembled a GC × GC chromatogram, despite the quite large modulation period, 30 seconds (18 seconds for the separation in the second-dimension and 12 seconds for equilibration of the low pressure in the second column followed by the transfer at low pressure). Indeed a very efficient focusing was demonstrated for heavy compounds (Petkovic et al. 2018), which explained why the peak width was reduced in the second dimension. Compared with 1D SFC or GC, these SFC × SFC results were very promising for gaining more

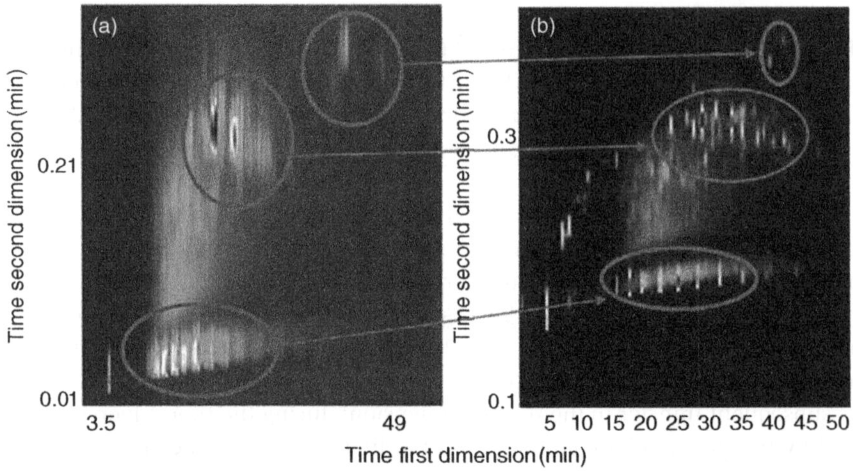

Figure 6.15 Comprehensive 2D SFC chromatogram of a coal tar vacuum distillate. (a) Without active modulation; (b) with active modulation. Mobile phase: carbon dioxide with pressure program, 60 °C; first dimension, three Ascentis Express fused core columns (each 15 cm × 4.6 mm, 2.7 μm); second-dimension column Syncronis Silica, 5 cm × 3/2.1 mm, 1.8 μm; CO_2 flow rate: 1.6 ml/min. Modulation period: 30 seconds. Source: Petkovic et al. (2018). Reproduced with permission, Copyright (2018), Elsevier.

detailed information on heavy samples through the improvement of peak capacity and selectivity offered by the combination of two different stationary phases. For hydrocarbon and low polarity compounds, SFC × SFC can mimic the performances of GC × GC, owing to the focusing effect permitted by the changes of the CO_2 density. It must be underlined that this focusing effect obtained via density changes would be reduced in the presence of polar modifiers in the mobile phase, where efficient focusing should be performed in a similar manner to LC × LC by varying the composition of the mobile phases (Sarrut et al. 2015b).

The complex system described earlier could be simplified by using new switching valves designed for LC × LC and pressure gradient capabilities of the very recent SFC systems, so very fast pressure gradients could be performed on the second column using only one pump.

References

Adam, F., Thiébaut, D., Bertoncini, F. et al. (2010). Supercritical fluid chromatography hyphenated with twin comprehensive two-dimensional gas chromatography for ultimate analysis of middle distillates. *Journal of Chromatography A* 1217 (8): 1386–1394. https://doi.org/10.1016/j.chroma.2009.11.092.

Andersson, P.E., Demirbüker, M., and Blomberg, L.G. (1993). Characterization of fuels by multi-dimensional supercritical fluid chromatography and supercritical fluid chromatography – mass spectrometry. *Journal of Chromatography A* 641 (2): 347–355. https://doi.org/10.1016/0021-9673(93)80151-W.

Anton, K. and Berger, C. (1998). *Supercritical Fluid Chromatography with Packed Columns* (eds. K. Anton and C. Berger). New York, NY: Marcel Dekker, Inc.

ASTM-D2887 (2019). *Standard Test Method for Boiling Range Distribution of Petroleum Fractions by Gas Chromatography*. West Conshohocken, PA: ASTM International.

ASTM-D2892 (2018). *Standard Test Method for Distillation of Crude Petroleum (15-Theoretical Plate Column)*. West Conshohocken, PA: ASTM International.

ASTM-D5186 (2019). *Standard Test Method for Determination of the Aromatic Content and Polynuclear Aromatic Content of Diesel Fuels and Aviation Turbine Fuels by Supercritical Fluid Chromatography*. West Conshohocken, PA: ASTM International.

ASTM-D5236 (2018). *Standard Test Method for Distillation of Heavy Hydrocarbon Mixtures*. West Conshohocken, PA: ASTM International.

ASTM-D6352 (2015). *Standard Test Method for Boiling Range Distribution of Petroleum Distillates in Boiling Range from 174 to 700 °C by Gas Chromatography*. West Conshohocken, PA: ASTM International.

ASTM-D6550 (2015). *Standard Test Method for Determination of Olefin Content of Gasolines by Supercritical-Fluid Chromatography*. West Conshohocken, PA: ASTM International.

ASTM-D7169 (2018). *Standard Test Method for Boiling Point Distribution of Samples with Residues Such as Crude Oils and Atmospheric and Vacuum Residues by High Temperature Gas Chromatography*. West Conshohocken, PA: ASTM International.

ASTM-D7347 (2015). *Standard Test Method for Determination of Olefin Content in Denatured Ethanol by Supercritical Fluid Chromatography*. West Conshohocken, PA: ASTM International.

ASTM-D7500 (2015). *Standard Test Method for Determination of Boiling Range Distribution of Distillates and Lubricating Base Oils – in Boiling Range from 100 to 735 °C by Gas Chromatography*. West Conshohocken, PA: ASTM International.

Berger, T.A. (1997). *Chemical basis for the instrumentation used in packed column supercritical fluid chromatography in: Supercritical fluid chromatography*, vol. 75 (eds. K. Anton and C. Berger). Marcel Dekker Inc.

Berger, T.A. (2010). Demonstration of high speeds with low pressure drops using 1.8 μm particles in SFC. *Chromatographia* 72: 597–602. https://doi.org/10.1365/s10337-010-1699-2.

Berger, T.A. and Wilson, W.H. (1993). Packed column supercritical fluid chromatography with 220,000 plates. *Analytical Chemistry* 65 (10): 1451–1455. https://doi.org/10.1021/ac00058a024.

Campbell, R.M. (1998). Design of a practical injector for capillary supercritical fluid chromatography that meets the needs of industry. Paper presented at the 8th International Symposium on Supercritical Fluid Chromatography and Extraction, St. Louis, Missouri, USA (12–16 July 1998).

Campbell, R.M., Cortes, H.J., and Green, L.S. (1992). Large-volume injection in capillary supercritical fluid chromatography. *Analytical Chemistry* 64 (22): 2852–2857. https://doi.org/10.1021/ac00046a033.

Campbell, R.M., Suszek, J., Oravetz, L., et al. (2013). High speed and high efficiency SFC separations of complex mixtures. Paper presented at the 7th International Conference on Packed-Column SFC, Boston, Massachusetts, USA (10–12 July 2013).

Campbell, R.M., Greziak, A., Bellinger, E., and Lehr, S. (2015). Software assisted characterization of complex materials by SFC/MS and determination of physical properties by SFC. Paper presented at the 9th International Conference on Packed-Column SFC, Philadelphia, PA, USA (22–24 July 2015).

Campbell, R.M., Grzesiak, A., Crosthwaite, J., et al. (2017). Physical properties by SFC: determination of CO_2-H_2O partition coefficients of surfactants. Paper presented at the SFC 2017, Rockville, Maryland, USA (15–17 October 2017). https://www.greenchemistrygroup.org/pastconferences/13-scientific-program-sfc-2017

Ciclet, O., Barron, D., Bajic, S. et al. (2018). Natural compounds analysis using liquid and supercritical fluid chromatography hyphenated to mass spectrometry: evaluation of a new design of atmospheric pressure ionization source. *Journal of Chromatography B: Analytical Technologies in the Biomedical & Life Sciences* 1083: 1–11. https://doi.org/10.1016/j.jchromb.2018.02.037.

Cortes, H.J., Campbell, R.M., Himes, R.P., and Pfeiffer, C.D. (1992). Online multidimensional chromatographic system with large-volume injection handling for supercritical-fluid chromatography. WO9205852A1, filed 10 October 1991. Dow Chemical Company, USA (28th July 1993).

Crepier, J., Le Masle, A., Charon, N. et al. (2017). Development of a supercritical fluid chromatography method with ultraviolet and mass spectrometry detection for the characterization of biomass fast pyrolysis bio oils. *Journal of Chromatography A* 1510: 73–81. https://doi.org/10.1016/j.chroma.2017.06.003.

Crepier, J., Le Masle, A., Charon, N. et al. (2018). Ultra-high performance supercritical fluid chromatography hyphenated to atmospheric pressure chemical ionization high resolution mass spectrometry for the characterization of fast pyrolysis bio-oils. *Journal of Chromatography B: Analytical Technologies in the Biomedical & Life Sciences* 1086: 38–46. https://doi.org/10.1016/j.jchromb.2018.04.005.

Cui, Y. and Olesik, S.V. (1991). High-performance liquid chromatography using mobile phases with enhanced fluidity. *Analytical Chemistry* 63 (17): 1812–1819. https://doi.org/10.1021/ac00017a028.

Dahan, L. (2005). *Chromatographie en phase supercritique capillaire: application à la distillation simulée des produits pétroliers lourds.* Université Pierre et Marie Curie, Paris 6,

Dulaurent, A., Dahan, L., Thiébaut, D. et al. (2007). Extended simulated distillation by capillary supercritical fluid chromatography. *Oil & Gas Science and Technology – Rev. IFP* 62 (1): 33–42. https://doi.org/10.2516/ogst:2007003.

Dutriez, T., Thiébaut, D., Courtiade, M. et al. (2012). Supercritical fluid chromatography hyphenated to bidimensional gas chromatography in comprehensive and heart-cutting mode: design of the instrumentation. *Journal of Chromatography A* 1255: 153–162. https://doi.org/10.1016/j.chroma.2012.03.075.

Dutriez, T., Thiébaut, D., Courtiade, M. et al. (2013). Application to SFC-GC × GC to heavy petroleum fractions analysis. *Fuel* 104: 583–592. https://doi.org/10.1016/j.fuel.2012.04.048.

Felix, D. (2000). *Reproducibility of Splitless Injection for SFC.* Selerity Technologies.

Francotte, E.R. (2017). Practical aspects and applications of preparative supercritical fluid chromatography. In: *Handbooks in Separation Science* (ed. C. Poole). Cambridge, MA: Elsevier.

Giddings, J.C., Myers, M.N., McLaren, L., and Keller, R.A. (1968). High pressure gas chromatography of nonvolatile species. *Science* 162 (3849): 67–73. https://doi.org/10.1126/science.162.3849.67.

Gouw, T.H. and Jentoft, R.E. (1972). Supercritical fluid chromatography. *Journal of Chromatography A* 68 (2): 303–323. https://doi.org/10.1016/S0021-9673(00)85722-2.

Grand-Guillaume Perrenoud, A., Veuthey, J.-L., and Guillarme, D. (2014). Coupling state-of-the-art supercritical fluid chromatography and mass spectrometry: From hyphenation interface optimization to high-sensitivity analysis of pharmaceutical compounds. *Journal of Chromatography A* 1339: 174–184. https://doi.org/10.1016/j.chroma.2014.03.006.

Guibal, P., Thiébaut, D., Sassiat, P., and Vial, J. (2012). Feasability of neat carbon dioxide packed column comprehensive two dimensional supercritical fluid chromatography. *Journal of Chromatography A* 1255: 252–258. https://doi.org/10.1016/j.chroma.2012.04.004.

Hirata, Y., Hashiguchi, T., and Kawata, E. (2003). Development of comprehensive two-dimensional packed column supercritical fluid chromatography. *Journal of Separation Science* 26 (6-7): 531–535. https://doi.org/10.1002/jssc.200390072.

Hoffman, B.J., Zheng, J., Taylor, L.T., et al. (2003). Meeting some of the separation and detection challenges in SFC: Alcohol polyethers and ionic compounds. Paper presented at the ACS National Meeting, New York, NY, USA (7–11 September 2003).

Hoffman, B.J., Taylor, L.T., Rumbelow, S. et al. (2004a). Determination of alcohol polyether average molar oligomer value/distribution via supercritical fluid chromatography coupled with UV and MS detection. *Journal of Chromatography A* 1043 (2): 285–290. https://doi.org/10.1016/j.chroma.2004.05.013.

Hoffman, B.J., Taylor, L.T., Rumbelow, S. et al. (2004b). Separation of derivatized alcohol ethoxylates and propoxylates by low temperature packed column supercritical fluid chromatography using ultraviolet absorbance detection. *Journal of Chromatography A* 1034 (1-2): 207–212. https://doi.org/10.1016/j.chroma.2004.02.006.

Hoffman, B.J., Taylor, L.T., Rumbelow, S., and Pinkston, J.D. (2004c). Increasing UV detection sensitivity in the supercritical fluid chromatographic analysis of alcohol polyethers. *Journal of Chromatography A* 1052 (1-2): 161–166. https://doi.org/10.1016/j.chroma.2004.08.106.

Huynh, V.-K. (1998). *Chromatographie en phase supercritique: application à la distillation simulée des produits pétroliers lourds*. Paris, France: Pierre et Marie Curie.

Jentof, R. and Gouw, T. (1970). Pressure programmed supercritical fluid chromatography of wide molecular weight range mixtures. *Journal of Chromatographic Science* 8: 138–142. https://doi.org/10.1093/chromsci/8.3.138.

Kaloinoski, H.T., Udseth, H.R., Wright, B.W., and Smith, R.D. (1987). Analytical applications of capillary supercritical fluid chromatography – mass spectrometry. *Journal of Chromatography* 400: 307–316. https://doi.org/10.1016/S0021-9673(01)81626-5.

Kelemidou, K. and Severin, D. (1996). Simulated distillation by chromatography with supercritical fluid on high boiling petroleum fractions. *Erdoel Erdgas Kohle/EKEP* 112 (1): 25–27.

Klesper, E., Corwin, A., and Turner, D. (1962). High pressure gas chromatography above critical temperatures. *Journal of Organic Chemistry* 27: 700–701. https://doi.org/10.1021/jo01049a069.

Lavison, G., Bertoncini, F., Thiébaut, D. et al. (2007). Supercritical fluid chromatography and two-dimensional supercritical fluid chromatography of polar car lubricant additives with neat CO_2 as mobile phase. *Journal of Chromatography A* 1161 (1): 300–307. https://doi.org/10.1016/j.chroma.2007.05.068.

Lavison-Bompard, G., Bertoncini, F., Thiébaut, D. et al. (2012). Hypernated supercritical fluid chromatography: Potential application for car lubricant analysis. *Journal of Chromatography A* 1270: 318–323. https://doi.org/10.1016/j.chroma.2012.10.065.

Lee, M.L. and Markides, K.E. (1990). *Analytical Supercritical Fluid Chromatography and Extraction* (eds. M.L. Lee and K.E. Markides). Provo, Utah: Chromatography Conferences, Inc.

Lesellier, E. (2012). Efficiency in supercritical fluid chromatography with different superficially porous and fully porous particles ODS bonded phases. *Journal of Chromatography A* 1228: 89–98. https://doi.org/10.1016/j.chroma.2011.11.058.

Lesellier, E. and West, C. (2015). The many faces of packed column supercritical fluid chromatography – A critical review. *Journal of Chromatography A* 1382, 2-2-46. doi:https://doi.org/10.1016/j.chroma.2014.12.083.

Losacco, G.L., Veuthey, J.-L., and Guillarme, D. (2019). Supercritical fluid chromatography – mass spectrometry: Recent evolution and current trends. *TrAC Trends in Analytical Chemistry* 118: 731–738. https://doi.org/10.1016/j.trac.2019.07.005.

Lübeck, J.S., Malmquist, L.M.V., and Christensen, J.H. (2019). Supercritical fluid chromatography for the analysis of oxygenated polycyclic aromatic compounds in unconventional oils. *Journal of Chromatography A* 1589: 162–172. https://doi.org/10.1016/j.chroma.2018.12.056.

Maftouh, M., Granier-Loyaux, C., Chavana, E. et al. (2005). Screening approach for chiral separation of pharmaceuticals: Part III. Supercritical fluid chromatography for analysis and purification in drug discovery. *Journal of Chromatography A* 1088 (1): 67–81. https://doi.org/10.1016/j.chroma.2004.12.038.

McLaren, L., Myers, M., and Giddings, J. (1968). Dense-gas chromatography of nonvolatile substances of high molecular weight. *Science* 159: 197–199. https://doi.org/10.1126/science.159.3811.197.

M'Hamdi, R., Thiébaut, D., Caude, M. et al. (1998). Packed column SFC of gas oils Part II: Experimental design optimization of the group-type analysis using CO_2-SF_6 as the mobile phase. *HRC Journal of High Resolution Chromatography* 21 (2): 94–102. https://doi.org/10.1002/(sici)1521-4168(19980201)21:2<94::aid-jhrc94>3.0.co;2-9.

Mourier, P.A., Eliot, E., Caude, M.H. et al. (1985). Supercritical and subcritical fluid chromatography on a chiral stationary phase for the resolution of phosphine oxide enantiomers. *Analytical Chemistry* 57 (14): 2819–2823. https://doi.org/10.1021/ac00291a017.

Norris, T.A. and Rawdon, M.G. (1984). Determination of hydrocarbon types in petroleum liquids by supercritical fluid chromatography with flame ionization detection. *Analytical Chemistry* 56 (11): 1767–1769. https://doi.org/10.1021/ac00275a003.

Novotny, M., Springston, S.R., Peaden, P.A. et al. (1981). Capillary supercritical fluid chromatography. *Analytical Chemistry* 53: 407A–414A. https://doi.org/10.1021/ac00226a002.

Peaden, P.A. and Lee, M.L. (1983). Theoretical treatment of resolving power in open tubular column supercritical fluid chromatography. *Journal of Chromatography* 259 (1): 1–16. https://doi.org/10.1016/S0021-9673(01)87974-7.

Petkovic, O., Guibal, P., Sassiat, P. et al. (2018). Active modulation in neat carbon dioxide packed column comprehensive two-dimensional supercritical fluid chromatography. *Journal of Chromatography A* 1536: 176–184. https://doi.org/10.1016/j.chroma.2017.08.063.

Pilařová, V., Plachká, K., Khalikova, M.A. et al. (2019). Recent developments in supercritical fluid chromatography – mass spectrometry: Is it a viable option for analysis of complex samples? *TrAC Trends in Analytical Chemistry* 112: 212–225. https://doi.org/10.1016/j.trac.2018.12.023.

Pinkston, J.D., Stanton, D.T., and Wen, D. (2004). Elution and preliminary structure-retention modeling of polar and ionic substances in supercritical fluid chromatography using volatile ammonium salts as mobile phase additives. *Journal of Separation Science* 27 (1-2): 115–123. https://doi.org/10.1002/jssc.200301672.

Poe, D.P. (2017). *Theories of Supercritical Fluid Chromatography in: Supercritical Fluid Chromatography*. Elsevier.

Potgieter, H., van der Westhuizen, R., Rohwer, E., and Malan, D. (2013). Hyphenation of supercritical fluid chromatography and two-dimensional gas chromatography–mass spectrometry for group type separations. *Journal of Chromatography A* 1294: 137–144. https://doi.org/10.1016/j.chroma.2013.04.020.

Prothmann, J., Sun, M., Spégel, P. et al. (2017). Ultra-high-performance supercritical fluid chromatography with quadrupole-time-of-flight mass spectrometry (UHPSFC/QTOF-MS) for analysis of lignin-derived monomeric compounds in processed lignin samples. *Analytical and Bioanalytical Chemistry* 409 (30): 7049–7061. https://doi.org/10.1007/s00216-017-0663-5.

Raynie, D.E., Markides, K.E., and Lee, M.L. (1991). Boiling range distribution of petroleum and coal-derived heavy ends by supercritical fluid chromatography. *Journal of Microcolumn Separations* 3 (5): 423–433. https://doi.org/10.1002/mcs.1220030507.

Reach. (2010). Annex XVII, Conditions of restriction, restrictions on the manufacture, placing on the market and use of certain dangerous substances, mixtures and articles. Retrieved from https://echa.europa.eu/documents/10162/176064a8-0896-4124-87e1-75cdf2008d59 (accessed 21 March 2020).

Sandra, P., Medvedovici, A., Kot, A., and David, F. (1998). Selectivity tuning in packed column SFC & the use of supercritical fluids in environmental analysis. In: *Supercritical Fluid Chromatography with Packed Columns* (eds. K. Anton and C. Berger). Marcel Dekker.

Sarazin, C., Thiébaut, D., Sassiat, P., and Vial, J. (2011). Feasibility of ultra high performance supercritical neat carbon dioxide chromatography at conventional pressures. *Journal of Separation Science* 34 (19): 2773–2778. https://doi.org/10.1002/jssc.201100332.

Sarrut, M., Corgier, A., Crétier, G. et al. (2015a). Potential and limitations of on-line comprehensive reversed phase liquid chromatography × supercritical fluid chromatography for the separation of neutral compounds: An approach to separate an aqueous extract of bio-oil. *Journal of Chromatography A* 1402: 124–133. https://doi.org/10.1016/j.chroma.2015.05.005.

Sarrut, M., D'Attoma, A., and Heinisch, S. (2015b). Optimization of conditions in on-line comprehensive two-dimensional reversed phase liquid chromatography. Experimental comparison with one-dimensional reversed phase liquid chromatography for the separation of peptides. *Journal of Chromatography A* 1421: 48–59. https://doi.org/10.1016/j.chroma.2015.08.052.

Sassiat, P.R., Mourier, P., Caude, M.H., and Rosset, R.H. (1987). Measurement of diffusion coefficients in supercritical carbon dioxide and correlation with the equation of Wilke and Chang. *Analytical Chemistry* 59 (9): 1164–1170. https://doi.org/10.1021/ac00135a020.

Schwartz, H.E. and Brownlee, R.G. (1986). Hydrocarbon group analysis of gasolines with microbore supercritical fluid chromatography and flame ionization detection. *Journal of Chromatography A* 353: 77–93. https://doi.org/10.1016/S0021-9673(01)87078-3.

Schwartz, H.E., Brownlee, R.G., Boduszynski, M.M., and Su, F. (1987). Simulated distillation of high-boiling petroleum fractions by capillary supercritical fluid chromatography and vacuum thermal gravimetric analysis. *Analytical Chemistry* 59 (10): 1393–1401. https://doi.org/10.1021/ac00137a005.

Shariff, S.M., Tong, D., and Bartle, K.D. (1994). Simulated distillation by supercritical fluid chromatography on packed capillary columns. *Journal of Chromatographic Science* 32 (12): 541–546. https://doi.org/10.1093/chromsci/32.12.541.

Sie, S.T., Beersum, W.v., and Rijnders, G.W.A. (1966). High-pressure gas chromatography with supercritical fluids. I. The effect of pressure on partition coefficients in gas-liquid chromatography with carbon dioxide as a carrier gas. *Separation Science* 1: 459–490. https://doi.org/10.1080/01496396608049460.

Skaar, H., Norli, H.R., Lundanes, E., and Greibrokk, T. (1990). Group separation of crude oil by supercritical fluid chromatography using packed narrow bore columns, column switching, and backflushing. *Journal of Microcolumn Separations* 2 (5): 222–228. https://doi.org/10.1002/mcs.1220020504.

Sotty, P. (1994). *Analyse de produits pétroliers lourds: mise en oeuvre de la distillation simulée avec les fluides supercritiques*. Lyon, Lyon France: Université Claude Bernard.

Sotty, P., Rocca, J.L., and Grand, C. (1993). Simulated distillation of petroleum compounds up to a boiling point of 750 °C using supercritical fluid chromatography with packed microcolumns. Paper presented at the 15th International Symposium on Capillary Chromatography, Riva del Garda (24–27 May 1993).

Sun, M., Lidén, G., Sandahl, M., and Turner, C. (2016). Ultra-high performance supercritical fluid chromatography of lignin-derived phenols from alkaline cupric oxide oxidation. *Journal of Separation Science* 39 (16): 3123–3129. https://doi.org/10.1002/jssc.201600169.

Sun, M., Sandahl, M., and Turner, C. (2018). Comprehensive on-line two-dimensional liquid chromatography × supercritical fluid chromatography with trapping column-assisted modulation for depolymerised lignin analysis. *Journal of Chromatography A* 1541: 21–30. https://doi.org/10.1016/j.chroma.2018.02.008.

Tang, Y., Sun, S., Zhang, L. et al. (2018). Determination of 16 polycyclic aromatic hydrocarbons in tire rubber by ultra-high performance supercritical fluid chromatography combined with atmospheric pressure photoionization-tandem mass spectrometry. *Analytical Methods* 10 (40): 4902–4908. https://doi.org/10.1039/c8ay01580e.

Tarafder, A. (2017). *Theories for preparative SFC. In: Supercritical Fluid Chromatography* (ed. C.F. Poole). Elsevier.

Thiébaut, D. (2012). Separations of petroleum products involving supercritical fluid chromatography. *Journal of Chromatography A* 1252: 177–188. https://doi.org/10.1016/j.chroma.2012.06.074.

Thiébaut, D. and Robert, E. (1999). Group-type separation and simulated distillation: A niche for SFC. *Analusis* 27 (8): 681–690. https://doi.org/10.1051/analusis:1999270681.

Thiébaut, D., Pinkston, J.D., Morin, P., and Lafosse, M. (1999). *Detection in SFC. In: Practical Supercritical Fluid Chromatography and Extraction* (eds. M. Caude and D. Thiébaut). Harwood Academic Publishers.

West, C. (2014). Enantioselective separations with supercritical fluids – review. *Current Analytical Chemistry* 10 (1): 99–120. https://doi.org/10.2174/1573411011410010009.

Wicker, A.P., JrCarlton, D.D., Tanaka, K. et al. (2018). On-line supercritical fluid extraction – supercritical fluid chromatography-mass spectrometry of polycyclic aromatic hydrocarbons in soil. *Journal of Chromatography B: Analytical Technologies in the Biomedical & Life Sciences* 1086, 82-82-88. doi:https://doi.org/10.1016/j.jchromb.2018.04.014.

Yoshioka, T., Nagatomi, Y., Harayama, K., and Bamba, T. (2018). Development of an analytical method for polycyclic aromatic hydrocarbons in coffee beverages and dark beer using novel high-sensitivity technique of supercritical fluid chromatography/mass spectrometry. *Journal of Bioscience & Bioengineering* 126 (1), 126-126-130. doi:https://doi.org/10.1016/j.jbiosc.2018.01.014.

Zheng, J., Glass, T., Taylor, L.T., and Pinkston, J.D. (2005a). Study of the elution mechanism of sodium aryl sulfonates on bare silica and a cyano bonded phase with methanol-modified carbon dioxide containing an ionic additive. *Journal of Chromatography A* 1090 (1-2): 155–164. https://doi.org/10.1016/j.chroma.2005.07.021.

Zheng, J., Pinkston, J.D., Mangels, M.L., and Taylor, L.T. (2005b). Effect of ionic additives on the elution of sulfonates and amine hydrochlorides in supercritical fluid chromatography. Paper presented at the ACS National Meeting, Washington DC, USA (28 August to 1 September 2005).

Zheng, J., Taylor, L.T., and Pinkston, J.D. (2005c). Supercritical fluid chromatography of ionic analytes. Paper presented at the ACS National Meeting, San Diego, CA, USA (13–17 March 2005).

Zheng, J., Taylor, L.T., Pinkston, J.D., and Mangels, M.L. (2005d). Effect of ionic additives on the elution of sodium aryl sulfonates in supercritical fluid chromatography. *Journal of Chromatography A* 1082 (2): 220–229. https://doi.org/10.1016/j.chroma.2005.04.086.

Zheng, J., Taylor, L.T., and Pinkston, J.D. (2006). Elution of cationic species with/without ion pair reagents from polar stationary phases via SFC. *Chromatographia* 63: 267–276. https://doi.org/10.1365/s10337-006-0731-z.

7

Online and In Situ Measurements for Environmental Applications in Oil and Gas

Eric Schmidt[1], *J.D. Tate*[2], *William L. Winniford*[1], *and Melissa N. Dunkle*[3]

[1] *Analytical Science, The Dow Chemical Company, 230 Abner Jackson Parkway, Lake Jackson, TX, USA*
[2] *ATC Analytical Operations, The Dow Chemical Company, 230 Abner Jackson Parkway, Lake Jackson, TX, USA*
[3] *Analytical Science, Dow Benelux, BV, Herbert H. Dowweg 5, Building 446, Hoek, The Netherlands*

7.1 Introduction

On-line analytical measurements have a long history of providing near real-time composition measurements in the refining and chemical industry (Kueppers and Haider 2003). However, the use of on-line analytical measurements for environmental applications in upstream oil and gas applications is somewhat limited. Measurements fit into two categories: regulated and unregulated. The most prevalent regulated on-line measurements are continuous emission monitoring systems (CEMSs) for combustion control in refineries. The United States has more requirements for environmental monitoring than other parts of the world. As a result, many nations use the requirements and methods from the United States as their starting point for regulation. Other regulated measurements are typically collected as grab samples and taken to a laboratory for analysis (e.g. water). In addition, there are a limited number of on-line water measurements, and typical flare measurements utilize gas flow meters as totalizers for flow into pipelines (Shaban and Tavoularis 2014). The unregulated measurements include gas detection for safety, process-related such as British thermal unit (BTU) or pipeline gas composition, and somewhat more recent applications of wireless sensors and satellite data for monitoring fugitive methane emissions. This chapter will provide a survey of the different on-line environmental measurements that are commonly practiced in the oil and gas industry as well as a discussion of the emerging remote monitoring techniques.

Analytical Techniques in the Oil and Gas Industry for Environmental Monitoring, First Edition.
Edited by Melissa N. Dunkle and William L. Winniford.

7.2 Characteristics of On-line Analyzers

7.2.1 Zone Classification

On-line analyzers differ from laboratory equipment in several aspects. Since their purpose is to run continuously, they must be engineered for the highest possible reliability. Maintenance in the field is more costly, difficult, and hazardous than in a laboratory. Robustness and simplicity are more important than flexibility. Typically, in the United States, they are certified for operation in Class I Div II environments – they will not provide a spark source if a flammable vapor is present. On-line analyzers are designed to do one specific analysis with minimal on-going maintenance. As an example, automated laboratory gas chromatographs (GCs) use syringe injection through a septum to introduce the sample to the separation column. A septum may start to leak considerably after 100 injections, requiring septa replacement. For an on-line GC, this is not practical to replace on this frequency, so the normal practice is to use valved injection to introduce the sample to the analytical column. Suppliers of injection valves will test the valves for a lifetime of over 1 million injections. On-line analyzers also differ from lab instruments in how the data is processed and transmitted. The volume of sample analysis is generally too large for regular inspection of the raw data. The raw data is converted to concentration values and transmitted via standard protocols to process control computers or certified continuous emission monitoring reporting systems. Calibration and check standards must be automatically run at regular intervals to verify the performance of an analyzer.

In Europe, on-line (or process) analytical equipment to be installed in a manufacturing environment must be ATEX certified for the correct zone where the installation will occur.

ATEX derives its name from the French title of the 94/9/EC directive of the European Commission (EC), *Appareils destinés à être Atmosphères Explosives*, which defines classifications, or zones, on what work or equipment is allowed in an environment with an explosive atmosphere (EC 1994). It is possible to have an explosive atmosphere with (i) gases, vapors, and mists, and (ii) solid dust. For gases, vapors, and mists, an explosive atmosphere is defined as an atmosphere where dangerous substances in the form of a gas, vapor, or mist mix with air. For solid dust, an explosive atmosphere is defined as a cloud of combustible dust in air.

Zone designations for gases, vapors, and mists:

- *Zone 0*: A location where an explosive atmosphere is present continuously over long periods of time.
- *Zone 1*: A location where an explosive atmosphere occasionally occurs during normal operation.

- *Zone 2*: A location where an explosive atmosphere is not likely to occur during normal operation; however, if it does occur, it will only persist for a short period of time.

Zone designations for solid dust:

- *Zone 20*: A location where an explosive atmosphere is present continuously over long periods of time.
- *Zone 21*: A location where an explosive atmosphere occasionally occurs during normal operation.
- *Zone 22*: A location where an explosive atmosphere is not likely to occur during normal operation; however, if it does occur, it will only persist for a short period of time.

7.2.2 Sampling Systems

A process or online GC analyzer is visually very different compared with a lab-scale GC instrument. State of the art in process GCs has been assessed recently (Waters 2020). A process GC is either mounted on a rack or on a wall with a display screen on the front panel. The gas and sample lines are typically plumbed into the top of the analyzer, and the analyzer will open to the front for easy access for maintenance. In many cases, a process GC is isothermal and utilizes packed columns; however, there are companies offering custom solutions for process GC. Another major difference between a process GC and a lab-scale instrument is the sample introduction, where process GCs are usually limited to a valve system because samples transferred from a given process through sample lines are potentially at high temperature and/or pressure. In such cases, the sample must be collected, cooled, and potentially de-pressurized prior to injection on the analyzer (Barrows et al. 1995; Waters 2013).

There are two general classes of sampling systems, extractive and non-extractive. Extractive sampling refers to a sample being taken from the process at regular intervals to be injected into the analyzer. Extractive sampling requires careful consideration of what is being sampled. Since it involves the transport of the sample to the analyzer, care must be taken that the sample actually represents what is in the process at the recorded time of analysis. A side stream of gas or liquid must be taken from the process at a flow rate that is sufficiently fast not to lag changes in the process stream being measured. Then there must be a way to return this sample stream to the process. The sample temperature must be carefully controlled. The particulate matter must be filtered out in such a way that it only requires occasional maintenance, i.e. using self-cleaning filters. All of these aspects introduce considerable expense and complexity to the analyzer. Non-extractive sampling eliminates many of these disadvantages but can only be used with techniques that do not alter the sample.

7.2.3 Detection

Optical absorption techniques and sensors are most commonly used with non-extractive sampling. Optical detectors are designed as pipe spool pieces so that the gas flows through unimpeded. A light source is located on one side and the detector on the other side. By choosing window material that is transparent at the relevant wavelengths, the electronic components are completely isolated and protected from the process stream. Sensors are attached via standard flanges that are used for temperature and pressure measurement. Consideration must be given to particulate matter or condensables that could foul the sensor, causing it to give erroneous readings.

One other type of monitoring is growing in use. Open path monitoring, most often using infrared, is a very effective way to do fenceline monitoring over wide areas. A single spectrometer, with a laser source and detector, is located at sufficient height above most processing equipment in a production facility. Mirrors are located at strategic positions along the perimeter of the facility to reflect the laser source back to the detector. One analyzer can be multiplexed to step through the positions where the mirrors are located sequentially. In this way, specific components such as H_2S or SO_2 can be monitored over a pathlength of hundreds of meters with relatively fast response time.

With process GC, the typical GC detectors can also be utilized, including the flame ionization detector (FID), the flame photoionization detector (FPD), and the thermal conductivity detector (TCD). Essentially, only the most stable GC detectors will be selected for process applications, as it is impractical and undesired to go into the field to maintain an analyzer continually.

Analytical laboratory-based gas chromatography–vacuum ultraviolet (GC-VUV) was discussed in Chapter 4; however, the authors would also like to mention that recently, an ATEX certified process GC-VUV system has become available. Wasson ECE offers the ability to integrate the VGA-100 and VGA-101 from VUV Analytics, Inc. into their Eclipse Process GC analyzer. To the best of our knowledge, Wasson ECE is the only vendor offering the possibility for a process GC-VUV. The intended application is currently for paraffins, isoparaffins, olefins, naphthenes, and aromatics (PIONA) analysis, according to ASTM D8071 (ASTM-D8071 2019; Wasson 2019).

7.3 Water Analysis

7.3.1 General Water Analysis

Water analysis in the oil and gas industry typically involves samples from reservoir formations and production that must be monitored to maintain the integrity of reservoirs and wells. Typical analysis includes pH, hydrocarbon content, specific gravity, benzene, toluene, ethylbenzene, and *m*-xylene (BTEX), and trace metals (Yang 2006). The purpose of the produced water analysis is for both process control of the well and regulatory compliance. For example, monitoring horizontal well integrity has been shown by monitoring the iron to manganese ratio of the produced water from the well to predict corrosion rates, which may lead to failures in the lateral sections of the well (Ofori et al. 2019).

These samples are typically taken at the source and then analyzed offsite at a specialized laboratory. Offline sample analysis can offer challenges to the oil and gas industry because of issues with sample integrity, whereas the composition of the sample changes before the lab analysis is performed, as well as the time required to transport, perform the analysis, and return the results. When possible, online analysis with real-time results is preferred.

One of the more recent on-line technologies to improve environmental compliance is an oil in water analyzer utilized to determine water contamination in the outfalls of process equipment (Brupbacher and Richerand 2013). This measurement uses laser-induced fluorescence to quantitate the hydrocarbon content in water for the purpose of diverting the outfall flow before reaching a non-compliance situation. Of particular interest is the automated ultrasonic cleaning mechanism, which improves measurement performance in contaminated water streams. This technology is useful for early warning of a potential environmental non-compliance for both offshore and onshore processes.

7.3.2 Application: Benzene in Drinking Water

The maximum contaminant levels (MCLs) established for benzene in drinking water vary slightly between the United States and Europe. The US Environmental Protection Agency (EPA) has set the MCL for benzene in water at 5 μg/l (EPA 2009), and the EU has set the MCL at 1 μg/l (EC 1998).

For real-time and online monitoring of benzene in water, the traditional technologies involving solid phase extraction (SPE) or solid phase microextraction (SPME) extraction prior to separation and detection on analytical instrumentation are not feasible, as the extraction step is time-consuming. A permeable membrane coupled to a photoacoustic sensor has been applied for online analysis. However, the response time was 40 minutes, and the limit of detection (LOD) was 350 μg/l (Mohácsi et al. 2001), which does not meet the MCL set by either the United States or Europe. Another technology developed was membrane inlet ion mobility spectrometry (MI-IMS) equipped with a radioactive source (^{63}Ni), offering a sensitive online monitoring of methyl *tert*-butyl ether (MTBE) and the monoaromatic compounds BTEX in water; here the LOD was 1 μg/l using the radioactive IMS source (Baumbach et al. 2003). Gas samples were also able

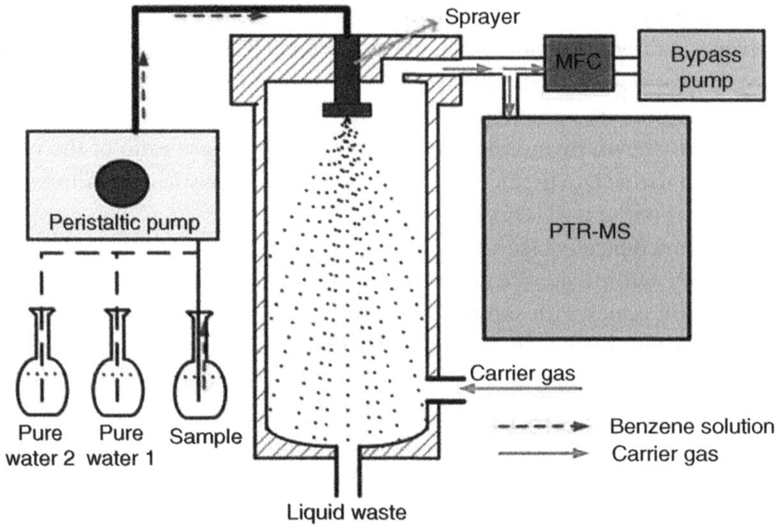

Figure 7.1 Schematic of the online SI-PTR-MS used for rapid and sensitive measurement of benzene dissolved in water. Source: Zou et al. (2016). Reproduced with permission, Copyright (2016), American Chemical Society.

to be analyzed using the developed MI-IMS system, which showed a significant improvement in the detection limits; 30 pg/l using the radioactive IMS source. While this technique reaches both the US and EU MCL for benzene in water, it is at the limit for the EU standards.

More recently, spray inlet proton transfer reaction mass spectrometry (SI-PTR-MS) has been applied for rapid and sensitive monitoring of trace levels of benzene in water (Zou et al. 2016). In Figure 7.1, a schematic is shown of the SI-PTR-MS. The droplets formed in the spray chamber are protonated via a proton transfer reaction from hydronium (H_3O^+) reagent ions (Eq. (7.1)).

$$H_3O^+ + C_6H_6 \rightarrow C_6H_6 \cdot H^+ + H_2O \tag{7.1}$$

SI-PTR-MS was able to detect benzene in water in 55 seconds, and the LOD was 0.14 μg/l, lower than the MCLs established by both the US EPA and the European Council.

7.4 Air Quality and Emissions Monitoring

Process gas analyzers are used for the continuous monitoring of one or more gases to check emissions, enabling process control and safety control, but also help to ensure compliance with governmental regulations.

In an industrial or manufacturing environment, in situ gas analyzers measure the flow of process gas in the actual process line and are available for a variety of gases, including oxygen, water, and greenhouse gases (GHGs), to name a few. With in situ gas analyzers, probes are installed in the duct at the top of the exhaust stack; both cross-duct or single-sided gas duct probes are available (Siemens 2018). Due to the compact design of in situ gas analyzers, the detector is typically a sensor or a diode laser, but no matter what, as in situ gas analyzers are installed at the top of an exhaust stack, the analyzer must be extremely stable and require very little maintenance.

7.4.1 Regulations

7.4.1.1 US Air Monitoring

Continuous air monitoring in the oil and gas industry is primarily done in the exhaust for any process that involves thermal oxidation/combustion to generate heat or burn unrecoverable components. In the United States, air monitoring is regulated by individual states as well as the federal government. The EPA has published the guidelines and requirements for on-line air monitoring. The primary EPA methods for on-line air monitoring are summarized in Table 7.1. Briefly, Methods 1–4 cover sampling procedures (EPA 2017a,b,c,d), 5F covers particulate matter for catalytic cracking (EPA 2017e), 6C focuses on SO_2 (EPA 2017f),7E measures NO_x (EPA 2018), 10 measures CO (EPA 2017g), 16C covers total reduced sulfur (EPA 2017i), and 25a focuses on volatile organic compounds (VOCs) (EPA 2017j).

7.4.1.2 European Union Air Monitoring

The European Union Offshore Oil and Gas Authorities Group (EUOAG) was established in 2012 (EU 2012). The purpose of the EUOAG is to discuss, assist, and give its opinion to the European Commission on issues and activities related to offshore oil and gas. Additionally, the EUOAG is to promote and ensure best practices as well as best safety standards in global offshore oil and gas operations. Best practices include regulatory reporting of major hazards and industrial risk assessments, prevention measures, and training requirements.

As per the Kyoto Protocol, the EU Member States are required to provide emissions reports to the United Nations (UN). On an annual basis, a report on their GHG emissions is required, and on a regular basis (e.g. biennial reports and national communications), reports on their climate change policies and measures, as well as their progress toward the targets are required (Kyoto 2008). The annual report must include emissions of seven GHGs from all sectors, details on how to cut GHG emissions (e.g. projections, policies, and measures), the measures taken to adapt to climate change, low carbon strategies, commitments

Table 7.1 Summary of EPA methods for on-line air monitoring.

Method name	Purpose	Instrumentation	References
Method 1	Sample traverses Velocity traverses	Directional probe	EPA (2017a)
Method 2	Stack gas velocity Volumetric flow rate	Type S pitot tube	EPA (2017b)
Method 3	Oxygen Carbon dioxide	Orsat analyzer Fyrite-type analyzer	EPA (2017c)
Method 4	Moisture content	Volumetric analysis Gravimetric analysis	EPA (2017d)
Method 5F	Particulate matter	Ion chromatography	EPA (2017e)
Method 6C	SO_2	UV NDIR Fluorescence	EPA (2017f)
Method 7E	NO_x	UV Electrochemical Chemiluminescence	EPA (2018)
Method 10	CO	NDIR Electrochemical Any that passes specs.	EPA (2017g)
Method 16A	Dimethyldisulfide Dimethylsulfide Hydrogen sulfide Methyl mercaptan	GC-FPD	EPA (2017h)
Method 16C	Total reduced S as SO_2 • Dimethyldisulfide • Dimethylsulfide • Hydrogen sulfide • Methyl mercaptan	UV NDIR Any that passes specs.	EPA (2017i)
Method 25A	Total gaseous organic	FID	EPA (2017j)

GC-FPD, gas chromatography–flame photometric detector.

to support developing countries both technically and financially, and a description of revenue usage. Regular reports are made on a biennial basis to the United Nations Framework Convention on Climate Change (UNFCCC) and every four years to the UN and must include information on emissions and removals, details on policies and measures to reduce emissions, details on financial, technological, and capacity-building support for developing countries, and other activities undertaken to implement the UNFCCC.

The European Environment Agency (EEA) gathers and summarizes all of the information presented to the UNFCCC on Air, Climate, Nature, Sustainability and Well-Being, and Economics for the European Countries. This includes the actual data on GHG emissions, but also the policies, finances, trends, and much more.

7.4.2 Proton Transfer Reaction Mass Spectrometry for Emission Monitoring

Proton transfer reaction mass spectrometry (PTR-MS) has also been used for VOC quantification from industrial flare gas. In an example by Knighton et al., flare gas was sampled by a probe situated near the flare and transferred to the analytical equipment in a mobile unit (Knighton et al. 2012). Direct injection to the PTR-MS, as well as gas chromatography–proton transfer reaction–mass spectrometry (GC-PTR-MS), was possible via a switching valve to select direct injection or separation prior to detection. It should be noted that with PTR-MS, the molecules of interest must have a proton affinity greater than that of water to react with H_3O^+. It is important with the analysis of air samples that the typical components of air (e.g. O_2, CO_2, N_2, Ar) will not react with the hydronium ion. Additional components more specific to flare gas that will not react with the hydronium ion include alkanes, acetylene, and ethylene; all other hydrocarbons from industrial flare gas should be detectable by PTR-MS. While GC-PTR-MS requires additional time to perform the separation, it does afford separation of the various components as well as confirmation of compound identification, which can then be used for interpretation of the direct injection PTR-MS data. Figure 7.2 shows an example of the GC-PTR-MS analysis of a flare gas stream, and as can be seen, displaying selected ions shows the peaks corresponding to different VOCs present in the flare gas.

In a study by Piel, the CHemical Analysis of AeRosol ONline (CHARON) inlet for particulate matter was utilized with proton transfer reaction coupled to time of flight mass spectrometry (PTR-ToF-MS) for the real-time analysis of sub-micron particulate matter from airborne sources (Piel et al. 2019). In this example, the instrument was installed on NASA's DC-8 aircraft to collect real-time data from a small wildfire and from a refinery plume, demonstrating the feasibility of this technology to chemically characterize airborne pollution from jet research aircraft (Figure 7.3). The CHARON PTR-ToF-MS combination has an average LOD of $<200\,pg/m^3$ for compounds with m/z 60–400.

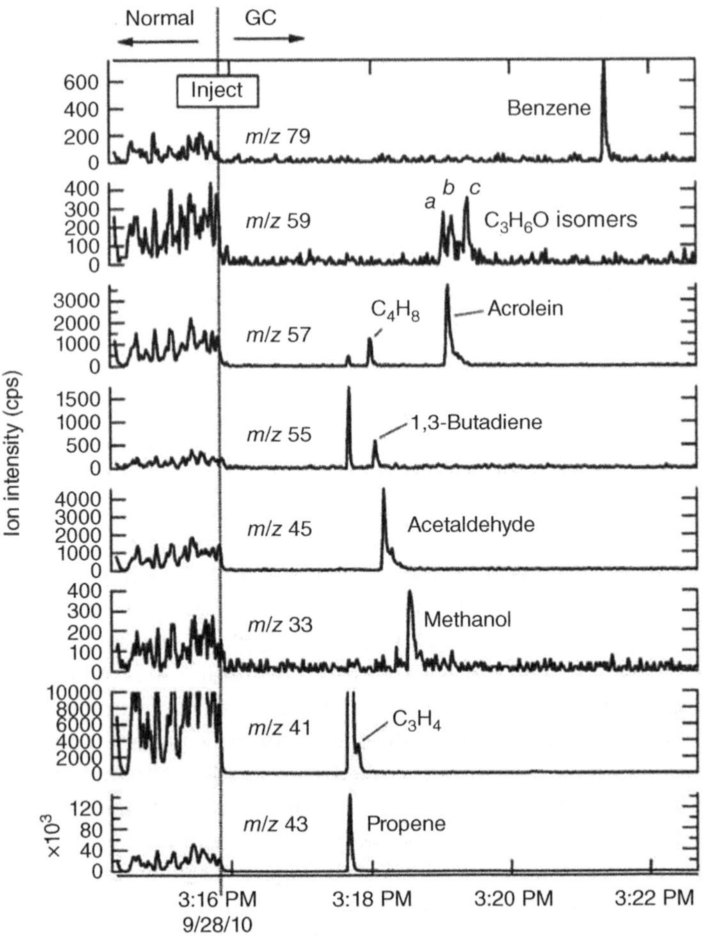

Figure 7.2 GC-PTR-MS selected ion chromatograms from the analysis of a flare stream. The first minute shows the trapping of the sample in the cryo-loop, and the dotted line shows the injection onto the GC-PTR-MS. Source: Knighton et al. (2012). Reproduced with permission, Copyright (2012), American Chemical Society.

The Houston Advanced Research Center initiated and implemented a "Benzene and other Toxics Exposure Study" (BEE-TEX) to demonstrate state of the art technologies for monitoring hazardous air pollutants. In one study, they used a van-mounted PTR-MS to do real-time monitoring of benzene near a marine terminal on the Houston ship channel (Olaguer et al. 2016). They were able to detect pipeline leaks and spikes due to transfer operations (i.e. truck, railcar, and ship loading).

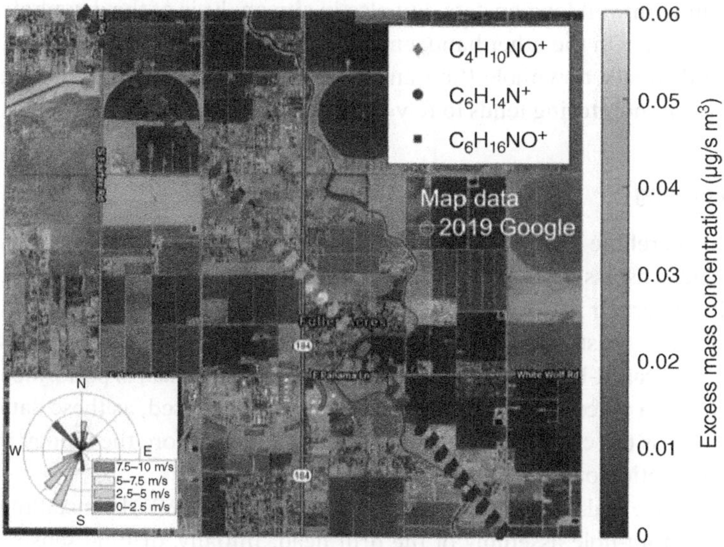

Figure 7.3 CHARON PTR-ToF-MS analysis of airborne particulate taken over a refinery (circled region). The plotted color-coded intensities of the observed $C_4H_{10}NO^+$, $C_6H_{14}N^+$, and $C_6H_{16}NO^+$ ions show a higher intensity nearer the refinery. Source: From Piel et al. (2019). License CC BY 4.0.

7.5 Sample Conditioning

For on-line combustion gas analysis, sample conditioning typically consists of diluting with dry air 50- to 200-fold to both cool the sample and reduce the water concentration. But this has the disadvantage that the O_2 concentration cannot be measured because the extracted sample is diluted with air. Also, the detection limits are impacted by dilution. If the dilution method cannot be used, then a drying technique may be used to remove water, typically a dryer that permeates water through a Nafion membrane and strips it away with a counter-current flow of dry nitrogen or air on the opposite side of the membrane. If only modest water removal is needed, then cooling the gas stream can be used to condense water. But whatever method is used for removal of water, it must not remove any of the components of interest. With either method, a filter may be needed to remove particulate matter.

7.6 Well Drilling and Production

In well drilling and production, real-time monitoring takes on different forms. On the one hand, the drilling process utilizes on-line analysis to provide information on where to drill with well logging data, but also in situ analysis of the downhole fluid during sourcing. On the other hand, emission monitoring is another aspect. Well logging and in situ downhole fluid analyses are well developed; however, currently, emission monitoring tends to rely on estimations.

7.6.1 Well Logging

Well logging, or borehole logging, allows for the evaluation of a formation; this evaluation includes assessing the resource size, aiding in the placement of wells, and interpreting reservoir performance (Fanchi and Christiansen 2017). Different types of well logging exist; open-hole logging is performed before the casing is set in the well, and cased-hole logging, which as the name suggests, is performed when the casing is in place. Open-hole logging is most widely used, as these data can be used to assess the commercial viability of a formation before the money is spent on the completion of the well.

It is possible to have real-time measurements while drilling (MWD) by installing sensors in the bottom-hole assembly of the drill head. Initially, drilling sensors only measured drilling direction; however, nowadays, drilling parameters such as weight on bit, torque, revolutions per minute (RPM), etc., as well as natural γ-ray logging sensors for the detection of shale features, are available. An example of real-time well logging data is given in Figure 7.4; as can be seen, for the Temsah-3 well, γ-ray, lithology, compressional wave delay time (DTCO), corrected bulk density (RHOB), neutron porosity (NPHI), and resistivity (RT) were all measured from the sensors mounted in the bottom-hole assembly of the drill head while drilling the borehole (Nabawy and Shehata 2015).

Once the borehole is drilled, the reservoir is tapped, and the material from the formation is collected. For downhole fluid analysis, it is possible to have an in situ fluid analyzer consisting of various sensors to provide quantitative information under reservoir conditions. Schlumberger offers an in situ fluid analyzer system providing quantitative measurements of the fluid properties during collection, such as hydrocarbon composition (C_1–C_6), gas/oil ratio, live-oil density and viscosity, CO_2, pH, color, free-gas detection, dew precipitation, flowline pressure, temperature, resistivity, and oil-based mud filtrate contamination (Schlumberger 2008). Halliburton has also developed many sensing technologies for downhole monitoring. H_2S is particularly problematic because of the corrosion that it causes to equipment. It is very difficult to take representative samples for laboratory analysis due to the tendency of H_2S to adsorb onto surfaces, causing the measured values to be lower than actual. Pressures downhole can be as high as 15 000 psi, which eliminates many techniques from consideration. Halliburton has developed a high-temperature/high-pressure cell that employs multivariate optical computing to quantify H_2S using the spectral range of 170–240 nm (Dai et al. 2018).

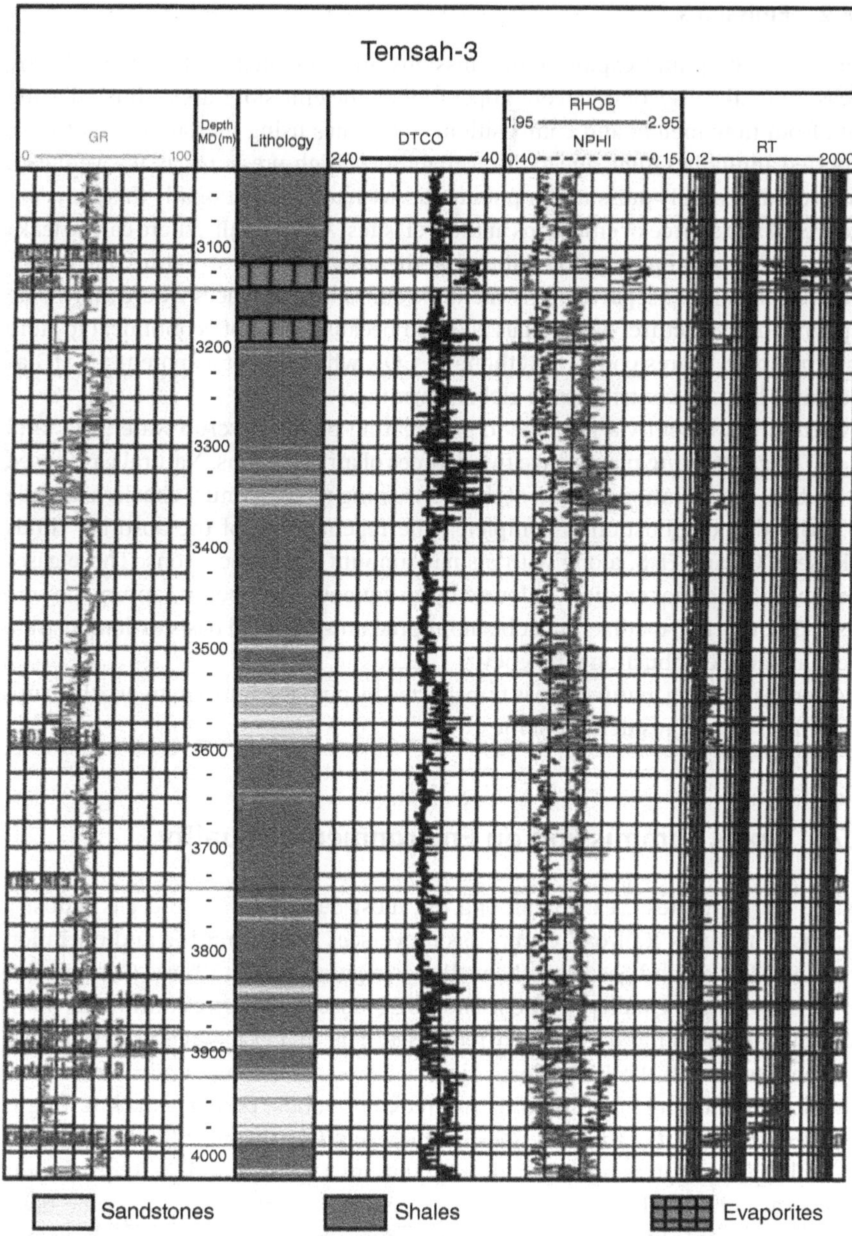

Figure 7.4 Conventional well log data from the Temsah-3 well, offshore Nile Delta, North Egypt. Source: Reprinted with permission from Nabawy and Shehata (2015). Copyright (2015), Elsevier.

7.6.2 Emissions

For well drilling and exploration, emissions are estimated, rather than directly measured. Likewise, once a well is operational, the emissions are estimated using data from flow meters and composition, sometimes using measured data that is reported annually. Often, due to the availability of high-pressure gas at a wellhead, gas is used to supply pressure for pneumatic actuators. Traditionally this gas is not recovered as a source of emissions at the well sites. Additionally, there can be leaks in the piping coming out of the ground or the above-ground structure that can be a source of emissions, particularly methane emissions. In places where there is no pipeline infrastructure, natural gas is flared because it is not economical to store and transport. Flares convert methane to CO_2, which is not as potent a GHG as methane, but still contributes to global warming.

Natural gas is sent via pipeline to gas treating plants to scrub water, CO_2 and H_2S. The gas treating plants have two sources of air emissions. Natural gas is used to heat the regeneration portion of the gas treating plant. Annual checks are made to verify that combustion is complete. The regeneration of the absorber fluids releases CO_2 and H_2S, along with a small amount of hydrocarbons. These gases are sent to a thermal oxidizer followed by an aqueous scrubber to capture the acid gases. The thermal oxidizer must be monitored on an annual basis to demonstrate the mandated combustion efficiency.

After gas treating, natural gas is transported by pipeline into a network that supplies industry and residential homes.

7.7 Texas Commission on Environmental Quality

In the United States, Texas is among the most developed oil and gas industry regions. The Texas Railroad Commission oversees the permitting and operation of wells. After oil and gas leave the well, it is regulated by the Texas Commission on Environmental Quality (TCEQ). The TCEQ does not require continuous monitoring until the oil and gas reach refineries. The TCEQ has requirements that are consistent with the EPA Clean Air Act. For certain compounds, continuous emission monitoring is required. For other compounds, periodic stack testing is required to certify that the emissions are not exceeding the mandated limits.

The TCEQ has a network of monitoring stations set up around the state of Texas to measure chemicals in the air. In five regions, automated GCs provide daily analysis of volatile hydrocarbons, including those that are considered to be significant ozone precursors such as ethylene, propylene, 1,3-butadiene, and butenes (*c*-2-butene, *t*-2-butene, and 1-butene). A 40-minute sample is collected and analyzed every hour. Table 7.2 lists the components monitored.

Table 7.2 List of components analyzed every hour via the TCEQ monitoring stations.

Ethane	*n*-Octane	Methylcyclopentane
Ethylene	*n*-Nonane	2-Methylhexane
Propane	*n*-Decane	1-Butene
Acetylene	Cyclopentane	2,3-Dimethylpentane
n-Butane	Isoprene	*n*-Undecane
Isobutane	2,2-Dimethylbutane	2-Methyheptane
trans-2-Butene	2,4-Dimethylpentane	*m*/*p*-Xylene
cis-2-Butene	Cyclohexane	Benzene
1,3-Butadiene	3-Methylhexane	Toluene
n-Pentane	2,2,4-Trimethylpentane	Ethylbenzene
Isopentane	2,3,4-Trimethylpentane	*o*-Xylene
1-Pentene	2-Methylheptane	1,3,5-Trimethylbenzene
trans-2-Pentene	Methylcyclohexane	1,2,4-Trimethylbenzene
cis-2-Pentene	2,2,4-Trimethylpentane	*n*-Propylbenzene
2-Methyl-2-butene	2,3,4-Trimethylpentane	Isopropyl benzene
n-Hexane	3-Methylheptane	Styrene
n-Heptane	Methylcyclohexane	1,2,3-Trimethylbenzene

Additionally, some sites will capture hourly data for CO, H_2S, NO_x, ozone, PM10, PM2.5, and SO_2. Figure 7.5 shows the location of monitoring stations. The data is available to the public through the TCEQ website.

Hourly data are acquired and averaged each day, and by the month, Figures 7.6 and 7.7. Meteorological conditions, wind speed, wind direction, wind gusts, temperature, and solar radiation are logged because the measured values of chemical species will be strongly affected by weather conditions.

7.8 Fenceline Monitoring

There are a variety of techniques that are used for fenceline monitoring. Fenceline monitoring is intended to identify and quantify fugitive emissions. These provide long-term data on hazardous air pollutants and also trigger an immediate response if a leak occurs that could impact the surrounding community. In 2016 the EPA initiated a fenceline monitoring program for refineries to verify benzene emission levels (EPA 2016). The threshold was set at 9 $\mu g/m^3$, averaged across the site annually. Six options have been proposed by the EPA.

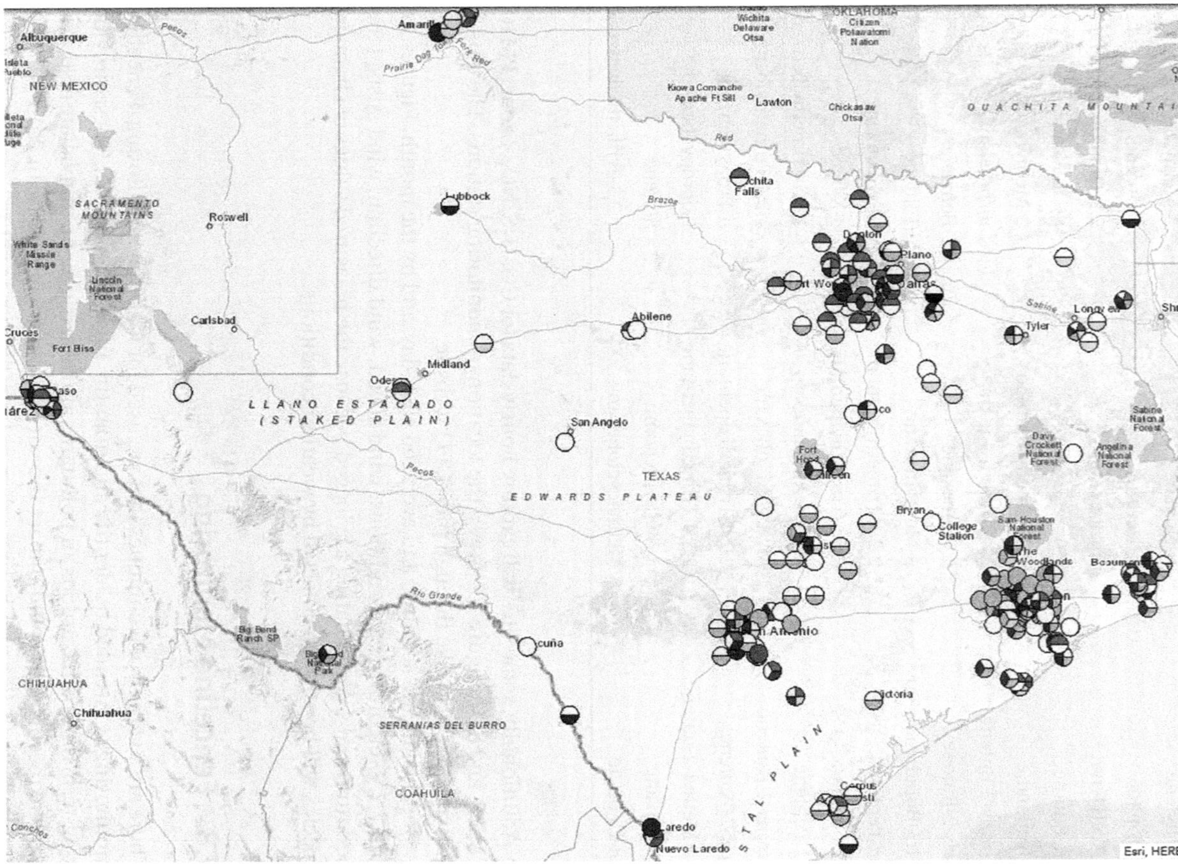

Figure 7.5 Location of on-line monitoring stations in Texas. Source: TCEQ (2020a).

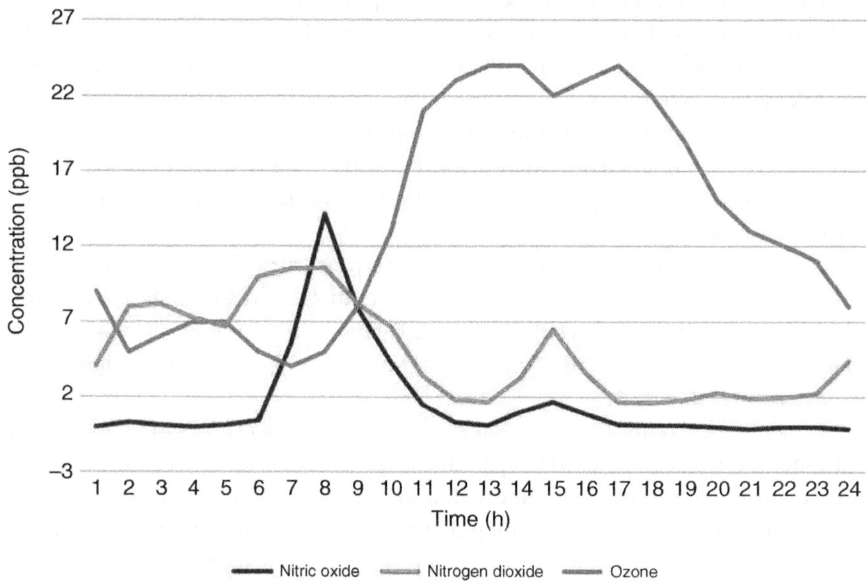

Figure 7.6 Hourly measurements for NO_x and ozone (ppb), example from a Houston monitoring station, 7 August 2019. Source: TCEQ (2020b).

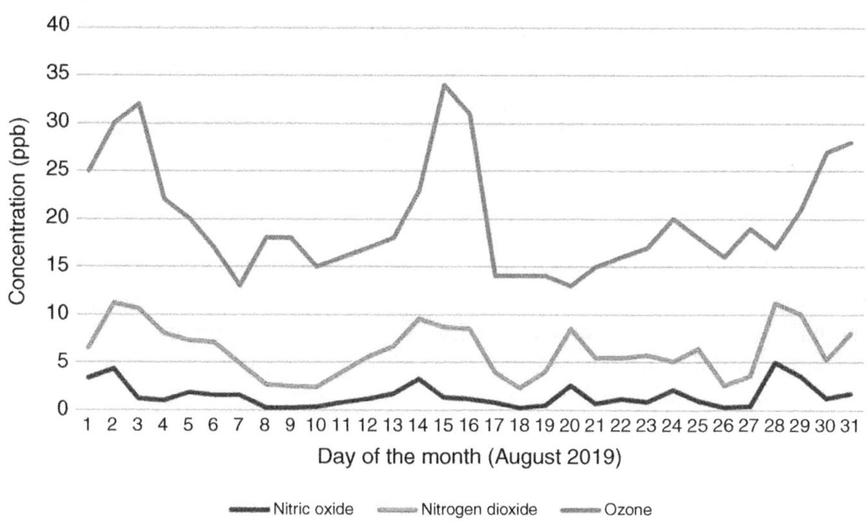

Figure 7.7 Daily averages for NO_x and ozone (ppb), example from a Houston monitoring station, August 2019. Source: TCEQ (2020b).

Passive diffusion tubes are the least costly option and easily implemented. The technology is based on benzene and other hazardous air pollutants diffusing through a special cap on a tube filled with an adsorbent. Periodically the tubes are taken to a laboratory for analysis. This approach is recommended as the minimum starting point for a fenceline monitoring program. Passive sampling has been well validated for measuring benzene. The disadvantage of passive sampling is that there is poor time resolution, and it is not useful for emergency response in the case of a spill or leak.

Active monitoring stations use a pump to draw air through sampling tubes. The advantage is that shorter sampling times are required. Thus there is a better time resolution. But the samples still are sent to a laboratory for analysis, which eliminates it as a tool for rapid response. And the cost can be 2–10 times higher than passive sampling.

Ultraviolet differential optical absorption spectroscopy (UV-DOAS) is an open path method that can be used to span distances up to 500 m between the UV light source and the detector. Fewer individual monitoring locations are needed since the response is averaged over a long distance. The response time can be almost instantaneous, which is an advantage for emergency response. Both adsorbent tube methods are accumulating organics at specific locations. Smoke and dust can interfere, giving false positive readings. The cost is substantially higher.

Open-path Fourier transform infrared spectroscopy (OP-FTIR) is similar to the UV method but uses infrared wavelengths. Since there are more bands that are specific to organic compounds in the infrared region versus the UV, it is possible to measure more compounds simultaneously. However, there can be interference from CO, CO_2, and water vapor. For benzene, there is less sensitivity in the infrared region versus the UV, and the EPA does not recommend its use for benzene monitoring.

Differential absorption lidar monitoring (DIAL) is another open path technique, but it differs from the previous two by using two wavelengths, one that is absorbed strongly by the monitored component and one that is absorbed weakly. The light source is pulsed, and the backscattered light is detected. Using the difference in the backscattered response between the two wavelengths, the concentration of the target component can be calculated. Also, by measuring the response time after the light pulse, the distance to the plume can be calculated. The EPA does not recommend this for long term measurement, but it is useful for short term mapping.

Solar occultation flux (SOF) monitoring is another open path variant that uses the sun's radiation as the light source. Either a UV or IR detector can be used, mounted in a van. Very high spatial resolution can be achieved, but the disadvantages are the staffing required for the operation of the van and the necessity of a sunny day.

Chevron recently published a comprehensive three-year study of the fenceline monitoring techniques that included data from nearby community monitoring stations in Richmond, California (Sanchez et al. 2019).

Installation of a network of sensors around the perimeter of a refinery can be very expensive due to the cost of installation that includes hardwire connections for power and data transmission. The EPA has begun to address this by developing a low-cost solar-powered sensor using a passive photoionization detector (PD) (described in Chapter 3) (EPA 2019). A wind anemometer is included. The data is processed by an on-board microprocessor and the results transmitted on a short-range wireless network. The combination of the concentration, wind speed, and direction is used with a gas dispersion model to map in real-time the location of the emission plume. The EPA has classified benzene, toluene, mixed xylenes, ethylbenzene, and *n*-hexane as air pollutants to be monitored. (Regulations for these compounds are found in section 112 of the US EPA Clean Air Act.) By using a 10.6 eV lamp, the PD is sensitive only to aromatic compounds.

7.9 Pipeline and Fugitive Emission Monitoring with Drones

Technology is rapidly developing and changing with respect to pipeline monitoring for leaks and fugitive emission monitoring with drones. Multiple technologies are converging to make significant advancements in capability. Infrared detectors, coupled with digital cameras and guidance from artificial intelligence algorithms, make it possible for drones to locate very small methane leaks. Flight times and distances are steadily increasing with the improvements in battery power density and reduction in weight in the analytical sensors. Southwest Research Institute has worked with the US Department of Energy to develop the technology for unmanned autonomous vehicle (UAV) leak detection using hyperspectral imaging cameras (Araujo et al. 2016; EPA 2019). An example of methane leak detection is shown in Figure 7.8. Many companies have now commercialized drones for methane leak detection.

7.10 Types of Continuous Emission Monitors

7.10.1 Nondispersive IR (NDIR)

Nondispersive IR (NDIR) is a very mature technology that has the advantages of simplicity and low cost (ABB 2007). Sensor assemblies can be purchased as cheaply as a few dollars. The cost is low enough that they are used widely in

Figure 7.8 Example of methane plume detection. Image on left without infrared overlay. Image on right with overlay. Source: Reproduced with permission from SwRI (2020); Image courtesy of the Southwest Research Institute (SwRI).

Table 7.3 Components measured by NDIR and corresponding ranges.

Component	Detection range (ppm)
CO	0–10
CO_2	0–5
NO	0–150
SO_2	0–25
N_2O	0–20
CH_4	0–50
C_3H_8	0–50
C_2H_4	0–300

heating, ventilation, and air conditioning (HVAC) systems. A broadband light source and a bandpass filter rather than a monochromator are used for the detection of specific components with either a thermal detector or photodetector. Typical components and ranges measured by NDIR using the range from 2.5 to 8 μm are shown in Table 7.3. Though NDIR has been used for many years, there are still advances in the technology that have been recently reviewed (Dinh et al. 2016). Since 2016, implementation with microelectromechanical systems (MEMS) has further reduced the size while improving sensitivity.

Gas filter correlation spectrometers have been used for over 40 years (Chaney and McClenny 1977). They differ from the earlier NDIR systems by using a rotating filter wheel between the light source and the gas sample cell. One half of the filter wheel is filled with the gas being monitored. The other half of the filter wheel consistently attenuates the light intensity at all wavelengths. As the gas being monitored enters the sample cell, there will not be a response when the gas-filled filter is between the light source and the cell. But there will be a change in detector response when the other filter rotates into place. The advantage of this type of detector is that components such as CO, CO_2, and water have enough fine structure to their spectra that the detector response is not biased when other components present if the spectral overlap is minimal (InstTools 2020a).

A Luft detector is configured with two parallel cells, a sample cell and a reference cell filled with a gas (Frant and LaButti 1980). Light from a single source is aimed through both cells using a parabolic mirror. The light is pulsed using a rotating chopper wheel between the cells and the light source. The detector has two compartments, aligned with the sample and reference cells, filled with the gas to be analyzed. When the gas composition in the sample and reference cells are the same, equal amounts of radiation impinge on the two detector cells. In the example of CO_2 analysis, the detector cells would be filled with CO_2. When CO_2 enters the sample cell, radiation from the light source is absorbed. Consequently, the sample side of the detector receives less radiation and the CO_2 molecules inside are not heated as much, resulting in less pressure relative to the reference side. In early versions, this change in pressure was detected using a diaphragm between the two detector cells. As the diaphragm deflected, the position was measured by parallel plates of a capacitor. More modern versions use a narrow channel connecting the two detector cells with a flow sensor that measures the change in flow direction as the chopper alternately passes light through the sample and reference cell (InstTools 2020b). The Luft detector works well for gas samples where there is no spectral overlap between the background components and the component to be analyzed. A spectral overlap will represent an interfering signal.

7.10.2 UV and Dispersive IR

These are both very mature technologies. The initial analyzers by ABB were introduced in the 1950s based on standard UV and IR absorption techniques; Table 7.4 contains a list of the components and corresponding range able to be detected (ABB 2007).

7.10.3 Chemiluminescent NO_x/SO_x Analyzers

The principle of detection is based on the reaction of ozone with the analytes to produce an excited state species that relaxes through the emission of light. The emitted light is detected with a photomultiplier tube. The technology is mature and well-characterized (Gluck et al. 2003). Compared with other analyzers, chemiluminescent analyzers are more complex. They require an ozone generator, a vacuum pump, catalyst systems to convert NO_x or SO_x to a single species

Table 7.4 ABB Limas UV and dispersive IR instrument range.

Limas 11 UV[a]		Limas 11 IR[b]		Limas 11 HW[a]	
Component	**Range (ppm)**	**Component**	**Range (ppm)**	**Component**	**Range (ppm)**
NO	0–10	CO	0–500	NO	0–25
SO_2	0–25	CO_2	0–150	NO_2	0–40
NO_2	0–50	HCl	0–2500	NH_3	0–25
H_2S	0–25	CH_4	0–1000		
Cl_2	0–125	C_2H_2	0–1250		
CS_2	0–50	C_2H_4	0–1500		
COS	0–250	C_2H_6	0–250		
		C_3H_6	0–500		
		C_3H_8	0–150		
		C_4H_{10}	0–250		

a) Limas 11 UV and Limas 11 HW: Ultraviolet and visible spectrum range ($\lambda = 200–600$ nm).
b) Limas 11 IR: Infrared spectrum range ($\lambda = 2.5–10$ μm).

to accurately quantify the sample, and a scrubber for the excess ozone. They are only used with extractive sampling, which adds expense, complexity, and maintenance. But they are used widely for SO_x and NO_x measurement with reliable detection down to a few ppm.

7.10.4 TDL Analyzers

Tunable diode lasers (TDLs), development driven by the telecommunications industry, have significantly advanced the reliability and robustness of instantaneous measurement for gas samples (Martin 2002). Analyzers for the following components are commercially available with single-digit detection limits for some components to a few hundred ppm for others.

- Oxygen
- Hydrogen chloride
- Hydrogen sulfide
- Carbon monoxide
- Moisture/water vapor
- Carbon monoxide (CO%)
- Carbon dioxide (CO_2%)
- Ammonia (NH_3)
- Methane (CH_4)

Because TDL analyzers are based on the absorption of near-infrared (NIR) radiation as it passes through a gas flow cell, the critical components are isolated from the sample stream. Multi-pass Herriot cells can be used to extend the path length for greater sensitivity while keeping the overall dimensions of the analyzer very compact.

7.10.5 QCL Analyzers

Quantum cascade lasers (QCLs) have the advantage of generating very narrow linewidth radiation in the mid-infrared wavelengths. They can be fabricated to operate over the range of 3–24 μm with high power. This makes them ideal for trace level detection of small molecules via infrared absorption. By folding the pathlength, it is possible to make a very compact gas sensor that is capable of ppb level detection or lower (Kosterev et al. 2007).

7.11 Portable GCs

Typically, in a manufacturing environment, analytical measurements are performed on-line, at-line, or in-line for a given process; however, in some cases, it is not always practical to install expensive process analytical equipment for a limited number of measurements, or in areas too far away from the plant. In these instances, a portable measuring device would be preferred. In addition to being portable, if used in a manufacturing environment, the potential for an explosive atmosphere needs to be considered. In Europe, such a device would require ATEX certification for the zone of use.

When comparing a portable analytical instrument, such as GC, to its lab-scale counterpart, it can be understood why there is not a huge offering of truly portable GC instrumentation. How can it be possible to fit all of the gases, electronics, temperature control, injection, etc. into something that is able to be carried by a person?

At the start of the miniaturization of GC equipment, transportable GCs able to be loaded into a cargo van were some of the first offerings. In an early example, Eckenrode used field-portable gas chromatography–mass spectrometry (GC-MS) loaded into a van for VOC monitoring of outside air (Eckenrode 2001). More recently, a portable GC-MS was used to identify and quantify VOCs from a petrochemical wharf; SPME was used to trap the VOCs prior to GC-MS analysis, for a total analysis time of less than five minutes, yielding detection limits in the low mg/m^3 range (Yun 2017). In this example, neither size nor weight for the portable GC was given, only that it "is quite small and convenient to carry out."

One of the smallest and lightest versions of a portable GC is the FROG-5000™ from Defiant Technologies.[1] The FROG-5000 is able to analyze VOCs in air,

1 ™ Trademark of Defiant Technologies.

Table 7.5 List of (trans)portable GCs.

Name	Analysis	Injection modes	Detector	LODs	Weight
TRIDION™-9[a)]	VOCs/SVOCs	SPME Needle-trap	Ion trap-MS	ppb–ppm	32 lb (14.5 kg)
Torion® T-9[b)]	VOCs/SVOCs	SPME Needle-trap	Ion trap-MS	sub-ppm	32 lb (14.5 kg)
Agilent 490 μ-GC	Hydrocarbon streams	Gas Liquid	Micro-TCD	ppm	14.4 lb (6.5 kg) 23.4 lb (10.6 kg)
Griffin™ G510[c)]	VOCs/SVOCs	Probe S/SL MIMS Direct injection SPME/twister	qMS	ppt–ppm	36 lb (16.3 kg)
Vernier Mini GC Plus	VOCs	Liquid	MEMS Chemicaptive sensor	0.1 ppm	3 lb (1.3 kg)
Seaport Mini GC	VOCs	Liquid	MEMS Chemicaptive sensor	0.1 ppm	3.1 lb (1.3 kg)
Hapsite® ER[d)]	VOCs/SVOCs	Headspace Probe SPME Thermal desorption	qMS	ppt–ppm	42 lb (19 kg)
Guardian	Vapor/gas/liquid/solid	SPME Thermal desorption	Ion trap-MS	ppt–ppm	32 lb (14.6 kg)

FROG-5000	VOCs	Gas Liquid Soil extraction	PID	sub-ppb	4.8 lb (2.2 kg)
CALIDUS™ GC[e)]	Fuel (up to C_{44})	Gas Liquid	FID TCD DBD FPD	Low wt%	25 lb (11.4 kg)
Model 312[f)]	VOCs/SVOCs	Gas/liquid	PID TCD FUV FID	ppb–ppm	12 lb (5.4 kg)
Compact GC[4.0]	VOCs/SVOCs/liquid/fuels	Gas/liquid	TCD FID PDD PED PID PFPD MS	ppt	50 lb (22 kg) (transportable)

MIMS, membrane inlet mass spectrometry; SVOC, semi-volatile organic compounds; qMS, quadrupole mass spectrometry; PID, photoionization detector; DBD, dielectric barrier discharge; FUV, far ultraviolet; PDD, pulsed discharge detector; PED, plasma emission detector; PFPD, pulsed flame photometric detector.

a) ™ Trademark of Torion Technologies, Inc.
b) ® Registered trademark of Perkin Elmer.
c) ™ Trademark of FLIR.
d) ™ Trademark of Inficon.
e) ™ Trademark of Teledyne-Falkon.
f) Produced by PID Analyzers, LLC.

water, and soil samples by GC coupled to a photoionization (PI) in a handheld device weighing only 5 lb (2.2 kg) (Defiant 2018). In a review of technologies able to meet the Air Quality Directive for benzene in water, the previous model, the FROG-4000 was able to meet the 5 μg/m^3 MCL (Spinelle et al. 2015), meaning that it is not only lightweight and portable, but also sensitive. Table 7.5 provides a list of (trans)portable GCs; it should be noted that this list is not exhaustive.

References

ABB (2007). Advance optima: Modular continuous gas analyzers. Retrieved from https://library.e.abb.com/public/ccda5d6159ba0989c1257bc00031735c/Advance_Optima_502401_EN_final.pdf (accessed 6 April 2020).

Araujo, M.S., Siebenaler, S.P., Dupont, E.M. et al. (2016). Near real-time automated detection of small hazardous liquid pipeline leaks using remote optical sensing and machine learning. *Paper presented at the International Pipeline Conference*, Calgary, Alberta, Canada.

ASTM-D8071 (2019). *Standard test method for determination of hydrocarbon group types and select hydrocarbon and oxygenate compounds in automotive spark-ignition engine fuel using gas chromatography with vacuum ultraviolet absorption spectroscopy detection (GC-VUV)*. West Conshohocken, PA: ASTM International.

Barrows, W.M., Foundos, P., and Liptak, B.G. (1995). Analytical instrumentation. In: *Instrument Engineer's Handbook: Process Measurement and Analysis*, 4e, vol. 1 (ed. B.G. Liptak), 1127–1764. New York: CRC Press.

Baumbach, J.I., Sielemann, S., Xie, Z., and Schmidt, H. (2003). Detection of the gasoline components methyl *tert*-butyl ether, benzene, toluene, and *m*-xylene using ion mobility spectrometers with a radioactive and UV ionization source. *Analytical Chemistry* 75: 1483–1490. https://doi.org/10.1021/ac020342i.

Brupbacher, B. and Richerand, F. Jr., (2013). Case history: Utilizing online oil in water analysis for process management and the prevention of environmental incidents. *Paper presented at the ISA 58th Analysis Division Symposium 2013*, Galveston, TX USA.

Chaney, L.W. and McClenny, W.A. (1977). Unique ambient carbon monoxide monitor based on gas filter correlation: Performance and application. *Environmental Science and Technology* 11 (13): 1186–1190. https://doi.org/10.1021/es60136a010.

Dai, B., Jones, C.M., Pearl, M. et al. (2018). Hydrogen sulfide gas detection via multivariate optical computing. *Sensors (Basel)* 18 (7) https://doi.org/10.3390/s18072006.

Defiant (2018). FROG-5000^{TM} chemical analysis system. Retrieved from www.defiant-tech.com/ (accessed 19 March 2020).

Dinh, T.V., Choi, I.Y., Son, Y.S., and Kim, J.C. (2016). A review on non-dispersive infrared gas sensors: Improvement of sensor detection limit and interference correction. *Sensors and Actuators B: Chemical* 231: 529–538. https://doi.org/10.1016/j.snb.2016.03.040.

EC (1994). Directive 94/9/EC of the European Parliament and the Council of 23 March 1994 on the approximation of the laws of the Member States concerning equipment and protective systems intended for use in potentially explosive atmospheres. *Official Journal of the European Union, L* 100: 1–39. Retrieved from https://eur-lex.europa.eu/.

EC (1998). Council Directive 98/83/EC of 3 November 1998 on the quality of water intended for human consumption. *Official Journal of the European Union, L* 330: 32–54.

Eckenrode, B.A. (2001). Environmental and forensic applications of field-portable GC-MS: An overview. *Journal of the American Society for Mass Spectrometry* 12: 683–693. https://doi.org/10.1016/S1044-0305(01)00251-3.

EPA (2009). *National Primary Drinking Water Regulations*. (EPA 816-F-09-004). Online. US Environmental Protection Agency (EPA). Retrieved from https://www.epa.gov/.

EPA (2016). *Petroleum Refinery Fenceline Monitoring Stakeholder Engagement*. Online. US Environmental Protection Agency (EPA). Retrieved from https://www.epa.gov/.

EPA (2017a). *Method 1 – Sample and Velocity Traverses for Stationary Sources*. Online. US Environmental Protection Agency (EPA). Retrieved from https://www.epa.gov/.

EPA (2017b). *Method 2 – Determination of Stack Gas Velocity and Volumetric Flow Rate (Type S Pitot Tube)*. Online. US Environmental Protection Agency (EPA). Retrieved from https://www.epa.gov/.

EPA (2017c). *Method 3 – Gas Analysis for the Determination of Dry Molecular Weight*. Online. US Environmental Protection Agency (EPA). Retrieved from https://www.epa.gov/.

EPA (2017d). *Method 4 – Determination of Moisture Content in Stack Gases*. Online. US Environmental Protection Agency (EPA). Retrieved from https://www.epa.gov/.

EPA (2017e). *Method 5F – Determination of Nonsulfate Particulate Matter Emissions from Stationary Sources*. Online. US Environmental Protection Agency (EPA). Retrieved from https://www.epa.gov/.

EPA (2017f). *Method 6C – Determination of Sulfur Dioxide Emissions from Stationary Sources*. Online. US Environmental Protection Agency (EPA). Retrieved from https://www.epa.gov/.

EPA (2017g). *Method 10 – Determination of Carbon Monoxide Emissions from Stationary Sources*. Online. US Environmental Protection Agency (EPA). Retrieved from https://www.epa.gov/.

EPA (2017h). *Method 16A – Determination of Total Reduced Sulfur Emissions from Stationary Sources*. Online. US Environmental Protection Agency (EPA). Retrieved from https://www.epa.gov/.

EPA (2017i). *Method 16C – Determination of Total Reduced Sulfur Emissions from Stationary Sources*. Online. US Environmental Protection Agency (EPA). Retrieved from https://www.epa.gov/.

EPA (2017j). *Method 25A – Determination of Total Gaseous Organic Concentration Using a Flame Ionization Analyzer*. Online. US Environmental Protection Agency (EPA). Retrieved from https://www.epa.gov/.

EPA (2018). *Method 7E – Determination of Nitrogen Oxides Emissions from Stationary Sources*. Online. US Environmental Protection Agency (EPA) Retrieved from https://www.epa.gov/.

EPA (2019). *SPod Fenceline Sensors Under Development*. Online. US Environmental Protection Agency (EPA). Retrieved from https://www.epa.gov/air-research.

EU (2012). Commission Decision 19 January 2012 on setting up of the European Union Offshore Oil and Gas Authorities Group (2012/C 18/07). *Official Journal of the European Union, C* 18: 8–10. Retrieved from https://euoag.jrc.ec.europa.eu/.

Fanchi, J.R. and Christiansen, R.L. (2017). *Well Logging. In: Introduction to Petroleum Engineering*, 1e. Hoboken, NJ: Wiley.

Frant, M.S. and LaButti, G. (1980). Process infrared measurements. *Analytical Chemistry* 52 (12): 1331A–1343A. https://doi.org/10.1021/ac50062a003.

Gluck, S., Glenn, C., Logan, T. et al. (2003). Evaluation of NO_x flue gas analyzers for accuracy and their applicability for low-concentration measurements. *Journal of the Air and Waste Management Association (1995)* 53 (6): 749–758. https://doi.org/10.1080/10473289.2003.10466208.

InstTools (2020a). Gas filter correlation (GFC) spectroscopy. Retrieved from https://instrumentationtools.com/gas-filter-correlation-gfc-spectroscopy/ (accessed 19 March 2020).

InstTools (2020b). Luft detector principle. Retrieved from https://instrumentationtools.com/luft-detector-principle/ (accessed 19 March 2020).

Knighton, W.B., Herndon, S.C., Franklin, J.F. et al. (2012). Direct measurement of volatile organic compound emissions from industrial flares using real-time online techniques: proton transfer reaction mass spectrometry and tunable infrared laser differential absorption spectroscopy. *Industrial and Engineering Chemistry Research* 51: 12674–12684. https://doi.org/10.1021/ie202695v.

Kosterev, A., Wysocki, G., Bakhirkin, Y. et al. (2007). Application of quantum cascade lasers to trace gas analysis. *Applied Physics B* 90 (2): 165–176. https://doi.org/10.1007/s00340-007-2846-9.

Kueppers, S. and Haider, M. (2003). Process analytical chemistry—future trends in industry. *Analytical and Bioanalytical Chemistry* 376 (3): 313–315. https://doi.org/10.1007/s00216-003-1907-0.

Kyoto (2008). *Kyoto Protocol Reference Manual*. Bonn, Germany: United Nations Framework Convention on Climate Change (UNFCCC).

Martin, P.A. (2002). Near-infrared diode laser spectroscopy in chemical process and environmental air monitoring. *Chemical Society Reviews* 31 (4): 201–210. https://doi.org/10.1039/b003936p.

Mohácsi, A., Bozóki, Z., and Niessner, R. (2001). Direct diffusion sampling-based photoacoustic cell for in situ and on-line monitoring of benzene and toluene concentrations in water. *Sensors and Actuators B: Chemical* 79: 127–131. https://doi.org/10.1016/S0925-4005(01)00857-7.

Nabawy, B.S. and Shehata, A.M. (2015). Integrated petrophysical and geological characterization for the Sidi Salem-Wakar sandstones, off-shore Nile Delta, Egypt. *Journal of African Earth Sciences* 110: 160–175. https://doi.org/10.1016/j.jafrearsci.2015.06.017.

Ofori, B. A., Wei, M., Bai, B., & Al-Kamil, E. H. K. (2019). Predicting mean time to failure in horizontal wells through metal loss analysis and produced water composition. *Paper presented at the Abu Dhabi International Petroleum Exhibition & Conference*, Abu Dhabi, UAE. https://doi.org/10.2118/197521-MS

Olaguer, E.P., Erickson, M.H., Wijesinghe, A., and Neish, B.S. (2016). Source attribution and quantification of benzene event emissions in a Houston ship channel community based on real-time mobile monitoring of ambient air. *Journal of the Air and Waste Management Association* 66 (2): 164–172. https://doi.org/10.1080/10962247.2015.1081652.

Piel, F., Müller, M., Mikoviny, T. et al. (2019). Airborne measurements of particulate organic matter by proton-transfer-reaction mass spectrometry (PTR-MS): A pilot study. *Atmospheric Measurement Techniques* 12: 5947–5958. https://doi.org/10.5194/amt-12-5947-2019.

Sanchez, N.P., Saffari, A., Barczyk, S. et al. (2019). Results of three years of ambient air monitoring near a petroleum refinery in Richmond, California, USA. *Atmosphere* 10 (7) https://doi.org/10.3390/atmos10070385.

Schlumberger (2008). InSitu Fluid Analyzer: Quantitative fluid measurements at reservoir conditions, in real time. Retrieved from https://www.slb.com/insitu (accessed 20 March 2020).

Shaban, H. and Tavoularis, S. (2014). Measurement of gas and liquid flow rates in two-phase pipe flows by the application of machine learning techniques to differential pressure signals. *International Journal of Multiphase Flow* 67: 106–117. https://doi.org/10.1016/j.ijmultiphaseflow.2014.08.012.

Siemens (2018). In situ continuous process gas analysis. Retrieved from https://cache.industry.siemens.com/ (accessed 20 March 2020).

Spinelle, L., Gerboles, M., Kok, G., and Sauerwald, T. (2015). Sensitivity of VOC sensors for air quality monitoring within the EURAMET key-VOC project. *Paper presented at the 4th Scientific Meeting EuNetAir*, Linkoping, Sweden.

SwRI (2020). Methane leak detection. Retrieved from https://www.swri.org/ (accessed 20 March 2020).

TCEQ (2020a). Geographical Texas air quality monitoring (GeoTAM) viewer. Retrieved from https://tceq.maps.arcgis.com/ (accessed 20 March 2020).

TCEQ (2020b). TCEQ data and records. Retrieved from https://www.tceq.texas.gov/agency/data (accessed 20 March 2020).

Wasson (2019). Can your online process gas chromatograph do this?! Retrieved from https://wasson-ece.com/ (accessed 20 March 2020).

Waters, T. (2013). *Industrial Sampling Systems: Reliable Design & Maintenance for Process Analyzers*. Solon, OH: Swagelok Company.

Waters, T. (2020). *Process Gas Chromatographs: Fundamentals, Design and Implementation*. Hoboken, NJ: Wiley.

Yang, M. (2006). Oil in produced water analysis and monitoring in the North Sea. *Paper presented at the SPE Annual Technical Conference and Exhibition*, San Antonio, Texas, USA. https://doi.org/10.2118/102991-MS

Yun, Z. (2017). The application of portable GC-MS on the petrochemical wharf. *Journal of Chromatography and Separation Techniques* 8: 1000374. https://doi.org/10.4172/2157-7064.1000374.

Zou, X., Kang, M., Li, A. et al. (2016). Spray inlet proton transfer reaction mass spectrometry (SI-PTR-MS) for rapid and sensitive online monitoring of benzene in water. *Analytical Chemistry* 88: 3144–3148. https://doi.org/10.1021/acs.analchem.5b04301.

Part IV

Special Cases and Examples Related to the Petroleum Industry

8

Tracers for Oil and Gas Reservoirs

William L. Winniford[1] *and Melissa N. Dunkle*[2]

[1] *The Dow Chemical Company, Analytical Science, 230 Abner Jackson Parkway, Lake Jackson, TX, USA*
[2] *Dow Benelux, BV, Analytical Science, Herbert H. Dowweg 5, Building 446, Hoek, The Netherlands*

8.1 Introduction

The use of tracers to track fluid flow in oil reservoirs dates back 60 years, driven by oil company geologists working to gain a better understanding of geologic formations. This practice has become increasingly important with the growth of fracking and enhanced oil recovery. Figure 8.1 depicts the strata in an oil and gas producing formation as well as water that is impacted by the operations (McIntosh et al. 2019).

For fracking, tracers can provide essential information on the effectiveness of the fracturing underground. In enhanced oil recovery, tracers provide information on channeling in water floods, CO_2 floods, or chemical recovery aids (i.e. surfactants); with each of these approaches, the aim is to maximize the yield of oil that can be recovered from a reservoir. Tracers can be water-soluble, oil-soluble, or gases depending on the geology of the reservoir and the intent of the tracing study. Modeling of tracer movement is an ever-growing technique and well beyond the scope of this chapter. Early tracers were radioactive due to the sensitivity needed and the cost of delivering a detectable amount of tracer. Zemel has published a comprehensive work on tracers that covers both the radioactive and non-radioactive versions (Zemel 1995). Safe handling of radioactive tracers is essential, but the use is so widespread that there have been inevitable cases of poor stewardship. Quality control has also been an issue in the industry. With the reputation of the oil and gas industry, it is not surprising that there is public concern over the use and disposal of radioactive tracers. The trend has been to use tracers that are viewed as environmentally benign, but the history of the chemical industry has taught us that what is considered safe at one point in time will likely be regulated or prohibited for certain uses in the future. The interaction of man-made

Analytical Techniques in the Oil and Gas Industry for Environmental Monitoring, First Edition.
Edited by Melissa N. Dunkle and William L. Winniford.

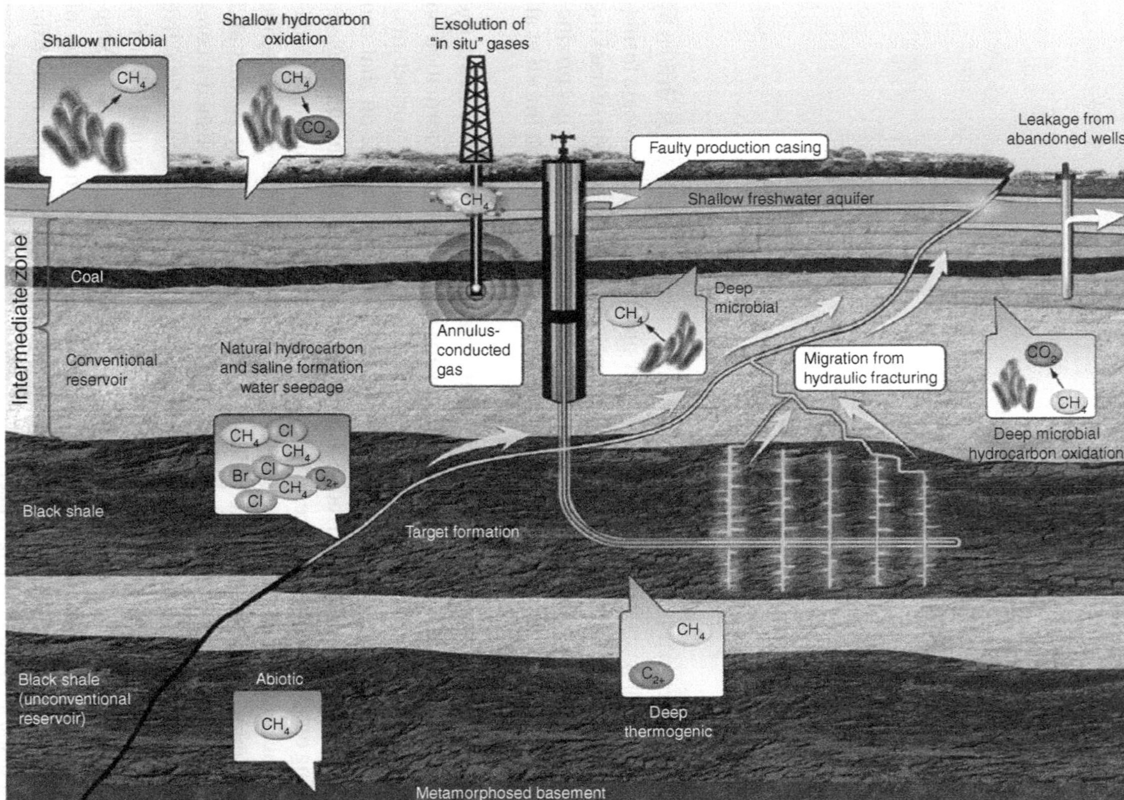

Figure 8.1 Conceptual model of sources and pathways of natural gas and associated formation water migration from multiple reservoirs at depth in *geologic basins* into *shallow groundwaters*, and biodegradation of hydrocarbons *via aerobic* and/or *anaerobic microbial* oxidation. Source: McIntosh et al. (2019). Reproduced with permission, Copyright (2019), American Chemical Society.

chemicals in the environment is difficult to assess comprehensively at the time of initial use. It is only in the retrospect of enough years of observation that the true impact may be known. Though petroleum reservoirs are typically much deeper than drinking water aquifers, there is growing public anxiety over the possibility of drinking water contamination from oil field operations. With the proliferation of oil exploration, transport, and processing, there is growing potential for contamination of surface water and aquifers that supply drinking water. Akstinat has given an excellent review of the differences in reservoir water, connate water, production water, etc. and how formations may be damaged through improper treatment (Akstinat 2019). Though practices for oil and gas production have improved over the years in first world nations, there is an ever-increasing infrastructure that is aging: pipelines, produced water ponds, and underground storage tanks, to name a few. For developing nations regulations are less stringent and resources to enforce them are limited, leading to significant abuse. The approach in this chapter is based on the likelihood that reservoir tracers will face increased environmental scrutiny or be used to validate interaction between oil reservoirs and drinking water aquifers over concern about other chemicals used in oil and gas production. There is already a growing parallel field of research on the use of tracers to study aquifers. Examples are given of tracer use to aid in oil field production and use in studying underground aquifers.

The types of tracers currently used and analytical methods will be described in the following sections.

A passive tracer that labels gas or water in a well-to-well tracer test must have several characteristics and must fulfill the following criteria (IAEA 2004).

- Have a very low detection limit
- Be stable under reservoir conditions
- Follow the phase that is being tagged and have a minimal partitioning into other phases
- Have no adsorption to rock material
- Have minimal environmental consequences

Partitioning tracers are commonly used to determine the oil in place in a reservoir, making use of the partitioning between the oil and water phase (Serres-Piole et al. 2012). They can be injected simultaneously with passive tracers to determine the difference in time arriving at the detection well. Partitioning tracers are subject to the same criteria mentioned earlier for the passive tracers but the chemical functionality required for partitioning makes adsorption on to rock more likely, depending on the geology of the reservoir. It is understood that cost is always a consideration for any tracer. Thus improving analytical detection limits can have a significant impact on the cost of such studies.

For much of the tracer use history, radioactive tracers were favored because the detection technology was much more sensitive than methods for non-radioactive chemical tracers. Thus, much smaller amounts of a tracer could be used. For example, several milliliter of tritiated water may be sufficient for a waterflood of closely spaced wells, while thousands of pounds of a non-radioactive anionic tracer would be needed for detection by ion chromatography (Zemel 1995). Scintillation counters and ionization detectors are relatively simple to use, and relatively little adjustment to methods are needed when switching between radioactive tracers. The National Lab programs initiated during World War II drove the investigation of radioactive isotopes and the development of sensitive detection methods. Since that time, medical applications have taken the lead in driving improvement in radiochemical technology. In more recent years, advances in detection for ion and liquid chromatography, as well as pre-concentration techniques, have made it possible to detect anions in the low ppb range. Advances in inductively coupled plasma–mass spectrometry (ICP-MS) have made it possible to measure part per trillion (ppt) levels of most transition metals, alkali, and alkaline earth elements. Also, advances in mass spectrometry have made possible low ppt detection limits, particularly with triple quad mass spectrometers becoming commonplace, meaning such instrumentation can be operated with a modest skill level. Thus far, there have been few published works using triple quad mass spectrometers for environmental monitoring of oil field contaminants, but they are routinely used for other environmental analyses, and it is likely that the service labs that do interwell testing are already using this technology.

Local regulations limit the amount of specific chemical and radioactive species that may be injected. There is a balance between injecting the smallest amount possible to minimize cost and injecting enough to ensure that the tracer can be quantified by the appropriate analytical technique. The minimum concentration in the diluted sample must be 10× the limit of detection for reliable quantitation, and often, higher concentrations are injected to limit the possibility that the tracer would not be detected.

8.2 Types of Tracers

8.2.1 Radioactive Water Tracers

Tritiated water (HTO) has some of the most desirable properties as a water tracer because it acts as a nearly perfect passive tracer with the same adsorption and partitioning properties of water. HTO has a half-life of 12 years, the cost is relatively low, and it is a low energy β emitter (does not penetrate human skin). The hazards of working with HTO are low if handled with appropriate personal protective equipment and engineering controls. The primary hazard is absorption through the skin; the half-life in the body is seven days. The only case where it would not be the best tracer choice is in reservoirs, where the natural abundance of tritium is already high. Using a liquid scintillation counter (LSC), HTO has a limit of detection of 2–4 Bq/l. Table 8.1 gives the most common radiolabeled water tracers. Tritiated water, thiocyanate, hexacyanocobalt, and iodide, historically, account for the majority of the tests.

Table 8.1 Overview of the most common radiolabeled water tracers in oil reservoir examinations.

Water tracer compound/ion	Half-life	Main radiation characteristics	Comments
HTO	12.32 a	β^- (18 keV)	Generally applicable
CH_2TOH[a)]	12.32 a	β^- (18 keV)	Caution at temperatures >100 °C (partition into gas phase)
CH_3CHTOH[b)]	12.32 a	β^- (18 keV)	Caution at temperatures >100 °C (partition into gas phase), some biodegradation below 70–80 °C
$CH_3CTOHCH_3$	12.32 a	β^- (18 keV)	Reasonably general application
$CH_3CH_2CH_2CHTOH$	12.32 a	β^- (18 keV)	Reasonably general application, some partition into oil
$S_{14}CN^-$	5730 a	β^- (156 keV)	For $T < 90–100$ °C, long-term experiments
$^{35}SCN^-$	87 d	β^- (167 keV)	For $T < 90–100$ °C, medium-term experiments
$^{36}Cl^-$	3×10^5 a	β^- (709 keV)	High-temperature reservoirs, long term, EMS analysis
$^{125}I^-$	60 d	γ (35.5 keV), e^-	Reducing chemical conditions, medium-term
$^{131}I^-$	8 d	β^- (606 keV), γ (364.5 keV)	Reducing chemical conditions, short term (fracture detection)
$^{56}Co(CN)_6{}^{3-}$	77.7 d	β^+ (1459 keV), γ (846.8 keV, 1238.29 keV)	Use with caution at $T < 90$ °C, medium term
$^{57}Co(CN)_6{}^{3-}$	271.8 d	γ (122.1 keV, 136.5 keV)	Use with caution at $T < 90$ °C, medium to long term
$^{58}Co(CN)_6{}^{3-}$	70.9 d	β^+ (470 keV), γ (810.8 keV)	Use with caution at $T < 90$ °C, medium term

Table 8.1 (Continued)

Water tracer compound/ion	Half-life	Main radiation characteristics	Comments
$^{60}Co(CN)_6^{3-}$	5.27 a	β^- (317.9 keV), γ (1173.2 keV, 1332.4 keV)	Use with caution at $T < 90\,°C$, long term
$CO(^{14}CN)(CN)_5^{3-}$	5730 a	β^- (156 keV)	Use with caution at $T < 90\,°C$, long term
$^{35}SO_4^{2-}$	87 d	β^- (167 keV)	Caution, not generally applicable, avoid alkaline earth containing waters
$^{22}Na^+$	2.6 a	β^+ (545 keV), γ (1274.5 keV)	High-temperature tracer in saline reservoirs, long term, slightly reversible sorption
$^{134}Cs^+$	2.065 a	β^- (658 keV), γ (604.7 keV, 795.8 keV)	Caution, not generally applicable, sorption on clays
$^{137}Cs^+$	30.2 a	β^- (512 keV), γ (661.6 keV)	Caution, not generally applicable, sorption on clays

EMS, electrochemically modulated separations.
a) The alcohol tracers may also be labeled with ^{14}C.
b) The position of tritium in the alcohol compounds heavier than methanol may vary.
Source: Reproduced with permission, Copyright (2004), International Atomic Energy Agency (IAEA) (2004).

8.2.2 Radioactive Gas Tracers

The common gas tracers are given in Table 8.2, with tritiated methane and ^{85}Kr being most commonly used. For studies where partitioning into the oil phase is desired, the tritiated higher alkanes are used.

8.2.3 Radioactive Measurement Techniques

The most common radioactive tracers for interwell studies emit β particles (tritium (^3H), ^{14}C, or ^{35}S). Liquid scintillation counting is used to measure the radiation. A scintillation cocktail (e.g. anthracene in a nonpolar solvent) is added to the sample. Emitted β particles excite the scintillation cocktail molecules, which in turn emit light as fluorescence as they return to the ground state. This light is detected by a photomultiplier tube (PMT), where the intensity of the light is proportional to the energy of the β particle. Both the intensity and rate are recorded. The energy is specific to the type of β emitter, and by monitoring specific portions of the energy distribution, noise from background radiation can be reduced. LSC have been optimized to improve sensitivity by using a pair of

Table 8.2 Overview of the most common radiolabeled gas tracers in oil reservoir examinations.

Gas tracer compound	Half-life	Main radiation characteristics	Comments
CH_3T[a]	12.32 a	β^- (18 keV)	First choice radioactive gas tracer, long-term experiments, high-temperature reservoirs
CH_2TCH_3[a]	12.32 a	β^- (18 keV)	Applicable gas tracer, long-term experiments, slight isotope exchange of the T label to H containing media
CH_3CHTCH_3[a]	12.32 a	β^- (18 keV)	Reasonable gas tracer, long-term experiments, unknown T exchange, larger partitioning into oil
$CH_3CHTCH_2CH_3$[a]	12.32 a	β^- (18 keV)	Can be used if necessary, long term, unknown T exchange, high partitioning into oil
^{85}Kr	10.76 a	β^- (687 keV)	Long history in reservoir examinations, chemically inert, high temperatures, long-term experiments, dynamic behavior close to that of methane at reservoir conditions
$^{14}CO_2$	5730 a	β^- (156 keV)	Applicable tracer for injection gas, ideal for CO_2 injection, long-term experiments, high temperatures
^{133}Xe	5.25 d	β^- (346 keV), γ (81 keV)	Chemically inert, high temperatures, dynamic reservoir movement rate in-between methane and ethane, usable for detection of high permeability streaks and "super" conducting flow channels, detectable until 25 d after injection
^{127}Xe	36.4 d	γ (202.9 keV, 172.1 keV, 375.0 keV)	Dynamic behavior as for ^{133}Xe, applications as for ^{133}Xe, but may be detected until 180 d after injection

a) These molecules may also be labeled by ^{14}C.
Source: Reproduced with permission, Copyright (2004), International Atomic Energy Agency (IAEA) (2004).

PMTs and surrounding the detectors with lead shielding to minimize background radiation. The detection limits are limited by the background signal and the length of counting time used (IAEA 2004). Though most commercial instruments are not portable and are intended to be used in the laboratory, however, field-portable LSCs are becoming available.

γ tracers are measured using solid scintillation detectors or semiconductors. In the former, incident γ rays strike a crystal (NaI doped with thallium). The interaction of the γ photon creates an excited state in the crystal, which emits light as it relaxes to the ground state. As in the LSC, a PMT is used to detect the light intensity (proportional to energy). These instruments are easily made into field-portable instruments, and they can also be used in downhole wells to provide real-time information on tracer studies. Semiconductor detectors (germanium) are doped with impurities so that incident γ rays will excite electrons from the valence band to the conduction band. The resulting electrical pulse in a high voltage field is proportional to the energy of the γ-ray. As with the NaI (Tl) detector, the signal is collected with a multichannel analyzer that can record the γ energy spectrum. The NaI (Tl) detector has high intrinsic counting efficiency but lower energy resolution. Semiconductor detectors have steadily improved to provide similar sensitivity with much higher energy resolution, but this comes at a higher cost, and the requirement for liquid nitrogen cooling precludes portability (IAEA 2004).

8.2.4 Example Studies of Radioactive Tracers

A recent study in California (McMahon et al. 2019) has shown that radium levels in produced water can be 100- to 1000-fold higher than the established limit by the US EPA of 0.185 Bq/l (5 pCi/l). Produced water that is pumped into unlined ponds can enter drinking water aquifers. The area studied was in the San Joaquin Valley of California, and similar issues can be found in unconventional oil plays in the Marcellus and Bakken formations. In this case, radium is both the tracer and the pollutant of concern.

8.2.5 Chemical Water Tracers

Thiocyanate (SCN^-) is the most common chemical water tracer, typically measured to 1 ppb via liquid chromatography with electrochemical detection (LC-EC) or ion chromatography with electrochemical detection (IC-EC).

Nitrate (NO_3^-), bromide (Br^-), iodide (I^-), hydrogen borate (HBO_3^-), and fluorobenzoic acids (2-fluoro, 4-fluoro, 2,6-difluoro) can all be measured in the sub-ppm range. With ion-exchange preconcentration, the detection limits can be extended to low ppb levels. DNA fragments, deuterated fatty acids, and fluorescent nanoparticles are newer tracers being explored to overcome problems with the incumbent technology.

Ugbena used NaCl at a concentration of 66 g/l as a tracer to monitor the groundwater movement affected by oil spills in the Niger Delta (Ugbena et al. 2019). Conductivity was used as the detection method; as such, background conductivity was measured in the test wells to establish the baseline conductivity prior to injection. With the rate of groundwater movement and size of the area measured, the peak in conductivity arrived at the outlying wells within four days of introduction to the injection well. The results were used to design the long-term groundwater monitoring program.

Eosin Y (3 g/l) was used in well to well tracer tests in Tatarstan along with a field-portable laser spectrofluorimeter (Shipaeva et al. 2019). The advantage of Eosin Y, in this case, is the increased fluorescent intensity as it partitioned into the oil phase. But fluorescent indicators are notoriously influenced by the matrix they are in, so users should carefully validate the fluorescent response in the respective fluids being evaluated.

A patent has been issued by Aramco on the use of substituted polystyrenes as tracers to more closely mimic the behavior of oil in the reservoirs (Cox 2019). The approach on this is unique because pyrolysis gas chromatography–mass spectrometry (GC-MS) was used to identify the specific substituted polystyrene markers. Water samples can be analyzed directly because the pyrolyzer can be operated to vaporize the water prior to analysis. Subsequently, the remainder of the sample is pyrolyzed, giving characteristic fragments of the substituted styrene monomer used to make the polystyrene marker.

Akondi used phospholipids and diglyceride fatty acids (DGFAs) as markers for studying deep subsurface microbial communities in black shale formations (Akondi 2019). Appropriate extraction and storage conditions are essential to achieving accurate results. After extraction, the DGFAs were converted to the fatty acid methyl esters (FAMEs) and then analyzed by GC-MS according to well-known procedures.

8.2.6 Chemical Gas Tracers

As with other tracers, sensitivity, cost, and stability are the most important attributes of chemical gas tracers. Perfluorocarbons, freons, and sulfur hexafluoride (SF_6) can be easily detected with GC-MS with selected ion monitoring or gas chromatography–electron capture detector (GC-ECD) for low or sub ppb detection. Freons were the first to come under scrutiny due to the potential for depleting ozone. There is growing public concern over the use of other fluorinated hydrocarbons. This leaves SF_6 as the primary option for an inert fluorinated gas marker. It is also desirable because it is easily separated from other permanent gases using a molecular sieve GC column. Deuterated methane, ethane, and propane have also been used, as they have the advantage of nearly identical physical properties to methane, ethane, and propane. While these tracers can be readily detected via GC-MS, their relatively high cost limits widespread use.

8.2.7 Naturally Occurring Tracers

Naturally occurring tracers must meet several requirements to be useful. Specifically, the production water and injection water must have statistically significant differences, and the naturally occurring tracers must be chemically stable during the measurement period. In addition, the fluid samples must be well characterized to verify whether interference exists.

8.2.7.1 Isotopes

Flude et al. (2016) has proposed that ^{13}C isotope ratios, along with natural gas levels, can be used to measure the source of CO_2 when it is reinjected into reservoirs. There is a growing interest in doing this as a way for CO_2 sequestration, but CO_2 floods are also used for enhanced oil recovery. Either way, there can be concern about a breakthrough to aquifers even if the projected risk is low. The authors show that there is enough variation in the relative amounts of ^{13}C with the source of CO_2 production – both natural and manmade – that the source can be identified using this and the noble gas concentrations. Another approach in measuring leakage of CO_2 from reservoirs is to measuring methane and ^4He that is released as the CO_2 dissolves in formation water and lowers the pH, thereby releasing permanent gases from carbonate or siltstone formations that are in the containment zone (Amonette et al. 2019). Mass spectrometry and gas chromatography were used to quantify the gases, including isotopes of the noble gases.

Frick et al. examined interconnectivity between aquifers at different depths under Berlin using naturally occurring isotopes (Frick et al. 2019). Though the focus of the work is not primarily on impact from the oil and gas industry, the findings are relevant because there can be discontinuities in geologic layers that are presumed to isolate deep aquifers from drinking water aquifers. Heavy pumping of drinking water can provide conditions that lead to an upwelling of deep aquifers and inter-aquifer flow.

Donner investigated whether arsenic and selenium transport into surface waters in the Athabasca River in Northern Alberta occurs as a natural process or as a result of oil sands mining (Donner 2019). Surface mining of tar sands can be considered as a reservoir and arsenic–selenium transport as a tracer for the indicator, as well as impacting on the environment. Steam assisted gravity drainage (SAGD) is growing rapidly as a technique to recover heavy oil from subsurface formations that can range from relatively shallow to very deep. The process of extracting oil via SAGD increases the permeability of the strata and the probability of impacting aquifers increases. ICP-MS was essential because the typical concentrations were sub ppb. The work concluded that arsenic and selenium in the Athabasca watershed was a result of natural leaching and not increased by the mining of bitumen. This work also detailed the importance of sampling protocol and quality control checks to establish accurate data.

Analysis of Ca–Mg versus Ca–Sr has been performed by cluster analysis to differentiate water produced from different levels in formations (Tisherman and Bain 2019). Water analysis data from the USGS Produced Water database was analyzed to distinguish water from unconventional production (e.g. fracking) and conventional sources. The authors were able to distinguish between conventional brines, unconventional brines, and non-impacted waters in four sedimentary formations: the Williston, Michigan, Appalachian, and Green River basin.

Lead isotopes ^{204}Pb, ^{206}Pb, ^{207}Pb, ^{208}Pb appear to provide a unique signature for the source of crude oils. In a study documenting the lead isotope ratios, 195 crude oils from around the world and 17 Northern European black shales were analyzed by ICP-MS (Fetter et al. 2019). By plotting the lead isotope data as ratios in three dimensions, $^{204}Pb/^{206}Pb$, $^{207}Pb/^{206}Pb$, $^{208}Pb/^{206}Pb$, each crude occupies a unique position in the 3D space. This is a substantial improvement over previous studies that use Pb, U, and Th ratios because they partition differently between oil and water phases; however, the lead isotope ratios remain constant. While the focus of the work was to improve the understanding of the formation of oil reservoirs, the same approach could be used in environmental forensic work to determine the source of crude oil spills.

Noble gases can be excellent tracers for subsurface fluid migration – water and oil. These have been used, particularly ^{4}He, to determine whether the water in oil fields has had contact with surface water based on the $^{3}He/^{4}He$ values or the $^{4}He/^{20}Ne$ values (Zhang et al. 2019).

8.2.7.2 Biomarkers

Demethylated hopanoids have been studied as biomarkers for nitrate dependent bacteria *Candidatus Methylomirabilis oxyfera* (Smit et al. 2019). Though the thrust of the work was to understand the aging of petroleum in reservoirs, these hopanoids can be readily measured via GC × GC-MS.

Akondi (2019) studied lipid biomarkers of the Marcellus shale that are produced as a result of microbial activity in deep formations (>7000 ft). Phospholipid fatty acids (PLFAs) are microbial lipid biomarkers providing evidence of living microbes, while DGFAs are evidence of nonviable microbes. The fatty acids can be readily determined after extraction and methylation to the FAME, followed by well-established GC-MS methods. Extraction methods have a significant influence on the distribution of fatty acids recovered, particularly substituted fatty acids.

8.3 Regulations

Current regulations primarily address the use of radioactive tracers and the naturally occurring radioactive materials (NORMs) released as a result of oil and gas production. The IAEA recommended exposure limit for the general public is 1 mSv/An (IAEA 2018a). Releases to the environment (solids, liquid, gas) must be limited according to the local regulations if the activity concentration of any radionuclide in the uranium decay chain or the thorium decay chain is greater than 1 Bq/g or if the activity concentration of ^{40}K is greater than 10 Bq/g (IAEA 2018b).

In North America, the oil and gas industry is subject to a combination of regulations from federal, state/provincial, and local governments. The Clean Air Act (1963), the Clean Water Act (1972), and the Safe Drinking Water Act (1974) provide the framework for most federal regulation of the oil and gas industry (Allison and Mandler 2018) in the United States. But within the US federal government, there are more than 10 agencies that have jurisdiction over specific locations. The Interstate Oil and Gas Compact Commission (IOGCC) was formed in 1935 to establish a common set of regulations among states. While in Canada, regulations are driven by the provincial governments, resulting in vastly different laws between Western and Eastern Canada (Millar 2019).

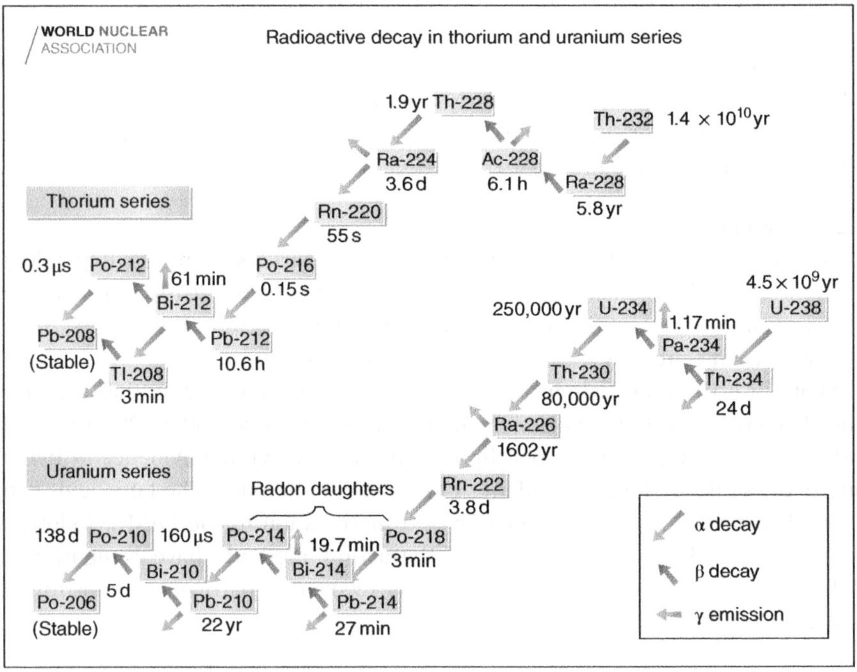

Figure 8.2 Radioactive decay pathway for thorium and uranium series. Source: WNA (2019). Reproduced with permission, Copyright (2019), World Nuclear Association.

In the United Kingdom, the regulations on NORM have become increasingly stringent for the oil and gas industry. Uranium and Thorium are not typically mobilized by oil and gas production, but radium isotopes and ^{210}Pb can be mobilized by produced water. These can be precipitated out as scale in pipelines and processing equipment. As shown in Figure 8.2, ^{226}Ra decays to ^{222}Rn, which as a gas, is readily transported. Since it rapidly decays to the much longer-lived ^{210}Pb, the daughter decay isotope can build up in films in gas processing equipment (WNA 2019).

In summary, current regulations are driven by drinking water standards, air emissions, and worker safety. In principle, most oil and gas production operates at depths far below aquifers used for drinking water, but as depicted in Figure 8.1, there are multiple pathways (fault lines, well casings, and gas transmission lines) that can introduce contaminants into drinking water aquifers. Proprietary tracers and reservoir modeling in the oil and gas industry continue to develop to improve production, while in parallel, there is a very active and growing field of research to use tracers to monitor the impact on water and air.

References

Akondi, R. (2019). Utilizing lipid biomarkers to understand the microbial community. PhD thesis. West Virginia University, Morgantown, WV, USA.

Akstinat, M. (2019). Chemical and physicochemical properties of formation waters of the oil and gas industry. *Journal of Hydrology* 578: 124011. https://doi.org/10.1016/j.jhydrol.2019.124011.

Allison, E. and Mandler, B. (2018). *Petroleum and the Environment* (A. G. Institute Ed.). American Geosciences Institute.

Amonette, J.E., Zhong, L., Darrah, T.H. et al. (2019). Noble and major gases released from rock core materials as intrinsic tracers for detecting carbon dioxide leakage – laboratory evaluation. *International Journal of Greenhouse Gas Control* 89: 76–88. https://doi.org/10.1016/j.ijggc.2019.05.010.

Cox, J.R. (2019). Polymeric tracers. US Patent US2019/0234200, filed 12 April 2019 and issued 10 September 2019.

Donner, M. W. (2019). Natural and anthropogenic inputs of arsenic and selenium to the environment in the Athabasca Bituminous Sands Region. PhD thesis. University of Alberta, Edmonton, Alberta, Canada.

Fetter, N., Blichert-Toft, J., Ludden, J. et al. (2019). Lead isotopes as tracers of crude oil migration within deep crustal fluid systems. *Earth and Planetary Science Letters* 525 (115747): 1–9. https://doi.org/10.1016/j.epsl.2019.115747.

Flude, S., Johnson, G., Gilfillan, S.M., and Haszeldine, R.S. (2016). Inherent tracers for carbon capture and storage in sedimentary formations: composition and applications. *Environmental Science and Technology* 50 (15): 7939–7955. https://doi.org/10.1021/acs.est.6b01548.

Frick, M., Scheck-Wenderoth, M., Cacace, M., and Schneider, M. (2019). Boundary condition control on inter-aquifer flow in the subsurface of Berlin (Germany) – new insights from 3-D numerical modelling. *Advances in Geosciences* 49: 9–18. https://doi.org/10.5194/adgeo-49-9-2019.

IAEA (2004). *Radiotracer Applications in Industry: A Guidebook*. Vienna: International Atomic Energy Agency (IAEA).

IAEA (2018a). *Radiation Protection for the Public and the Environment*. Vienna, Austria: International Atomic Energy Agency (IAEA).

IAEA (2018b). *Regulatory Control of Radioactive Discharges to the Environment* (GSG-9). Vienna, Austria: International Atomic Energy Agency (IAEA).

McIntosh, J.C., Hendry, M.J., Ballentine, C. et al. (2019). A critical review of state-of-the-art and emerging approaches to identify fracking-derived gases and associated contaminants in aquifers. *Environmental Science and Technology* 53 (3): 1063–1077. https://doi.org/10.1021/acs.est.8b05807.

McMahon, P.B., Vengosh, A., Davis, T.A. et al. (2019). Occurrence and sources of radium in groundwater associated with oil fields in the Southern San Joaquin Valley, California. *Environmental Science and Technology* 53 (16): 9398–9406. https://doi.org/10.1021/acs.est.9b02395.

Millar, H.L. (2019). *Managing Uncertainty: Risk Narratives and Learning in Provincial Hydraulic Fracturing Regulation in Canada 2006–2016*. Toronto, Canada: University of Toronto.

Serres-Piole, C., Preud'homme, H., Moradi-Tehrani, N. et al. (2012). Water tracers in oilfield applications: Guidelines. *Journal of Petroleum Science and Engineering* 98–99: 22–39. https://doi.org/10.1016/j.petrol.2012.08.009.

Shipaeva, M.S., Sudakov, V.A., and Akhmadullin, R.R. (2019). Analysis of the results of tracer tests for the monitoring of the development of super-viscous oil deposit. *IOP Conference Series: Earth and Environmental Science* 282: 012042. https://doi.org/10.1088/1755-1315/282/1/012042.

Smit, N.T., Rush, D., Sahonero-Canavesi, D.X. et al. (2019). Demethylated hopanoids in 'Ca. Methylomirabilis oxyfera' as biomarkers for environmental nitrite-dependent methane oxidation. *Organic Geochemistry* 137: 103899. doi: 10.1016/j.orggeochem.2019.07.008.

Tisherman, R. and Bain, D.J. (2019). Alkali earth ratios differentiate conventional and unconventional hydrocarbon brine contamination. *Science of the Total Environment* 695 (133944): 1–9. https://doi.org/10.1016/j.scitotenv.2019.133944.

Ugbena, K.G., Akpokodje, E.G., and Tse, A.C. (2019). Radial flow tracer test to investigate coefficient of transverse dispersion in an aquifer in the Eastern Niger Delta, Nigeria. *World Scientific News* 132: 233–243.

WNA (2019). Naturally-occurring radioactive materials (NORM). Retrieved from https://www.world-nuclear.org/information-library/safety-and-security/radiation-and-health/naturally-occurring-radioactive-materials-norm.aspx (accessed 20 March 2020).

Zemel, B. (1995). *Tracers in the Oilfield*. Amsterdam: Elsevier.

Zhang, W., Li, Y., Zhao, F. et al. (2019). Quantifying the helium and hydrocarbon accumulation processes using noble gases in the North Qaidam Basin, China. *Chemical Geology* 525: 368–379. https://doi.org/10.1016/j.chemgeo.2019.07.020.

9

Environmental Impact of Emissions Originating from the Petroleum Industry

Melissa N. Dunkle[1] *and William L. Winniford*[2]

[1] *Dow Benelux, BV, Analytical Science, Herbert H. Dowweg 5, Building 446, Hoek, The Netherlands*
[2] *The Dow Chemical Company, Analytical Science, 230 Abner Jackson Parkway, Lake Jackson, TX, USA*

9.1 Global Warming

In this chapter, the authors want to provide real-world and relatable examples of how various aspects of the petroleum industry are responding to the strict emission standards implemented globally. Detailed examples related to improvements in diesel emissions and fossil-based power production will be provided, and the link to selected analytical techniques described earlier in the book, together with their use for environmental impact monitoring, will be provided.

Global warming is a serious phenomenon affecting our planet. Surface air temperature is the primary measure of global climate change, and as can be seen from Figure 9.1, the global surface temperature of the planet has risen by 0.8 °C since 1980.

9.1.1 Causes of Global Warming

The primary cause of global warming has been attributed to air pollution and greenhouse gases (GHGs). GHGs are those gases that trap heat in the atmosphere and include carbon dioxide (CO_2), methane (CH_4), nitrous oxide (N_2O), and fluorinated gases such as hydrofluorocarbons (HFCs) and chlorofluorocarbons (CFCs). Figure 9.2 shows the 2011 global GHG emissions (IPCC 2014).

Air pollution occurs when harmful or excessive amounts of materials are introduced into the atmosphere and can include gases, particulate, and biological matter. Air pollutants include the GHGs and many other harmful substances, including sulfur oxides (SO_x), nitrogen oxides (NO_x), volatile organic compounds

Analytical Techniques in the Oil and Gas Industry for Environmental Monitoring, First Edition.
Edited by Melissa N. Dunkle and William L. Winniford.

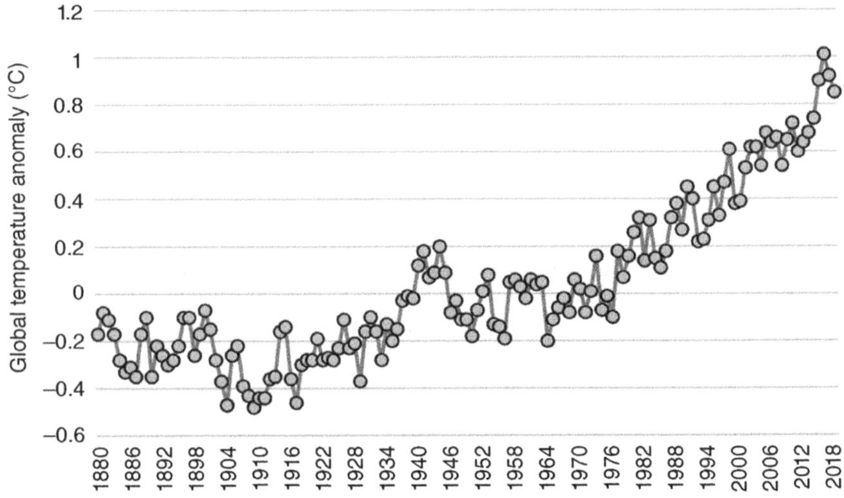

Figure 9.1 Global temperature from 1880 to 2018, normalized to the 1951–1980 average. Source: Graph created using data courtesy of NASA/JPL-Caltech (Shaftel et al. 2019).

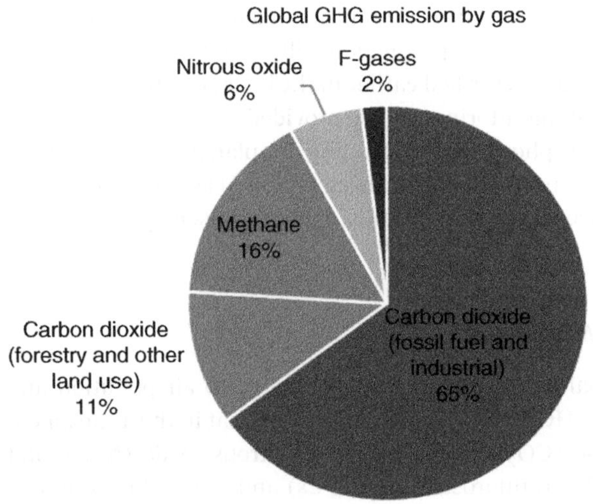

Figure 9.2 2011 Global GHG emissions. Source: Reproduced with permission of IPCC (2014).

(VOCs), particulate matter (PM), persistent free radicals, ammonia, odors, and radioactive pollutants. When SO_x and NO_x are present in the atmosphere, reaction with air forms sulfuric acid (H_2SO_4), sulfurous acid (H_2SO_3), and nitrous acid (HNO_2), which are contributors to acid rain. In addition, NO_2 and PM contribute to smog.

Air pollution has many sources, both natural and man-made, or anthropogenic. Natural sources include dust from areas of little to no vegetation, methane from livestock (cattle are one example), radon gas that is naturally decaying in the Earth's crust, and the release of sulfur, chlorine, and ash particulate through volcanic activity. Anthropogenic sources include manufacturing facilities or factories that contain furnaces for burning fossil fuels (e.g. burning coal and natural gas to create electricity), the transportation industry (personal vehicles, transport vehicles, marine vessels, and aircraft), the controlled burning of fields and forests, methane generation from landfills, NO_x from the fertilization of farmlands, and the military.

9.1.2 Combatting Global Warming

In 1970, the US Congress passed the Clean Air Act, which was meant to reduce harmful emissions from automobiles and industry (CAA 1990). While this act was not focused on combatting global warming, it was focused on improving air quality. Based on this act, standardized emission limits were developed, allowing for measurement and enforcement. In 1990, an amendment to the Clean Air Act was passed, which called for enforcement on acid rain, specifically focusing on the strict reduction of hydrocarbons (HCs), SO_x, CO, NO_x, and PM in the United States.

The United Nations Framework Convention on Climate Change (UNFCCC) is an international environmental treaty initiated in 1992, with the intention of stabilizing GHG concentrations in the atmosphere to levels where anthropogenic sources would not affect the climate. Many countries participate in the UNFCCC, and regularly updated information can be found on their webpage (https://unfccc.int). This framework, which still exists today, does not enforce the set limitations, rather it helps to outline specific legislation. Since 1995, the parties involved in the UNFCCC have met annually at the Conference of the Parties (COP) to assess global progress in dealing with climate change.

At the third such COP meeting in Kyoto, Japan, in 1997, the Kyoto Protocol was drafted and signed, calling for industrialized nations to limit or reduce their GHG emissions by up to 8% (Kyoto 2008). This was the first time that binding commitments were agreed upon, and these financial commitments would be used to finance projects and programs in developing nations. The commitment period ran from 2008 to 2012.

In a continuing effort after the end of the Kyoto Protocol, when the COP met in Paris in 2015, a new agreement was drawn up and signed. The Paris Agreement aimed to strengthen the global response against climate change while requiring sustainable efforts to prevent the global surface temperature from rising further (Paris 2015).

One area where governments are looking to reduce air pollution is from motor vehicles, and especially diesel engines, as motor vehicles add considerably to the levels of GHGs and air pollution in the atmosphere (OECD 2002). Diesel engines are used globally to transport goods, to power much of the world's equipment, and to generate economical electricity; however, diesel-powered engines are among the largest contributors to environmental pollution worldwide (Lloyd and Cackette 2011). Diesel engines have evolved throughout the years in order to reduce emissions, and improvements to diesel fuel have also been made. The evaluation of some of these improvements will be discussed.

Another area of focus is the global burning of fossil fuels to generate electricity. As a fuel source, coal and natural gas are abundant and relatively inexpensive to produce and convert into energy; however, sourcing the fossil fuels as well as combusting them into usable energy have negative effects on the environment. Advances have been made throughout the years to reduce emissions and the environmental impact from the sourcing and burning of fossil fuels; a detailed discussion will be included later in this chapter.

9.2 Environmental Impact of Diesel Emissions

9.2.1 Diesel Engine

The diesel engine is a type of compression – ignition engine that was developed by Rudolf Diesel (1913). Diesel engines are internal combustion engines, where the ignition of fuel occurs in the presence of elevated temperature of the air in the cylinder, which is caused by mechanical compression. In terms of the energy produced per unit fuel, diesel engines are one of the most efficient engines available; it is for this reason that they are so widely implemented. Diesel engines have been used since the early 1900s in submarines and ships (1920s), trucks (1930s), locomotives (1950s), heavy equipment (1910s), and to generate electricity in power plants (Bryant 1976). In the 1930s, diesel engines first appeared in automobiles; however, it was not until the 1970s, with the incorporation of direct fuel injection into the engine design, that the implementation of diesel engines in larger vehicles started to take off (Zima 2001).

9.2.2 Diesel Exhaust

Even before the US Government passed the Clean Air Act in 1970, there was a growing concern about diesel exhaust emissions and their contribution to air pollution. In 1960, the state of California established the Motor Vehicle Pollution Control Board, as well as a program to control air pollution from motor vehicles

(Clarkson and Middleton 1962). However, such a program requires accurate quantification of exhaust from motor vehicles, both gasoline and diesel. As gasoline engines were more popular at this point in time, extensive data were available in the literature on exhaust from gasoline engines; however, this was not the case for diesel engines. One of the first in-depth comparison studies of gasoline and diesel engines was made by McKee, Clark, and Wheeler, which showed that the emission profiles between combustion engines and conventional gasoline engines differed (McKee et al. 1962). Over the years, standardized methods have been developed for the measurement of automotive exhaust; the International Organization for Standardization (ISO) has 41 published methods available for the measurement of transport exhaust emissions, which includes methods for agricultural tractors, road vehicles, motorcycles, and mopeds, to name a few. While ISO has several standardized methods for diesel engines, ISO 16183 details the standardized methodology for the measurement of diesel exhaust (ISO:16183 2002).

Through better measurements of diesel exhaust, a better understanding of the composition, and thus potential health effects became clearer. In 1988, the International Agency for Research on Cancer (IARC), which is part of the World Health Organization (WHO), classified diesel exhaust as probably carcinogenic to humans; however, after re-evaluation in 2012, re-classified diesel exhaust as carcinogenic to humans (IARC 2012). Thus, it is clear to see why governments are looking to reduce air pollution and GHGs created by diesel engines.

The typical composition of the exhaust from modern diesel engines is described in detail in Table 9.1; briefly, 99% is made up of CO_2, H_2O, N_2, and O_2, and the remaining 1% is made up of CO, HC, NO_x, and PM (CIMAC 2008; Zheng et al. 2004). The components making up the 1% of diesel exhaust are the four main pollutants of concern (Resitoglu et al. 2015), and through engine and fuel modifications, the amount of these pollutants in diesel exhaust has been reduced.

9.2.3 Diesel Engine Modifications

Improvements in performance and reduction in emissions can be accomplished through engine and exhaust modification. With respect to diesel engines, PM and NO_x are the main concerns; therefore, efforts have been focused on reducing these without significantly affecting fuel economy.

Diesel engines introduce fuel into the combustion chamber of the engine by direct injection. With direct injection, the fuel is injected into hot compressed air where it ignites, which is more energy and fuel-efficient than gasoline engines that pre-mix the fuel and air prior to heating and ignition. In early diesel engines, mechanical injection of the fuel was the only option; however, with advancements in electronics in the 1950s, it was soon realized that electronic control of the fuel injection could optimize engine performance. An electronic control unit monitors

Table 9.1 Typical diesel exhaust composition.

Major component	Typical concentration in diesel exhaust (vol%)	Typical concentration in ambient air (vol%)
Nitrogen (N_2)	75–77	78.08
Oxygen (O_2)	11.5–15.5	20.95
Carbon dioxide (CO_2)	4–6.5	0.038
Water (H_2O)	4–6	
Argon (Ar)	0.8	0.934
Total	>99.7	
Minor components	**Typical concentration in diesel exhaust**	**Comment**
Nitrogen oxides (NO_x)	1000–1500 ppm-vol	
Sulfur oxides (SO_x)	30–900 ppm-vol	Related to fuel composition
Carbon monoxide (CO)	20–150 ppm-vol	
Total hydrocarbons (THC)	20–100 ppm-vol	
Volatile organic compounds (VOC)	20–100 ppm-vol	
Particulates (PM)	20–100 mg/Nm^3	Based on ISO 9096 or similar

Source: CIMAC (2008).

engine performance through sensor responses and alters the fuel injection rate to provide exactly the right amount of fuel at the perfect timing. With electronic control of the fuel into the combustion chamber, diesel engines afford higher engine performance, lower fuel consumption, and reduced noise pollution and pollutant emissions (Zima 2001).

Exhaust gas recirculation (EGR) recirculates a portion of the exhaust gases back into the engine. Different technologies are available for EGR (Zheng et al. 2004), but the basic concept involves mixing the recirculated exhaust gas with the incoming air, which in turn dilutes the combustion mixture in the engine. This lowers the peak combustion temperatures and the amount of excess oxygen, reducing NO_x formation; however, the downside is that PM is increased (Rizvi 2009). As PM in diesel exhaust is of great concern, the addition of a particulate filter was mandated to be installed on all heavy-duty diesel vehicles built since 2007 (EPA 2006). As filters tend to become blocked with the material they are filtering out, it was necessary to install a filter that could self-regenerate. The PM from diesel exhaust has a combustion temperature between 500 and 600 °C, while diesel exhaust has a temperature around 200–500 °C (Rizvi 2009), meaning that under these conditions, a filter will not be able to burn off the PM for regeneration. As such, one option employed coating the filter with an oxidative catalyst (diesel oxidative catalyst, DOC), which not only lowers the activation energy needed for the combustion of the PM, but also lowers the amount of HCs and CO present in the exhaust through conversion to water and CO_2. This technology is known as selective catalytic reduction (SCR), and the different options for SCR have been reviewed previously in the literature (Guan et al. 2014; Shelef 1995).

In order to further reduce NO_x emissions, starting in 2006, a diesel exhaust fluid (DEF) was required when SCR was installed in diesel vehicles (ISO:22241-1 2019). DEF is a non-hazardous fluid consisting of 32.5% high purity urea + 67.5% deionized (DI) water and is mixed with the diesel exhaust prior to entering the PM filter containing SCR. Figure 9.3 shows the inclusion of DEF in a diesel engine; in this example, AdBlue, a commercially available DEF solution, is shown. The result is diesel exhaust comprising water and nitrogen exiting the SCR, which occurs via the overall equations shown below (Eqs. (9.1)–(9.5)) (Resitoglu et al. 2015). The water is evaporated off, leaving the urea to thermally decompose to ammonia and isocyanic acid (Eq. (9.1)). The isocyanic acid hydrolyzes to form carbon dioxide and ammonia (Eq. (9.2)). Then in the presence of oxygen and the catalyst, ammonia reduces NO species to N_2 gas and water vapor (Eqs. (9.3) through (9.5)). SCR equipped in combination with DEF can conservatively reduce NO_x emissions by up to 90%, HC and CO emissions from 50% to 90%, and PM emissions by 30–50% (Hansen 2015).

$$(NH_2)_2CO \rightarrow NH_3 + HNCO \text{ (Thermolysis)} \tag{9.1}$$

$$HNCO + H_2O \rightarrow NH_3 + CO_2 \text{ (Hydrolysis)} \tag{9.2}$$

$$4NO + 4NH_3 + O_2 \rightarrow 4N_2 + 6H_2O \tag{9.3}$$

$$2NO + 2NO_2 + 4NH_3 \rightarrow 4N_2 + 6H_2O \tag{9.4}$$

$$6NO_2 + 8NH_3 \rightarrow 7N_2 + 12H_2O \tag{9.5}$$

To compensate for the performance loss incurred from the EGR, variable geometry turbochargers (VGTs) are installed on diesel engines. The various technologies for VGT have been recently reviewed (Feneley et al. 2017). No matter the design, the purpose of the VGT is to provide improved fuel efficiency, engine response, torque, and emissions. Improved acceleration is a result of enhanced response and cleaner combustion while reducing emissions through EGR (Figure 9.3).

All sulfur entering the combustion chamber of the diesel engine is oxidized, forming SO_x, which is then released into the atmosphere via the diesel exhaust. As the only source of sulfur originates from the diesel fuel, SO_x emissions from a diesel engine are proportional to the amount of sulfur present in the fuel. Therefore, rather than implementing engine modifications to reduce SO_x emissions, modifications to diesel fuel are desired.

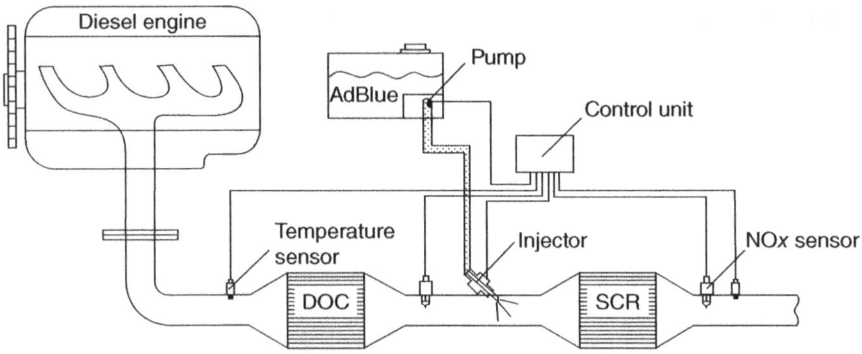

Figure 9.3 Schematic of DEF inclusion (AdBlue) in a diesel engine. Source: Resitoglu et al. (2015). Reproduced with permission of Springer Nature, License CC BY.

9.2.4 Diesel Fuel Modifications

While modifications to the diesel engine can improve emissions, it should also be noted that improvements to diesel fuel can also have a positive impact on the emissions. It is known that sulfur in diesel fuel forms SO_x and PM during combustion (Di et al. 2009); therefore, by reducing the level of sulfur in the fuel, emissions will also be reduced.

In an article from 1966, a process is detailed for converting high sulfur crude into diesel, where the resulting diesel had up to 1 wt% (10,000 ppm) sulfur (Bugai et al. 1966). However, based on ASTM methods, the sulfur level for diesel fuel in the United States was limited to 0.5 wt% (5000 ppm) (ASTM-D975 2019).

Hydrotreatment, or hydrodesulfurization, is applied to reduce sulfur in diesel. For this process, hydrogen is added, cleaving the carbon–heteroatom bond; an example is shown in Eq. (9.6), where ethanethiol, a common sulfur component found in petroleum, is hydrogenated to ethane and hydrogen sulfide.

$$C_2H_5SH + H_2 \rightarrow C_2H_6 + H_2S \tag{9.6}$$

On an industrial scale, hydrodesulfurization takes place in a fixed bed reactor in the presence of a catalyst at elevated temperatures and pressures (El-Gendy and Speight 2016). Typical refinery feedstocks undergoing hydrodesulfurization contain several classes of sulfur compounds, including thiols, thiophenes, organic sulfides, and disulfides, among others. While reaction rates may differ between the different sulfur compound classes, hydrodesulfurization is the most effective manner to reduce the sulfur content of diesel fuel. Several review articles are available for more information on how hydrodesulfurization works as well as advancements in the field (Babich and Moulijn 2003; Song 2003; Takatsuka et al. 1997; Tanimu and Alhooshani 2019; Vrinat 1983).

9.2.4.1 Low Sulfur Diesel

In order to meet the PM emission standard set forth in 1994 by the EPA under the Clean Air Act, a maximum limit of 500 ppm sulfur was implemented on diesel fuel, which is a 10-fold decrease in the sulfur content for diesel fuel. In 1996, Europe implemented the same maximum sulfur content for diesel, in what is also known as Euro 2 diesel fuel. Diesel with a total sulfur content below 500 ppm is classified as low sulfur diesel, and high sulfur diesel is when the sulfur content is above 500 ppm. The implementation of low sulfur diesel in the United States accounted for a 95% reduction in measured SO_2 and PM levels (Darlington and Kahlbaum 1999).

9.2.4.2 Ultra-Low Sulfur Diesel

Significant improvements in diesel emissions were observed with low sulfur diesel, and as a result, a further reduction in diesel sulfur content was required. In 2004, Europe set a maximum limit of 50 ppm sulfur in highway diesel (Euro 4 diesel fuel) but required ultra-low sulfur diesel containing only 10 ppm (Euro 5 diesel fuel) be made available. Then in 2009, Europe changed the maximum sulfur content in diesel to 10 ppm. In the United States, ultra-low sulfur diesel containing a maximum of 15 ppm sulfur replaced low sulfur diesel in 2006. This is an improvement of around 97% less sulfur in diesel fuel compared with low sulfur diesel.

9.2.4.3 Biodiesel

The term biodiesel encompasses any vegetable oil or animal fat-based diesel fuel and consists of fatty acid esters (methyl, ethyl, and propyl). In Europe, rapeseed oil is the predominant natural oil used to produce biodiesel, while in the United States, soybean oil is utilized; however, other natural oils can also be used to produce biodiesel. Biodiesel has many advantages in terms of fuel properties; however, there are still several parameters that require improvements, such as low calorific value, low horsepower output, and the higher emission of NO_2.

One example of biodiesel is found in waste cooking oil, which, upon transesterification, was evaluated as an alternative to fossil-based diesel (Dorado et al. 2003). In this example, used olive oil was transesterified, and the exhaust emissions from a diesel engine using this biofuel were compared with commercial diesel. For the exhaust gasses evaluated, a significant reduction in CO, CO_2, SO_2, NO, and NO_x was observed with the biodiesel. However, even though the NO_x value decreased, which is the sum of $NO + NO_2$, the NO_2 values showed an increase with the biodiesel. An additional effect from the biodiesel was observed due to the increase in oxygen present in the fuel compared to fossil-based diesel. This increase means that there is more oxygen available for combustion; however, no significant effect was observed on the combustion efficiency of the engine. In terms of engine performance, biodiesel produced from the transesterification of used olive oil was similar to fossil-based diesel. In addition to the improvements in the exhaust emission of the diesel engine, used cooking oil could be processed and reused for another purpose.

An example of a vegetable oil-based biofuel was found in transesterified soybean oil. In this instance, blends of biodiesel, as well as 100% biodiesel, were compared with petroleum-based diesel (Sheehan et al. 2000). CO_2 emissions from urban buses were significantly reduced when biodiesel was used compared with petroleum diesel; B100 (100% biodiesel) showed a 78% decrease in CO_2 emissions, and B20 (20% biodiesel blend) showed nearly a 16% decrease in CO_2 emissions compared with petroleum-based diesel. As for the other air pollutants measured from the tailpipe, B100 showed a 68% reduction in PM less than 10 μm in particle size, a 46% reduction in CO emissions, a 37% reduction in HC, and complete elimination in SO_x emissions compared with petroleum-based diesel. For B20 blends, reduction in PM, CO, and SO_x emissions were proportional to the B100, with values approximately 20% that of the B100 reductions. As was mentioned earlier, NO_x values were increased when biodiesel was compared with petroleum-based diesel. The tailpipe NO_x values for B100 showed an increase of nearly 9% compared with petroleum-based diesel. In this example, only NO_x values were monitored; there was no distinction made between NO and NO_2 values.

In an example from Buyukkaya, transesterified rapeseed oil was blended with petroleum-based diesel to study the effects of diesel engine emission and performance (Buyukkaya 2010). The blend ratios studied were 5% (B5), 20% (B20), 70% (B70), and 100% (B100) biodiesel. As was observed for the soybean-based biodiesel and the waste cooking oil examples described earlier, the exhaust gas emissions were significantly reduced for CO, CO_2, SO_2, NO, and NO_x when the transesterified rapeseed oil was used as a bio-fuel; likewise, the NO_2 values were increased. Overall, engine performance was reduced when the biodiesel was blended or at 100%. In terms of engine performance and environmental impact, only low concentrations of biodiesel blended into fossil-based diesel could be recognized as potential candidates to be certified for wide-spread use in unmodified diesel engines.

In 2009, the European Union mandated that biofuels would be implemented to provide a GHG reduction of 45% by the end of 2012 (EU Reg 2009). Then in 2015, the European Union directed each Member State to have 10% of transport-based energy consumption to originate from renewable sources by the end of 2020 (EU Reg 2015). The United States has not yet required biodiesel to be offered when fueling up, but in the EU, one will find B7 as an offering across the various Member States, and occasionally, "Blue Diesel," or 100% biodiesel can be found.

9.2.4.4 Modification of Diesel and Biodiesel with Oxygenates

9.2.4.4.1 Diesel

Alcohol-based fuels have a higher-octane number than petroleum-based fuels, which can be a benefit for gasoline engines, as the higher-octane value offers higher compression ratios, and therefore, higher performance. Instead of compressing the fuel, in a diesel engine, the air is compressed, and the fuel is injected into the air that was heated by compression; therefore, a higher-octane value has no benefit to performance. In fact, fuels with a higher-octane value typically have a lower cetane number. The cetane number is an indicator of the combustion speed of diesel fuel, and a typical diesel engine operates well when the cetane number is around 50. The lower the cetane number of the fuel, the longer it takes for the combustion cycle to complete; therefore, with higher speed diesel engines, a fuel with a higher cetane number is preferred.

Low amounts of alcohol can be added to diesel fuel without the need to modify the diesel engine; several alcohols have been studied, including methanol, ethanol, butanol, and dodecanol. It was previously shown that alcohol–fuel blends could improve some diesel exhaust emissions, such as smoke density, PM, CO, and NO_x (Babich and Moulijn 2003; Lapuerta et al. 2008; Merritt et al. 2005). As alcohols have an inherently low cetane number, engine performance is affected, alcohol–diesel blends suffer from an ignition delay and a lower rate of heat release (He et al. 2003).

9.2.4.4.2 Biodiesel

In a 2011 study, the effect of diethyl ether and ethanol added to a biodiesel blend was evaluated (Qi et al. 2011). A B30 blend (30% biodiesel/70% petroleum-based diesel (v/v)) was used as a reference; the amount of petroleum-based diesel was held constant for all blends, meaning that 25% biodiesel and 5% diethyl ether or 5% ethanol were blended. In both the diethyl ether and ethanol blends, smoke emissions were reduced at low engine loads and eliminated at high engine loads. NO_x levels were slightly increased when ethanol was used and were similar to the reference when diethyl ether was used, and CO levels were reduced with both the diethyl ether and ethanol blends. Both the diethyl ether and ethanol blends showed better stability than the biodiesel reference, and combustion characteristics were nearly identical to the reference. Both can be used in diesel engines without any modifications to the engine.

In another example, a comparison of the particle size distribution from PM was made between low sulfur fossil-based diesel, a blend of ethanol (7.7% v/v) in low sulfur fossil-based diesel, and a blend of ethanol (10% v/v) and biodiesel (30% v/v) in low sulfur fossil-based diesel (Armas et al. 2013). Here, a city bus following a

13 km route (Line 34 of the Seville public bus service) was used instead of a test engine. Compared with the low sulfur diesel reference, an 80% reduction in particulate emission, as well as a decrease in the particle diameter, was observed with both the ethanol and ethanol/biodiesel blends.

9.2.5 Sulfur Monitoring of Diesel Fuels

As the amount of sulfur in diesel decreased from percentage levels to low parts per million levels, the analytical techniques used to monitor sulfur content were required to improve in sensitivity. As was discussed in Chapter 3, for diesel engine fuel, ASTM D5453 is the standardized method for sulfur analysis utilizing ultraviolet fluorescence to provide a total sulfur value for samples containing between 1 and 8000 mg/kg (ppmw) sulfur (ASTM-D5453 2019).

A different option for total sulfur quantification of high sulfur diesel includes a carbon, hydrogen, nitrogen, sulfur (CHNS) analyzer, which simultaneously quantifies C, H, N, and S in a sample. CHNS analyzers use a combustion oven to convert all carbon to CO_2, all hydrogen to water, all nitrogen to NO_2, and all sulfur to SO_2; then a carrier gas is used to transfer the combusted sample to traps to remove all components except for the CO_2, water, NO_2, and SO_2, which are then detected using a thermal conductivity detector. Limits of detection vary depending on the instrument manufacturer, but typically, hundreds of ppms to low percentage levels are reachable with a CHNS analyzer.

Inductively coupled plasma (ICP) coupled to an optical emission spectrophotometer (OES) is an analytical technique able to quantify various elements, including total sulfur, in a wide range of samples. Inductively coupled plasma-optical emission spectrophotometer (ICP-OES) uses plasma to excite atoms in the sample, which then release photons upon relaxation that are detected by the OES detector and used for quantification. ICP-OES has an limit of detection (LOD) of around 10 ppb for many elements. However, detection limits vary significantly by element. ICP can also be coupled to mass spectrometric (MS) detection for further enhancement in sensitivity; inductively coupled plasma–mass spectrometry (ICP-MS) has an LOD of low ppt levels for most elements. Recent examples of ICP analysis of crude and diesel for low-level total sulfur quantification are available (Evans et al. 2001; Maryutina and Soin 2009).

While a total sulfur analysis can give a quick measurement of the amount of sulfur present in a sample, it gives no indication of what compound classes are present. By coupling a GC separation to a sulfur chemiluminescence detector (SCD), speciation of the different sulfur components is possible. Figure 9.4 is an example of just such a separation of petroleum-based diesel with different levels of sulfur content (Bacaud et al. 2002). It is clear to see the complexity of the different sulfur species present in the analyzed diesel. In addition to one-dimensional gas

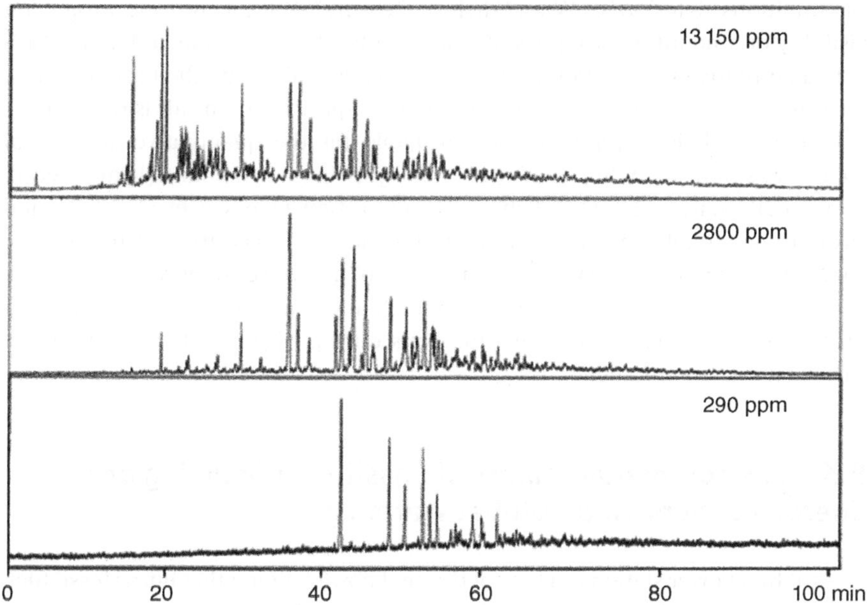

Figure 9.4 GC-SCD chromatograms of diesel as a function of total sulfur content. Source: Bacaud et al. (2002). Reproduced with permission, Copyright (2002), American Chemical Society.

chromatography–sulfur chemiluminescence detector (GC-SCD), it is also possible to couple the SCD to a two-dimensional GC separation, which has the added benefit of offering class type separation of the sulfur species (Wang et al. 2003). For a detailed description of GC and SCD detection, the reader is referred to Chapter 3.

9.2.6 Monitoring Air Pollution/Haze

As was mentioned earlier, sulfur in petroleum-based diesel forms SO_x and PM during combustion, and methods to monitor sulfur levels and even sulfur speciation in diesel were discussed. However, it is also important to monitor emissions from diesel, as SO_x and PM will contribute to air pollution and haze.

Visibility is a perceptive measurement and can vary from person to person, which is why, in 1994, a standard visibility index was proposed by Mark Pitchford and William Malm (1994). Using traditional metrics, such as visible range and extinction coefficients, a linear scale, expressed in deciview (dv), respective to fractional changes in extinction coefficients as a function of sensitive scenic targets, was developed to be perceptible regardless of the baseline conditions. With the deciview scale, pristine conditions are at near-zero levels, and the scale increases as conditions deteriorate. While such a measurement can provide insight on visibility or amount of haze present at a given location and time, it does not provide quantitative information on the components making up the haze. For that, quantitative elemental composition and PM composition of the air is required.

Various analytical techniques are available for the quantitative analysis of air composition, including gravimetric analysis for PM amount (Kinney et al. 2000), ICP analysis for the elemental composition (Janssen et al. 1997), and gas chromatography–mass spectrometry (GC-MS) for speciation of the organic components comprising PM (Phuleria et al. 2007). In the examples listed, the link between air measurements taken during high and low traffic conditions showed increased PM values from diesel emissions during higher traffic conditions in various urban environments.

9.3 Environmental Impact of Fossil Fuel Sourcing and Energy Conversion on Global Warming

A fossil fuel power station is a type of thermal power station that burns fossil fuel, such as coal or natural gas, to produce electricity. Upon combustion of the fossil fuel, heat energy is produced, which in turn operates an electrical generator, thereby converting heat energy into mechanical energy. Fossil fuel power stations dominate the landscape globally, with around 38% of all energy production coming from coal and around 23% coming from natural gas. Figure 9.5 shows the 2016 global breakdown of electricity production by source (IEA 2018).

Using coal as an energy source accounts for around one-third of the world's energy production, meaning that coal is still in high demand. Table 9.2 shows the demand for coal as an energy source along with the 2017–2018 growth rate (IEA 2019). Based on the negative values reported, the United States and Europe are decreasing the use of coal as an energy source; however, with China and the United States being the first and second-largest coal producers, respectively, mining coal is still prevalent in the United States. Both surface and underground mining are utilized to source coal, and both have negative effects on the environment.

While the demand for coal is decreasing, as can be seen from Table 9.3, 2018 saw an almost 5% global increase in the demand for natural gas (IEA 2019). Only Europe showed a decrease in natural gas demand.

9.3.1 Coal Mining, Natural Gas Wells, and Methane Release

Both coal mining and sourcing of natural gas are responsible for methane release. Methane is a powerful GHG that can trap 28 times more heat than that of CO_2.

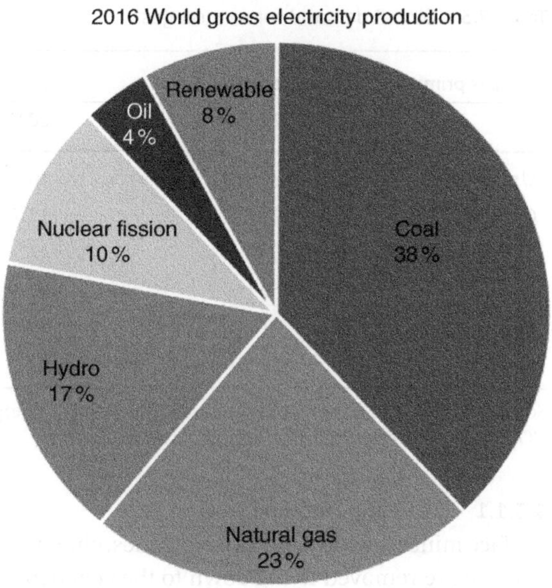

Figure 9.5 2016 Global breakdown of electricity production by source. Source: IEA (2018), World Energy Outlook.

Table 9.2 Global demand for coal.

Coal primary energy demand (Mtce)		Growth rate (%)
	2018	2017–2018
United States	451	−4.5
China	2771	1.0
India	600	5.0
Europe	462	−2.6
Rest of the world	1113	1.4
World	5397	0.7

Source: Based on IEA data from the IEA (2019), Global Energy and CO_2 Status Report (IEA 2019).

Methane has a lifetime of 12 years in the atmosphere, and it is responsible for around 25% of the observed global warming. The UN Climate Summit held in September 2018 has called for countries to reduce methane emissions by 45% in 2025 and between 60% and 75% in 2030, or to a near-zero target (UN & CCAC 2019). In 2017, methane emission from the oil and gas industry accounted for 6% of the global energy sector GHG emissions (McGlade et al. 2019).

Table 9.3 Global demand for natural gas.

Gas primary energy demand (Bcm)		Growth rate (%)
	2018	2017–2018
United States	854	10.5
China	279	17.7
India	60	4.7
Europe	599	−1.9
Rest of the world	2137	2.8
World	3928	4.6

Source: Based on IEA data from the IEA (2019), Global Energy and CO_2 Status Report (IEA 2019).

9.3.1.1 Coal Mine Methane

Surface mines, also known as strip mines, change the landscape as the water, rock, and soil are removed to dig down to the coal deposit. As a result of growing environmental concerns, the Surface Mining Control and Reclamation Act (SMCRA) of 1977 was passed into law in the United States (SMCRA 2017). The SMCRA set forth environmental standards that mines were required to follow, as well as regulations that must be followed when reclaiming lands. As the materials removed during mining can potentially contain contaminants, these contaminants must be dealt with to ensure that mining areas are returned to a state as close as possible to the original condition.

Underground mining, as the name suggests, has less of an effect on the landscape. However, the collapse of the ground above the mine is a significant risk. An additional hazard originating from underground coal mines is methane release, which, based on 1998 global statistics, accounted for 90% of the global methane emissions (Bibler et al. 1998). Methane is the second most abundant GHG, making it an important opportunity for mitigating global warming, which is why the Global Methane Initiative (GMI) was formed. Through the implementation of coal mine methane recovery and use projects, the amount of methane released from coal mines in 2010 was reduced to 8% of the global methane emissions (GMI 2011), a significant reduction from the 1998 levels. A review of methane capture techniques and usage was recently published (Ozgen Karacan et al. 2011). Despite methane capture and usage, global methane emission is expected to remain around this level (9%) through 2030 (EPA 2012), which is in part due to the challenges associated with the implementation of recovery and use projects (e.g. policy framework, cost, technology, etc.). If a full implementation of methane recovery and usage technologies were achieved, then a potential reduction of 10% of non-CO_2 GHG could be achieved by 2030 (EPA 2014).

9.3.1.2 Natural Gas Methane

Natural gas is essentially comprised of methane, but small amounts of natural gas liquids (HCs) and non-hydrocarbon gases, such as CO_2 and water vapor, are also observed. Natural gas is a fuel used to create energy, materials, and chemicals. Unlike coal, natural gas is sourced through drilling, in a similar manner as was discussed for petroleum sourcing in Chapter 2. Natural gas wells can be drilled vertically or horizontally into the natural gas deposit, which is essentially an open pocket or reservoir deep in the ground that contains only natural gas. With a conventional natural gas deposit, the natural gas typically flows easily to the surface through the well by naturally occurring pressure in the ground or, in some cases, by a pumping mechanism. In the United States, along with a few other countries, natural gas can be found in shale sedimentary rock formations. Sourcing shale natural gas deposits require a technique called hydraulic fracturing, or fracking, which involves forcing water, chemicals, and sand into the rock formation at high pressures once the well is drilled. Once the rock formation is broken up, the natural gas easily flows to the surface through the well.

As natural gas is mostly methane, any leak in the well piping, transport piping, connections, and valves will result in methane release to the environment. In 2016, methane release from leaks in the natural gas drilling and transport infrastructure accounted for 1.3% of the total GHG emissions (GIE & MARCOGAZ 2019). While the solution, plugging the leaks, sounds simple enough, locating and measuring methane leaks are more challenging. To compound matters, some pipelines cross state-lines in the United States and international borders in Europe, where cross-jurisdictional policies may be different.

9.3.2 Fossil Fuel Power Stations

The combustion of coal and natural gas to produce electricity occurs in a power station. The major difference between a coal-fired power station and a gas-fired power station is how the fossil fuel is transported to the combustion chamber, which will be discussed in more detail below. Even though the fuels are different, the emissions from each type of power station are quite similar. The exceptions come from SO_x, PM, heavy metal, and the ash produced from coal combustion, which are not produced in gas-fired power stations.

9.3.2.1 Coal-Fired Power Station

Coal is converted to energy in a coal-fired power station, also known as a coal power plant, which is a type of thermal power station (Figure 9.6). The sourced coal is first pulverized or milled into smaller pieces before entering the combustion chamber of a boiler where it is burned at high temperatures. Burning the coal releases hot gases and heat energy that converts water in tubes lining the boiler into steam; the high-pressure steam then turns the blades in a turbine, creating

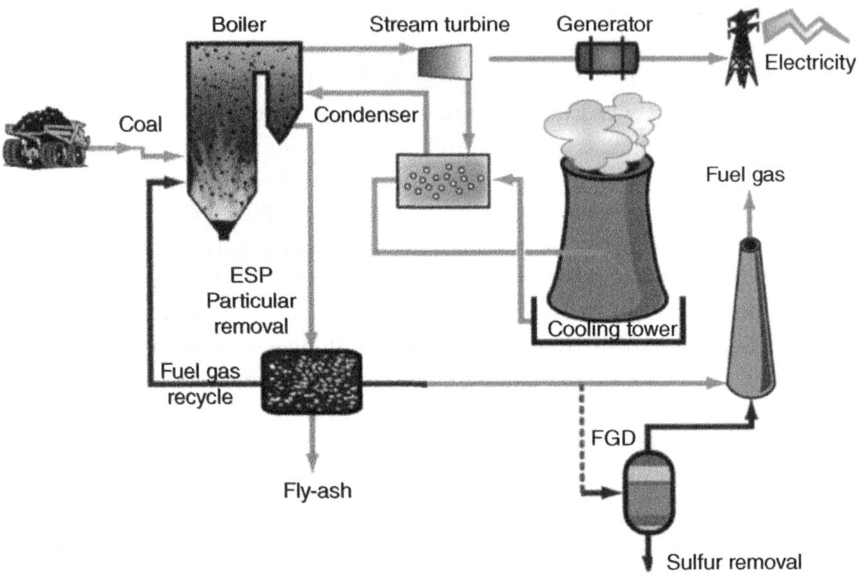

Figure 9.6 Simplified schematic of a coal-fired power station. Source: Guttikunda and Jawahar (2014). Reproduced with permission, Copyright (2014), Elsevier.

energy. After passing through the turbine, the steam is condensed and directed back to the boiler to be re-heated. Coal-fired power stations generate over a third of the world's electricity.

9.3.2.2 Gas-Fired Power Station

Natural gas is converted to energy in a gas-fired power station, which is another type of thermal power station. Figure 9.7 shows an example of a combined cycle gas turbine, which produces energy from the combustion gases of natural gas, but also the production of steam converted to energy. The sourced natural gas is transported to the power station via pipelines, and as the fuel is a gas, it can be piped directly to the combustion chamber. As the natural gas is combusted, hot gas and heat energy are released, which are converted to mechanical energy by turning the turbine. The hot gases are then directed to the recovery boiler and chimney, where the remaining heat is used to convert water in piping into steam, which also creates mechanical energy by turning the steam turbine. The steam is then condensed and recovered again to be turned into steam.

9.3.3 Emissions from Fossil Fuel Power Stations

The combustion or burning of fossil fuels produces energy but also results in harmful emissions, including NO_x and CO_2. More specific to the combustion of coal are SO_2, PM, mercury and other heavy metals, and fly and bottom ash.

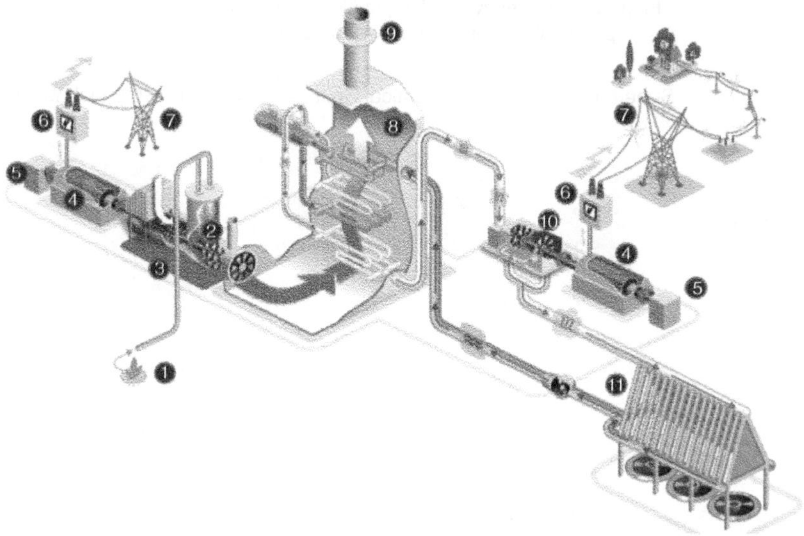

Figure 9.7 Gas-fired power station: (1) natural gas supply, (2) combustion chamber, (3) gas turbine, (4) alternator, (5) exciter, (6) transformer, (7) high voltage cable, (8) recovery steam boiler, (9) chimney, (10) steam turbine, and (11) air condenser. Source: Engie (2019). Reproduced with permission from Engie Electrabel.

9.3.3.1 Carbon Dioxide

Improvements to fossil-fuel power stations have been made to improve the thermal efficiency while simultaneously reducing CO_2 emissions. Specific to coal-fired power stations, different phases of improvement occurred over the years (Figure 9.8): first subcritical, then supercritical, ultra-supercritical, advanced ultra-supercritical, and finally, carbon capture and storage (CCS) technology (Campbell 2013). The first coal-fired power stations were subcritical, having an energy conversion efficiency of around 30%. Supercritical combustion technology, which operates at higher temperatures and pressures to obtain supercritical water, was commercialized in the 1960s and showed a conversion efficiency close to 40%. Ultra-supercritical combustion technology operates at even higher temperatures and pressures in order to push the energy conversion to 45%, while advanced ultra-supercritical technology attempts to reach an energy conversion of 50%. For every 1% increase in efficiency, a 2–3% reduction in CO_2 emissions was observed.

With CCS technology, around 90% of the carbon released from fuel (both coal and natural gas) combustion is captured (de Queiroz Fernandes Araújo and de Medeiros 2017; Herzog 2001); however, the drawback is a loss in conversion efficiency by 7–12%. Research is continuing to improve the conversion efficiency when CCS technology is installed.

Table 9.4 shows the energy-related CO_2 emissions from fuel combustion (IEA 2019). As can be seen, there was a global increase in CO_2 emissions from 2017 to

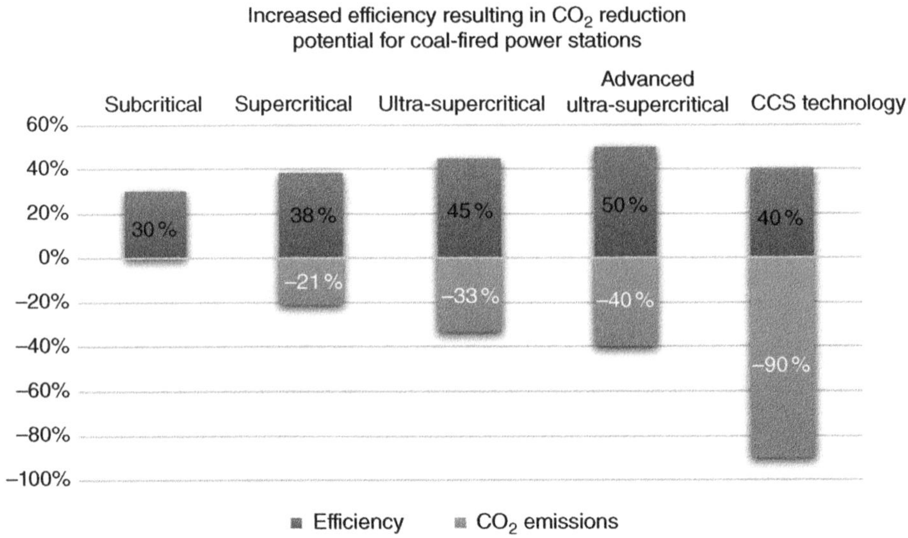

Figure 9.8 Improvements to coal-fired power stations to increase efficiency and decrease CO_2 emissions. Source: Graph constructed using data from Campbell and Herzog (Campbell 2013; de Queiroz Fernandes Araújo and de Medeiros 2017; Herzog 2001).

Table 9.4 Total energy-related CO_2 emissions.

Total CO_2 emissions (Mt CO_2)		Growth rate (%)
	2018	**2017–2018**
United States	4888	3.1
China	9481	2.5
India	2299	4.8
Europe	3956	−1.3
Rest of the world	11 249	1.1
World	33 143	1.7

Source: Based on IEA data from the IEA (2019), Global Energy and CO_2 Status Report (IEA 2019).

2018 by nearly 2%. This is not surprising when considering that from 2017 to 2018, there was a global increase in fossil fuel demand, from both coal and natural gas.

9.3.3.2 Sulfur Dioxide

Sulfur oxides are harmful in terms of their ability to react in the atmosphere to create acid rain. SO_2 produced from coal combustion will react with air to create

H_2SO_3 and H_2SO_4. In addition, SO_x emissions have been related to respiratory issues (Moretti and Jones 2012).

In terms of SO_2 emissions, 65% of US SO_2 emissions were nearly all (90%) originating from coal-fired power stations, and 70% of European SO_2 emissions came from electricity production (Lin et al. 2018). Advancements in technology to filter the stack exhaust from coal-fired power stations have dramatically reduced SO_2 emissions by up to 80%. These technologies include wet flue gas desulfurization (FGD), dry FGD utilizing a spray dryer absorber (SDA), circulating dry scrubbers, and dry sorbent injection (Moretti and Jones 2012). The most predominant technology implemented for coal-fired power stations is wet FGD utilizing limestone, as this covers the widest range of fuel coal; a review on FGD technologies was recently published (Cordoba 2015). Briefly, the wet FGD process involves wet limestone with in situ oxidation to remove SO_2, forming calcium sulfate (gypsum) via (Eq. (9.7)). This process can remove SO_2 at efficiencies of 92–98%. Instead of needing to be disposed of as waste, the gypsum produced can be used to replace the need for naturally sourced gypsum in cement plants.

$$CaCO_3 + SO_2 + {}^1/_2O_2 \rightarrow CaSO_4 \cdot 2H_2O + CO_2 \tag{9.7}$$

9.3.3.3 Nitrogen Oxides

Nitrogen oxides are harmful to the environment, because, like sulfur oxides, nitrogen oxides also contribute to acid rain, as NO generated from coal and natural gas combustion will react with air to form HNO_2. Additionally, excess NO_x can result in a brown color and is attributed to respiratory issues (Moretti and Jones 2012), and NO_2 contributes to smog.

Controlling the fuel–air mixing ratios of coal combustion is an effective way to reduce NO_x formation; however, additional technologies are required to reduce NO_x to acceptable levels. One such technology is SCR using ammonium and urea to catalytically reduce the NO_x to N_2 and H_2O (Moretti and Jones 2012). If this technology seems familiar, it is because the same technology is used on a smaller scale to reduce NO_x emissions from automotive diesel engines. On the industrial scale, SCR technology can reduce NO_x emissions from fossil-fuel power stations between 70% and 90% depending on the operating conditions (Baig and Yousaf 2017).

9.3.3.4 Particulate Matter (PM)

PM emissions from the fly and bottom ash of coal-fired power stations can range in size from 1 to 100 μm and have a multitude of environmental issues. PM contributes to smog and haze issues, soil contamination, stunted plant growth, and human respiratory issues.

Electrostatic precipitators (ESPs), fabric filters, mechanical collectors, and venturi scrubbers are technologies available to reduce PM emissions from fly ash

(Moretti and Jones 2012); however, ESP and fabric filters are able to remove the highest percentage of PM, with fabric filters being preferred over ESP (Baig and Yousaf 2017). ESP devices contain electrode plates that generate an electric field that the flue gases must pass through; the PM present in flue gas is charged, and as it passes through the electric field, it is attracted to electrodes of the opposite charge. As the name suggests, a fabric filter is made up of thousands of small fabric filter bags laid out in a matrix that collects PM from dry flue gas as it passes through the filter. Both ESP and well-made filters can capture up to 99.9% of PM, meeting the most stringent emission standards; however, as fabric filters are available at a fraction of the cost of ESP, they are more preferred.

9.3.3.5 Coal Ash and Heavy Metals

The combustion of coal produces energy; however, it also produces coal ash. Coal ash is made up of both fly ash and bottom ash. Only about 10–20% of the ash produced is bottom ash, meaning that the rest is fly ash that gets caught in the flue gas.

Fly ash quality depends on many factors including coal type, particle fineness, and combustion conditions; however, based on scanning electron microscopy–energy dispersive spectroscopy (SEM-EDS) analysis, it is made up mostly of amorphous alumino-silicate spheres and a smaller amount of iron-rich spheres that is released upon the combustion of finely pulverized coal in the combustion chamber (Kutchko and Kim 2006).

The major heavy metal concern from coal-fired power stations is mercury. Coal mercury is converted during the combustion process into gaseous mercury (Hg^0) and subsequently into oxidized states as the combustion gases cool (Pavlish et al. 2003). Mercury air emissions from coal-fired power stations are very harmful to the environment because when reacted with water, the mercury forms water-soluble compounds that accumulate in fish and other aquatic life. An additional risk to humans arises from the potential to consume contaminated fish.

Powdered activated carbon (PAC) is effective at removing up to 95% of mercury in the various possible forms from the flue gas of coal-fired power stations. PAC can remove elemental mercury, oxidized mercury, and particulate mercury from PM. For coal-fired power stations with wet FGD technology already implemented, nearly 90% of mercury(II) is removed; unfortunately, wet FGD technology is unable to remove elemental mercury (Hg^0), meaning that an additional removal technique is required (Pavlish et al. 2003).

Bottom ash is made up of the larger ash particles that are too heavy to be carried up the flue stack, and as the name insinuates, the bottom ash accumulates at the bottom of the combustion chamber. Mercury was discussed earlier, as it can be volatilized; however, mercury, along with other heavy metals, including arsenic (As), selenium (Se), and lead (Pb), among others, can be found in coal ash.

There is no shortage of analytical techniques for the determination of the elemental and heavy metal composition of coal ash. An example of the comparison of flue- and bottom-ash from three different coal-fired power stations in India was performed using flame atomic absorption spectrometry (AAS) (Sushil and Batra 2006). In a study around a Canadian bituminous coal-fired power station, a battery of analytical techniques was utilized to provide comparative quantitative information on heavy metal emissions; elemental composition was determined using neutron activation analysis (NAA), inductively coupled plasma combined with optical emission spectroscopy (ICP-OES) and mass spectroscopy (ICP-MS), and various modes of AAS (Goodarzi et al. 2008). Review articles are also available, describing additional advanced characterization techniques (Gupta 2007; Ward 2016).

Currently, less than 30% of coal ash produced is reused in agriculture and in construction as cement, concrete, structural fill, and soil stabilization (Chindaprasirt et al. 2009; Jayaranjan et al. 2014; Rajakumar and Meenambal 2015). Much of the coal ash produced ends up in either landfills or mixed with water and disposed of in isolated ponds near the coal-fired power station. When using wet ponds, the lining of the pond is extremely important; in the heavy metal study from India briefly described earlier, two of the three power plants used wet ponds for disposal of bottom coal ash; however, neither pond was lined. Based on the AAS results, heavy metals were detected in the bottom ash, and without a pond liner, the risk exists for these heavy metals to leach into the surrounding environment (Sushil and Batra 2006).

9.3.4 Wastewater from Fossil Fuel Power Stations

Nearly 90% of the water requirements in fossil-fuel power stations is for cooling purposes; the remainder is used for ash handling and flue-gas desulfurization, which is specific to coal-fired power stations (Delgado 2012). It is in part for this reason that gas-fired power stations have lower water consumption than coal-fired power stations. Additionally, as coal-fired power stations decline in number and as gas-fired power stations increase in number, the overall water requirement for fossil-fuel power stations has and is declining.

In terms of cooling, most modern fossil-fuel power stations employ water recycling technology and non-pristine water is used for cooling the produced steam back to water that can then be reused (EIA 2018). The use of non-pristine water means that instead of taking on fresh water or drinking water, treated wastewater originating from another source is used. Ash handling and flue-gas desulfurization are the main sources for contamination of wastewater for fossil-fuel power stations, more specifically for coal-fired power stations.

Table 9.5 2018 Coal-fired power station scrubber wastewater limits.

	Unit	Daily limit	Monthly limit
Arsenic (As)	μg/l	8	11
Mercury (Hg)	μg/l	0.356	0.788
Selenium (Se)	μg/l	12	23
Nitrates	mg/l	4.4	17

Source: Data source EIA (2018).

Coal-fired power stations account for one-third of all toxic industrial pollution discharged into waterways (EIP 2014). The Clean Water Act was first drafted in 1972, and throughout the years, additions have been made to protect waterways from industrial pollution, including pollution from coal-fired power stations (CWA 2002). In 2018, new limits were established for scrubber wastewater from coal-fired power stations, which are presented in Table 9.5 (EIA 2018). While it is recommended to implement strategies to reach these limits as soon as possible, the limits will not be enforced until 2023.

Water is typically used in ash transport in coal-fired power stations; some possibilities for ash handling include wet ash systems, semi-dry systems, and dry ash systems. In a dry ash system, no water is used for ash transport, but instead, air is used. By using semi-dry systems, which include an ash de-watering step, and recycling the water, a zero-liquid discharge from the ash handling system can be achieved (Carpenter 2017). However, wet ash systems are the most commonly used for the disposal of coal ash. With wet ash systems, fly and bottom ash are mixed with water to form what are called ash ponds. These ash ponds are isolated from the surrounding environment to prevent contamination while the larger particulates from the ash settle out. Unfortunately, water from these ash ponds is not treated to remove pollutants. One way to purify the wastewater from ash handling is the construction of wetlands. This long-term method used for water purification has already been implemented at coal mines and is a potential option for coal-fired power stations utilizing ash ponds (Wood 2012).

In the Clean Water Act, it is recognized that water sources are for the public. Clean water is used to support life, and our demand for clean water is continually growing. In 2050, global water demand is expected to increase by 55% compared with 2020 (Romano and Akhmouch 2019), making the protection of freshwater sources increasingly important. Liquid chromatography–mass spectrometry (LC-MS) and elemental analysis are most commonly used to confirm water purity.

9.3.5 Analysis of Ground Water

As was mentioned earlier, ash ponds are used to dispose of coal ash. Coal ash is made up of both fly and bottom ash; however, fly ash makes up the bulk of the coal ash composition. Both major and trace elements are enriched on the surface of fly ash, making leaching into the ash pond a distinct possibility. There are several factors that affect leaching, including the number of elements present and their respective concentrations, the pH of the pond water, the ash source, and the leaching time (Singh et al. 2010). While undesired, coal ash contaminants can migrate from the ash pond to the environment, where a higher risk of environmental contamination is possible when ash ponds are not properly designed, such as the example mentioned earlier of the unlined ash ponds observed in India (Singh et al. 2010; Sushil and Batra 2006). Therefore, it is not only important to analyze ash pond water, but also groundwater surrounding ash ponds, pondwater from natural lakes and ponds, and also the aquatic species present.

In a 2016 study, ash samples taken from Parichha, a coal-fired power station in India, and groundwater samples taken from villages around the ash pond were analyzed for heavy metal content (Verma et al. 2016). It should be noted that the ash pond in this example was not lined, increasing the risk for heavy metal contamination into the surrounding environment. AAS was used for heavy metal quantification; Pb, Ni, Fe, Cr, and Mn were detected in all the groundwater samples, with some of the heavy metals showing values well above the WHO recommended levels for drinking water. Another study taking place in 2016 in the Southeastern United States also showed groundwater sources around various coal-fired power stations to be contaminated with heavy metals above drinking water and ecological standards in 10 of 24 samples (Harkness et al. 2016). In this study, direct current plasma optical emission spectroscopy (DCP-OES) and ICP-MS were used for heavy metal identification and quantification.

The few examples of groundwater analysis described earlier are specific for coal-fired power stations. A more detailed explanation of water analysis and environmental impact drawn from other examples from the oil and gas industry is described in Chapter 10.

References

Armas, O., Gomez, A., Mata, C., and Ramos, A. (2013). Particle size distributions from a city bus fuelled with ethanol–biodiesel–diesel fuel blends. *Fuel* 111: 393–400. https://doi.org/10.1016/j.fuel.2013.03.036.

ASTM-D5453 (2019). *Standard Test Method for Determination of Total Sulfur in Light Hydrocarbons, Spark Ignition Engine Fuel, Diesel Engine Fuel, and Engine Oil by Ultraviolet Fluorescence*. West Conshohocken, PA: ASTM International.

ASTM-D975 (2019). *Standard Specification for Diesel Fuel Oils*. West Conshohocken, PA: ASTM International.

Babich, I.V. and Moulijn, J.A. (2003). Science and technology of novel processes for deep desulfurization of oil refinery streams: A review. *Fuel* 82: 607–631. https://doi.org/10.1016/S0016-2361(02)00324-1.

Bacaud, R., Cebolla, V.L., Membrado, L. et al. (2002). Evolution of sulfur compounds and hydrocarbons classes in diesel fuels during hydrodesulfurization. Combined use of thin-layer chromatography and GC-sulfur-selective chemiluminescence detection. *Industrial and Engineering Chemistry Research* 41: 6005–6014. https://doi.org/10.1021/ie020231+.

Baig, K.S. and Yousaf, M. (2017). Coal fired power plants: Emission problems and controlling techniques. *Journal of Earth Science and Climatic Change* 8: 404–413. https://doi.org/10.4172/2157-7617.1000404.

Bibler, C.J., Marshall, J.S., and Pilcher, R.C. (1998). Status of worldwide coal mine methane emissions and use. *International Journal of Coal Geology* 35: 283–310. https://doi.org/10.1016/S0166-5162(97)00038-4.

Bryant, L. (1976). The development of the diesel engine. *Technology and Culture* 17: 432–446. https://doi.org/10.2307/3103523.

Bugai, E.A., Selivanov, T.I., Akhmetshim, M.I. et al. (1966). Experience in producing gasoline and diesel fuel from high-sulfur crudes. *Chemistry and Technology of Fuels and Oils* 2: 373–376.

Buyukkaya, E. (2010). Effects of biodiesel on a DI diesel engine performance, emission and combustion characteristics. *Fuel* 89: 3099–3105. https://doi.org/10.1016/j.fuel.2010.05.034.

CAA (1990). *Clean Air Act*. Washington, DC: Environmental Protection Agency (EPA). Retrieved from https://www.epa.gov.

Campbell, R.J. (2013). *Increasing the Efficiency of Existing Coal-Fired Power Plants*. Washington, DC: Congressional Research Service. Retrieved from https://www.crs.gov.

Carpenter, A. (2017). *Water Conservation in Coal-Fired Power Plants*. London: EIA Clean Coal Center.

Chindaprasirt, P., Jaturapitakkul, C., Chalee, W., and Rattanasak, U. (2009). Comparative study on the characteristics of fly ash and bottom ash geopolymers. *Waste Management* 29: 539–543. https://doi.org/10.1016/j.wasman.2008.06.023.

CIMAC (2008). *Guide to Diesel Exhaust Emissions Control of NO_x, SO_x, Particulates, Smoke, and CO_2: Seagoing Ships and Large Stationary Diesel Power Plants*. Frankfurt, Germany: International Council on Combustion Engines (CIMAC).

Clarkson, D. and Middleton, J.T. (1962). The California control program for motor vehicle created air pollution. *Journal of the Air Pollution Control Association* 12: 22–28. https://doi.org/10.1080/00022470.1962.10468042.

Cordoba, P. (2015). Status of flue gas desulphurization (FGD) systems from coal-fired power plants: Overview of the physic-chemical control processes of wet limestone FGDs. *Fuel* 144: 274–286. https://doi.org/10.1016/j.fuel.2014.12.065.

CWA (2002). *Federal Water Pollution Control Act*. Washington, DC: Environmental Protection Agency (EPA).

Darlington, T. and Kahlbaum, D. (1999). *Nationwide Emission Benefits of a Low Sulfur Diesel Fuel*. Michigan, US: Air Improvement Resource, Inc. Retrieved from https://www.epa.gov.

Delgado, A. (2012). Water footprint of electric power generation: Modeling its use and analyzing options for a water-scarce future. Master's thesis. Massachusetts Institute of Technology, Cambridge, MA, USA.

Di, Y., Cheung, C.S., and Huang, Z. (2009). Experimental study on particulate emission of a diesel engine fueled with blended ethanol–dodecanol–diesel. *Journal of Aerosol Science* 40: 101–112. https://doi.org/10.1016/j.jaerosci.2008.09.004.

Diesel, R. (1913). Die Entstehung des Dieselmotors. In: *Jahrbuch Schiffbautechnichsen Gesellschaft*, vol. 14 (ed. R. Diesel), 267–355. Berlin, Germany: Springer.

Dorado, M.P., Ballesteros, E., Arnal, J.M. et al. (2003). Exhaust emissions from a Diesel engine fueled with transesterified waste olive oil. *Fuel* 82: 1311–1315. https://doi.org/10.1016/S0016-2361(03)00034-6.

EIA. (2018). Water withdrawals by U.S. power plants have been declining. Retrieved from https://www.eia.gov.

EIP (2014). *Toxic Wastewater from Coal Plants*. Washington, DC: Environmental Integrity Project. Retrieved from https://www.environmentalintegrity.org.

El-Gendy, N.S. and Speight, J.G. (2016). *Handbook of Refinery Desulfurization*. Boca Raton, FL, USA: Taylor & Francis.

Engie (2019). Gas: How does a CCGT work? Retrieved from https://corporate.engie-electrabel.be/.

EPA (2006). *Direct Final Rule for Amendments to Regulations for Heavy-Duty Diesel Engines*. US Environmental Protection Agency (EPA). Retrieved from https://www.epa.gov.

EPA (2012). *Global Anthropogenic Emissions of Non-CO_2 Greenhouse Gases: 1990–2030*. US Environmental Protection Agency (EPA). Retrieved from https://www.epa.gov.

EPA (2014). *Global Mitigation of Non-CO_2 Greenhouse Gases: 2010–2030 – Executive Summary*. US Environmental Protection Agency (EPA). Retrieved from https://www.epa.gov.

EU Reg (2009). Regulation (EC) no 443/2009 of the European Parliament and of the Council of 23 April 2009: Setting emission performance standards for new passenger cars as part of the community's integrated approach to reduce CO_2 emissions from light-duty vehicles. *Official Journal of the European Union* 52: 1–152. https://doi.org/10.3000/17252555.L_2009.140.eng.

EU Reg (2015). Directive (EU) 2015/1513 of the European Parliament and of the Council of 9 September 2015: Amending Directive 98/70/EC relating to the quality of petrol and diesel fuels and amending Directive 2009/28/EC on the promotion of the use of energy from renewable sources. *Official Journal of the European Union* 239: 1–29.

Evans, P., Wolff-Briche, C., and Fairman, B. (2001). High accuracy analysis of low level sulfur in diesel fuel by isotope dilution high resolution ICP-MS, using silicon for mass bias correction of natural isotope ratios. *Journal of Analytical Atomic Spectrometry* 16: 964–969. https://doi.org/10.1039/B102288C.

Feneley, A.J., Pesiridis, A., and Andwari, A.M. (2017). Variable geometry turbocharger technologies for exhaust energy recovery and boosting – A review. *Renewable and Sustainable Energy Reviews* 71: 959–975. https://doi.org/10.1016/j.rser.2016.12.125.

GIE & MARCOGAZ (2019). Potential ways the gas industry can contribute to the reduction of methane emissions. Report for the Madrid Forum. Paper presented at the European Gas Regulatory Forum, Madrid, Spain (5–6 June 2019).

GMI (2011). *Coal Mine Methane: Reducing Emissions, Advancing Recovery and Use Opportunities*. Online. Global Methane Initiative. Retrieved from https://www.globalmethane.org.

Goodarzi, F., Huggins, F.W., and Sanei, H. (2008). Assessment of elements, speciation of As, Cr, Ni and emitted Hg for a Canadian power plant burning bituminous coal. *International Journal of Coal Geology* 74: 1–12. https://doi.org/10.1016/j.coal.2007.09.002.

Guan, B., Zhan, R., Lin, H., and Huang, Z. (2014). Review of state of the art technologies of selective catalytic reduction of NO_x from diesel engine exhaust. *Applied Thermal Engineering* 66: 395–414. https://doi.org/10.1016/j.applthermaleng.2014.02.021.

Gupta, R. (2007). Advanced coal characterization: A review. *Energy & Fuels* 21: 451–460. https://doi.org/10.1021/ef060411m.

Guttikunda, S.K. and Jawahar, P. (2014). Atmospheric emissions and pollution from the coal-fired thermal power plants in India. *Atmospheric Environment* 92: 449–460. https://doi.org/10.1016/j.atmosenv.2014.04.057.

Hansen, S. (2015). Clean diesel: What is SCR (selective catalytic reduction), how does it work . . . & why is it important? [press release]. Retrieved from https://www.dieselforum.org/ (accessed 19 March 2020).

Harkness, J.S., Sulkin, B., and Vengosh, A. (2016). Evidence for coal ash ponds leaking in the Southeastern United States. *Environmental Science and Technology* 50: 6583–6592. https://doi.org/10.1021/acs.est.6b01727.

He, B., Wang, J., Yan, X., et al. (2003). Study on combustion and emission characteristics of diesel engines using ethanol blended diesel fuels. 01-0762: 1–6. https://doi.org/10.4271/2003-01-0762.

Herzog, H.J. (2001). What future for carbon capture and sequestration? *Environmental Science and Technology* 35: 148A–153A. https://doi.org/10.1021/es012307j.

IARC (2012). IARC: diesel engine exhaust carcinogenic [press release]. Retrieved from https://www.iarc.fr (accessed 19 March 2020).

IEA (2018). Electricity information: Overview. Retrieved from https://www.iea.org (accessed 19 March 2020).

IEA (2019). Global energy and CO_2 status report: The latest trends in energy and emissions in 2018. Retrieved from https://www.iea.org (accessed 19 March 2020).

IPCC (2014). *Climate Change 2014: Mitigation of Climate Change. Working Group III Contribution to the Fifth Assessment Report of the Intergovernmental Panel on Climate Change*. New York: Cambridge University Press. Retrieved from https://ipcc.ch.

ISO:16183 (2002). *Heavy Duty Engines – Measurement of Gaseous Emissions from Raw Exhaust Gas and of Particulate Emissions Using Partial Flow Dilution Systems Under Transient Test Conditions*. Geneva, Switzerland: International Organization for Standardization (ISO).

ISO:22241-1 (2019). *Diesel Engines – NO_x Reduction Agent AUS 32*. Geneva, Switzerland: International Organization for Standardization.

Janssen, N.A.H., van Mansom, D.F.M., van der Jagt, K. et al. (1997). Mass concentration and elemental composition of airborne particulate matter at street and background locations. *Atmospheric Environment* 31: 1185–1193. https://doi.org/10.1016/S1352-2310(96)00291-9.

Jayaranjan, M.L.D., Hullebusch, E.D.v., and Annachhatre, A.P. (2014). Reuse options for coal fired power plant bottom ash and fly ash. *Reviews in Environmental Science and Bio/Technology* 13: 467–486. https://doi.org/10.1007/s11157-014-9336-4.

Kinney, P.L., Aggarwal, M., Northridge, M.E. et al. (2000). Airborne concentrations of PM(2.5) and diesel exhaust particles on Harlem sidewalks: a community-based pilot study. *Environmental Health Perspectives* 108: 213–218. https://doi.org/10.1289/ehp.00108213.

Kutchko, B.G. and Kim, A.G. (2006). Fly ash characterization by SEM–EDS. *Fuel* 85: 2537–2544. https://doi.org/10.1016/j.fuel.2006.05.016.

Kyoto (2008). *Kyoto Protocol Reference Manual*. Bonn, Germany: United Nations Framework Convention on Climate Change (UNFCCC).

Lapuerta, M., Armas, O., and Herreros, J.M. (2008). Emissions from a diesel–bioethanol blend in an automotive diesel engine. *Fuel* 87: 25–31. https://doi.org/10.1016/j.fuel.2007.04.007.

Lin, C.K., Lin, R.T., Chen, P.C. et al. (2018). A global perspective on sulfur oxide controls in coal-fired power plants and cardiovascular disease. *Scientific Reports* 8: 2611–2620. https://doi.org/10.1038/s41598-018-20404-2.

Lloyd, A.C. and Cackette, T.A. (2011). Diesel engines: Environmental impact and control. *Journal of the Air and Waste Management Association* 51: 809–847. https://doi.org/10.1080/10473289.2001.10464315.

Maryutina, T.A. and Soin, A.V. (2009). Novel approach to the elemental analysis of crude and diesel oil. *Analytical Chemistry* 81: 5896–5901. https://doi.org/10.1021/ac900615t.

McGlade, C., Sondak, G., and Gould, T. (2019). *Methane Emissions from Oil and Gas: Tracking Clean Energy Progress*. The International Energy Agency (IEA). Retrieved from https://www.iea.org.

McKee, H.C., Clark, J.M., and Wheeler, R.J. (1962). Measurement of diesel engine emissions. *Journal of the Air Pollution Control Association* 12: 516–546. https://doi.org/10.1080/00022470.1962.10468122.

Merritt, P.M., Ulmet, V., McCormick, R.L. et al. (2005). Regulated and unregulated exhaust emissions comparison for three tier II non-road diesel engines operating on ethanol-diesel blends. *Journal of Fuels and Lubricants* 114: 1111–1122.

Moretti, A.J. and Jones, C.S. (2012). *Advanced Emissions Control Technologies for Coal-Fired Power Plants*. (BR-1886). Barberton, OH: Babcock & Wilcox Power Generation Group, Inc.

OECD (2002). *Strategies to Reduce Greenhouse Gas Emissions from Road Transport: Analytical Methods*. Paris, France: Organization for Economic Co-Operation and Development (OECD).

Ozgen Karacan, C., Ruiz, F.A., Cote, M., and Phipps, S. (2011). Coal mine methane: A review of capture and utilization practices with benefits to mining safety and to greenhouse gas reduction. *International Journal of Coal Geology* 86: 121–156. https://doi.org/10.1016/j.coal.2011.02.009.

Paris (2015). *Paris Agreement*. Paris, France: United Nations.

Pavlish, J.H., Sondreal, E.A., Mann, M.D. et al. (2003). Status review of mercury control options for coal-fired power plants. *Fuel Processing Technology* 82: 89–165. https://doi.org/10.1016/S0378-3820(03)00059-6.

Phuleria, H.C., Sheesley, R.J., Schauer, J.J. et al. (2007). Roadside measurements of size-segregated particulate organic compounds near gasoline and diesel-dominated freeways in Los Angeles, CA. *Atmospheric Environment* 41: 4653–4671. https://doi.org/10.1016/j.atmosenv.2007.03.031.

Pitchford, M.L. and Malm, W.C. (1994). Development and applications of a standard visual index. *Atmospheric Environment* 28: 1049–1054. https://doi.org/10.1016/1352-2310(94)90264-X.

Qi, D.H., Chen, H., Geng, L.M., and Bian, Y.Z. (2011). Effect of diethyl ether and ethanol additives on the combustion and emission characteristics of biodiesel-diesel blended fuel engine. *Renewable Energy* 36: 1252–1258. https://doi.org/10.1016/j.renene.2010.09.021.

de Queiroz Fernandes Araújo, O. and de Medeiros, J.L. (2017). Carbon capture and storage technologies: present scenario and drivers of innovation. *Current Opinion in Chemical Engineering* 17: 22–34. https://doi.org/10.1016/j.coche.2017.05.004.

Rajakumar, C. and Meenambal, T. (2015). Effect of coal ash in the stabilization of expansive soil for the pavement. *International Journal of ChemTech Research* 8: 170–177.

Resitoglu, I.A., Altinisik, K., and Keskin, A. (2015). The pollutant emissions from diesel-engine vehicles and exhaust after treatment systems. *Clean Technologies and Environmental Policy* 17: 15–27. https://doi.org/10.1007/s10098-014-0793-9.

Rizvi, S.Q.A. (2009). Emissions in an internal combustion engine. In: *A Comprehensive Review of Lubricant Chemistry, Technology, Selection, and Design* (ed. S.Q.A. Rizvi), 657. Baltimore, MD: ASTM International.

Romano, O. and Akhmouch, A. (2019). Water governance in cities: current trends and future challenges. *Water* 11: 500–509. https://doi.org/10.3390/w11030500.

Shaftel, H., Jackson, R., & Callery, S. (2019). Global temperature (12 July 2019). Retrieved from https://climate.nasa.gov/vital-signs/global-temperature/.

Sheehan, J., Camobreco, V., Duffield, J. et al. (2000). *An Overview of Biodiesel and Petroleum Diesel Life Cycles*. Washington DC, United States: US Department of Agriculture and US Department of Energy.

Shelef, M. (1995). Selective catalytic reduction of NO_x with N-free reductants. *Chemical Reviews* 95: 209–225. https://doi.org/10.1021/cr00033a008.

Singh, R., Singh, R.K., Gupta, N.C., and Guha, B.K. (2010). Assessment of heavy metals in fly ash and groundwater – a case study of NTPC Badarpur Thermal Power Plant, Delhi, India. *Pollution Research* 29: 685–689.

SMCRA (2017). *Surface Mining Control and Reclamation Act of 1977*. Washington, DC: US Congress.

Song, C. (2003). An overview of new approaches to deep desulfurization for ultra-clean gasoline, diesel fuel and jet fuel. *Catalysis Today* 86: 211–263. https://doi.org/10.1016/S0920-5861(03)00412-7.

Sushil, S. and Batra, V.S. (2006). Analysis of fly ash heavy metal content and disposal in three thermal power plants in India. *Fuel* 85: 2676–2679. https://doi.org/10.1016/j.fuel.2006.04.031.

Takatsuka, T., Inoue, S., and Wada, Y. (1997). Deep hydrodesulfurization process for diesel oil. *Catalysis Today* 39: 69–75. https://doi.org/10.1016/S0920-5861(97)00089-8.

Tanimu, A. and Alhooshani, K. (2019). Advanced hydrodesulfurization catalysts: A review of design and synthesis. *Energy & Fuels* 33: 2810–2838. https://doi.org/10.1021/acs.energyfuels.9b00354.

UN & CCAC (2019). *A Global Alliance to Significantly Reduce Methane Emissions in the Oil and Gas Sector by 2030*. United Nations Environmental Programme, Climate & Clean Air Coalition (CCAC).

Verma, C., Madan, S., and Hussain, A. (2016). Heavy metal contamination of groundwater due to fly ash disposal of coal-fired thermal power plant, Parichha, Jhansi, India. *Cogent Engineering* 3: 1179243–1179251. https://doi.org/10.1080/23311916.2016.1179243.

Vrinat, M.L. (1983). The kinetics of the hydrodesulfurization process – A review. *Applied Catalysis* 6: 137–158. https://doi.org/10.1016/0166-9834(83)80260-7.

Wang, F.C., Robbins, W.K., Sanzo, F.P.D., and McElroy, F.C. (2003). Speciation of sulfur-containing compounds in diesel by comprehensive two-dimensional gas chromatography. *Journal of Chromatographic Science* 41: 519–523. https://doi.org/10.1093/chromsci/41.10.519.

Ward, C.R. (2016). Analysis, origin and significance of mineral matter in coal: An updated review. *International Journal of Coal Geology* 165: 1–27. https://doi.org/10.1016/j.coal.2016.07.014.

Wood, H. (2012). *Disasters and Minewater: Good Practice and Prevention*. London: IWA Publishing.

Zheng, M., Reader, G.T., and Hawley, J.G. (2004). Diesel engine exhaust gas recirculation – A review on advanced and novel concepts. *Energy Conversion and Management* 45: 883–900. https://doi.org/10.1016/S0196-8904(03)00194-8.

Zima, S. (2001). Historical review. In: *Internal Combustion Engine Handbook – Basics, Components, Systems, and Perspectives* (eds. R.v. Basshuysen and F. Schafer), 1–8. Warrendale, PA: SAE International.

Part V

Environmental Analysis

10

Environmental Analysis of Soil, Water, and Air

Paige Teehan[1], Kyra A. Murrell[2], Romano Jaramillo[2], A. Paige Wicker[3], Robert Parette[4], Kevin A. Schug[3,5], and Frank L. Dorman[1]

[1] *Department of Biochemistry and Molecular Biology, The Pennsylvania State University, University Park, PA, USA*
[2] *Department of Chemistry, The Pennsylvania State University, University Park, PA, USA*
[3] *Department of Chemistry and Biochemistry, The University of Texas at Arlington, Arlington, TX, USA*
[4] *Maston and Associates, Inc., State College, PA, USA*
[5] *Collaborative Laboratories for Environmental Analysis and Remediation (CLEAR), The University of Texas at Arlington, Arlington, TX, USA*

10.1 Water and Soil Monitoring

The determination of the identity and quantity of hydrocarbon composition in water and soil matrices is critical in several applications. These determinations may be undertaken in the form of a "class" or group type analysis (gasoline range organics [GROs], total petroleum hydrocarbon [TPH], etc. as defined in the following text), or they may be "detailed" hydrocarbon analyses, where the ultimate goal is the complete separation of each specific compound that makes up a more complex mixture. While it is currently impractical to achieve a complete separation of each individual compound in complex hydrocarbon analyses, significant strides have been made toward this goal of complete analytical profiling, through sophisticated separation, detection, and data processing advancements. Often, the decision on which approach to undertake is based on the desired goals from the analysis. If organic range, or class, is desired, then the analytical demands are less rigorous than if detailed trace composition or hydrocarbon (forensic) "fingerprinting" is required.

When compounds are released into the environment and may potentially impact the health and welfare of living organisms, this release generally first impacts one or more of the common environmental matrices: air, water, or soil. Generally, living systems are impacted by either direct interaction with these environmental systems, or matrices, or indirect interaction (ingestion of other materials in direct

Analytical Techniques in the Oil and Gas Industry for Environmental Monitoring, First Edition. Edited by Melissa N. Dunkle and William L. Winniford.

contact with these matrices, for example). In order to determine the source and the extent of a potential impact, it is often desired to achieve a detailed separation, so that both the toxicological impact may be estimated as well as source identification and apportionment may be determined. Typically, in cases where environmental impact is the reason for the analytical determination, increasingly, the use of more sophisticated fingerprinting techniques is becoming the norm, as class-type separations often do not yield enough data quality for source identification. This chapter discusses the analytical approaches of both class-type separations, as well as detailed hydrocarbon determinations as applied to cases where addressing potential environmental impact is the concern.

10.2 Total Petroleum Hydrocarbons in Soil

10.2.1 Introduction

Crude oil and refined petroleum products are complex mixtures possessing a wide variety of physical and chemical properties. These products consist of hundreds of hydrocarbon compounds ranging in length from C_1 to C_{40} or higher. TPH are a measure of all extractable hydrocarbons in a sample by a given analytical method. It is often used to identify and monitor environmental samples for oil and gas contamination.

Due to the complex nature of petroleum products, there is no single analytical method capable of detecting the entire range of petroleum-derived hydrocarbons. The distillation of refined petroleum products from crude oil depends on the different boiling point properties of each product, resulting in characteristic hydrocarbon ranges for each. TPH analyses can be tailored to fit these hydrocarbon ranges based on the suspected type of contamination. For example, gasoline is mostly composed of hydrocarbons ranging from C_6 to C_{12}, known as gasoline range organics (GRO), while diesel fuel is composed of higher molecular weight hydrocarbons ranging from C_{11} to C_{25}, or diesel range organics (DROs) (Farrell-Jones 2003; Speight 2005). Consequently, GRO analyses will utilize methods specific for volatile compounds, while DRO analyses will focus on semivolatile compounds. Refined petroleum products often have overlapping characteristic hydrocarbon ranges, making the exact identification of a product more challenging.

Analysis of petroleum-derived hydrocarbons is further complicated by the type of contaminated material or matrix. The monitoring of soil, water, or air all require different approaches, even when the same type of contamination is suspected. This section will focus on the challenges associated with soil contamination and present an overview of current methodologies that are approved by the United States Environmental Protection Agency (EPA).

10.2.2 Soil as a Matrix

Soil composition, like that of petroleum products, varies greatly in physical and chemical properties. A defining feature of soils is their texture, which is determined by the mineral particle composition. Soil mineral particles are divided into three classes, based on size and chemical properties: sand, silt, and clay. Sand particles are the largest, ranging from 2.0 to 0.5 mm, and are usually composed of quartz or other silicates. Silt particles are smaller, ranging from 0.5 to 0.002 mm. These particles are granular, like sand, and are often composed of quartz or feldspars. Clay particles are the smallest, with diameters less than 0.002 mm, and are composed of silicates from three main categories: kaolinites, illites, or montmorillonites. Soil textural classes are defined by the percent abundance of each type of mineral particle present. Most soils are a type of loam, or a combination of all three mineral particles (Brady and Weil 2014; Varona-Torres et al. 2017). Soil structure is determined by the aggregation of mineral particles and the network of pores that form. The pore network allows for the movement of air and water throughout the soil, as well as plant and organism growth. Contamination of these spaces with petroleum products changes the natural pathways available for air and water, altering the chemical and physical properties of the soil. Characteristics such as pH, total organic carbon (TOC), and permeability can be drastically changed due to petroleum product contamination (Devatha et al. 2019).

Soil texture and structure influence the impact of weathering on an area. Weathering is defined as the change in the composition of a soil or contaminant due to naturally occurring processes. Adsorption, volatilization, dissolution, biotransformation, photolysis, oxidation, and hydrolysis are all capable of altering the composition and distribution of oil and gas in the environment (Douglas et al. 2018). Highly volatile, low molecular weight hydrocarbons, such as those found in GRO fractions, are more susceptible to the effects of weathering. These compounds are often lost to volatilization unless samples are taken immediately following a contamination event. Mischaracterization of the weathered GRO fraction may lead to incorrect identification of the petroleum contaminants. It is the responsibility of the analyst to ensure that the weathering of samples is accounted for at the start of an analysis.

10.2.3 Sample Preparation

Sample preparation can be divided into four categories:

- Collection and preservation
- Extraction
- Concentration
- Cleanup

The steps taken in each category will depend greatly on both the suspected type of contamination and the matrix, and it is important to note that not all soil samples will require concentration and/or cleanup. The following information is provided to help guide the preparation of a contaminated soil sample for analysis of petroleum-derived hydrocarbons.

10.2.3.1 Collection and Preservation

Soil sample collection and preservation should always be performed in consultation with the analytical lab or regulatory agency guidelines of the area, to ensure valid analytical results and to prevent the loss of analytes due to improper collection containers or storage conditions. Samples should be representative of the area of interest and maintained under conditions that preserve the chemical and physical properties at the time of collection (Weisman 1998). Spills involving gasoline should be sampled as close to the time of the spill as possible, to prevent loss of volatile compounds through weathering. In cases where multiple contaminants are suspected, several different samples may be needed for complete analysis of volatile, semivolatile, and nonvolatile compounds.

10.2.3.2 Extraction

Table 10.1 presents a summary of EPA methods for the analysis of hydrocarbons in soil. It is important to remember that no single extraction method can cover the entire range of petroleum-derived hydrocarbons. Therefore, extraction methods should be matched to the suspected type of contamination.

10.2.3.3 Concentration

The concentration of extracts ensures that the lower detection limits required from an extraction method can be achieved, but this is also a point where there is a high risk for analyte loss. Volatile compounds are concentrated by either sorbent or cryogenic trapping, usually during purge-and-trap (EPA-5030B 1996; EPA-5035 1996). Semivolatile and nonvolatile compounds are concentrated by a Snyder column or Kuderna-Danish concentrator, nitrogen evaporation, or vacuum evaporation. Kuderna-Danish concentrators and nitrogen evaporation are the most commonly recommended concentration methods for semivolatile compounds, while the less common vacuum evaporation method is best suited to nonvolatile compounds.

10.2.3.4 Cleanup

Cleanup methods remove interferences from non-petroleum substances; cleanups allow many extraction methods to report a TPH value instead of a gas and oil (G and O) value. In soils, high molecular weight organic acids and microbial or plant breakdown materials may interfere with the analysis of petroleum-derived

Table 10.1 Extraction and screening methods for petroleum-derived hydrocarbons in soils.

EPA method	Title	Comments	References
3540C	Soxhlet extraction	Extracts nonvolatile and semivolatile compounds. Long extraction times and specialized glassware required.	EPA-3540C (1996)
3541	Automated Soxhlet extraction	Extracts nonvolatile and semivolatile compounds. Employs commercially available apparatus to reduce extraction times and produce comparable results to Method 3540C.	EPA-3541 (1994)
3545A	Pressurized fluid extraction (PFE)	Extracts semivolatile organic compounds and DRO. Employs automated extraction system with elevated temperature and pressure to produce comparable results to Method 3540C, with less solvent and shorter extraction times.	EPA-3545A (2007)
3546	Microwave extraction	Extracts semivolatile organic compounds. Employs microwave energy to create elevated temperature and pressure, producing comparable results to Method 3540C, with less solvent and shorter extraction times.	EPA-3546 (2007)
3550C	Ultrasonic extraction	Extracts nonvolatile and semivolatile compounds. Caution: method has limited contact time between sample and solvent and may be less rigorous than other extractions.	EPA-3550C (2007)
3560	Supercritical fluid extraction (SFE) of total recoverable petroleum hydrocarbons (TRPHs)	Extracts TRPHs from soils/solids. Employs supercritical fluids as extraction solvent. Caution: not suitable for low boiling point TRPHs like GRO.	EPA-3560 (1996)
3561	Supercritical fluid extraction of polynuclear aromatic hydrocarbons (PAHs)	Extracts PAHs from soils/solids. Employs supercritical fluids as extraction solvent. Caution: not suitable for direct extraction of liquid samples.	EPA-3561 (1996)

(*continued*)

Table 10.1 (Continued)

EPA method	Title	Comments	References
3815	Screening solid samples for volatile organics	In field screening for known VOC contamination detectable by photoionization detector. Used to screen known contaminated sites to determine if low concentration or high concentration VOC analysis should be used.	EPA-3815 (2007)
5021A	Volatile organic compounds in various sample matrices using equilibrium headspace analysis	Static headspace extraction for low concentration VOCs in samples. Can be applied to higher concentration VOCs in solid samples after solvent extraction.	EPA-5021A (2014)
5030B	Purge-and-trap for aqueous samples	Combined with Method 5035 for analysis of high concentration VOCs in soils/solids.	EPA-5030B (1996)
5035	Closed-system purge-and-trap and extraction for volatile organics in soil and waste samples	Used for analysis of low concentration VOCs. High concentration VOCs in solid samples can be extracted using this procedure when combined with sample introduction outlined in Method 5030B.	EPA-5035 (1996)
9074	Turbidimetric screening method for total recoverable petroleum hydrocarbons	In field screening for broad range of hydrocarbons in soil. May determine lack of hydrocarbon contamination at a site.	EPA-9074 (2007)

hydrocarbons. The most common cleanup method passes the extract through a sorbent packed column to remove the non-petroleum compounds. EPA Methods 3611B and 3630C detail sample cleanup methods using alumina and silica gel columns for samples containing petroleum wastes or polyaromatic hydrocarbons (PAHs), respectively (EPA-3611B 1996; EPA-3630C 1996). Additional cleanup procedures can be found in EPA Method 3600C (EPA-3600C 1996).

10.2.4 Sample Analysis

The most commonly used analytical methods for TPH analysis in soils include gravimetric, immunoassay, infrared (IR) spectroscopy, and gas chromatography (GC) applications. A broad overview of each analytical method, including the advantages and disadvantages of each, is presented in this section.

Using gravimetric methods, every component in an extract that was not removed during the solvent evaporation stage is weighed to determine a G and O value (EPA-9071B 1998). If a sample cleanup is performed, then a TPH value is measured instead. Gravimetric methods are simple, quick, inexpensive, and useful for contamination by oily sludges. Due to the heating of the extracts during solvent evaporation, they cannot be used to measure low molecular weight hydrocarbons that volatilize below 85 °C. Gravimetric methods do not give any information on individual hydrocarbons, though it is possible to gain some knowledge of the chemical classes present by varying the cleanup procedures used.

Immunoassay test kits, such as those listed in EPA Methods 4030 and 4035 (PETRO RISc Soil Test and PAH RISc Soil Test from EnSys Inc. and EnviroGard™ Petroleum Fuels in Soil from Millipore Inc.[1]), are commercially available for use in both laboratory and field analyses, providing estimates of the TPH or PAH concentration (EPA-4030 1996; EPA-4035 1996). An immunoassay is a biochemical test that measures the presence or concentration of a molecule, either a macromolecule or small molecule, in a solution through the use of an antibody or antigen. The test kits mentioned earlier employ antibodies against specific TPH/PAH compounds that are immobilized in a test cell or membrane. Soil samples are extracted, per kit directions, and added to the test cell with an enzyme conjugate reagent. The enzyme conjugate competes with the sample to bind to the antibodies, and the test results are interpreted by comparing the response of the sample with a simultaneously run standard. Immunoassays are simple, quick, and inexpensive; however, they do not provide specific information about the compounds present and are prone to interference from other components that may bind the test kit's antibodies. All manufacturer guidelines should be followed when selecting an appropriate test kit.

IR methods measure TPH values by including every compound extracted by the solvent that was not removed during silica gel cleanup. For analysis by EPA Method 8440, samples should be prepped following EPA Method 3560 guidelines (EPA-3560 1996; EPA-8440 1996). The samples should be extracted using supercritical carbon dioxide and collected in tetrachloroethylene or a similar solvent with no C—H bonds present. IR spectroscopy measures vibrations that occur when molecules absorb infrared energy, and each bond type and functional group has a specific absorption frequency. For TPH analysis, C—H bonds at frequency $2950\,cm^{-1}$ are measured, which is why the extraction solvent should not contain any C—H bonds that may skew the results. TPH concentration is calculated by comparing the sample response with a calibration plot. IR methods are quick, but the method is prone to interferences from compounds that are not completely

1 ™ Trademark of Millipore Inc.

removed during the silica gel cleanup. This method is also not recommended for highly volatile compounds that may be lost during sample preparation.

Gas chromatography (GC), which was extensively discussed in Chapters 3 and 4, separates extracts into individual compounds based on volatility and stationary phase affinity. For TPH analysis, GC may be combined with a flame ionization detector (FID), photoionization detector (PID), or mass spectrometer (MS). FIDs detect all carbons in a sample that will burn in a hydrogen flame. Conditions for detecting GROs and DROs using GC/FID are found in EPA Method 8015D (EPA-8015D 2003). EPA Method 8021B outlines the conditions for measuring aromatic and halogenated volatiles in soil using GC/PID (EPA-8021B 2014). PIDs use ultraviolet (UV) lamps to ionize a small portion of the analyte molecules, meaning it is a nondestructive method. Therefore PIDs can be paired with other detectors for confirmatory analyses and are more amenable for use in portable field GCs. Several EPA methods (EPA-8260D 2018; EPA-8270E 2018; EPA-8275A 1996) outline the analysis of hydrocarbons in soil using GC/MS, where MS detectors provide mass spectral fragmentation data in addition to retention time data for each compound in a mixture.

A major advantage of GC analyses is the detailed data obtained on the individual components of the petroleum product contamination in an area. This enables the analyst to determine not only the amount of weathering a sample has undergone but also the source of the contamination by matching the spilled product with the original. This process is known as fingerprinting, and more details are available at the end of this chapter. Knowing the individual compounds present at a site also aids in risk assessment and remediation for the impacted area. Drawbacks for these analyses are the cost, time, and complexity of the data collected, which often requires an experienced user to interpret the results.

While GC is a powerful separation tool, gasoline and diesel products often contain many compounds that cannot be fully resolved, resulting in a "hump" on the chromatogram, as seen in the elution pattern obtained for motor oil in Figure 10.1. Though specific components that make up this unresolved complex mixture may not be clear, these compounds still contribute to the total petroleum-derived hydrocarbons of the sample, and therefore, must be accounted for in the TPH calculation. This requires a baseline-to-baseline integration as opposed to a peak-to-peak integration, which only includes fully resolved peaks (Weisman 1998). DRO concentration is determined by summing the areas of all peaks eluting between C_{10} and C_{28}, after the subtraction of column bleed. GRO concentration is determined by summing the areas of all peaks eluting between 2-methylpentane and 1,2,4-trimethylbenzene. GRO analysis does not normally require column bleed subtraction due to the early elution of compounds. More information on the calculations of analyte concentrations and calibration verifications can be found in EPA Method 8000D (EPA-8000D 2018).

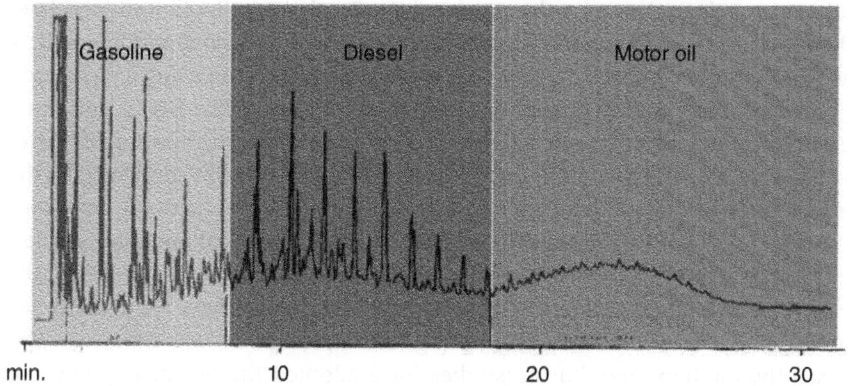

Figure 10.1 Oil fractions as determined by gas chromatography. Source: From Sadler and Connell (2003). License CC 4.0.

10.3 Volatile Organic Compound Analysis

10.3.1 Introduction

Concern around volatile organic compound (VOC) analysis of unconventional oil and gas developments has been increasing due to the rapid growth of horizontal drilling and hydraulic fracturing industries. Methane release from unconventional oil and gas development facilities, due to inefficiencies in collecting, processing, and transporting natural gas, has warranted discussions on whether unconventional oil and gas development industries provide cleaner energy sources compared with conventional oil and gas industry products. Furthermore, unconventional oil and gas development facilities have been shown to release other hazardous VOCs, albeit at significantly lower levels than other oil and gas refinement processes. The proximity of unconventional oil and gas development facilities to residential areas has led to community involvement and advocacy to conduct more VOC monitoring and health studies. Studies monitoring VOC contamination often focus on two areas: first, the release of the greenhouse gas methane primarily from unconventional oil and gas development facilities, and second, the release of other hazardous VOCs, such as the aromatic compounds benzene, toluene, ethylbenzene, and xylene (BTEX), nitrogen oxides, sulfur oxides, and other hydrocarbons from petroleum refinement sites.

10.3.2 Methane Monitoring

Natural gas is composed of 60–90% methane, meaning there are ample opportunities for methane escape during natural gas production. An estimated 13 Tg of methane are released per year throughout the natural gas production supply chain (Alvarez et al. 2018). Methane emissions have traditionally been monitored by leak detection using an infrared camera, followed by quantification using a Bacharach Hi-Flow Sampler (BHFS), which was introduced in 2001. The BHFS samples a measured volume of air around a leak and determines the concentration of hydrocarbons present. However, issues with the BHFS that have been confirmed by the manufacturer Bacharach have led to severe underreporting of methane emissions (Connolly et al. 2019; Howard 2015). This has affected several emission studies; for example, a recent survey of 190 natural gas sites used the BHFS for 17% of the measurements taken (Allen et al. 2013). These discrepancies have led to increased interest in alternative methane monitoring techniques.

Recently, methane monitoring studies have adopted the use of top-down or bottom-up approaches for emission estimates (Allen 2014; Alvarez et al. 2018; Brandt et al. 2014; Harriss et al. 2015). Top-down studies determine aggregate emissions over large areas using measurements typically obtained through aircraft or satellites. Bottom-up studies perform multiple emission measurements across several unconventional oil and gas development facilities and use this data to extrapolate average emissions over larger geographic areas. Top-down and bottom-up studies involve similar sampling and analytical methods; however, they often result in different estimates, with bottom-up approaches routinely reporting lower emissions. This has been attributed to faulty collection equipment, such as the previously discussed BHFS, as well as the omission of contributions from abnormally high emission sites, deemed super emitters. Conversely, top-down studies are not optimized for point source analysis, leading to overestimates caused by the emissions of super emitters. As a result, monitoring campaigns often incorporate both approaches when collecting data.

10.3.2.1 Cavity Ring-Down Laser Spectrometry Techniques

The Picarro cavity ring-down laser spectrometer (CRDS) is the most commonly used instrument for methane monitoring in both top-down and bottom-up approaches. It operates by directing a laser tuned to the spectral wavelength of methane into an optical cavity composed of at least two mirrors. The light intensity builds through constructive interference. Afterward, the laser is turned off. The light intensity decays exponentially while it is reflected off the mirrors producing a "ring-down" time. The degree to which light is absorbed by the sample in the cavity reduces the ring-down time, allowing for the quantification of the sample in the cavity. During the ring down measurement, the light is reflected off the mirrors thousands of times, resulting in a path length of approximately 20 km. The extended path length results in highly sensitive measurements of <1 ppb methane (Crosson 2008; Payne et al. 2017). The CRDS can be tuned to compounds other than methane, such as carbon dioxide and water vapor, and several novel monitoring techniques, described in the following text, have arisen from the use of CRDS.

Two studies have derived an inverse Gaussian method to model methane emissions rates based on EPA Other Test Method 33A (EPA 2014). Here CRDS was used to determine the mixing ratio of methane in air, while a weather station attached to a mobile platform determined wind speed and direction. Plotting methane mixing ratios against wind direction results in a Gaussian distribution and provides peak methane concentration. The methane emission rate can then be determined as a function of the peak concentration, wind speed, and dispersion parameters (Lan et al. 2015; Robertson et al. 2017). This technique is notable for circumventing the mobile platform's restriction of only being able to take measurements at ground level by modeling a methane plume dispersion using the collected wind data. A similar study using the inverse Gaussian method removed the need for atmospheric transport modeling by equipping vehicles with a mast sporting six air inlets, at varying heights, all routed to a single CRDS. The methane concentration readings were correlated to the height of the inlet to provide an image of the plume flux. Because plume dispersion does not need to be modeled, it was reported that emission rates could be calculated in six minutes following sample collection (Rella et al. 2015).

Another technique now considered for both top-down and bottom-up approaches uses tracer flux to determine facility-level methane emissions (Alvarez et al. 2018; Omara et al. 2016; Swarthout et al. 2015). This technique involves releasing tracer gases at a known emission rate from the target facility; then the plumes of both the tracers and methane are measured downwind using CRDS. The tracer flux principle depends on the downwind concentration of a gas being directly proportional to the flow rate at its source. By using a tracer, the ratio of the two downwind concentrations is equal to the ratio of the known emission flow rate of the tracer and the emission flow rate of methane, eliminating the need for collecting the meteorological information necessary in the inverse Gaussian plume modeling (Roscioli et al. 2015).

10.3.2.2 Mobile Platforms for Bottom-Up Analyses

A unique bottom-up method utilizes car-mounted mobile platforms to measure methane emissions. The automobiles are equipped with compact, rugged instrumentation, typically gas chromatographs or spectroscopic equipment, to allow for direct, point source ambient air analysis. These mobile platforms can provide both on-site and downwind measurements of unconventional oil and gas development facilities, and this approach provides a more comprehensive analysis of methane emissions at individual facilities and over larger geographic areas. Variations on mobile platforms may also be coupled with CRDS for methane quantification. Vehicles are normally equipped with anemometers and global positioning system

(GPS) to account for wind conditions and to correlate readings to geographic coordinates, respectively (Lan et al. 2015; Payne et al. 2017; Rella et al. 2015; Robertson et al. 2017; Swarthout et al. 2015).

10.3.2.3 Aircraft-Based Top-Down Analysis

Aircraft-based mass balance experiments are the most common approach for top-down methane monitoring. Geographic areas of interest are outlined, and numerous measurements of methane flux are recorded throughout the area. The measurements are associated with GPS coordinates and are often presented as color gradients imposed on satellite photography of the area. As such, emissions can be attributed to specific facilities, while still collecting an overall analysis of a larger area (Karion et al. 2015; Karion et al. 2013; Lavoie et al. 2015; Pétron et al. 2014; Smith et al. 2015). CRDS is another heavily used method to determine methane mixing ratios. In a recent study, a 2-m remote-control aircraft mounted with a custom laser-based spectroscopic instrument was able to collect measurements much closer to emission sources compared to standard airborne platforms (Nathan et al. 2015).

10.3.3 Non-Methane VOC Monitoring

The greenhouse gas methane is not the only concerning compound that may be released into the environment during oil and gas production. Harmful VOCs, such as nitrogen oxides, sulfur oxides, hydrogen sulfide, carbon dioxide, and a variety of hydrocarbons, are also routinely monitored at production sites and contamination events. Several of these compounds are known to be toxic or carcinogenic, like BTEX, and others are also greenhouse gases, like CO_2. While monitoring of toxic VOCs is typically performed alongside methane monitoring, the collection methods and analytical techniques used to determine their concentration vary.

10.3.3.1 Air Sampling

Air samples are often collected using evacuated, stainless steel canisters (Prenni et al. 2016; Su et al. 2019; Swarthout et al. 2015), which prevent the sample from escaping or being degraded by light. Additionally, the interior of the canister is treated to be chemically inert. Sampling can be performed by either passive, active, or grab methods. Passive sampling involves opening an inlet in a previously evacuated canister, generating a negative pressure, which draws the sample into the canister at a controlled flow rate. Active sampling, as opposed to passive sampling, uses a pump to pull ambient air through a sorbent tube where VOCs are collected. The flow rate of the pump is calibrated so that the concentration of VOCs can be determined in the sample volume. Grab sampling is similar to passive sampling, except the sample is introduced into a canister at an uncontrolled rate (Macey et al. 2014). Grab sampling is typically performed when a strong odor is observed, and samples must be collected rapidly.

10.3.3.2 Analysis of Air Samples

A variety of analytical methods are utilized for VOCs, and analyses can be tailored to specific compounds of interest. For example, a recent study evaluated samples for nitrogen oxides using chemiluminescence detectors, while sulfur oxides were monitored with UV fluorescence (Prenni et al. 2016). Gas chromatography-flame ionization detection (GC-FID), gas chromatography-photoionization detection (GC-PID), and gas chromatography-mass spectrometry (GC-MS) are also used in hydrocarbon analysis, especially in risk assessment studies monitoring BTEX (de Blas et al. 2011; Spinelle et al. 2017; Varona-Torres et al. 2017). GC analyses require the use of reference standards (external calibration standards or internal standards) for accurate quantification, but qualitative analyses may be performed without these. Air samples are often concentrated before GC analysis using cryogenic or sorbent trapping to achieve lower detection limits. Another method involves the use of proton transfer reaction mass spectrometry (PTR-MS), which involves using hydronium ions to ionize VOC samples, resulting in reduced fragmentation compared to typical electron ionization (EI). With PTR-MS, analytes are thus identified by their protonated mass (Goetz et al. 2017; Goetz et al. 2015; Prenni et al. 2016; Swarthout et al. 2015). PTR-MS instruments can analyze ambient air samples directly, in real time, and are considerably smaller than traditional GC-MS systems, making them ideal for fieldwork. Direct sampling sacrifices the enrichment steps associated with GC-MS sample preparation; however, many PTR-MS systems are equipped with quadrupole analyzers, providing sufficient detection limits down to the parts per billion range (Majchrzak et al. 2018).

10.4 Water Analysis

10.4.1 Introduction

Most oil and gas companies these days will acknowledge that they are involved just as much in the water business as they are in the oil and gas business. The transition of technologies to accommodate resource extraction from conventional to more unconventional oil and gas (UOG) reservoirs has raised additional concerns about water, especially its sourcing, its quality, and its handling/use. The penetration of aquifers during drilling, through completion stages, and during resource recovery is a concern for groundwater quality; meanwhile, the sourced petroleum will return to the surface alongside a large amount of wastewater, which needs to be treated or disposed of. Spills above-ground can also threaten surface water quality. From the monitoring of water quality in proximity to oil and gas extraction activities, to the determination of the effectiveness of different wastewater treatment technologies, there exists a significant need for a variety of analytical methodologies to measure the prevalence of organic, inorganic, and biological constituents in complex aqueous samples. These measurements can help ensure the safety of drinking water, as well as the presence or absence of constituents that can impact reuse or repurposing of treated or untreated waste.

UOG extraction is facilitated by the injection of large volumes of fluid at high pressures to fracture low porosity shale formations. The fluids used in well stimulation are aqueous solutions comprised of approximately 1% sand and other additives, with water making up the remaining 99%. The complex mixture of hydraulic fracturing fluid may contain various additives (Table 10.2) to enable efficient oil and gas extraction; these include proppants or sand, which aid in opening the shale fractures for constant flow of commodity; gellants for increased viscosity; friction reducers for reduced friction; biocides to limit microbial growth; oxygen scavengers for corrosion prevention; and surfactants to reduce surface tension and break emulsions (Vidic et al. 2013). Due to the complexity of aqueous mixtures/suspensions and the often-proprietary nature of the identities and

Table 10.2 Typical chemical additives used in hydraulic fracturing. FracFocus Chemical Disclosure Registry.[a)]

Additive type	Use	Main compound	% v/v
Acid	Helps dissolve minerals and initiate cracks in shale	Hydrochloric acid	0.07
Corrosion inhibitor	Prevent pipe corrosion	*N,N*-dimethylformamide	0.05
Friction reducer	Decrease pumping friction	Polyacrylamide	0.05
Clay stabilizer	Prevent clay from swelling	Potassium chloride	0.034
Cross-linker	Maintain fluid viscosity as temperature increases	Borate salts	0.032
Scale inhibitor	Prevent scale deposits in pipe	Ethylene glycol	0.023
Breaker	Promote breakdown of gelling agent	Ammonium persulfate	0.02
Iron control	Prevent metal oxide precipitation	Ammonium bisulfite	0.004
Biocide	Control bacterial growth	Glutaraldehyde	0.01
Gellant	Improve proppant placement	Guar gum	0.5

a) This table was created for FracFocus and has not been updated since 2011.
Source: FracFocus (2015).

composition of the additives used, analysis of water associated with UOG can be quite challenging.

Groundwater is the water found below the Earth's surface, and as it moves through geologic formations, inorganic materials such as ions dissolve into the water. Groundwater is a source of drinking water for many landowners, and as such its quality is of paramount importance. This view and concern were highlighted in a recent article, where Stigler-Granados et al. evaluated water quality and health in high fracking regions, examining the attitudes and perceptions associated with private water wells near UOG activity (Granados et al. 2019). Maximum contamination limits for potable water have been set as part of the US Environmental Protection Agency's Clean Water Act; these values are often used as a reference for establishing groundwater quality. Some groundwater is naturally non-potable, often due to its high salinity (total dissolved solids, TDS). Groundwater near UOG activities has been extensively studied and characterized by a variety of analytical methods (Burton et al. 2016; Fontenot et al. 2013; Hildenbrand et al. 2015; Hildenbrand et al. 2016).

During well stimulation, an average of 12 million gallons of water is pumped downhole, and due to the high pressure in the well, a fluid, known as flowback water, will return to the surface carrying with it high levels of salts, dissolved solids, and additives. Additionally, produced water is recovered following flowback; produced water containing additives and high levels of salts, may also contain potentially harmful amounts of metals and naturally occurring radioactive materials (NORMs). These two waste streams are either disposed of in saltwater disposal wells or treated for possible reuse. Produced water reuse was thoroughly reviewed by Liden et al. (2018) with a focus on the challenges associated with reusing this water in hydraulic fracturing and the types of available technologies.

Comparatively, produced water forms a much more complex matrix than what is seen in groundwater. In addition to the additives used by the oil and gas companies, natural components of hydrocarbons are present, such as light distillates, middle distillates, and residues (wax hydrocarbons). The salt level in produced water may be as low as 500 ppm TDS to as high as supersaturated, at greater than 40,000 ppm TDS (Liden et al. 2017). The high level of salts poses a significant challenge in sample preparation and analysis.

10.4.2 Sample Preparation

Prior to sample preparation and analysis, it is important to consider proper sample collection and preservation. In the case of water samples, four types of sample blanks are typically used to monitor any potential contamination during sampling, transport, and sample handling. An analytical blank, which mimics the sample matrix containing no analytes, is used to expose contamination that may occur during sample preparation, including effects from solvent impurities and other consumables. A field blank, prepared in the lab prior to sampling, is comprised of a sample matrix with no analytes. It is opened at the site of collection to account for any contamination from the environment at the time of sampling. A travel blank is prepared in the lab prior to sampling. It also travels to the collections site; however, it is not opened during sampling. The travel blank is used to represent any contamination potentially caused by the travel environment and/or the sampling container. A fourth blank is prepared post-sample collection in the lab to account for any contamination associated with sample storage. The storage blank is stored along with the collected samples to be analyzed (Carlton et al. 2017).

Water samples associated with UOG are complex due to their varying composition; analytes can be present at trace levels in a highly concentrated brine solution. This makes sample cleanup an imperative step in removing potential matrix interferences while preconcentrating or isolating the analytes of interest (Chirila et al. 2006). Several modes of extraction may be applied to water samples, including liquid–liquid extraction (LLE), solid phase extraction (SPE), solid phase microextraction (SPME), and headspace (HS) analysis.

LLE provides a cost-effective approach to isolating target analytes. LLE separates analytes based on their relative solubilities in two different immiscible liquids. An organic solvent (non-polar) is added to an aqueous sample (polar), and the desired components partition into the organic layer, depending on their relative degree of affinity for the organic versus aqueous solutions. The solvent can then be evaporated to concentrate the analytes further. Although LLE is laborious and may use large volumes of solvent, it is widely used as it allows for high throughput at low cost and increases sensitivity for target analytes. In one of the first efforts to characterize oilfield wastewater, Thacker et al. (2015) demonstrated the use of LLE to extract organics from flowback and produced water samples for analysis by GC-MS.

A method for the extraction and identification of pesticides and organohalogens from wastewater was developed by Dorman and coworkers (Prebihalo et al. 2015). Here they used various sample preparation approaches to extract emerging contaminants of concern from a local water treatment facility. A modified US EPA 3510C extraction technique was used to extract and concentrate a starting sample volume of 500 ml to 2 ml. Samples were serially extracted three times under acidic conditions (pH = 2) and three times under basic conditions (pH = 11) using dichloromethane as the extraction solvent. If an emulsion was observed, the resulting sample was then centrifuged for three minutes at 3000 rpm to separate the phases. A Kuderna-Danish evaporation apparatus was used to concentrate the samples to a final volume of 2 ml. The extracted compounds were analyzed by comprehensive gas chromatography coupled (GC × GC) to time-of-flight mass spectrometry (TOFMS), a comprehensive multidimensional analysis that provides enhanced resolution of complex samples.

SPE was used to extract and concentrate semivolatile and nonvolatile analytes from environmental matrices; here the aqueous sample was passed through a packed bed in a cartridge, similar to the stationary phases found in liquid chromatography columns. The analytes were extracted and concentrated based on their physical and chemical properties. SPE is also useful in sample clean up by either trapping impurities on the packed bed of sorbent, or by allowing impurities to pass through prior to the analytes of interest. In contrast to LLE, SPE requires less solvent, but it is relatively more expensive due to the cost of the SPE cartridges and the associated apparatus (vacuum or positive pressure manifolds) needed. Prior to analysis by liquid chromatography-mass spectrometry (LC-MS), Cluff et al. (Cluff et al. 2014) demonstrated the use of a C18 SPE cartridge to extract ethoxylated surfactants from produced water.

SPME is a type of SPE that requires no solvent to extract target analytes. In this case, volatile or semivolatile analytes are trapped on a fiber coated with a selective sorbent. The fiber is then transferred to the injection port of the analytical instrument where the analytes are desorbed onto the analytical column, explaining why SPME is usually coupled to GC. Another solvent-less extraction method used with GC is HS analysis, which is well suited for highly volatile target analytes. The HS is the "empty" gas space above the sample in the top portion of a sealed vial. The volatile components in the sample diffuse into the gas space upon heating and agitation of the vial; this space is then sampled using a syringe needle through a septum in the vial cap for analysis. By only sampling the HS, nonvolatile interferences remain in the sample matrix. Both SPME and HS have the advantage of very little sample handling, limiting the opportunity for sample contamination. These methods of extraction may also be automated to increase throughput. Schenk et al. analyzed the effects of simulated downhole conditions on leachates from proppants by HS coupled to GC with both mass spectrometry and vacuum ultraviolet (VUV) spectroscopy detection (Schenk et al. 2019). In general, for the measurement of volatiles, such as light alcohols in water, a protocol similar to that used for forensics blood alcohol determinations can be used. This workflow ideally includes a range of positive and negative quality control steps, as well as the use of a dual column set-up for simultaneous duplication of analysis from a single injection.

10.4.3 Sample Analysis

The US EPA has approved many analytical methods (Figure 10.2) for the analysis of drinking water and wastewater; however, there are no approved methods that have been validated for either flowback water or produced water with dissolved

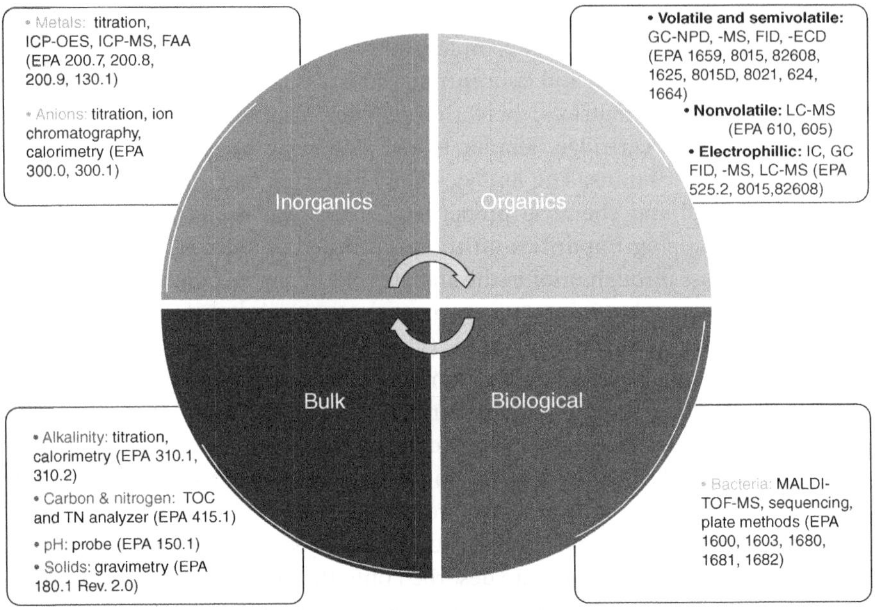

Figure 10.2 Analytical and EPA approved methods for the analysis of wastewater. Source: Santos et al. 2019. Reproduced with permission, Copyright (2019), American Chemical Society.

salt content as high as 350,000 mg/l TDS (Liden et al. 2019, Santos et al. 2019). Recently, several additional analytical methods have been applied to water related to UOG activities; while these methods make use of best practices from EPA methods for sample collection and handling, new methods have been created to account for more varied and unknown analytes. To adequately characterize water, many parameters should be monitored, including basic water quality (pH, TDS, ion composition), organic compounds (volatile and semivolatile organic compounds), inorganic compounds, and microbial species.

Bulk measurements allow for a quick and relatively simple assessment to evaluate overall water quality. These measurements include testing pH, temperature, alkalinity, salinity, total suspended solids, TDS, TOC, and total nitrogen. While these measurements do not give evidence of specific analytes, they do provide important information as to the general water quality. pH measurements provide critical information for improving the fracturing fluid and may be useful in determining environmental impacts. Measurements of alkalinity and water hardness may provide evidence of potential scale buildup in equipment. Some of these basic measurements may be taken in the field by use of a probe, while others require instrumentation in an analytical lab. Alkalinity levels may be monitored by titration as in EPA method 310.1 or by a colorimetric method as in EPA method 310.2 (EPA-310.1 1978; EPA-310.2 1974). TOC is typically determined by a benchtop organic carbon analyzer. Due to the high salinity of produced water, it is common practice to dilute samples for bulk measurements to diminish any potential interferences. Measurements of total nitrogen can be useful to help distinguish compromises in water quality arising from UOG operations versus agricultural operations (e.g. run-off from fertilizers, pesticides, etc.).

Both produced water and groundwater contain ions that present several concerns to the environment, as well as the equipment and infrastructure associated with UOG. The halogens, chloride and bromide, interfere with treatment at water processing facilities, with bromide ions forming toxic disinfection byproducts (DBPs). Hydrogen sulfide, a source of corrosion, can form from sulfate ions, while other anions of concern include strontium, barium, and calcium. These ions can cause significant damage to infrastructure through scaling (Hildenbrand et al. 2018). SM 4110 and EPA methods 300.0 and 300.1 describe the standard methods of analysis for these ions (NEMI-4110B 2017; Pfaff 1993; Pfaff et al. 1997). Ion chromatography is preferred (Ferrer and Thurman 2015); however, in order to accurately analyze produced water for anions, it is necessary to dilute the sample as described by Coday (Coday et al. 2015), where samples were diluted by a factor 30 to a final concentration below 500 mg/l.

Metal ions also pose a concern to the environment and UOG infrastructure. A critical concern is given to the high metal ion concentration in flowback and produced waters, and the effect they could have on drinking water supplies if they contaminate surface or groundwater. Analysis of metals is typically carried out by inductively coupled plasma-optical emission spectroscopy (ICP-OES) or inductively coupled plasma-mass spectrometry (ICP-MS). Standard methods include USGS I-4020-05, ASTM D1976, ASTM 5673, EPA 200.7, EPA 200.8, SM 3120B, and SM 3125 (ASTM-D1976 2018; ASTM-D5673 2016; Creed et al. 1994; Garbarino et al. 2006; Martin et al. 1994; NEMI-3120B 2017; NEMI-3125 2017). Inductively coupled plasma (ICP) utilizes a torch at high temperatures that diminishes many interferences from the complex matrix. Determining which method to use, ICP-MS versus ICP-OES, is analyte dependent; ICP-MS is more sensitive than ICP-OES and allows for better detection of analytes at trace quantities. EPA methods 265.2 and 267.2 may also be used for metal analysis; however, these utilize flame atomic absorption (FAA) or graphite furnace atomic absorption (GFAA), which have limited throughput and variable sensitivity (EPA-265.2 1978; EPA-267.2 1978).

For groundwater contamination, the analysis of volatile and semivolatile organic compounds, such as DRO and organic solvents, is of interest. Other VOCs of interest include natural gas constituents (methane, ethane, etc.) and BTEX. Due to analyte volatility, GC is the preferred method of VOC determination, where the GC may be coupled to with a variety of detectors. Mass spectrometry is the most sensitive detector, using electron ionization (EI) to ionize the gas molecules for analysis. EI is a hard ionization technique that produces fragment ions that can be used to identify analytes through a mass spectral library. Other detectors include a thermal conductivity detector, an electron capture detector, and an FID. The FID cannot provide qualitative data as can the MS, but it can be used for quantitative data based on the retention time of the target analyte.

Worthy of note, a relatively new, universal detector has been developed based on absorption in the VUV spectral region (125–240 nm) (Schug et al. 2014). This detector has been used with GC to monitor dissolved gases in water (Bai et al. 2015). Since analytes produce their own unique VUV spectrum, it is possible to use deconvolution for coeluting analytes. VUV is also highly complementary to MS in that it can well distinguish isomeric species that might be difficult to separate chromatographically, and/or might exhibit identical MS fragmentation patterns (Schenk et al. 2016).

Nonvolatile analytes such as surfactants, fatty amines, and polyacrylamide friction reducers are well suited to analysis by LC-MS. Few methods currently exist that use LC-MS for the analysis of water related to UOG activities. Among them, polypropylene glycols and polyethylene glycols have been studied in produced water by LC-MS (Heyob et al. 2017; Thurman et al. 2017), as well as heterocyclic compounds including benzothiazole in wastewater near gas production (Orem et al. 2014).

Liquid chromatography (LC) separations for produced water may be expanded through the use of tandem mass spectrometry. Tandem mass spectrometry offers structural information and can improve confidence in identification. Low resolution mass spectrometers are limited in their ability to perform nontargeted analysis because they lack either sensitivity or mass accuracy. The use of high-resolution mass spectrometry (HRMS) allows for the identification of untargeted analytes or unknowns and has been used to characterize produced water (Carlton et al. 2017; Ferrer and Thurman 2015; Luek et al. 2017; Thurman et al. 2014, 2017). Luek et al. (2017) used an ultra-HRMS to identify halogenated organic compounds in flowback water.

Another critical aspect of produced and flowback water is the presence of bacteria. Bacteria may cause corrosion of infrastructure or sour the product of UOG, and it can also lead to disease if found in drinking water. The identification of microorganisms is typically carried out by traditional biochemical plating and genotypic methods. Recently, the use of matrix assisted laser/desorption ionization (MALDI) has been applied to the analysis and identification of intact bacterial colonies based on their ionized protein profiles (Havlicek et al. 2013; Lay 2001; Schenk et al. 2016). Martin et al. (2018) and Santos et al. (2018) successfully used the matrix assisted laser/desorption ionization-time-of-flight (MALDI-TOF) MS to identify the microbial community in produced water and groundwater. The workflow used in this work is depicted in Figure 10.3.

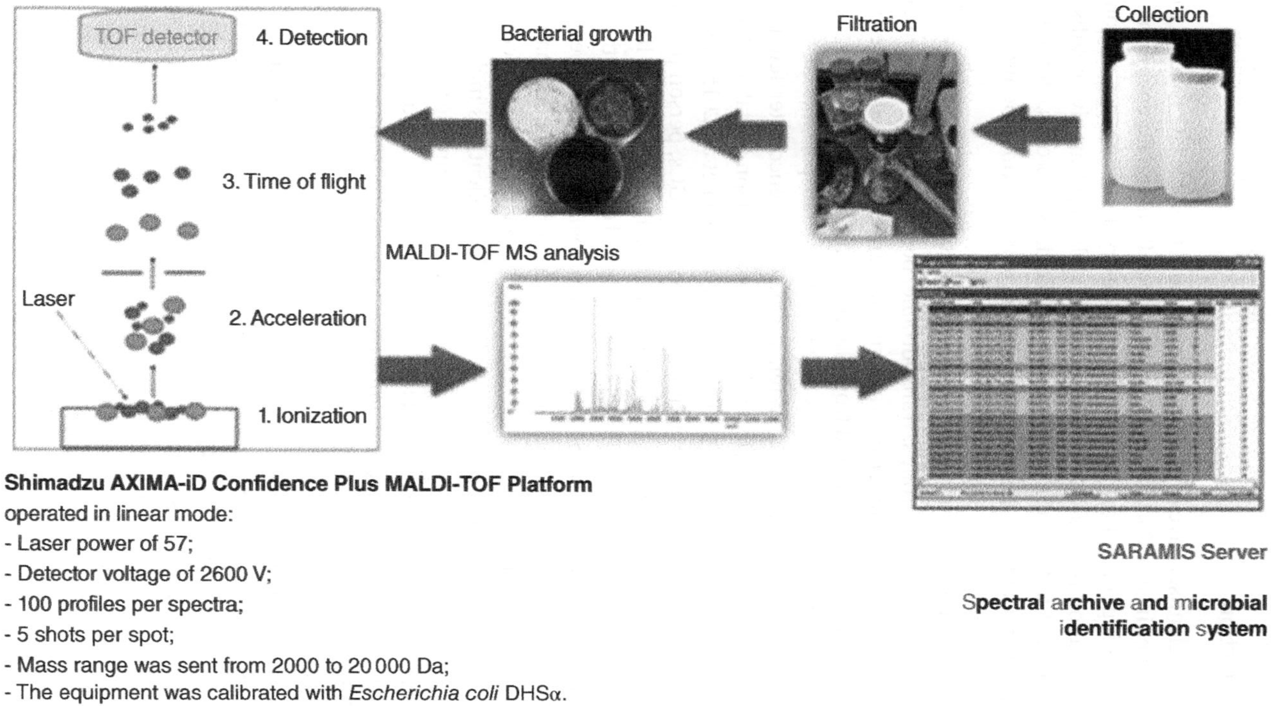

Figure 10.3 Workflow for bacterial identification using MALDI-TOF MS. Source: Santos et al. (2019). Reproduced with permission, Copyright (2019), American Chemical Society.

10.5 Portable GCs for Field Monitoring

10.5.1 Introduction

Since the 1970s, scientists have investigated developing portable GC systems that can analyze samples in situ. Portable GCs can be especially useful for environmental impact monitoring related to the oil and gas industry. In emergency response situations, information is required quickly, and portable GCs can give results in minutes as compared to laboratory analyses that take days or weeks. When connected to a laptop, portable GC systems offer online, near real-time monitoring of environmental samples.

Many categories of portable GCs exist, but for the purposes of this text, they have been divided into three major groups: transportable, person portable, and micro. Transportable GCs consist of larger instruments that are transported to the field by vehicle and should be kept sheltered from the elements. They also require a stable power supply, as well as gas cylinders or generators. Transportable GCs most resemble conventional laboratory GC systems; hence they encounter few problems when given a steady source of electricity and gases. Person portable GCs are smaller, typically weighing less than 50 lb (23 kg) (Anderson et al. 2015), and do not require an external power source. Unlike transportable GCs, person portable systems are often limited by their power demands and small size, having a limited battery life. They must also use gas cylinders that are small enough to fit in the carrying case. Due to their compact, rugged design, person portable GCs can be taken places that laboratory or transportable GCs cannot access. Even smaller than person portable GCs are the micro-fabricated systems, micro-GCs, also referred to as chip-based GCs. The advantages of micro-GCs for field applications include fast analysis time, low energy consumption, and less use of consumables. Although commercial micro-GC systems exist, much of this field is still in the development phase. The two main areas still needing improvement are the creation of uniform coatings on the column or channel and robust connections from the micro-column to an inlet/detector (Ghosh et al. 2018).

Field samples, including air, water, and soil, are most commonly monitored for VOCs and a range of petroleum hydrocarbons. Extensive sample preparation and extraction are often not feasible while in the field, therefore sampling methods should be portable, quick, and solvent-free. Frequently used environmental sampling methods include SPME, active air sampling on thermal desorption tubes, and purge-and-trap or headspace sampling. These methods allow for quick sample collection, followed by direct injection and analysis by the portable GC system. Portable GCs can potentially be applied in the oil and gas industry for refinery gas analysis, mud-logging characterization, remediation monitoring, and well plume investigation.

10.5.2 Analyzing Field Samples

There are many differences between laboratory scale GC systems and person portable systems, apart from their size. Person portable GCs typically use resistive heating techniques for temperature control of a low thermal mass (LTM) separation column. This method requires less energy and space and is faster than the air-bath oven utilized in laboratory GCs. Person portable systems not only use less power to operate but also require less carrier gas. Some systems can even use ambient air as the carrier gas, eliminating the need for onboard gas cylinders. In terms of performance and capabilities, both systems are comparable when analyzing targeted analytes in specific types of samples, such as VOCs in air. Person portable systems are good at analyzing smaller molecular weight, more volatile compounds. For the analysis of BTEX in soils, a portable GC-PID was found to yield comparable results to a laboratory GC-MS in both detection limits and precision. The main difference was a smaller linear range for the portable system (Zhou et al. 2013). As compounds increase in molecular weight and decrease in volatility, higher temperatures are required to elute them from the column. Many portable GCs lack the power capabilities to achieve high enough temperatures for these separations, limiting their effective separation range.

To determine the quality of field portable GC systems, the EPA tested four systems in 1997 through their Environmental Technology Verification (ETV) Program. The GCs were tested at two sites with large subsurface plumes of chlorinated VOCs near Aiken, South Carolina, and Sacramento, California (Einfeld 1998a,b,c; Staples et al. 1998). Each system was a portable GC, but employed a different detector, including a surface acoustic wave detector, dual electron capture and PID, argon ion/electron capture detector, and a mass spectrometer. Although the ETV program ended in 2014, it demonstrated the capabilities of commercial instruments in the analysis of environmental samples. Table 10.3 shows a wide array of commercially available, portable GC systems from a variety of instrument manufacturers, although it is not an exhaustive list.

Portable GCs in the oil and gas industry are commonly used for characterizations of natural gas and petroleum hydrocarbon VOCs, as well as pinpointing hydrocarbon leaks in storage tanks, but few recent peer-reviewed studies are available. However, there are many examples of both transportable and person portable GCs in other types of VOC field analyses of air, soil, and water samples. Micro-GC systems are less established and have not been used for field analysis of oil and gas samples. Laboratory versions of micro-GCs have tested similar compounds to those expected to be found in the field, where one study demonstrated a high temperature (up to 350 °C) separation with a micromachined column using a simulated distillation mix (ASTM2887) (ASTM-D2887 2019; Gaddes et al. 2014). Both stainless steel and silicon microcolumns have been tested for high boiling

Table 10.3 Commercially available portable GC systems and their distinguishing information.

Instrument, Company	Detector(s)	Weight (kg)	Carrier gas	Inlet type
490 Micro GC, Agilent	Micro-machined thermal conductivity detector (TCD)	6.5 (dual channel) 10.6 (quad channel)	He, H_2, N_2, or Ar	Micro-machined injector, can be heated to 110 °C
HAPSITE ER, Inficon	Mass spectrometer (MS)	19 (with battery)	N_2	Air probe, other options can be purchased
FROG-5000, Defiant Technologies	Photoionization detector (PID)	2.2	Ambient air	Air probe, other option can be purchased
CALIDUS GCs, Teledyne- Falcon	TCD and flame ionization detector (FID), other options available	11 (max weight)	He, H_2	Split/splitless injection port
Torion T-9, PerkinElmer	Toroidal ion trap mass spectrometer (TMS)	14.5 (with battery and case)	He	Heated injection port with removable liner

alkanes and even more complex mixtures, such as gasoline by headspace and commercial diesel samples (Ghosh et al. 2017, 2019). In addition to high temperature separations, fast separation times are an important aspect of micro-GC; temperature-programmed silica-sputtered micromachined columns separated a light alkane mix (C_1–C_5) in 9 seconds and heavier alkanes (C_1–C_9) in 15 seconds (Haudebourg et al. 2013). Micro-GCs offer a combination of fast separation, low power consumption, extremely small size, and temperature programming that offers a promising future in environmental impact monitoring of oil and gas development.

10.6 Fingerprinting in the Oil and Gas Industry

10.6.1 Introduction

Environmental forensic investigations involve determining the chemical nature, source, and age of site contamination. In environmental forensic investigations, a chemical fingerprint, or the unique pattern of a source, can be utilized to support expert opinions regarding sources of environmental contamination. Chemical fingerprinting has been used extensively in the oil and gas industry. It allows one to distinguish between contamination and background sources, for example, to distinguish between petrogenic and pyrogenic sources, or to identify and/or eliminate a potential source. In some cases, chemical fingerprinting can be utilized to determine the contamination timeframe.

10.6.2 Hydrocarbon Fingerprinting

Hydrocarbon fingerprinting of petroleum products is performed using one- or two-dimensional (1D or 2D) GC coupled with an FID or MS detector. 2D GC has superior separation power to 1D GC, leading to increased identification of the individual components present in a sample. Additionally, the use of an MS over an FID further enhances the identification of the compounds by providing mass spectral fragmentation data for each. This wealth of knowledge about the individual components of a contaminant enables the analyst to account for weathering in the sample. Weathering, as previously discussed in this chapter, changes the hydrocarbon composition through naturally occurring processes, such as volatilization. When determining the source of a contaminant, it is important to understand the extent of weathering it has undergone. Otherwise, an incorrect match, or no match, may be made.

An often-cited example of a fingerprinting application is the Deepwater Horizon oil release in the Gulf of Mexico. On 20 April, 2010, an explosion at the drilling platform released an estimated 470–690 thousand metric tons of oil into the surrounding ocean. Oil continued to wash up along gulf shores for years after this accident, prompting the need to identify the source of these contaminants and differentiate between naturally occurring oil seeps. Comprehensive 2D GC was employed to analyze samples that had washed ashore months after the release event and compare them to oil from the original Macondo well to determine the source and examine the effects of biodegradation on the samples (Gros et al. 2014; Gruber et al. 2018).

10.6.3 Additional Texts on Fingerprinting Oil Spills and Petroleum Products

Several textbooks on environmental forensics are available in print. Listed in the following are recommendations for additional information or case studies.

- *Environmental Forensics: Contaminant Specific Guide* (Morrison and Murphy 2006).
- *Introduction to Environmental Forensics* (3rd ed.) (Murphy and Morrison 2015).

- *Environmental Forensics Fundamentals: A Practical Guide* (Petrisor 2014).
- *Standard Handbook of Oil Spill Environmental Forensics: Fingerprinting and Source Identification* (2nd ed.) (Stout and Wang 2016).
- *Oil Spill Environmental Forensics Case Studies* (Stout and Wang 2018).

References

Allen, D.T. (2014). Methane emissions from natural gas production and use: Reconciling bottom-up and top-down measurements. In: *Current Opinion in Chemical Engineering*, vol. 5, 78–83. Elsevier Ltd.

Allen, D.T., Torres, V.M., Thomas, J. et al. (2013). Measurements of methane emissions at natural gas production sites in the United States. *Proceedings of the National Academy of Sciences of the United States of America* 110: 17768–17773. https://doi.org/10.1073/pnas.1213080110.

Alvarez, R.A., Zavala-Araiza, D., Lyon, D.R. et al. (2018). Assessment of methane emissions from the U.S. oil and gas supply chain. *Science* 361: 186–188. https://doi.org/10.1126/science.aar7204.

Anderson, J.L., Berthod, A., Pino, V., and Stalcup, A.M. (2015). *Analytical Separation Science*. Weinheim, Germany: Wiley-VCH.

ASTM-D1976 (2018). *Standard test method for elements in water by inductively-coupled plasma atomic emission spectroscopy*. West Conshohocken, PA: ASTM International.

ASTM-D2887 (2019). *Standard test method for boiling range distribution of petroleum fractions by gas chromatography*. West Conshohocken, PA: ASTM International.

ASTM-D5673 (2016). *Standard test method for elements in water by inductively coupled plasma – mass spectrometry*. West Conshohocken, PA: ASTM International.

Bai, L., Smuts, J., Walsh, P. et al. (2015). Permanent gas analysis using gas chromatography with vacuum ultraviolet detection. *Journal of Chromatography A* 1388: 244–250. https://doi.org/10.1016/j.chroma.2015.02.007.

de Blas, M., Navazo, M., Alonso, L. et al. (2011). Automatic on-line monitoring of atmospheric volatile organic compounds: gas chromatography–mass spectrometry and gas chromatography–flame ionization detection as complementary systems. *Science of the Total Environment* 409: 5459–5469. https://doi.org/10.1016/j.scitotenv.2011.08.072.

Brady, N.C. and Weil, R.R. (2014). *The Nature and Properties of Soils*, 14e. Essex, England: Pearson Education.

Brandt, A.R., Heath, G.A., Kort, E.A. et al. (2014). Methane leaks from North American natural gas systems. *Science* 343: 733–735. https://doi.org/10.1126/science.1247045.

Burton, T.G., Rifai, H.S., Hildenbrand, Z.L. et al. (2016). Elucidating hydraulic fracturing impacts on groundwater quality using a regional geospatial statistical modeling approach. *Science of the Total Environment* 545–546: 114–126. https://doi.org/10.1016/j.scitotenv.2015.12.084.

Carlton, D.D., Hildenbrand, Z.L., and Schug, K.A. (2017). Analytical approaches for high-resolution environmental investigations of unconventional oil and gas exploration. In: *Environmental Issues Concerning Hydraulic Fracturing* (eds. K.A. Schug and Z.L. Hildenbrand), 193–226. Cambridge, MA: Elsevier.

Chirila, E.S., Draghici, C., and Dobrinas, S. (2006). *Sampling and Sample Pretreatment for Environmental Analysis*. Dordrecht: Springer.

Cluff, M.A., Hartsock, A., Macrae, J.D. et al. (2014). Temporal changes in microbial ecology and geochemistry in produced water from hydraulically fractured Marcellus shale gas wells. *Environmental Science and Technology* 48 (11): 6508–6517. https://doi.org/10.1021/es501173p.

Coday, B.D., Almaraz, N., and Cath, T.Y. (2015). Forward osmosis desalination of oil and gas wastewater: Impacts of membrane selection and operating conditions on process performance. *Journal of Membrane Science* 488: 40–55. https://doi.org/10.1016/j.memsci.2015.03.059.

Connolly, J.I., Robinson, R.A., and Gardiner, T.D. (2019). Assessment of the Bacharach Hi Flow® Sampler characteristics and potential failure modes when measuring methane emissions. *Measurement: Journal of the International Measurement Confederation* 145: 226–233. https://doi.org/10.1016/j.measurement.2019.05.055.

Creed, J.T., Brockhoff, C.A., and Martin, T.D. (1994). *EPA Method 200.8: Determination of Trace Elements in Waters and Wastes by Inductively Coupled Plasma-Mass Spectrometry (Revision 5.4)*. Cincinnati, OH: United States Environmental Protection Agency (USEPA).

Crosson, E.R. (2008). A cavity ring-down analyzer for measuring atmospheric levels of methane, carbon dioxide, and water vapor. *Applied Physics B: Lasers and Optics* 92: 403–408. https://doi.org/10.1007/s00340-008-3135-y.

Devatha, C.P., Vishnu Vishal, A., and Purna Chandra Rao, J. (2019). Investigation of physical and chemical characteristics on soil due to crude oil contamination and its remediation. *Applied Water Science* 9: 10. https://doi.org/10.1007/s13201-019-0970-4.

Douglas, R.K., Nawar, S., Cipullo, S. et al. (2018). Evaluation of vis-NIR reflectance spectroscopy sensitivity to weathering for enhanced assessment of oil contaminated soils. *Science of the Total Environment* 626: 1108–1120. https://doi.org/10.1016/j.scitotenv.2018.01.122.

Einfeld, W. (1998a). *Environmental Technology Verification Report Field-Portable Gas Chromatograph Sentex Systems, Inc. Scentograph Plus II*. (EPA/600/R-98/145). Washington, DC: US Environmental Protection Agency (EPA).

Einfeld, W. (1998b). *Environmental Technology Verification Report Field-Portable Gas Chromatograph*. Perkin-Elmer Photovac, Voyager (EPA-VS-SCM-24). Washington DC: US Environmental Protection Agency.

Einfeld, W. (1998c). *Environmental Technology Verification Report Field-Portable Gas Chromatograph/ Mass Spectrometer Inficon, Inc., HAPSITE*. (EPA-VS-SCM-25). Washington, DC: US Environmental Protection Agency (EPA).

EPA (2014). *OTM33A Geospatial Measurement of Air Pollution-Remote Emissions Quantification-Direct Assessment (GMAP-REQ-DA)*. US Environmental Protection Agency (EPA).

EPA-265.2 (1978). *Rhodium (Atomic Absorption, Furnace Technique)*. Online. US Environmental Protection Agency (EPA). Retrieved from https://www.epa.gov.

EPA-267.2 (1978). *Ruthenium (Atomic Absorption, Furnace Technique)*. Online. US Environmental Protection Agency (EPA). Retrieved from https://www.epa.gov.

EPA-310.1 (1978). *Alkalinity (Titrimetric, pH 4.5)*. Online. US Environmental Protection Agency (EPA). Retrieved from https://www.epa.gov.

EPA-310.2 (1974). *Alkalinity (Colorimetric, Automated, Methyl Orange) by Autoanalyzer*. Online. US Environmental Protection Agency (EPA). Retrieved from https://www.epa.gov.

EPA-3540C (1996). *Soxhlet Extraction*. Online. US Environmental Protection Agency (EPA). Retrieved from https://www.epa.gov.

EPA-3541 (1994). *Automated Soxhlet Extraction*. Online. US Environmental Protection Agency (EPA). Retrieved from https://www.epa.gov.

EPA-3545A (2007). *Pressurized Fluid Extraction (PFE)*. Online. US Environmental Protection Agency (EPA). Retrieved from https://www.epa.gov.

EPA-3546 (2007). *Microwave Extraction*. Online. US Environmental Protection Agency (EPA). Retrieved from https://www.epa.gov.

EPA-3550C (2007). *Ultrasonic Extraction*. Online. US Environmental Protection Agency (EPA). Retrieved from https://www.epa.gov.

EPA-3560 (1996). Supercritical Fluid Extraction of Total Recoverable Petroleum Hydrocarbons (TRPHs). *Online: US Environmental Protection Agency (EPA)*. Retrieved from https://www.epa.gov.

EPA-3561 (1996). *Supercritical Fluid Extraction of Polynuclear Aromatic Hydrocarbons (PAHs)*. Online. US Environmental Protection Agency (EPA). Retrieved from https://www.epa.gov.

EPA-3600C (1996). *Cleanup*. Online. US Environmental Protection Agency (EPA). Retrieved from https://www.epa.gov.

EPA-3611B (1996). *Alumina Column Cleanup and Separation of Petroleum Wastes*. Online. US Environmental Protection Agency (EPA). Retrieved from https://www.epa.gov.

EPA-3630C (1996). *Silica Gel Cleanup*. Online. US Environmental Protection Agency (EPA). Retrieved from https://www.epa.gov.

EPA-3815 (2007). *Screening Solid Samples for Volatile Organics*. Online. US Environmental Protection Agency (EPA). Retrieved from https://www.epa.gov.

EPA-4030 (1996). *Soil Screening for Petroleum Hydrocarbons by Immunoassay*. Online. US Environmental Protection Agency (EPA). Retrieved from https://www.epa.gov.

EPA-4035 (1996). *Soil Screening for Polynuclear Aromatic Hydrocarbons by Immunoassay*. Online. US Environmental Protection Agency (EPA). Retrieved from https://www.epa.gov.

EPA-5021A (2014). *Volatile Organic Compounds (VOCs) in Various Sample Matrices Using Equilibrium Headspace Analysis*. Online. US Environmental Protection Agency (EPA). Retrieved from https://www.epa.gov.

EPA-5030B (1996). *Purge-and-Trap for Aqueous Samples*. Online. US Environmental Protection Agency (EPA). Retrieved from https://www.epa.gov.

EPA-5035 (1996). *Closed-System Purge-and-Trap and Extraction for Volatile Organics in Soil and Waste Samples*. Online. US Environmental Protection Agency (EPA). Retrieved from https://www.epa.gov.

EPA-8000D (2018). *Determinative Chromatographic Separations*. Online. US Environmental Protection Agency (EPA). Retrieved from https://www.epa.gov.

EPA-8015D (2003). *Nonhalogenated Organics Using Gas Chromatography/Flame Ionization Detection (GC/FID)*.Online. US Environmental Protection Agency (EPA).

EPA-8021B (2014). *Aromatic and Halogenated Volatiles by Gas Chromatography Using Photoionization and/or Electrolytic Conductivity Detectors*. Online. US Environmental Protection Agency (EPA). Retrieved from https://www.epa.gov.

EPA-8260D (2018). *Volatile Organic Compounds by Gas Chromatography/Mass Spectrometry (GC/MS)*. Online. US Environmental Protection Agency (EPA). Retrieved from https://www.epa.gov.

EPA-8270E (2018). *Semivolatile Organic Compounds by Gas Chromatography/Mass Spectrometry (GC/MS)*. Online. US Environmental Protection Agency (EPA). Retrieved from https://www.epa.gov.

EPA-8275A (1996). *Semivolatile Organic Compounds in Soil/Sludges and Solid Wastes Using Thermal Extraction/Gas Chromatography/Mass Spectrometry (TE/GC/MS)*. Online. US Environmental Protection Agency (EPA). Retrieved from https://www.epa.gov.

EPA-8440 (1996). *Total Recoverable Petroleum Hydrocarbons by Infrared Spectrophotometry*. Online. US Environmental Protection Agency (EPA). Retrieved from https://www.epa.gov.

EPA-9071B (1998). *n-Hexane Extractable Material (HEM) for Sludge, Sediment, and Solid Samples*. Online. US Environmental Protection Agency (EPA). Retrieved from https://www.epa.gov.

EPA-9074 (2007). *Turbidimetric Screening Method for Total Recoverable Petroleum Hydrocarbons in Soil*. Online. US Environmental Protection Agency (EPA). Retrieved from https://www.epa.gov.

Farrell-Jones, J. (2003). Petroleum hydrocarbons and polyaromatic hydrocarbons. In: *Chemical Analysis of Contaminated Land* (eds. K.C. Thompson and C.P. Nathanail), 45. Boca Raton, FL: CRC Press LLC.

Ferrer, I. and Thurman, E.M. (2015). Analysis of hydraulic fracturing additives by LC/Q-TOF-MS. *Analytical and Bioanalytical Chemistry* 407 (21): 6417–6428. https://doi.org/10.1007/s00216-015-8780-5.

Fontenot, B.E., Hunt, L.R., Hildenbrand, Z.L. et al. (2013). An evaluation of water quality in private drinking water wells near natural gas extraction sites in the Barnett shale formation. *Environmental Science and Technology* 47 (17): 10032–10040. https://doi.org/10.1021/es4011724.

FracFocus. (2015). Why chemicals are used. Retrieved from http://fracfocus.ca/en/chemical-use/why-chemicals-are-used

Gaddes, D., Westland, J., Dorman, F.L., and Tadigadapa, S. (2014). Improved micromachined column design and fluidic interconnects for programmed high-temperature gas chromatography separations. *Journal of Chromatography A* 1349: 96–104. https://doi.org/10.1016/j.chroma.2014.04.087.

Garbarino, J.R., Kanagy, L.K., and Cree, M.E. (2006). Determination of elements in natural-water, biota, sediment, and soil samples using collision/reaction cell inductively coupled plasma–mass spectrometry. In: *U.S. Geological Survey Techniques and Methods Book 5, Ch. B1*. Reston, VA: United States Geological Survey (USGS).

Ghosh, A., Johnson, J.E., Nuss, J.G. et al. (2017). Extending the upper temperature range of gas chromatography with all-silicon microchip columns using a heater/clamp assembly. *Journal of Chromatography A* 1517: 134–141. https://doi.org/10.1016/j.chroma.2017.08.036.

Ghosh, A., Vilorio, C.R., Hawkins, A.R., and Lee, M.L. (2018). Microchip gas chromatography columns, interfacing and performance. *Talanta* 188: 463–492. https://doi.org/10.1016/j.talanta.2018.04.088.

Ghosh, A., Foster, A.R., Johnson, J.C. et al. (2019). Stainless-steel column for robust, high-temperature microchip gas chromatography. *Analytical Chemistry* 91: 792–796. https://doi.org/10.1021/acs.analchem.8b04174.

Goetz, J.D., Floerchinger, C., Fortner, E.C. et al. (2015). Atmospheric emission characterization of Marcellus shale natural gas development sites. *Environmental Science and Technology* 49: 7012–7020. https://doi.org/10.1021/acs.est.5b00452.

Goetz, J.D., Avery, A., Werden, B. et al. (2017). Analysis of local-scale background concentrations of methane and other gas-phase species in the Marcellus Shale. *Elementa* 5: 1. https://doi.org/10.1525/elementa.182.

Granados, H., Mata, H., Martin, J. et al. (2019). Attitudes, perceptions, and geospatial analysis of water quality and individual health status in a high-fracking region. *Water* 11 (8): 1633. https://doi.org/10.3390/w11081633.

Gros, J., Reddy, C.M., Aeppli, C. et al. (2014). Resolving biodegradation patterns of persistent saturated hydrocarbons in weathered oil samples from the Deepwater Horizon disaster. *Environmental Science and Technology* 48: 1628–1637. https://doi.org/10.1021/es4042836.

Gruber, B., Weggler, B.A., Jaramillo, R. et al. (2018). Comprehensive two-dimensional gas chromatography in forensic science: a critical review of recent trends. *TrAC – Trends in Analytical Chemistry* 105: 292–301. https://doi.org/10.1016/j.trac.2018.05.017.

Harriss, R., Alvarez, R.A., Lyon, D. et al. (2015). Using multi-scale measurements to improve methane emission estimates from oil and gas operations in the Barnett Shale region, Texas. *Environmental Science and Technology* 49: 7524–7526. https://doi.org/10.1021/acs.est.5b02305.

Haudebourg, R., Vial, J., Thiebaut, D. et al. (2013). Temperature-programmed sputtered micromachined gas chromatography columns: An approach to fast separations in oilfield applications. *Analytical Chemistry* 85: 114–120. https://doi.org/10.1021/ac3022136.

Havlicek, V., Lemr, K., and Schug, K.A. (2013). Current trends in microbial diagnostics based on mass spectrometry. *Analytical Chemistry* 85: 790–797. https://doi.org/10.1021/ac3031866.

Heyob, K.M., Blotevogel, J., Brooker, M. et al. (2017). Natural attenuation of nonionic surfactants used in hydraulic fracturing fluids: Degradation rates, pathways, and mechanisms. *Environmental Science and Technology* 51 (23): 13985–13994. https://doi.org/10.1021/acs.est.7b01539.

Hildenbrand, Z.L., Carlton, D.D., Fontenot, B.E. et al. (2015). A comprehensive analysis of groundwater quality in the Barnett Shale region. *Environmental Science and Technology* 49 (13): 8254–8262. https://doi.org/10.1021/acs.est.5b01526.

Hildenbrand, Z.L., Carlton, D.D., Fontenot, B.E. et al. (2016). Temporal variation in groundwater quality in the Permian Basin of Texas, a region of increasing unconventional oil and gas development. *Science of the Total Environment* 562: 906–913. https://doi.org/10.1016/j.scitotenv.2016.04.144.

Hildenbrand, Z.L., Santos, I.C., Liden, T. et al. (2018). Characterizing variable biogeochemical changes during the treatment of produced oilfield waste. *Science of the Total Environment* 634: 1519–1529. https://doi.org/10.1016/j.scitotenv.2018.03.388.

Howard, T. (2015). Comment on “Methane Emissions from Process Equipment at Natural Gas Production Sites in the United States: Pneumatic Controllers”. *Environmental Science & Technology* 49: 3981–3982. https://doi.org/10.1021/acs.est.5b00507.

Karion, A., Sweeney, C., Pétron, G. et al. (2013). Methane emissions estimate from airborne measurements over a western United States natural gas field. *Geophysical Research Letters* 40: 4393–4397. https://doi.org/10.1002/grl.50811.

Karion, A., Sweeney, C., Kort, E.A. et al. (2015). Aircraft-based estimate of total methane emissions from the Barnett Shale region. *Environmental Science and Technology* 49: 8124–8131. https://doi.org/10.1021/acs.est.5b00217.

Lan, X., Talbot, R., Laine, P., and Torres, A. (2015). Characterizing fugitive methane emissions in the Barnett Shale area using a mobile laboratory. *Environmental Science and Technology* 49: 8139–8146. https://doi.org/10.1021/es5063055.

Lavoie, T.N., Shepson, P.B., Cambaliza, M.O.L. et al. (2015). Aircraft-based measurements of point source methane emissions in the Barnett Shale Basin. *Environmental Science and Technology* 49: 7904–7913. https://doi.org/10.1021/acs.est.5b00410.

Lay, J.O. (2001). MALDI-TOF mass spectrometry of bacteria. *Mass Spectrometry Reviews* 20: 172–194. https://doi.org/10.1002/mas.10003.

Liden, T., Clark, B.G., Hildenbrand, Z.L., and Schug, K.A. (2017). Unconventional oil and gas production: Waste management and the water cycle. In: *Environmental Issues Concerning Hydraulic Fracturing*, vol. 1 (eds. K.A. Schug and Z.L. Hildenbrand), 17–45. Cambridge, MA: Elsevier.

Liden, T., Santos, I.C., Hildenbrand, Z.L., and Schug, K.A. (2018). Treatment modalities for the reuse of produced waste from oil and gas development. *Science of the Total Environment* 643: 107–118. https://doi.org/10.1016/j.scitotenv.2018.05.386.

Liden, T., Santos, I.C., Hildenbrand, Z.L., and Schug, K.A. (2019). Analytical methods for the comprehensive characterization of produced water. *Separation Science and Technology* 11: 199–217. https://doi.org/10.1016/B978-0-12-815730-5.00009-0.

Luek, J.L., Schmitt-Kopplin, P., Mouser, P.J. et al. (2017). Halogenated organic compounds identified in hydraulic fracturing wastewaters using ultrahigh resolution mass spectrometry. *Environmental Science and Technology* 51 (10): 5377–5385. https://doi.org/10.1021/acs.est.6b06213.

Macey, G.P., Breech, R., Chernaik, M. et al. (2014). Air concentrations of volatile compounds near oil and gas production: A community-based exploratory study. *Environmental Health* 13: 82. https://doi.org/10.1186/1476-069X-13-82.

Majchrzak, T., Wojnowski, W., Lubinska-Szczygeł, M. et al. (2018). PTR-MS and GC-MS as complementary techniques for analysis of volatiles: A tutorial review. *Analytica Chimica Acta* 1035: 1–13. https://doi.org/10.1016/j.aca.2018.06.056.

Martin, T.D., Brockhoff, C.A., Creed, J.T., and Group, E.M.W. (1994). *EPA Method 200.7: Determination of Metals and Trace Elements in Water and Wastes by Inductively Coupled Plasma-Atomic Emission Spectrometry (Revision 4.4).* Cincinnati, OH: US Environmental Protection Agency (EPA).

Martin, M.S., Santos, I.C., Carlton, D.D. et al. (2018). Characterization of bacterial diversity in contaminated groundwater using matrix-assisted laser desorption/ionization time-of-flight mass spectrometry. *Science of the Total Environment* 622-623: 1562–1571. https://doi.org/10.1016/j.scitotenv.2017.10.027.

Morrison, R.D. and Murphy, B.L. (2006). *Environmental Forensics: Contaminant Specific Guide* (eds. R.D. Morrison and B.L. Murphy). San Diego, CA: Academic Press.

Murphy, B.L. and Morrison, R.D. (2015). *Introduction to Environmental Forensics*, 3e (eds. B.L. Murphy and R.D. Morrison). San Diego, CA: Academic Press.

Nathan, B.J., Golston, L.M., O'Brien, A.S. et al. (2015). Near-field characterization of methane emission variability from a compressor station using a model aircraft. *Environmental Science and Technology* 49: 7896–7903. https://doi.org/10.1021/acs.est.5b00705.

NEMI-3120B (2017). *Standard Method 3120B: Metals in Water by Plasma Emission Spectroscopy*. Online. National Environmental Methods Index (NEMI). Retrieved from https://www.nemi.gov.

NEMI-3125 (2017). *Standard Method 3125: Metals in Water by ICP/MS*. Online. National Environmental Methods Index (NEMI). Retrieved from https://www.nemi.gov.

NEMI-4110B (2017). *Standard Method 4110B: Ion Chromatography with Chemical Suppression of Eluent Conductivity*. Online. National Environmental Methods Index (NEMI). Retrieved from https://www.nemi.gov.

Omara, M., Sullivan, M.R., Li, X. et al. (2016). Methane emissions from conventional and unconventional natural gas production sites in the Marcellus Shale Basin. *Environmental Science and Technology* 50: 2099–2107. https://doi.org/10.1021/acs.est.5b05503.

Orem, W., Tatu, C., Varonka, M. et al. (2014). Organic substances in produced and formation water from unconventional natural gas extraction in coal and shale. *International Journal of Coal Geology* 126: 20–31. https://doi.org/10.1016/j.coal.2014.01.003.

Payne, B.F., Ackley, R., Paige Wicker, A. et al. (2017). Characterization of methane plumes downwind of natural gas compressor stations in Pennsylvania and New York. *Science of the Total Environment* 580: 1214–1221. https://doi.org/10.1016/j.scitotenv.2016.12.082.

Petrisor, I.G. (2014). *Environmental Forensics Fundamentals: A Practical Guide*. Boca Raton, FL: CRC Press, Taylor and Francis Group.

Pétron, G., Karion, A., Sweeney, C. et al. (2014). A new look at methane and nonmethane hydrocarbon emissions from oil and natural gas operations in the Colorado Denver-Julesburg Basin. *Journal of Geophysical Research* 119: 6836–6852. https://doi.org/10.1002/2013JD021272.

Pfaff, J.D. (1993). *EPA Method 300.0: Determination of Inorganic Anions by Ion Chromatography*. Cincinnati, OH: United States Environmental Protection Agency (USEPA).

Pfaff, J.D., Hautman, D.P., and Munch, D.J. (1997). *EPA Method 300.1: Determination of Inorganic Anions in Drinking Water by Ion Chromatography*. Cincinnati, OH: United States Environmental Protection Agency (USEPA).

Prebihalo, S., Brockman, A., Cochran, J., and Dorman, F.L. (2015). Determination of emerging contaminants in wastewater utilizing comprehensive two-dimensional gas-chromatography coupled with time-of-flight mass spectrometry. *Journal of Chromatography A* 1419: 109–115. https://doi.org/10.1016/j.chroma.2015.09.080.

Prenni, A.J., Day, D.E., Evanoski-Cole, A.R. et al. (2016). Oil and gas impacts on air quality in federal lands in the Bakken region: An overview of the Bakken Air Quality Study and first results. *Atmospheric Chemistry and Physics* 16: 1401–1416. https://doi.org/10.5194/acp-16-1401-2016.

Rella, C.W., Tsai, T.R., Botkin, C.G. et al. (2015). Measuring emissions from oil and natural gas well pads using the mobile flux plane technique. *Environmental Science and Technology* 49: 4742–4748. https://doi.org/10.1021/acs.est.5b00099.

Robertson, A.M., Edie, R., Snare, D. et al. (2017). Variation in methane emission rates from well pads in four oil and gas basins with contrasting production volumes and compositions. *Environmental Science and Technology* 51: 8832–8840. https://doi.org/10.1021/acs.est.7b00571.

Roscioli, J.R., Yacovitch, T.I., Floerchinger, C. et al. (2015). Measurements of methane emissions from natural gas gathering facilities and processing plants: Measurement methods. *Atmospheric Measurement Techniques* 8: 2017–2035. https://doi.org/10.5194/amt-8-2017-2015.

Sadler, R. and Connell D. (2003). Analytical methods for the determination of total petroleum hydrocarbons in soil. Online: National Environment Protection Council (NEPC). Retrieved from http://www.nepec.gov.au.

Santos, I.C., Martin, M.S., Reyes, M.L. et al. (2018). Exploring the links between groundwater quality and bacterial communities near oil and gas extraction activities. *Science of the Total Environment* 618: 165–173. https://doi.org/10.1016/j.scitotenv.2017.10.264.

Santos, I.C., Hildenbrand, Z.L., and Schug, K.A. (2019). A review of analytical methods for characterizing the potential environmental impacts of unconventional oil and gas development. *Analytical Chemistry* 91: 689–703. https://doi.org/10.1021/acs.analchem.8b04750.

Schenk, J., Mao, J.X., Smuts, J. et al. (2016). Analysis and deconvolution of dimethylnaphthalene isomers using gas chromatography vacuum ultraviolet spectroscopy and theoretical computations. *Analytica Chimica Acta* 945: 1–8. https://doi.org/10.1016/j.aca.2016.09.021.

Schenk, J., Carlton, D.D., Smuts, J. et al. (2019). Lab-simulated downhole leaching of formaldehyde from proppants by high performance liquid chromatography (HPLC), headspace gas chromatography-vacuum ultraviolet (HS-GC-VUV) spectroscopy, and headspace gas chromatography-mass spectrometry (HS-GC-MS).

Environmental Science: Processes and Impacts 21 (2): 214–223. https://doi.org/10.1039/c8em00342d.

Schug, K.A., Sawicki, I., Carlton, D.D. et al. (2014). Vacuum ultraviolet detector for gas chromatography. *Analytical Chemistry* 86: 8329–8335. https://doi.org/10.1021/ac5018343.

Smith, M.L., Kort, E.A., Karion, A. et al. (2015). Airborne ethane observations in the Barnett Shale: Quantification of ethane flux and attribution of methane emissions. *Environmental Science and Technology* 49: 8158–8166. https://doi.org/10.1021/acs.est.5b00219.

Speight, J.G. (2005). *Environmental Analysis and Technology for the Refining Industry*. Hoboken, NJ: Wiley.

Spinelle, L., Gerboles, M., Kok, G. et al. (2017). Review of portable and low-cost sensors for the ambient air monitoring of benzene and other volatile organic compounds. *Sensors* 17: 1520. https://doi.org/10.3390/s17071520.

Staples, E.J., Matsuda, T., and Viswanathan, S. (1998). Real Time Environmental Screening of Air, Water and Soil Matrices Using a novel Field Portable GC/SAW System. Paper presented at the Environmental Strategies for the 21st Century, Asia Pacific Conference, Singapore (8–10 April 1998).

Stout, S.A. and Wang, Z. (2016). *Standard Handbook of Oil Spill Environmental Forensics: Fingerprinting and Source Identification*, 2e (eds. S.A. Stout and Z. Wang). San Diego, CA: Academic Press.

Stout, S.A. and Wang, Z. (2018). *Oil Spill Environmental Forensics Case Studies* (eds. S.A. Stout and Z. Wang). Cambridge, MA: Butterworth-Heinemann.

Su, Y.-C., Chen, W.-H., Fan, C.-L. et al. (2019). Source apportionment of volatile organic compounds (VOCs) by positive matrix factorization (PMF) supported by model simulation and source markers – using petrochemical emissions as a showcase. *Environmental Pollution* 254: 112848. https://doi.org/10.1016/j.envpol.2019.07.016.

Swarthout, R.F., Russo, R.S., Zhou, Y. et al. (2015). Impact of Marcellus shale natural gas development in southwest Pennsylvania on volatile organic compound emissions and regional air quality. *Environmental Science and Technology* 49: 3175–3184. https://doi.org/10.1021/es504315f.

Thacker, J., Carlton, D., Hildenbrand, Z. et al. (2015). Chemical analysis of wastewater from unconventional drilling operations. *Water* 7 (12): 1568–1579. https://doi.org/10.3390/w7041568.

Thurman, E.M., Ferrer, I., Blotevogel, J., and Borch, T. (2014). Analysis of hydraulic fracturing flowback and produced waters using accurate mass: Identification of ethoxylated surfactants. *Analytical Chemistry* 86 (19): 9653–9661. https://doi.org/10.1021/ac502163k.

Thurman, E.M., Ferrer, I., Rosenblum, J. et al. (2017). Identification of polypropylene glycols and polyethylene glycol carboxylates in flowback and produced water from hydraulic fracturing. *Journal of Hazardous Materials* 323: 11–17. https://doi.org/10.1016/j.jhazmat.2016.02.041.

Varona-Torres, E., Carlton, D.D., Payne, B. et al. (2017). The characterization of BTEX in variable soil compositions near unconventional oil and gas development. In: *Advances in Chemical Pollution, Environmental Management and Protection*, vol. 1 (ed. D. Barcelo), 321–351. New York: Elsevier.

Vidic, R.D., Brantley, S.L., Vandenbossche, J.M. et al. (2013). Impact of shale gas development on regional water quality. *Science* 340: 1235009. https://doi.org/10.1126/science.1235009.

Weisman, W. (1998). *Analysis of Petroleum Hydrocarbons in Environmental Media*, vol. 1. Amherst, MA: Amherst Scientific Publishers.

Zhou, Y.Y., Yu, J.F., Yan, Z.G. et al. (2013). Application of portable gas chromatography-photo ionization detector combined with headspace sampling for field analysis of benzene, toluene, ethylbenzene, and xylene in soils. *Environmental Monitoring and Assessment* 185: 3037–3048. https://doi.org/10.1007/s10661-012-2771-1.

Part VI

Future Trends in the Petroleum Industry

11

Future Trends

William L. Winniford[1] *and Melissa N. Dunkle*[2]

[1] *The Dow Chemical Company, Analytical Science, 230 Abner Jackson Parkway, Lake Jackson, TX, USA*
[2] *Dow Benelux, BV, Analytical Science, Herbert H. Dowweg 5, Building 446, Hoek, The Netherlands*

11.1 Introduction

Predicting future trends is an inherently risky business. Some years from now it will be obvious that we overestimated how quickly some technology would evolve and be implemented. Conversely, there will be trends that we underestimated or missed completely. Nonetheless, we offer our insight based on personal experience and observations on technology development relative to human behavior. And that begins by looking back at how public concern for the environment has developed over the past 70 years.

The beginning of a widespread environmental movement was signaled by Rachel Carson's "Silent Spring." Prior to the publication of that book in 1962, public opinion of the petrochemical industry was generally positive. "Science" was bringing many new products to market that were improving the way of life for first-world nations. "Silent Spring" was the tipping point of the optimism that followed the rapid expansion of the petrochemical industry. It came at a time when the cold war had created fear of a nuclear holocaust that would end life on earth. A new fear was introduced, extinction of life due to the overuse of chemicals, specifically pesticides. Fear can be a useful tool for those that are seeking funding for research work. The 1960s gave rise to the rapid development of new analytical techniques that were used to study environmental contamination. For example, polarography and atomic absorption techniques were extensively studied and developed during this time. Another powerful force at work at the same time was the growth in semiconductor electronics that made it possible to commercialize affordable instruments with orders of magnitude greater sensitivity than those made in the previous decade. The electron capture detector

Analytical Techniques in the Oil and Gas Industry for Environmental Monitoring, First Edition.
Edited by Melissa N. Dunkle and William L. Winniford.

for gas chromatography was invented in 1957 but the real impact was in the 1960s when it became commercially available, providing parts per billion detection for halogenated pesticides. Suddenly researchers, transitioning from classical wet chemical methods to instrumental methods, were finding contamination everywhere. With these findings, more and more funding became available for further research. Environmental analysis became one of the most significant forces for developing analytical chemistry techniques in the 1960s and 1970s. If the environmental movement had not occurred at that time, it seems likely that the development of trace analytical methods would have been much slower. In the 1980s and 1990s, funding began to grow for biochemical and biomedical research and commensurately funding for related analytical techniques. Liquid chromatography-mass spectrometry (LC-MS) development was driven by needs in drug discovery, proteomics, and metabolomics. Interfacing liquid chromatography to mass spectrometry is inherently difficult, supplying liquid to a high vacuum detector. The development was relatively gradual compared with some other techniques that have been marked by a new invention that made a major step forward. The development of nuclear magnetic resonance (NMR) over the last 40 years has been similar. It is, likewise, a critical technique for bioanalytical research. Analytical NMR has benefitted greatly from the research money poured into developing magnetic resonance imaging (MRI) for health care.

In the mid-1980s, the IBM PC was introduced, which was the next major step in the electronics revolution. Prior to this, minicomputers were being used for instrument data acquisition. These were essential for data processing in mass spectrometry, NMR, and Fourier transform infrared spectroscopy (FTIR). The price of computing dropped by orders of magnitude while computing power increased by orders of magnitude. The net effect was computer-based acquisition for all types of analytical measurements, which also spawned wide-scale automation of sample preparation and introduction to instruments. High throughput screening became possible, first for biological samples and then to other areas such as catalyst screening. With this capability in hand, the amount of collected data has increased exponentially. Traditional approaches to evaluating data could not keep pace with the volume of data generated. Analysis of huge data sets is now commonplace in many fields. The learning from this retrospective is that technology development is strongly influenced by co-current trends that supply the money for development. It is tempting to think that the money follows public perception, but the ideas that grip public perception can change quickly. The global economy is so complex that it is difficult, if not impossible, to predict the events that will quickly seize public attention. But the factors that influence the flow of money for funding research tend not to change as quickly. By studying these, one can make more intelligent estimations on future developments.

11.2 Climate Change

"Climate change" is currently one of the hottest topics in science and politics, at least, in the United States and Europe. As such, the funding was originally pushed to study "global warming," and it has been helpful to change the title to include more areas of research. There is significant political pressure to reduce the release of carbon dioxide and related greenhouse gases (GHGs) generated from the use of fossil fuels. In 2017 the International Energy Agency (IEA) estimated that 80 million metric tons of methane were released to the air by the oil and gas sector (McGlade et al. 2019). Methane has more than 80 times the warming potential as carbon dioxide. The IEA has proposed a Sustainable Development Scenario that would achieve a 45% reduction in methane emissions by 2025 with little cost to the oil and gas sector. Since the lifetime of methane in the atmosphere is only 12 years versus centuries for CO_2, reducing methane emissions is one of the more effective ways in the short term to impact GHG emissions. But it is important to note that this is not viewed as equally important around the globe. In developing nations fossil fuels are the cheapest way to produce energy, and energy production is closely related to improving the standard of living. People who live on day to day subsistence do not have time to worry about the possible impact of climate change 50 years from now. Thus we have seen, and it is likely to continue, that only nations with sufficient affluence will pass laws and take action on the environmental impact of poor stewardship by some parts of the oil and gas industry. We should be encouraged by the progress that has been made in the United States and Europe with regard to air and water pollution. Air pollution is not nearly as bad in many urban areas as it was in the 1960s. Though the industry has protested the introduction of environmental laws, the last 60 years have shown that the industry can adapt to regulations and still be profitable. In many cases, improvements in waste disposal have saved money by turning more feedstock into products that can be sold. Some developing nations have been able to bypass pollution from industrialization by implementing laws to prevent previous mistakes. China implemented the use of liquified natural gas (LNG) as fuel in taxis to reduce air emissions in urban areas. Bangladesh has done likewise. For some people, though, this is not enough. They urge immediate conversion to electric vehicles. In doing so, they fail to take many things into account. Due to the global infrastructure in place and the energy density of petroleum, all other solutions are more costly, and the CO_2 footprint is not necessarily better. This is where economic solutions tend to prevail in the long run. Solar energy is viewed as clean, and the price per kilowatt-hour inches its way toward being cost-competitive with natural gas, but it takes significant energy to produce solar cells. And solar cells have a finite lifetime. Wind energy is clean but can only be applied in certain areas. Both solar and wind energies require the development of battery technology that can efficiently store energy during low load hours and release at peak times. General public awareness of the technical issues involved with reducing GHG emissions is low. Indeed the long-term consequences of making energy policy decisions are difficult for highly educated people to agree on.

11.3 Likely Scenarios

11.3.1 Gas Emissions

Natural gas as a power source has enjoyed a significant boom and has all but eliminated the construction of new coal-fired power plants in the United States and Europe. China does not have abundant access to oil and gas. It is much more dependent on a blend of imports and domestic coal for energy. In the past, there was considerable research on "clean coal" combustion but the supply of natural gas available through fracking in many parts of the world has significantly reduced the funding. China is most likely to lead future research in this area. CO_2 capture and reuse are receiving funding now in many of the developed nations. Capturing CO_2 from power plants and cement kilns has the greatest potential for an immediate reduction in emissions, but it is not trivial to implement because CO_2 comes out in a dilute, low-pressure gas stream. Many solutions are being proposed, but the most likely routes involve the use of membranes to separate and pressurize the CO_2 for reuse in oilfields. The technical challenges are sufficient enough that this research will go on for a number of years. Because of the concern about climate change, monitoring for CO_2 release (Figure 11.1) and other GHGs will continue to grow. As discussed in Chapter 7, there are several ways for monitoring CO_2 levels, ranging from point sensors to open-path infrared measurements that can monitor average levels over long distances. Since CO_2 sensors are relatively cheap and 5G communication technology is currently being implemented, it is very likely that oil and gas companies will implement networks of wireless gas sensors to provide instantaneous feedback on leaks. Laws already exist in the petrochemical industry requiring continuous emission monitoring of processes for targeted compounds. Over time these requirements will be added to gas treating plants and production wells. Industry leaders recognize that they need to be proactive in addressing these issues. BP has announced the use of drones to provide real-time feedback on methane emissions (Blum 2019) and will use them globally. They tested with a pilot project at their Khazzan field in Oman and now have that as a permanent installation. Shell is using drones in Australia to reduce the manpower required for pipeline inspections. Many of the major aircraft companies have begun ventures related to the use of drones for inspection in oil and gas production.

In the summer of 2016, the GHGSat-D (Claire) was launched as a service to pinpoint methane leaks. Satellite imaging is used to locate leaks initially, and then

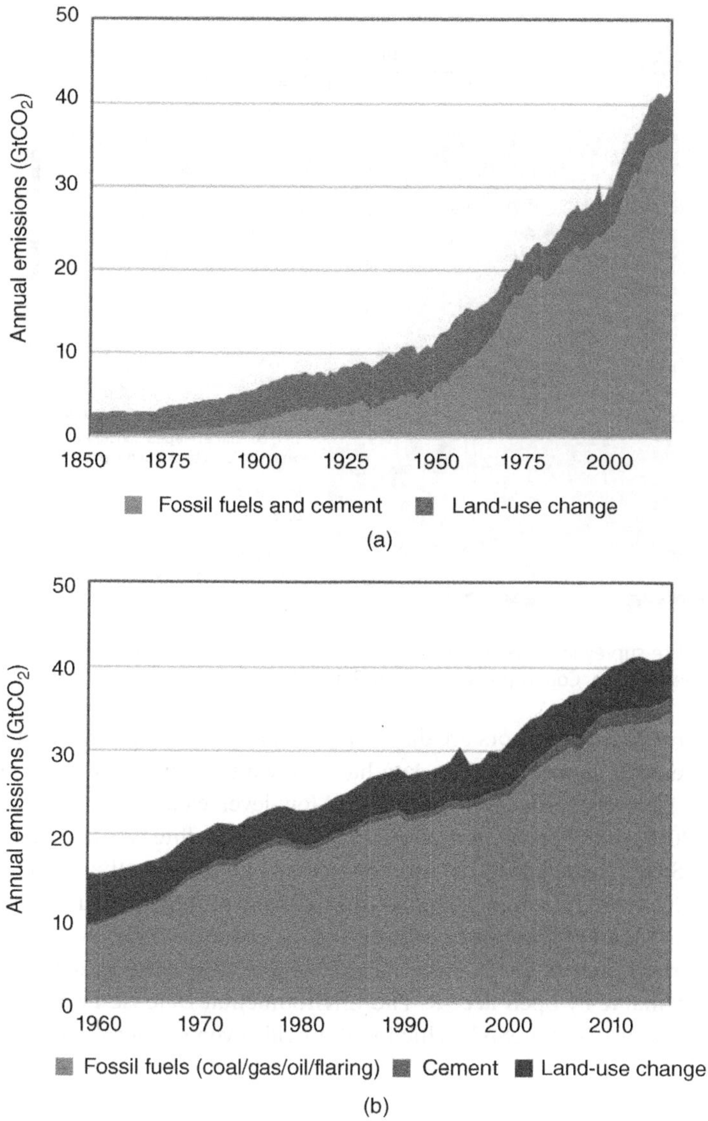

Figure 11.1 Growth in CO_2 emissions from fossil fuel use. Source: Friedlingstein et al. (2019). Licenced under CC BY 4.0.

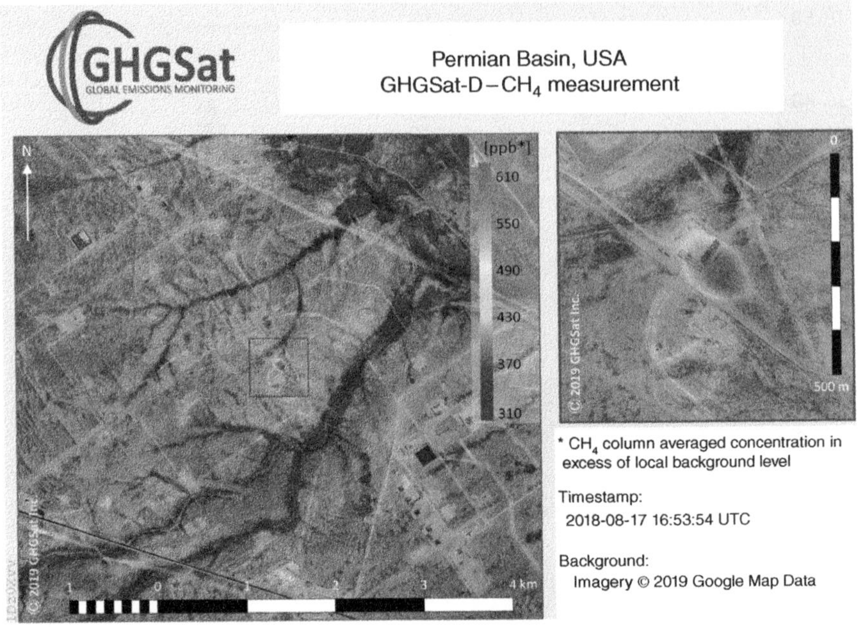

Figure 11.2 Satellite survey showing methane plume. Source: Chaturvedi (2019). Reproduced with permission, Copyright (2019), GHGSat and Google Maps.

aircraft with high sensitivity sensors are dispatched to provide even greater accuracy in locating leaks (Figure 11.2). The data have shown that emissions tend to be due to specific gas wells rather than widespread low-level leakage.

In October of 2017, the Copernicus Sentinel-5 Precursor satellite was launched by the European Space Agency with the mission to map plumes of methane, CO_2, nitrogen oxides, CO, ozone, formaldehyde, SO_2 and aerosols around the Earth. The instrument onboard, the TROPOspheric Monitoring Instrument (TROPOMI), currently has a pixel resolution of 3.6×5.6 km. It can map the entire earth in 24 hours and the data is available by open access. The Environmental Defense Fund has commissioned work on MethaneSAT, which will provide even greater sensitivity and resolution. MethaneSAT is reported to have a sensitivity of 2 ppb versus 12 ppb for TROPOMI. Figure 11.3 shows a comparison of the attributes of each satellite. Both TROPOMI and MethaneSAT make their data available to the public; this will put significant pressure on offending emitters of GHGs to reduce. Never before has the capability existed for the entire Earth to be mapped on a daily/weekly basis. It is likely that a combination of these monitoring techniques will grow in usage. Satellites can provide the big picture mapping of hot spots for GHG emissions. Drones or manned aircraft can then be used to provide much higher lateral resolution in locating point sources. Then on-line instruments or sensors can be used

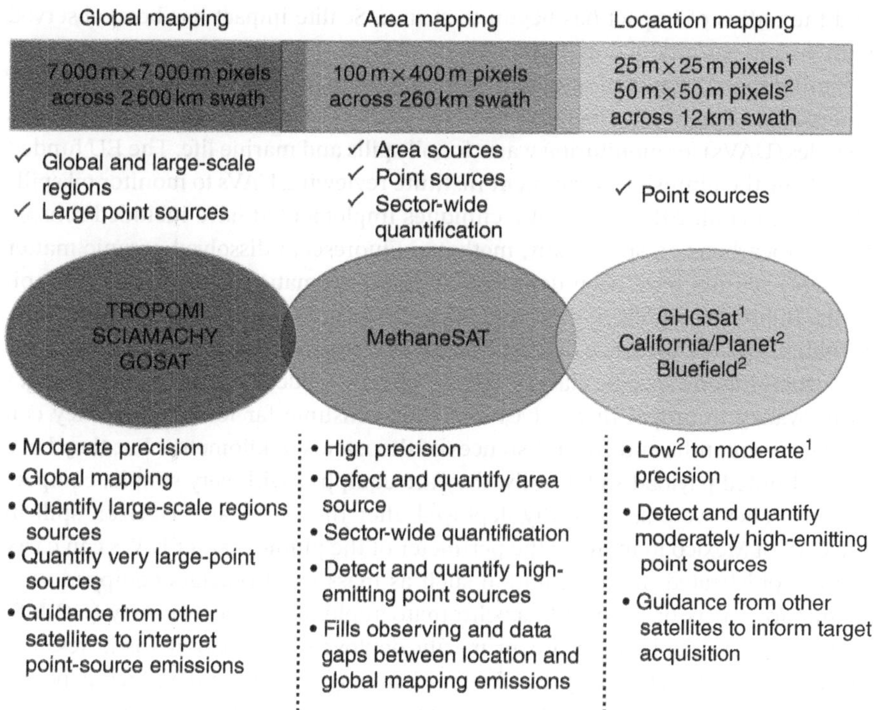

Figure 11.3 Comparison of satellites for mapping methane emissions. Source: Reproduced with permission, Copyright (2020), MethaneSAT (2019).

for real-time feedback on point sources. In some cases, leaks will be repaired and the on-line sensors will provide data for verification of compliance. In other cases, sensors will be used to identify operational issues that lead to unacceptable GHG releases.

11.3.2 Water Emissions

The oil and gas industry relies heavily on recycling water to have sufficient water for drilling and production operations. In many parts of the world, oil is produced in relatively arid areas. In these regions, it is increasingly unlikely that water would be discharged to the environment because it is too valuable for use in drilling and production. Disposal of water is often done by deep well injection, and there are not significant requirements for monitoring deep wells. There are some special cases where water contamination is currently monitored and studied. Oil production in the North Sea has resulted in the accumulation of related compounds in the tissue of marine life (Bakke et al. 2013). Oil companies have already responded, and the radius of impact has begun to shrink. Sealife impact has been observed with oil and gas production in the Gulf of Mexico.

Similar to the growth of satellites for air emission monitoring, there is a parallel emergence of remotely operated vehicles (ROVs) and underwater autonomous vehicles (UAVs) for monitoring water for oil spills and marine life. The EU funded a study by the Finnish Environment Institute reviewing UAVs to monitor oil spills (Jørgensen et al. 2019). Current techniques implemented have included sensors for hydrocarbons or, specifically, methane. Fluorescent dissolved organic matter (FDOM) sensors have been deployed to detect aromatic hydrocarbons from oil spills. Indirect methods of detecting oil spills include conductivity, temperature, turbidity, and the presence of dissolved gases (oxygen). The indirect methods are important for one specific class of UAVs – gliders. Gliders use very small changes in buoyancy to propel themselves and thus consume far less power. They can operate for months and cover distances of thousands of kilometers but they have a very limited payload so typically, they are equipped with very simple, compact sensors. They were successfully deployed after the Deepwater Horizon spill in the Gulf of Mexico to monitor the perimeter of the plume. Larger UAVs can carry more sophisticated instrumentation such as mass spectrometers equipped with membrane inlets or even full gas chromatography-mass spectrometry (GC-MS) systems. At present, these are primarily deployed by research institutes, but portable mass spectrometry has been advancing rapidly and reducing power requirements, which are particularly important for implementation in UAVs and ROVs. State of the art in underwater mass spectrometers has been recently reviewed (Chua et al. 2016). The growth in artificial intelligence (AI) and machine learning have had a significant influence on the development of UAVs. Ship-based sensors and towed probes have been used since World War II, but UAVs require the ability to sense their environment and change course based on that feedback. Developments in UAV adaptive sampling methods have been reviewed recently (Hwang et al. 2019). As the age of oil and gas structures in marine environments grow, many are approaching decommissioning. Much of the North Sea basin will be decommissioned in the coming years. Marine autonomous systems (MASs) may be the most cost-effective way to monitor such structures as described in a recent review (Jones et al. 2019). Structures, pipelines, the near-field seabed, water column, and far-field seabed all are under consideration by the European Union and particularly the United Kingdom. Sensors for these studies include geo-acoustic, visual (cameras), and the sensors mentioned earlier for UAVs. At present, there are few specific regulations for decommissioning, but many are under discussion. There is not a simple answer to this. Off-shore platforms that have been in place for many years have developed their own ecosystem. The platforms provide structure for a wide variety of marine life. Removal and remediation would certainly bring short-term disruption. Drilling-related sediments and cuttings piles could be remediated but it is not clear yet whether that can be done without creating more damage to marine life than leaving them in place. Either way, it is certain that decommissioning will require more monitoring, and regulations are likely to grow.

For Example:

- "Surface water monitoring monitors the pH of surface water, potassium permanganate index (CODMn), biochemical oxygen demand (BOD5), petroleum, ammonia nitrogen, sulfide, cyanide, and volatile phenol pollutants" (Zou 2017).
- "Underground water monitoring monitors the total hardness of underground water, petroleum, fluoride, cyanide, volatile phenol, ammonia nitrogen, pH, CODMn, number of bacteria, and coil group" (Zou 2017).

11.3.3 Oil Sands

The tailings ponds for oil sands mining water in Alberta, Canada, have become an icon for those that decry "dirty oil." The Clark hot water extraction (CHWE) method is used to separate bitumen from the mined tar sands. The Syncrude Mildred Lake Settling Basin is the largest earthen dam structure on Earth, and the Syncrude Southwest Sand Storage is the third-largest earthen dam structure. Collectively there are over a trillion liters of tailings pond water impounded by Syncrude, Suncor, and the other operators in Alberta. The process of extracting bitumen requires a settling pond to separate the sand and clay from the process water. Quickly the sand drops out, but a layer of suspended colloidal clay persists (mature fine tails or MFTs) with a clear water layer on top. The MFT layer is 70% water and stable for years. The water contains naphthenic acids and other soluble hydrocarbons from the bitumen as well as dissolved minerals. The chief concern is that the naphthenic acids are known to be acutely toxic to aquatic life, and many of the settling ponds are located very close to the Athabasca River which flows into the Arctic Ocean. If there were ever a breach in one of the settling ponds, there would be an enormous environmental impact. Thus there have been numerous studies to determine whether the retention ponds are leaking and whether they are impacting surface waters or groundwater in the surrounding area. Over 150 publications were reviewed (Fennell and Arciszewski 2019) to summarize what is known about seepage from the settling ponds. There are naphthenic acids detected in the groundwater but there is conflicting data on whether they are due to seepage from the ponds or due to natural interaction of groundwater with unmined bitumen. Because this is such a controversial subject, it is certain that studies will continue. It is important to note that the majority of the studies on this have been published in the last 10 years. The growth in commercial availability of high-resolution LC-MS correlates well with this and has become the preferred method of analysis for naphthenic acids.

11.3.4 Food Contact – MOSH/MOAH

Another area of emerging concern is over mineral oil saturated hydrocarbons (MOSHs) and mineral oil aromatic hydrocarbons (MOAHs). Currently, there are no specific regulations on MOSH/MOAH, but they are being scrutinized. In February 2017, the German government amended its consumer goods ordinance to stipulate that barriers be used so that MOSH/MOAH could not be transferred from food packaging into food. The detection limit was specified as 0.5 mg/kg of foodstuff (IFP 2019). This is challenging because MOSH/MOAH is comprised of hundreds, if not thousands of individual components. Chromatographically, these components elute as a broad hump in a chromatogram, which makes it difficult to achieve low detection limits. Mineral oil is used widely in industry in the manufacturing of packaging materials, in printing inks, as a lubricant for machinery, and as a food additive. Thus far, mineral oil has been considered benign and there has not been toxicological evidence to suggest a human health risk from MOSH. But MOAH is suspect due to chemical similarity to other carcinogenic aromatic hydrocarbons, particularly those with three to seven rings. Technical grade mineral oil can contain 65–85% MOSH and 15–35% MOAH (EPEA 2018). Food grade mineral oil should have MOAH minimized. It is assumed that MOSH is present in nearly all processed foods. To distinguish MOAH from MOSH, either GC × GC-MS is required or the use of a selective detector such as the vacuum ultraviolet (VUV) or the photoionization detector (PID), which will respond selectively to aromatic compounds. Further work is needed in sample preparation techniques to selectively extract these compounds and concentrate them to reach the desired detection limits.

11.3.5 Industry 4.0 and the 4th Wave of Environmentalism

Industry 4.0 was initiated in 2011 when the German government laid out a blueprint for advanced manufacturing. In the simplest terms, it can be considered the logical next step after the third industrial revolution of the 1980s that ushered in the widespread use of computers, development of the internet, industrial automation, and significant advances in communications. Industry 4.0 brings together a number of technologies to create what is referred to as the "smart factory." It is now possible to do real-time monitoring, analysis of big data sets, modeling, and optimization to achieve breakthroughs in productivity, cost reduction, and sustainability. The tools commonly cited are Internet of things (IoT), machine learning, AI, blockchain, and autonomous systems (Manoukian 2018). It is expected that Manufacturing 4.0 will bring major changes to the oil and gas industry (Lobene 2019). The ability to store and process massive amounts of data makes it possible to create autonomous systems that can control processes to maximize yield and minimize waste.

There has been a significant shift in the relationship between the oil and gas industry and environmental groups. The 1960s, as referred to at the start of this chapter, was the second wave of environmentalism, mostly focused on the negative impact of industry on the environment. The EPA was formed at this time. The Clean Air Act and the Clean Water Act were enacted as well. Industry was often in the position of defending itself and lobbying to control legislation. Environmental regulations were the primary driver for environmental monitoring. The third wave of environmentalism began in the 1990s and marked a reduction in tension between environmentalists and industry. Market-based incentives were introduced to encourage sustainability. There was a strong push for recycling at this time. The biggest impact on the oil and gas industry came from cap and trade policies introduced at this time. The fourth wave of environmentalism is marked by innovations in technology that both improve profitability and sustainability. This is a significant shift in mindset from the 1960s, where environmental regulations were always viewed by industry as increasing costs. The Environmental Defense Fund (EDF) has defined seven technologies that can drive sustainability (Murray 2018).

- Data Analytics
- Blockchain
- Automation Technologies
- Sharing Technologies
- Sensors
- Dematerialization
- Mobile Ubiquity

As an example, in 2014, the EDF issued the Methane Detectors Challenge to identify low-cost sensors that could be widely deployed to monitor for methane leaks. This raised awareness of the need to detect and fix leaks in the oil and gas industry in days instead of months. It is estimated that 1–2% of production is lost due to leaks, and a small number of leaks contribute to the majority of the emissions (EPA 2019; Zavala-Araiza et al. 2015). This requires robust sensors capable of measuring methane down to 5 ppm or less. Five submissions were received and four were evaluated by the Southwest Research Institute in 2014–2015. This resulted in two laser-based sensors being selected for pilot trials with Statoil, PG&E, and Shell (Nowlan 2016). The significance of this example is that it illustrates a profound shift. Rather than environmental groups working as adversaries to the oil and gas industry, the EDF has shown that it is possible to bring together the objectives of the environmental movement with tech startups and the oil and gas industry. Also, companies in the oil and gas industry no longer deny responsibility for climate change. They recognize that compliance with environmental regulations is no longer sufficient. The social contract with

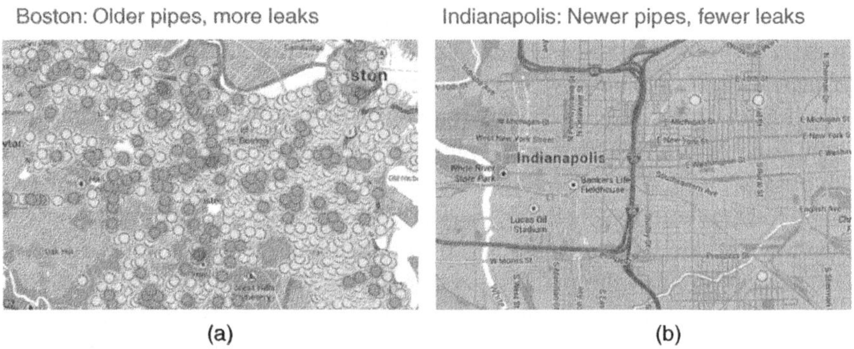

Figure 11.4 Comparison of natural gas leaks in Boston and Indianapolis. Source: Reproduced with permission of Environmental Defense Fund (EDF 2019).

society is now playing an ever more significant role in the decisions that investors make. In turn, this applies pressure to companies in the oil and gas industry to demonstrate that they are actively working toward sustainability.

Another example is the partnership between EDF and Google Earth to equip the Google Earth Streetview cars with methane sensors to map leaks in natural gas pipelines. This began in 2014 and grew to include collaboration with Colorado State University for data analysis and local utility companies to validate leaks. Now, 13 urban areas have been mapped in this pilot study. As would be expected, cities with older infrastructure have significantly more leaks than cities that have grown more recently, as shown in Figure 11.4. Boston averaged 1 leak for every mile driven while Indianapolis averaged 1 leak for every 200 miles driven.

11.4 Summary

In summary, recent years have shown an explosion in the amount of data that can be collected and stored. Sophisticated analytical techniques will grow in capability as driven by other fields such as biotechnology, medicine, electronics, and nanotechnology. Environmental monitoring in the oil and gas industry will reap this technology and benefit. But the greater impact is likely to come from the proliferation of simple, inexpensive sensors that can be placed everywhere to provide real-time data. We are at the threshold of one of the most profound changes in the industry in all of history.

References

Bakke, T., Klungsøyr, J., and Sannicd, S. (2013). Environmental impacts of produced water and drilling waste discharges from the Norwegian offshore petroleum industry. *Marine Environmental Research* 92: 154–169. https://doi.org/10.1016/j.marenvres.2013.09.012.

Blum, J. (2019). BP launches gas cloud imaging, drones to monitor methane emissions. *Houston Chronicle*. Retrieved from https://www.houstonchronicle.com/ (accessed 20 March 2020).

Chaturvedi, A. (2019). Canadian startup aims to become the global reference for greenhouse gas monitoring. Retrieved from https://www.geospatialworld.net/ (accessed 20 March 2020).

Chua, E.J., Savidge, W., Short, R.T. et al. (2016). A review of the emerging field of underwater mass spectrometry. *Frontiers in Marine Science* 3: 209. https://doi.org/10.3389/fmars.2016.00209.

EDF. (2019). Local leaks impact global climate. Retrieved from https://www.edf.org/ (accessed 20 March 2020).

EPA (2019). *Inventory of U.S. Greenhouse Gas Emissions and Sinks: 1990–2017* (EPA 430-R-10-001). Online. US Environmental Protection Agency (EPA). Retrieved from https://www.epa.gov/.

EPEA (2018). *EPEA-Factsheet on MOSH/MOAH*. Online. EPEA Internationale Umweltforschung GmbH. Retrieved from www.healthyprinting.eu/.

Fennell, J. and Arciszewski, T.J. (2019). Current knowledge of seepage from oil sands tailings ponds and its environmental influence in northeastern Alberta. *Science of the Total Environment* 686: 968–985. https://doi.org/10.1016/j.scitotenv.2019.05.407.

Friedlingstein, P., Jones, M.W., O'Sullivan, M. et al. (2019). Global carbon budget 2019. *Earth System Science Data* 11: 1783–1838. https://doi.org/10.5194/essd-11-1783-2019.

Hwang, J., Bose, N., and Fan, S. (2019). AUV adaptive sampling methods: A review. *Applied Sciences* 9 (15): 3145 (1–30). https://doi.org/10.3390/app9153145.

IFP (2019). *Mineral Oil Hydrocarbons (MOSH/MOAH)*. Online. Institut für Produktqualität GmbH. Retrieved from https://www.produktqualitaet.com/.

Jones, D.O.B., Gates, A.R., Huvenne, V.A.I. et al. (2019). Autonomous marine environmental monitoring: Application in decommissioned oil fields. *Science of the Total Environment* 668: 835–853. https://doi.org/10.1016/j.scitotenv.2019.02.310.

Jørgensen, K.S., Kreutzer, A., Lehtonen, K.K. et al. (2019). The EU Horizon 2020 project GRACE: Integrated oil spill response actions and environmental effects. *Environmental Sciences Europe* 31: 1–10. https://doi.org/10.1186/s12302-019-0227-8.

Lobene, M. (2019). The Impact of Industry 4.0 on the Oil and Gas Industry. Retrieved from https://medium.com/ (accessed 20 March 2020).

Manoukian, J. G. (2018). Industry 4.0 Meets the 4th Wave of Environmentalism. Retrieved from https://enablon.com/ (accessed 20 March 2020).

McGlade, C., Gould, T., and Sondak, G. (2019). *Tracking Fuel Supply*. Paris: International Energy Agency (IEA). Retrieved from https://www.iea.org/reports/tracking-fuel-supply-2019.

MethaneSAT (2019). *MethaneSAT Mission: How it Fits into the Methane Measurement Eco-system*. Online. MethaneSAT, LLT. Retrieved from https://www.methanesat.org.

Murray, T. (2018). *Business and the Fourth Wave of Environmentalism*. Online. Environmental Defense Fund (EDF). Retrieved from https://www.edf.org/.

Nowlan, A. (2016). *Methane Detector Challenge: An Unexpected Partnership*. Online. Environmental Defense Fund (EDF). Retrieved from https://www.edf.org/.

Zavala-Araiza, D., Lyon, D.R., Alvarez, R.A. et al. (2015). Reconciling divergent estimates of oil and gas methane emissions. *Proceedings of the National Academy of Sciences of the United States of America* 112: 15597–15602. https://doi.org/10.1073/pnas.1522126112.

Zou, C. (2017). Oil shale. In: *Unconventional Petroleum Geology*, 2e (ed. C. Zou), 371–390. Cambridge, MA: Elsevier.

Index

Analytical Techniques in the Oil and Gas Industry for Environmental Monitoring, First Edition.
Edited by Melissa N. Dunkle and William L. Winniford.

d

e

f

g

h

i

k

l